Lecture Notes in Electrical Engineering

Volume 968

The book series *Lecture Notes in Electrical Engineering* (LNEE) publishes the latest developments in Electrical Engineering—quickly, informally and in high quality. While original research reported in proceedings and monographs has traditionally formed the core of LNEE, we also encourage authors to submit books devoted to supporting student education and professional training in the various fields and applications areas of electrical engineering. The series cover classical and emerging topics concerning:

- Communication Engineering, Information Theory and Networks
- Electronics Engineering and Microelectronics
- Signal, Image and Speech Processing
- Wireless and Mobile Communication
- Circuits and Systems
- Energy Systems, Power Electronics and Electrical Machines
- Electro-optical Engineering
- Instrumentation Engineering
- Avionics Engineering
- Control Systems
- Internet-of-Things and Cybersecurity
- Biomedical Devices, MEMS and NEMS

For general information about this book series, comments or suggestions, please contact leontina.dicecco@springer.com.

To submit a proposal or request further information, please contact the Publishing Editor in your country:

China

Jasmine Dou, Editor (jasmine.dou@springer.com)

India, Japan, Rest of Asia

Swati Meherishi, Editorial Director (Swati.Meherishi@springer.com)

Southeast Asia, Australia, New Zealand

Ramesh Nath Premnath, Editor (ramesh.premnath@springernature.com)

USA, Canada

Michael Luby, Senior Editor (michael.luby@springer.com)

All other Countries

Leontina Di Cecco, Senior Editor (leontina.dicecco@springer.com)

** This series is indexed by EI Compendex and Scopus databases. **

Anupam Shukla · B. K. Murthy · Nitasha Hasteer ·
Jean-Paul Van Belle
Editors

Computational Intelligence

Select Proceedings of InCITe 2022

Editors
Anupam Shukla
Sardar Vallabhbhai National Institute
of Technology
Surat, India

B. K. Murthy
Innovation and Technology Foundation
Indian Institute of Technology, Bhilai
Bhilai, India

Nitasha Hasteer
Department of Information Technology
Amity School of Engineering
and Technology
Amity University Uttar Pradesh
Noida, India

Jean-Paul Van Belle
Department of Information Systems
University of Cape Town
Rondebosch, South Africa

ISSN 1876-1100 ISSN 1876-1119 (electronic)
Lecture Notes in Electrical Engineering
ISBN 978-981-19-7348-2 ISBN 978-981-19-7346-8 (eBook)
https://doi.org/10.1007/978-981-19-7346-8

This Springer imprint is published by the registered company Springer Nature Singapore Pte Ltd.
The registered company address is: 152 Beach Road, #21-01/04 Gateway East, Singapore 189721, Singapore

Preface

Computational intelligence is the study of design, application and development of intelligent agents. Enormous success has been achieved in solving a variety of challenging problems through the modelling of biological and natural intelligence, resulting in so-called intelligent systems. These systems encompass paradigms such as artificial neural networks, evolutionary computation, swarm intelligence, artificial immune systems and fuzzy systems. Over the last few years, there has been an explosion of research in this area as it is a promising technology capable of bringing drastic economic impact. The recent technological advancements in this field have led us to discuss and deliberate on emerging novel discoveries and important insights in all areas of computational intelligence design and applications.

This book is a compilation of research work carried out by esteemed experts in the area of computational intelligence. The objective of this compilation is to provide relevant and timely information to those who are striving to contribute to this domain by utilizing the latest technology. The content of this book is based on the research papers accepted during the 2nd International Conference on Information Technology (InCITe 2022) held at Amity University, Uttar Pradesh, Noida, India, on 3 and 4 March 2022. It is a conglomeration of research papers covering interdisciplinary research and in-depth applications of computational intelligence, deep learning, machine learning, artificial intelligence, data science, enabling technologies for IoT, blockchain and other futuristic computational technologies. It covers interdisciplinary and innovative research on neural networks for medical imaging, object detection using deep learning, modelling for disease analysis, prediction of financial data sets using genetic algorithms, stock price prediction, smart farming, sentiment analysis, etc. The content would serve as a rich knowledge repository on information and communication technologies, neural networks, fuzzy systems, natural language processing, data mining and warehousing, big data analytics, cloud computing, security, social networks and intelligence, decision-making and modelling, information systems and IT architectures. We are hopeful that it will prove to be of high value to graduate students, researchers, scientists, practitioners and policymakers in the field of information technology.

We thank the management of Amity University, Uttar Pradesh, who believed in us and provided the opportunity for publishing the book. We are grateful to Science and Engineering Research Board, Department of Science and Technology (DST), Council of Scientific and Industrial Research (CSIR) and Defence Research and Development Organization (DRDO), Government of India, for their support. We are thankful to the advisory committee of InCITe 2022 for the continuous support and guidance. Special thanks to all the national and international reviewers who helped us in selecting the best of the works as different chapters of the book. We wish to acknowledge and thank all the authors and co-authors of different chapters who cooperated with us at every stage of publication and helped us to sail through this mammoth task. We owe our sincere gratitude to all our family members and friends who helped us through this journey of publishing a book. Our appreciation goes to each and every individual who supported us in this endeavour. Last but not least, we are grateful to the editing team of Springer who provided all guidance and support to us in the compilation of the book and also shaped it into a marketable product.

Surat, India   Anupam Shukla
Bhilai, India   B. K. Murthy
Noida, India   Nitasha Hasteer
Rondebosch, South Africa   Jean-Paul Van Belle

Contents

About the Editors

Dr. Anupam Shukla is the Director of Sardar Vallabhbhai National Institute of Technology, Surat. Prior to this he was Director at IIIT Pune and Professor in the Department of Information and Communication Technology (ICT) at ABV-Indian Institute of Information Technology and Management (ABV-IIITM), Gwalior. Prof. Shukla has over 35 years of teaching, research, and administrative experience. He is globally renowned for his research in the field of artificial intelligence and has won several academic accolades. He has been awarded 'Distinguished Professor' by Computer Society of India, Mumbai in 2017 and 'Dewang Mehta National Education Award' for best Professor in 2016. He is the author of patents, books, 190 peer-reviewed publications and mentor of 18 Doctorate and 116 Post Graduate theses. He has successfully completed 13 Government-sponsored projects aimed at developing IT applications for Indian farmers, skill development of the Indian youth and infrastructure development at parent institutes.

Dr. B. K. Murthy is the CEO of Innovation and Technology Foundation at IIT Bhilai. He was Scientist G and Group Coordinator (R&D in IT) in the Ministry of Electronics and IT (MeitY), Government of India where he was responsible for promotion of R&D in the area of IT. He has been conferred the prestigious VASVIK Industrial Research Award for the year 2020 in the category of Information and Communication Technology. Dr. Murthy spearheaded for bringing out the National Strategy on Blockchain Technology from the Ministry of Electronics and IT which was released in December 2021. His research interests include Artificial Intelligence, Software Defined Networking, Cloud Computing, Quantum Computing and Blockchain Technologies. He was awarded Ph.D. degree by IIT Delhi. He has published and presented more than 70 papers in various journals and conferences and is a regular speaker at International forums. He has served as member of Board of Governors Indira Gandhi Technological University for Women, Delhi, IIITs at Allahabad, Gwalior, Kurnool and Kanchipuram. He also served as Guest faculty at IIT Roorkee, IIIT Gwalior and IP University.

Dr. Nitasha Hasteer is the Head of the Information Technology Department and Dy. Director (Academics) at Amity School of Engineering and Technology, Amity University Uttar Pradesh. She has 21 Years of teaching, research and administrative experience in academics and industry and is a Ph.D. in Computer Science and Engineering. Her area of interests includes Machine Learning, Cloud Computing, Crowdsourced Software Development, Software Project Management and Process Modeling through Multi-Criteria Decision-Making Techniques. She has published more than 50 papers in various journals and international conferences in these areas and has guided many postgraduate students for their dissertation. She has been on the editorial board of many international conferences and obtained funding from government agencies such as Science and Engineering Research Board, Department of Science and Technology (DST), Defence Research and Development Organization (DRDO), Indian National Science Academy (INSA) and Council of Scientific and Industrial Research (CSIR) for organizing conferences in the area of Information Technology.

Dr. Jean-Paul Van Belle is the Professor in the Department of Information Systems, University of Cape Town, South Africa and Director of the Centre for IT and National Development in Africa (CITANDA). His particular research focus and passion is the adoption and appropriation of emerging ICTs in a development context i.e. Development Informatics, ICT4D and Mobile for Development (M4D). He is the Prime Investigator of FOWIGS (Future of Work in the Global South) Fair work network project, and the Fairwork South Africa project. His other research and teaching specializations are Social Networking, Decision Support, Business Analytics, Open Government Data, E- and M-commerce, E- and M-government, Organizational Impacts and Adoption of IS, Open-Source Software, IT/IS Architectures and Artificial Intelligence. He has active research collaborations with researchers in India, UK, Ethiopia, Kenya, Ecuador and Chile.

Computational Modeling of Multilevel Organizational Learning: From Conceptual to Computational Mechanisms

Gülay Canbaloğlu, Jan Treur, and Anna Wiewiora

Abstract This paper addresses formalization and computational modeling of multi-level organizational learning, which is one of the major challenges for the area of organizational learning. It is discussed how various conceptual mechanisms in multi-level organizational learning as identified in the literature, can be formalized by computational mechanisms which provide mathematical formalizations that enable computer simulation. The formalizations have been expressed using a self-modeling network modeling approach.

Keywords Organizational learning · Mechanisms · Computational modeling · Self-modeling networks

1 Introduction

Multilevel organizational learning is a complex adaptive process with multiple levels and nested cycles between them. Much literature is available analyzing and describing in a conceptual manner the different conceptual mechanisms involved, e.g., [18, 9,

G. Canbaloğlu (✉)
Department of Computer Engineering, Koç University, Istanbul, Turkey
e-mail: gcanbaloglu17@ku.edu.tr

J. Treur
Social AI Group, Department of Computer Science, Vrije Universiteit Amsterdam, Amsterdam, The Netherlands
e-mail: j.treur@vu.nl

G. Canbaloğlu · J. Treur
Center for Safety in Healthcare, Delft University of Technology, Delft, The Netherlands

A. Wiewiora
School of Management, QUTBusinessSchool, Queensland University of Technology, Brisbane, Australia
e-mail: a.wiewiora@qut.edu.au

1

26, 27, 16]. However, mathematical or computational formalization of organizational learning in a systematic manner is a serious challenge. Successfully addressing this challenge requires:

- An overall mathematical and computational modeling approach able to handle the interplay of the different levels, adaptations and mechanisms involved
- For the conceptual mechanisms involved mathematical and computational formalization as computational mechanisms.

In this paper, for the first bullet, the self-modeling network modeling approach described in [20] is chosen, this approach has successfully been applied to the use, adaptation and control of mental models in [21]. For the second bullet, for many of the identified conceptual mechanisms from the literature, it is discussed how they can be modeled mathematically and computationally as computational mechanisms within a self-modeling network format.

In Sect. 2, an overview is given of conceptual mechanisms and how they can be related to computational mechanisms. Section 3 briefly describes the self-modeling network modeling approach used for computational formalization. In Sect. 4, a few examples of computational mechanisms for the level of individuals are described more in detail. Section 5 discusses more complex examples of computational models for feedforward and feedback learning that form bridges between the levels.

2 Overview: From Conceptual to Computational Mechanisms

In this section, a global overview of conceptual organizational learning mechanisms is described and supported by relevant references. Some of these conceptual mechanisms do not have pointers yet to computational mechanisms and can be considered items for a research agenda. Organizations operate as a system or organism of interconnected parts. Similarly, organizational learning is considered a multilevel phenomenon involving dynamic connections between individuals, teams and organization [12, 9], see Fig. 1. Due to the complex and changing environment within which organizations operate, the learning constantly evolves, and some learning may become obsolete. Organizational learning is a vital means of achieving strategic renewal and continuous improvement, as it allows an organization to explore new possibilities as well as exploit what they have already learned (March, 1991). Organizational learning is a dynamic process that occurs in feedforward and feedback directions. Feedforward learning assists in exploring new knowledge by individuals and teams and institutionalizing this knowledge at the organizational level [9]. Feedback learning helps in exploiting existing and institutionalized knowledge, making it available for teams and individuals. The essence of organizational learning is best captured in the following quote:

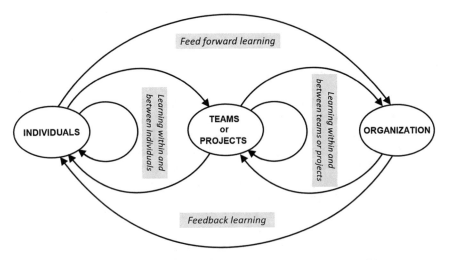

Fig. 1 Organizational learning: multiple levels and nested cycles (with depth 3) of interactions

'organizations are more than simply a collection of individuals; organizational learning is different from the simple sum of the learning of its members. Although individuals may come and go, what they have learned as individuals or in groups does not necessarily leave with them. Some learning is embedded in the systems, structures, strategy, routines, prescribed practices of the organization, and investments in information systems and infrastructure.'Crossan et al. [9], p. 529).

Individuals can learn by reflecting on their own past experiences, learn on the job (learning by doing), by observing others and from others, or by exploiting existing knowledge and applying that knowledge to other situations and contexts. Individuals can also learn by exploring new insights through pattern recognition, deep evaluation of a problem at hand, materialized in the 'aha' moment, when a new discovery is made. It is highly subjective and deeply rooted in individual experiences [9]. Teams learn by interpreting and integrating individual learnings by interpreting and sharing knowledge through the use of language, mind maps, images, or metaphors and thus jointly developing new shared mental models [9]. Team-level learning encompasses integration of possibly diverse, conflicting meanings, in order to obtain a shared understanding of a state or a situation. The developed shared understanding results in taking coordinated action by the team members [9]. Eventually, these shared actions by individuals and teams are turned into established routines, deeply embedded into organizational cultures, and/or captured in new processes or norms. From a computational perspective, such a process of shared mental model formation out of a number of individual mental models may be considered to relate to some specific form of knowledge integration or (feedforward) aggregation of the individual mental models.

In order for the organizational learning to occur, it has to be triggered by learning mechanisms, which are defined as apparatus for enabling learning. A recent review by

Wiewiora et al. [26] and their subsequent empirical investigation [27] identified organizational and situational mechanisms affecting project learning in a project-based context. Organizational learning mechanisms include culture, practices, systems, leadership and structure. From the organizational learning perspective, *organizational systems* are designed to capture knowledge, which was developed locally by teams or projects and captured into manuals or guidelines. These can represent knowledge management systems such as centralized knowledge repositories or specialized software used to collect, store and access organizational knowledge. Future learnings of individuals based on this in general may be considered another specific form of (feedback) aggregation, this time of the organization-level mental model with the already available individual mental model,an extreme form of this is fully replacing the own mental model by the organization mental model.

Organizational practices include *coaching and mentoring* sessions for building competencies. Coaching and mentoring occurs on individual level, where individuals have opportunities to learn from more experienced peers or teachers. Coaching and mentoring can also facilitate organizational to individual-level learning. These experts have often accumulated, through the years, a vast of organizational knowledge and experiences, in which case they can be also sharing organizational learnings. Furthermore, *training sessions* provide opportunities for developing soft and technical skills. During the sessions, a facilitator shares their own soft or technical skills or organizational knowledge with individuals or teams. *Leaders* have been described as social architects and orchestrators of learning processes (Hannah and Lester 2009). Leaders who limit power distances and encourage input and debate promote an environment conducive to openness and sharing, hence facilitating individual to team learning [10]. Meanwhile, self-protected leaders are more likely to use their position of power and impose control, hence restricting collective learning opportunities. When it comes to the *organizational structure*, decentralized structures promote rapid diffusion of ideas and encourage the exploration of a more diverse range of solutions [1]. The ideal structure appears to be the one that is loosely coupled, providing some degree of team separation, while ensuring weak connections between teams and the organization [11]. A situational mechanism affecting multilevel learning is *occurrence of major events* [26]: significant situations, positive or negative, that trigger immediate reaction (Madsen 2009).

There is limited research that systematically and empirically investigates mechanisms that trigger learning flows within and between levels. Tables 1, 2 and 3 synthesize existing research into multilevel learning and offers (in the first three columns of Tables 1, 2 and 3) a list of learning mechanisms facilitating multilevel learning flows. This paper demonstrates one of the first attempts to translate these (conceptual) mechanisms into computational mechanisms (in the last three columns of Tables 1, 2 and 3) and proposes a new computational modeling approach (briefly summarized in Sect. 3) that can handle the interplay between the levels and consider learning mechanisms that trigger learning flows between the levels.

Table 1 addresses the learning at the individual level, in Fig. 1 indicated by the circular arrow from individuals to individuals. A number of examples of conceptual mechanisms are shown in the different rows, and for each of them, it is indicated

Table 1 Conceptual and computational mechanisms for learning at the level of individuals

Conceptual mechanisms	Examples	Relevant references	Computational mechanisms	Examples	Relevant references
Individual: within persons					
Learning by internal simulational	Mental simulation (sometimes called visualization) of individual mental models to memorize them better		Hebbian learning during internal simulation of an individual mental model	Mental simulation of individual mental models for surgery in a hospital before shared mental model formation and after learning from a shared mental model	[6, 7]
Learning by observing oneself during own task execution	Individuals are observing themselves while they are performing a task	Iftikhar and Wiewiora (2020)	Hebbian learning for mirroring and internal simulation of an individual mental model	Observation of own task execution by nurses and doctors in a hospital operation room	[2, 24, 25]
Learning from past experiences	Individuals are learning by reflecting on their own past experiences	Iftikhar and Wiewiora (2020)	Learning based on counterfactual thinking	Counterfactual internal what-if simulation of nearest alternative scenarios	[3]
Individual: between persons					
Learning by observing others during their task execution	Individuals are observing how their peers are performing a task	Iftikhar and Wiewiora (2020)	Hebbian learning for mirroring and internal simulation of an individual mental model	Observation of each other's task execution by nurses and doctors in a hospital operation room	[2, 24, 25]
Coaching and mentoring	Learning from more senior and experienced people their individual 'tricks of the trade' via coaching and mentoring	[26, 27]	Learning from internal communication channels and aggregation	Doctor explains own mental model to nurse as preparation for surgery	[2, 24]

(continued)

Table 1 (continued)

Conceptual mechanisms	Examples	Relevant references	Computational mechanisms	Examples	Relevant references
Training sessions	Training sessions in which a facilitator shares their own soft and technical skills with individuals	[26, 27]	Learning from internal communication channels and aggregation	Experienced doctor explains own mental model to team	[2, 24]

Table 2 Conceptual and computational mechanisms for feedforward learning: from individual to teams or projects or to the organization and from teams or projects to the organization

Conceptual mechanisms	Examples	Relevant references	Computational mechanisms	Examples	Relevant references
Feedforward learning					
From individuals to organization					
Shared organization mental model formation and improvement based on individual mental models	Individuals share their mental models and institutionalize a shared mental model for the organization	[26, 27]	Feedforward aggregation of individual mental models for formation or improvement of shared mental models	Aggregating individual mental models from a nurse and a doctor to form a shared mental model of an intubation	[4, 5], Canbaloğlu et al. (2021a)
Occurrence of major events	Individuals mobilize to react and find solutions to major events. The best solution is selected and institutionalized by the organization	[26]	Feedforward aggregation of individual mental models for formation or improvement of shared mental models	Aggregating individual mental models from a nurse and a doctor to form a shared mental model of an intubation	[4, 5], Canbaloğlu et al. (2021a)
From individuals to teams or projects					
Training sessions	Training sessions in which a facilitator shares their own soft and technical skills with teams or projects	[26, 27]	Feedforward aggregation of individual mental models (and perhaps an existing shared organization mental model)	Aggregating individual mental models for surgery by hospital teams to form a shared team or project mental model	Canbaloğlu et al. (2021b)

(continued)

Table 2 (continued)

Conceptual mechanisms	Examples	Relevant references	Computational mechanisms	Examples	Relevant references
Learning by working together and joint-problem solving	Individuals of a team while working together are sharing their individual mental models and creating a new shared mental model by discussing and jointly solving a problem in hand	[18] Iftikhar and Wiewiora, (2020)	Feedforward aggregation of individual mental models (and perhaps an existing shared organization mental model)	Aggregating individual mental models for surgery by hospital teams to form a shared organization mental model	Canbaloğlu et al. (2021b)
From teams or projects to organization					
Formalizing team learnings	Teams capture their learnings into manuals or guidelines, which then inform new organizational practices	[9], Iftikhar and Wiewiora (2020)	Feedforward aggregation of team or project mental models to obtain a shared organization mental model	Aggregating team or project mental models for surgery by hospital teams to form a shared team or project mental model	Canbaloğlu et al. (2021b)
Occurrence of major events	Teams mobilize to react and find solutions to major events. The best solution is selected and institutionalized by the organization	[26]	Feedforward aggregation of team or project mental models to obtain a shared organization mental model	Aggregating team or project mental models for surgery by hospital teams to form a shared organization mental model	Canbaloğlu et al. (2021b)

which computational mechanisms have been found that can be associated to them. These computational mechanisms are based on findings from neuroscience, such as Hebbian learning [13] and mirroring, e.g., [15, 17, 19, 22]. Tables 2 and 3 address the mechanisms behind the arrows from left to right and vice versa connecting different levels in Fig. 1. Here, the arrows from left to right indicate feedforward learning (see Table 2), and the arrows from right to left indicate feedback learning (Table 3). The three different sections in Table 2 relate to arrows from individuals to teams or projects, from teams or projects to the organization, and from individuals directly to the organization level. Similarly, the three different sections in Table 3 relate to the

Table 3 Conceptual and computational mechanisms for feedback learning: from the organization to individuals and to teams or projects and from teams or projects to individuals

Conceptual mechanisms	Examples	Relevant references	Computational mechanisms	Examples	Relevant references
Feedback learning					
From organization to individuals					
Instructional learning from a shared organization mental model					
Coaching and mentoring	Learning from experienced people who have through years accumulated organizational knowledge	[26, 27]	Learning from internal communication channels and feedback aggregation	Individuals (doctors and nurses) learning an own mental model from a shared organization mental model for intubation	Canbaloğlu et al. (2021a)
Organizational systems	Individuals access organizational knowledge management systems, policies and procedures to inform their practices	[9], Iftikhar and Wiewiora (2020)	Learning from internal communication channels and feedback aggregation	Individuals (doctors and nurses) learning an own mental model from a shared organization mental model for intubation	Canbaloğlu et al. (2021a)
Training sessions	Provision of courses during which a facilitator shares organizational knowledge to individuals	[26, 27]	Learning from internal communication channels and feedback aggregation	Individuals (doctors and nurses) learning an own mental model from a shared organization mental model for intubation	Canbaloğlu et al. (2021a)
From organization to teams or projects					
Training sessions	Provision of courses during which a facilitator shares organizational knowledge to teams or projects	[26, 27)	Learning from internal communication channels and feedback aggregation	Teams of doctors and nurses learning a team mental model from a shared organization mental model for intubation	Canbaloğlu et al. (2021b)

(continued)

Table 3 (continued)

Conceptual mechanisms	Examples	Relevant references	Computational mechanisms	Examples	Relevant references
From teams or projects to individuals					
Learning by working together and joint-problem solving	Individuals of a team while working together are sharing their individual mental models and creating a new shared mental model by discussing and jointly solving a problem in hand	[18] Iftikhar and Wiewiora (2020)	Learning from internal communication channels and feedback aggregation	Individuals (doctors and nurses) learning an own mental model from a shared team mental model for intubation	Canbaloğlu et al. (2021b)
Training sessions	Provision of courses during which a facilitator shares team knowledge to individuals	[26, 27]	Learning from internal communication channels and feedback aggregation	Individuals (doctors and nurses) learning an own mental model from a shared team mental model for intubation	Canbaloğlu et al. (2021a)

arrows in Fig. 1 from the organization to teams or projects, from teams or projects to individuals and from the organization level directly to individuals. After introducing the computational modeling approach based on self-modeling networks in Sect. 3, in the subsequent Sects. 4 and 5 for a number of the computational mechanisms indicated in Tables 1, 2 and 3. more details will be given.

3   The Self-modeling Network Modeling Approach Used

In this section, the network-oriented modeling approach used is briefly introduced. A temporal-causal network model is characterized by the following; here X and Y denote nodes of the network that have activation levels that can change over time, also called states [20]:

- *Connectivity characteristics*: Connections from a state X to a state Y and their weights $\omega_{X,Y}$
- *Aggregation characteristics*: For any state Y, some combination function $c_Y(..)$ defines the aggregation that is applied to the impacts $\omega_{X,Y}X(t)$ on Y by its incoming connections from states X
- *Timing characteristics*: Each state Y has a speed factor η_Y defining how fast it changes for given causal impact.

The following canonical difference (or related differential) equations are used for simulation; they incorporate these network characteristics $\omega_{X,Y}$, $\mathbf{c}_Y(..)$, η_Y in a standard numerical format:

$$Y(t + \Delta t) = Y(t) + \eta_Y \left[\mathbf{c}_Y \left(\omega_{X_1,Y} X_1(t), \ldots, \omega_{X_k,Y} X_k(t) \right) - Y(t) \right] \Delta t \qquad (1)$$

for any state Y, where X_1 to X_k are the states from which Y gets incoming connections.

Modeling and simulation are supported by a dedicated software environment described in [20], it comes with a combination function library with currently around 65 combination functions. Some examples of these combination functions that are used here can be found in Table 4.

Applying the concept of *self-modeling network*, the network-oriented approach can also be used to model *adaptive* networks; see (Treur 2020a, b). By the addition of new states to the network which represent certain network characteristics, such characteristics become adaptive. These additional states are called self-model states, and they are depicted at the next level, distinguished from the base level of the network. For instance, the weight $\omega_{X,Y}$ of a connection from one state X to another state Y is represented by an additional self-model state $\mathbf{W}_{X,Y}$. In such a way, by including self-model states, any network characteristic can be made adaptive. An adaptive speed factor η_Y can be modeled by a self-model state $\mathbf{H}_Y$. The self-modeling network

Table 4 Examples of combination functions for aggregation available in the library

Name	Formula	Parameters
Advanced logistic sum $\mathbf{alogistic}_{\sigma,\tau}(V_1, \ldots, V_k)$	$\left[\frac{1}{1+e^{-\sigma(V_1+\ldots+V_k-\tau)}} - \frac{1}{1+e^{\sigma\tau}} \right] (1 + e^{-\sigma\tau})$	Steepness $\sigma > 0$; excitability threshold τ
Scaled maximum $\mathbf{smax}_\lambda(V_1, \ldots, V_k)$	$\max(V_1, \ldots, V_k)/\lambda$	Scaling factor λ
Euclidean $\mathbf{eucl}_{n,\lambda}(V_1, \ldots, V_k)$	$\sqrt[n]{\frac{V_1^n+\ldots+V_k^n}{\lambda}}$	Order n; scaling factor λ
Scaled geometric mean $\mathbf{sgeomean}_\lambda(V_1, \ldots, V_k)$	$\sqrt[k]{\frac{V_1*\ldots*V_k}{\lambda}}$	Scaling factor λ
Hebbian learning $\mathbf{hebb}_\mu(V_1, V_2, V_3)$	$V_1 * V_2(1 - V_3) + V_3$	V_1, V_2 activation levels of the connected states; V_3 activation level of the self-model state for the connection weight; persistence factor μ
Maximum composed with Hebbian learning $\mathbf{max\text{-}hebb}_\mu(V_1, \ldots, V_k)$	$\max(\mathbf{hebb}(V_1, V_2, V_3), V_4, \ldots, V_k)$	V_1, V_2 activation levels of the connected states; V_3 activation level of the self-model state for the connection weight; persistence factor μ

concept can be applied iteratively thus creating multiple orders of self-models (Treur 2020b). A second-order self-model can model an adaptive speed factor $\eta \mathbf{W}_{X,Y}$ by a second-order self-model state $\mathbf{HW}_{X,Y}$. Moreover, a persistence factor $\mu \mathbf{W}_{X,Y}$ of a first-order self-model state $\mathbf{W}_{X,Y}$ used for Hebbian learning can be modeled by a second-order self-model state $\mathbf{MW}_{X,Y}$.

4 Some Examples of Computational Mechanisms

In this section and the next one, for a number of conceptual mechanisms, it will be shown in more detail how they can be related to computational mechanisms in terms of the self-modeling network format. Some examples from (Canbaloğlu and Treur 2021a, b; Canbaloğlu et al. 2021a, b) will be briefly discussed, e.g., (see also Fig. 1 and Tables 1, 2 and 3):

- Learning by internal simulation: individual mental model learning based on internal simulation (conceptual) modeled by Hebbian learning (computational)
- Learning by observation: individual mental model learning based on observation (conceptual) and mirroring combined with Hebbian learning (computational)
- Learning by communication: individual mental model learning based on communication with another individual (conceptual) modeled by aggregation of communicated information with already available information (computational)
- Feedforward learning: shared team or organization mental model learning based on an individual or shared team mental model (conceptual) modeled by aggregation of multiple individual mental models
- Feedback learning: individual mental model learning based on a shared team or organization mental model (conceptual) modeled by aggregation of multiple shared team mental models.

In the current section, the first three bullets (all relating to Table 1) are addressed as mechanisms. In Sect. 5, the last two bullets (relating to Tables 2 and 3, respectively) are addressed, and it is also shown how the mechanisms involved can play their role in an overall multilevel organizational learning process. The mental models used as examples are kept simple; they concern tasks a, b, c, d which are assumed to be linearly connected.

Learning by internal simulation: Hebbian learning for an individual mental model

In Fig. 2a (see also Table 1) it is shown how internal simulation of a mental model by person B (triggered by context state con_1) activates subsequently the mental model states a_B to d_B of B and these activations in turn activate Hebbian learning of their mutual conection weights. Here for the Hebbian learning [13], the self-model state $\mathbf{W}_{X,Y}$ for the weight of the connection from X to Y uses the combination function $\mathbf{hebb}_\mu(V_1, V_2, W)$ shown in Table 4. More specifically, this function $\mathbf{hebb}_\mu(V_1, V_2, W)$ is applied to the activation values V_1, V_2 of X and Y and the current value

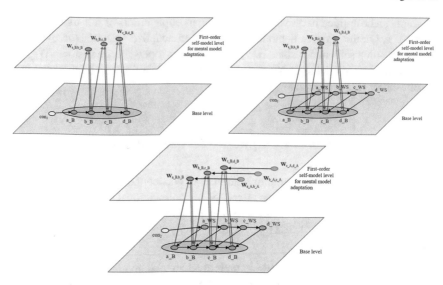

Fig. 2 **a** Upper, left. Learning by internal simulation: Hebbian learning during internal simulation. **b** Upper, right. Learning by observation: Hebbian learning after mirroring of the world states, **c** Lower. Learning by communication and by observation combined: learning by communication from person A to person B combined with Hebbian learning based on mirroring within person B

W of $\mathbf{W}_{X,Y}$. To this end upward (blue) connections are included in Fig. 2a (also a connection to $\mathbf{W}_{X,Y}$ itself is assumed but usually such connections are not depicted). The (pink) downward arrow from $\mathbf{W}_{X,Y}$ to Y depicts how the obtained value of $\mathbf{W}_{X,Y}$ is actually used in activation of Y. Thus, the mental model is learnt. If the persistence parameter μ is 1, the learning result persists forever, if $\mu < 1$, then forgetting takes place. For example, when $\mu = 0.9$, per time unit 10% of the learnt result is lost.

Learning by observation: observing, mirroring and Hebbian learning of an individual mental model

For learning by observation, see Fig. 2b (see also Table 1). Here, mirror links are included: the (black) horizontal links from World States a_WS to d_WS to mental model states a_B to d_B within the base (pink) plane. When the world states are activated, through these mirror links, they in turn activate B's mental model states which in their turn activate Hebbian learning like above; this is modeled, e.g., in [2].

Learning by communication: receiving communication and aggregation in an individual mental model

See Fig. 2c for a combined form of learning by communication and by observation as modeled, e.g., in [2]; see also Table 1. The horizontal links within the upper (blue) plane model communication from A to B. This communication provides input from the mental model self-model states $\mathbf{W}_{a\_A,b\_A}$ to $\mathbf{W}_{a\_A,b\_A}$ of A to the mental model self-model states $\mathbf{W}_{a\_B,b\_B}$ to $\mathbf{W}_{a\_B,b\_B}$ of B; this input is aggregated within these self-model states of B's mental model using the **max-hebb**$_\mu$ combination function (see Table 4). This function takes the maximum of the communicated value originating from $\mathbf{W}_{x\_A,y\_A}$ and the current value of $\mathbf{W}_{x\_B,y\_B}$ that is being learnt by B through Hebbian learning.

More complex examples covering multiple mechanisms for feedforward and feedback learning relating to Table 2 and 3 are shown in Sect. 5.

5 Computational Models for Feedforward and Feedback Learning

In this section, feedforward and feedback learning mechanisms (see also Fig. 1) in computational models in self-modeling network format will be explained with two examples from (Canbaloğlu et al. 2021b) and [5]; see the network pictures in Fig. 3 and Fig. 4. These are the mechanisms listed as the last two bullets of the list of building blocks of organizational learning process in the first paragraph of the Sect. 4.

Feedforward learning: formation of a shared mental model by individuals or teams

In Fig. 3, self-model **W**-states representing the weights of connections between mental model states at the base level and horizontal **W**-to-**W** connections between them are depicted in the first-order self-model level (blue plane). The rightward connections from **W**-states of individuals' mental models to **W**-states of teams'

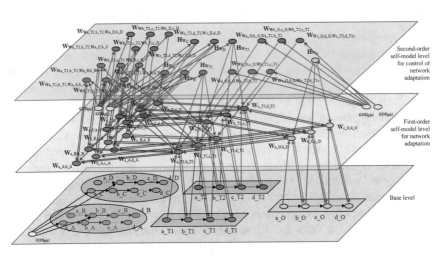

Fig. 3 Connectivity within the first-order self-model level for the adaptation of the mental models by formation of shared team and organization mental models (links from left to right: feedforward learning) and by instructional learning of individual mental models and shared team mental models from these shared mental models (links from right to left: feedback learning). (Canbaloğlu et al. 2021b)

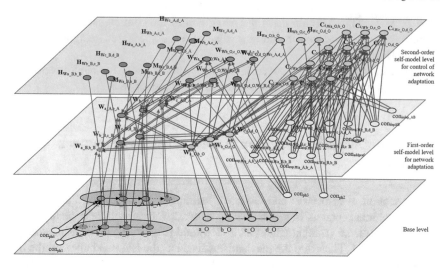

Fig. 4 Example involving context-sensitive control of aggregation in the process of shared mental model formation based on 16 context states (gray ovals) and four options of combination functions for aggregation [5]

shared mental models and from **W**-states of teams' shared mental models to **W**-states of the organization's shared mental model trigger the formation (by a form of aggregation) of shared mental models for teams and for the organization by feedforward learning.

Feedback learning: learning of individuals from shared mental models

In Fig. 3, self-model **W**-states of individuals' mental models (on the left) have connections coming from self-model **W**-states of their corresponding teams' shared mental models, and these team **W**-states have connections coming from self-model **W**-states of the organization's shared mental model. These leftward connections are used for individuals' improvements on their knowledge with the help of the shared mental models: the aggregated knowledge returns to the individuals by feedback learning.

Feedforward learning requires a combination function for aggregation of separate individual mental models to form a team's shared mental model, and a combination function for aggregation of different teams' mental models to form the organization's shared mental model. This aggregation can take place always according to one and the same method (modeled by one combination function), like in Fig. 3, or it can be adaptive according to the context. For real-life cases, the formation of a shared mental model is not same for different scenarios. Thus, making the aggregation adaptive improves the model in terms of applicability and accuracy.

In Fig. 4, context factors placed in the first-order self-model level (gray ovals in the blue plane) determine the choice of combination function during the aggregation of different mental models. Here, the combination function is dynamically chosen according to the activation status of the context factors that by their activation values characterize the context. In the second-order self-model level, C-states represent the choice of combination function for different mental model connections (between tasks a to d). Each C-state has (1) an incoming connection from each of the relevant context factors for the corresponding task connection it addresses (upward connections), and (2) one (downward) outward connection to the corresponding W-state. Thus, the control of the selection of the combination function is realized by the connections between context factors and C-states. Therefore, this approach makes the choice of combination function for the aggregation context sensitive. This makes the aggregation adaptive.

6 Discussion

Formalization and computational modeling of multilevel organizational learning is one of the major challenges for the area of organizational learning. The current paper addresses this challenge. Various conceptual mechanisms in multilevel organizational learning as identified in the literature were discussed. Moreover, it was shown how they can be formalized by computational mechanisms. For example, it has been discussed how formation of a shared mental model on the basis of a number of individual mental models, from a computational perspective can be considered a form of (feedforward) aggregation of these individual mental models.

The formalizations have been expressed using the self-modeling network modeling approach introduced in [20] and used as a basis for modeling dynamics, adaptation and control of mental models in [21]. The obtained computational mechanisms provide mathematical formalizations that form a basis for simulation experiments for the area of organizational learning, as has been shown in [4, 5], Canbaloğlu et al. (2021a, b). For example, in [4, 5], it is shown how specific forms of context-sensitivity of feedforward aggregation to obtain shared mental models can be modeled by second-order adaptive self-modeling networks according to the self-modeling network modeling approach applied to mental models from [20, 21].

The different types of mechanisms addressed cover almost all of the overall picture of multilevel organizational learning shown in Fig. 1, but by no means cover all relevant mechanisms. For example, for the sake of shortness factors that affect all levels, such as leaders, organization structure and culture, have been left out of consideration here. However, the modeling approach described here provides a promising basis to address in the future also the ones that were not addressed yet.

References

1. Benner MJ, Tushman ML (2003) Exploitation, exploration, and process management: the productivity dilemma revisited. Acad Manag Rev 28:238–256

2. Bhalwankar R, Treur J (2021) Modeling learner-controlled mental model learning processes by a second-order adaptive network model. PLoS ONE 16(8):e0255503

3. Bhalwankar R, Treur J (2021b) If only i would have done that…': a controlled adaptive network model for learning by counterfactual thinking. In: Proceedings of the 17th international conference on artificial intelligence applications and innovations, AIAI'21 advances in information and communication technology, vol 627. Springer, pp 3–16

4. Canbaloğlu G, Treur J (2021a) Context-sensitive mental model aggregation in a second-order adaptive network model for organisational learning. In: Proceedings of the 10th international conference on complex networks and their applications. studies in computational intelligence, vol 1015. Springer, pp 411–423

5. Canbaloğlu G, Treur J (2021b). Using boolean functions of context factors for adaptive mental model aggregation in organisational learning. In: Proceedings of the 12th international conference on brain-inspired cognitive architectures, BICA'21. Studies in computational intelligence, vol 1032. Springer, pp 54–68

6. Canbaloğlu G, Treur J, Roelofsma PHMP (2022) Computational modeling of organisational learning by self-modeling networks. Cognitive Syst Res J 73:51–64

7. Canbaloğlu G, Treur J, Roelofsma PHMP (2022b) An adaptive self-modeling network model for multilevel organisational learning. In: Proceedings of the 7th international congress on information and communication technology, ICICT'22, vol 2. Lecture notes in networks and systems, vol 448, Springer, pp 179–191

8. Chang A, Wiewiora A, Liu Y (2021) A socio-cognitive approach to leading a learning project team: a proposed model and scale development. Int J Project Manage

9. Crossan MM, Lane HW, White RE (1999) An organizational learning framework: from intuition to institution. Acad Manag Rev 24:522–537

10. Edmondson AC (2002) The local and variegated nature of learning in organizations: a group-level perspective. Organ Sci 13:128–146

11. Fang C, Lee J, Schilling MA (2010) Balancing exploration and exploitation through structural design: The isolation of subgroups and organizational learning. Organ Sci 21:625–642

12. Fiol CM, Lyles MA (1985) Organizational learning. Acad Manag Rev 10:803–813

13. Hebb DO (1949) The organization of behavior: a neuropsychological theory. Wiley, New York

14. Hogan KE, Pressley ME (1997) Scaffolding student learning: instructional approaches and issues. Brookline Books

15. Iacoboni M (2008) Mirroring people: the new science of how we connect with others. Farrar, Straus & Giroux, New York

16. Iftikhar R, Wiewiora A (2021) Learning processes and mechanisms for interorganizational projects: insights from the Islamabad-rawalpindi metro bus project. IEEE Trans Eng Manage. https://doi.org/10.1109/TEM.2020.3042252

17. Keysers C, Gazzola V (2014) Hebbian learning and predictive mirror neurons for actions, sensations and emotions. Philos Trans R Soc Lond B Biol Sci 369:20130175

18. Kim DH (1993) The link between individual and organisational learning. In: Sloan management review, fall 1993, pp 37–50. Reprinted in: Klein DA (ed) The strategic management of intellectual capital. Routledge-Butterworth-Heinemann, Oxford

19. Rizzolatti G, Sinigaglia C (2008) Mirrors in the brain: how our minds share actions and emotions. Oxford University Press

20. Treur J (2020) Network-oriented modeling for adaptive networks: designing higher-order adaptive biological, mental and social network models. Springer, Cham

21. Treur J, Van Ments L (eds) (2022) Mental models and their dynamics, adaptation, and control: a self-modeling network modeling approach. Springer

22. Van Gog T, Paas F, Marcus N, Ayres P, Sweller J (2009) The mirror neuron system and observational learning: implications for the effectiveness of dynamic visualizations. Educ Psychol Rev 21(1):21–30

23. Van Ments L, Treur J (2021) Reflections on dynamics, adaptation and control: a cognitive architecture for mental models. Cogn Syst Res 70:1–9

24. Van Ments L, Treur J, Klein J, Roelofsma PHMP (2021) A second-order adaptive network model for shared mental models in hospital teamwork. In: Nguyen NT et al (eds) Proceedings of the 13th international conference on computational collective intelligence, ICCCI'21. Lecture notes in AI, vol 12876. Springer, pp 126–140

25. Van Ments L, Treur J, Klein J, Roelofsma PHMP (2022) Are we on the same page: a controlled adaptive network model for shared mental models in hospital teamwork. In: J Treur, L Van Ments (2022), Ch 14

26. Wiewiora A, Smidt M, Chang A (2019) The 'How' of multilevel learning dynamics: a systematic literature review exploring how mechanisms bridge learning between individuals, teams/projects and the organization. Eur Manag Rev 16:93–115

27. Wiewiora A, Chang A, Smidt M (2020) Individual, project and organizational learning flows within a global project-based organization: exploring what, how and who. Int J Project Manage 38:201–214

Deep Learning-Based Black Hole Detection Model for WSN in Smart Grid

Korra Cheena, Tarachand Amgoth, and Gauri Shankar

Abstract Integration of wireless sensor network (WSN) in smart grid (SG) facilitates power distribution. The transfer of data in the sensor nodes (SN) is affected by malicious nodes in WSN at the same time, which leads to a black hole (BH) attack in the system. The BH attacks block the data instead of forwarding it to the destination. In order to overcome this, the proposed method adopts deep learning-based scheme to detect BH attacks in WSN. The proposed method used Bayesian theory and deep recurrent neural network (DRNN) for recognizing the malicious nodes in the network by corresponding route discovery time. After eliminating hostile nodes from the network, an optimum destination pathway is discovered by the grasshopper optimization algorithm (GOA). From the result analysis, the proposed approach can be seen to give higher throughput whereby reduced delay and increased average residual energy.

Keywords Wireless sensor network · Malicious nodes · Black hole attack · Route discovery time · Deep recurrent neural network · Grasshopper optimization algorithm

1 Introduction

The smart grid (SG) is an advanced bi-directional power flow system, which has the capability of self-healing, adaptive, and foresight detection of uncertainties [1]. Most importantly, the SG has the ability of self-monitoring, self-healing, remote control by analyzing the system by use of high numerous sensors. The SG is aimed to customize

K. Cheena (✉)
Electrical and Electronics Engineering, University College of Engineering, Kakatiya University, Kothagudem, India
e-mail: korrachinna@gmail.com

T. Amgoth
Computer Science and Engineering, Indian Institute of Technology (ISM), Dhanbad 826004, India

G. Shankar
Electrical Engineering, Indian Institute of Technology (ISM), Dhanbad 826004, India

© The Author(s), under exclusive license to Springer Nature Singapore Pte Ltd. 2023   19
A. Shukla et al. (eds.), *Computational Intelligence*, Lecture Notes in Electrical
Engineering 968, https://doi.org/10.1007/978-981-19-7346-8_2

the power demand based on the availability of power sources [2]. The WSN has a significant role in managing SG power distribution, whereas it is a risky task due to the features of electric grid and traffic patterns.

The sensor node (SN) comprises a microcontroller unit for processing and a transceiver unit, and these are powered by a battery. The SN in WSN is the units that overlook the data gathering and transmission then data gathered by sensors are transfer to the sink nodes [3]. Moreover, the data in WSN is routed via several routing approaches [4].

The black hole (BH) attack is a dangerous safety threats that degrades the WSN reliability. At that time, the malicious node replies to the request to capture more data packets false route to destination [5]. Malicious nodes in WSN can be held responsible and are the leading cause for BH occurrence. To accomplish this, a hidden Markov model (HMM) detection along with a decision model is suggested by Hanane et al. [6]. Whereas, this method was not providing a satisfactory outcome in terms of delivering the packet. A hybrid method is adopted for identifying both abnormality and BH, which includes the K-medoid tailored clustering algorithm and data set [7]. The approach, on the other hand, takes longer to send data to its target. The distributed coverage hole detection (DCHD) technique detects coverage holes by analyzing intersection points without sinks [8]. Whereas, this method fails to cope with the monitoring of irregular regions and to recover the hole with minimum mobility.

The suspicious router identification is an essential task for identifying the BH attack that affects the customized setting of the nodes for dropping the data packets without further forwarding. Researches in this field developed several methods to overcome this kind of attacks. Unfortunately, this kind of attack was still existed in real-world applications. In a technological view, artificial intelligence (AI) techniques gaining more attention to solve real-world problem. Thus, our research question is whether the AI can be employed in order to address the elimination of BH attacks; this motivated us to develop a deep learning-based BH attack detection and optimum path routing approach.

The contribution of this work is listed below,

- To detect the route discovery time of the SNs for examining the most probable malicious nodes containing path based on the Bayesian theory.
- To discover the BH attacks, the data of the SNs in the discovered malicious path is given as an input to the DRNN. Whereas the DRNN, find out the BH attack by sending examining its capability of data transmission. It can be considered that the affected nodes cannot transfer the data to the destination.
- Route the data packets via optimum path by using GOA, thereby clearing network by removing the harmful nodes.

The organization of the work is given as; Sect. 2 discusses the several methods suggested by the authors to detect the BH attacks. The method suggested in the paper is explained in Sect. 3 that comprises the Bayesian theory, DRNN, and GOA. The output of the present research in results is validated in Sect. 4. The conclusions observed from the system are given in Sect. 5.

2   Literature Review

Data transmission without delay by consuming minimum power is the primary considerations of WSN. Thus, various authors were suggested different approaches to detect the anomaly in WSN. Some of the methods to detect and prevent BH in WSN are listed below.

For finding and removing BH in WSN, Bisen et al. [9] proposed trust and secure routing (TSR) in WSN. In that proposed method, the source node, node number, and destination node were given as input. That proposed method facilitates the data transmission through an alternative route through identification nodes. That the methods suggested in the research offered higher throughput by minimizing energy utilization and packet loss.

To detect the wormhole (WH) as well as BH attack in WSN, Umarani and Kannan et al. [10] proposed a hybrid tissue growing algorithm (HTGA). The suggested model was used in two different models: networked tissue growth (NTG) and swarm tissue growth (STG). After the WSN is built, the proposed technique includes numerous processes such as neighbor evaluation, tissue growth, node grouping, route finding, and so on.

Alqahtani et al. [11] proposed an extreme gradient boosting model based on genetics in order to protect data from a range of network risks such as BH, gray hole, WH, and floods (GXGboost). Furthermore, a GB model was improved in order to identify tiny network intrusions. The suggested model outperforms the competition in terms of gray hole, BH, and flooding schedule detection rates.

To examine and predict the intrusions in WSN, Rezvi et al. [12] proposed a data mining approach. In that proposed method, the types of attack were expressed by numeric values, then it was divided for training as well as testing data by employing onefold cross-validation. In addition, in the training, Naive Bayes, logistic regression, support vector machine (SVM), ANN, and the KNN approach were utilized. That model was made for real-time use. Moreover, in terms of managing imbalance issues, the suggested method performs well than synthetic minority over-sampling technique (SMOTE).

In order to expand the lifetime of WSN, Vimalarani et al. [13] suggested an enhanced particle swarm optimization (PSO) for centralized clustering. Moreover, the head of the cluster was chosen by the PSO algorithm in a distributed manner based on minimized energy utilization. The result of that proposed method was taken in terms of several performances for a varying numbers of rounds.

To enhance the performance of cluster-based data transmission in WSN, Sefati et al. [14] proposed an optimized BH algorithm and an ant colony optimization (ACO) algorithm. In iterations of the BH detection algorithm, the greatest solution was chosen as BH; after that, it pulls out other candidates around it to form a star.

3   Proposed Methodology

The model deployed 50 SNs randomly in 100 m × 100 m area of WSN. The HELLO packets are sent by the source nodes to all the SNs in the network, including malicious nodes, to detect the routing attacks. Based on the route discovery time (RT), the high probability SN to become malicious is identified by Bayesian theory. Here, the route discovery time is evaluated with the threshold route discovery time. If the duration of RT is too large, then it is regarded as a BH attacked path.

The lower RT duration indicates that the route is not affected by any kind of attacks. The Bayesian theory identifying the path where the interruption occurs. The DRNN is used for spotting the BH attacks in the WSN network. After the identification of BH attacked malicious nodes, they are subjected to elimination in order enhance the data transmission. The DRNN categorizes the nodes with respect to their malicious probability. For routing the data to an optimum path, the grasshopper optimization algorithm (GOA) is adapted in this work. By this algorithm, each node in WSN expects malicious nodes to route the destination by the optimized path. Figure 1 depicts the process of suggested methodology.

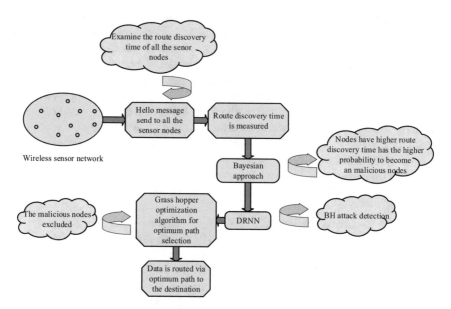

Fig. 1 Flow of proposed methodology

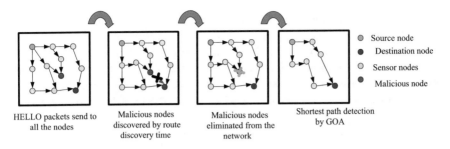

HELLO packets send to all the nodes | Malicious nodes discovered by route discovery time | Malicious nodes eliminated from the network | Shortest path detection by GOA

○ Source node
● Destination node
○ Sensor nodes
● Malicious node

Fig. 2 Overall working of the proposed method

3.1 Proposed Network Architecture

Initially, the source node sent the HELLO packets to all nodes to know about its status. By using this manner, the route discovery time of the SNs is examined. The RT is defining as the duration of the initial node discovers the message to the duration it obtains the resultant route response. Consider the source node (SN) transmits the data (HELLO), which is gathered by a node (AN) at the time, 'T_1' and the 'AN' respond at the time 'T_2'. The route discovery time is expressed as

$$RT = T_2 - T_1 \tag{1}$$

If the RT is larger or smaller, then the data may be corrupted. Moreover, the considerable duration of RT indicates that the particular path is affected by an intrusion. The duration of HELLO data packets obtained by the SNs is measured to check the nodes for malicious activity. The steps used in the proposed work are depicted in Fig. 2.

3.2 Bayesian Theory

The Bayes theorem is applied with RNN for prediction of malicious nodes presented in the WSN. The input to DRNN is number of nodes $N = (n_1, n_2, \ldots n)$ along with its route discovery, time $RT = (t_1, t_2, t_3, \ldots t)$ represents neuron input. For example, A and B represent events and that are not equal to zero $P(Y)$ is not zero. Then, the Bayes theorem is given by

$$P(X \cap Y) = P(X|Y)P(Y) = P(Y|X)P(X) \tag{2}$$

$$P(X|Y) = \frac{P(Y|X)P(X)}{P(Y)} \tag{3}$$

here, $P(X|Y)$ represents the probability of event A happens assumed that B is true, and if A is true, $P(Y|X)$ reflects the likelihood that event B will occur. It is like A given a fixed B because $P(Y|X) = L(Y|X)$, $P(X)$, and $P(Y)$ resemble the probability of observing A and B, correspondingly. It should be noted that X and Y must be different events. The proposed work adopted BRNN, in which the RNN is trained by Bayesian theory. By this process, the BH occurrence is subjected to elimination from network.

3.3 Deep Recurrent Neural Network (DRNN)

After finding the affected route, the malicious nodes and BH attack are identified by the DRNN. The input to RNN is through a time sequence 't' which is represented as $(\ldots x_{t-1}, x_t, x_{t+1} \ldots)$, and the connection is assigned with the weight matrix A_{1h}. If the hidden layers having 'n' units, they are represented as $p_t = (p_1, p_2, p_3, \ldots p_n)$. The hidden layers are connected through a recurrent network. By incurring p_t as current state, p_{t-1} as the preceding state and x_t denoting input state. The equation of the state at 't' time is given by

$$p_t = \tan p(A_{hh} p_{t-1} + A_{xh} x_t) \tag{4}$$

where the weight at a recurrent neuron is A_{hh}, A_{xh} is the weight at input neuron, $\tan p$ is the activation function. The output parameter produced after examining last state was calculated. The output calculation was done by

$$y_t = A_{hy} p_t \tag{5}$$

where y_t is output, A_{hy} is the weight at output layer, and f_0 is output function. The set of 'N' training sequence is given by

$$S = \left\{ \left((x_1^n, y_1^n) \ldots (x_T^n, y_T^n) \right) \right\}_{n-1}^N \tag{6}$$

An optimum path is built to send data to destination. Hence to optimize the path without including malicious nodes, the GOA is used. The BRNN needs few data for training; thus, it is suitable for classification problems. The BRNN for detecting malicious nodes is classified by using following equation.

$$\theta \approx N(0, I) \tag{7}$$

$$\pi = f_{nn}(\text{seq}; \theta) \tag{8}$$

$$y \approx \text{cat}(\text{logistic}(\pi)) \tag{9}$$

where θ is the potential feature between the node, f_{nn} is the neural network, π is the non-normalized log parameter to detect the classes, y denotes whether the data forwarded further or not. cat represents the classification distribution.

3.4 Grasshopper Optimization Algorithm (GOA)

The optimum path for transferring the data is chosen by the GOA. The mathematical representation formulated for the suggested algorithm in terms of finding an optimum path in WSN is given below.

$$X_i = D_i + B_i + R_i \tag{10}$$

here, X_i is SN's current position, D_i refers to the distance in between neighboring nodes, B_i represents the distance from BS, and R_i is the node's remaining energy. $d_{ij} = |x_j - x_i|$ and $d_{ij} = x_J - x_i/d_{ij}$ represent unit vectors from ith SN to jth SN. c is the maximum and minimum value. The position update of SN can be obtained by the following equation.

$$X_i^d = c \left(\sum_{\substack{j=1 \\ j \neq i}}^{N} c \frac{ub_d - lb_d}{2} s(|x_j - x_i|) \frac{x_j - x_i}{d_{ij}} \right) + T_d^\wedge \tag{11}$$

In the optimal solution, ub_d symbolizes the upper bound of dth dimension, lb_d denotes the lower bound of dth dimension, and $T_d^\wedge$ specifies the value of dth dimension.

4 Result and Discussion

The efficacy of suggested model for spotting malicious nodes and selecting the optimum route routing is described in this section. Here, we are using 50 sensor nodes in WSN for data communication. The layout of the network is depicted in Fig. 3. The suggested proposition is evaluated in a network having dimension of 100 m × 100 m. The topology used in our proposed method is depicted in Fig. 3.

In this proposed method, malicious node is detected based on the route discovery time. Based on its capacity to send data farther, the corrupt node can be identified. The SNs are represented by little circles (blue-unaffected sensor node, red-malicious node, green-best nodes to route the data) in Fig. 3. Moreover, the suggested approach is equated with the prevailing approaches such as competitive clustering (CC), enhanced PSO clustering (PECC), cuckoo search (CS) optimization, firefly

Fig. 3 WSN topology with 50 SNs

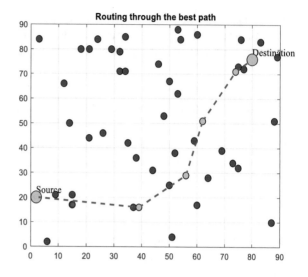

algorithm (FA), BH-ACO, and ant colony optimization (ACO) for increasing number of rounds.

The packet transmission delay in accordance with the increasing number of rounds is depicted in Fig. 4. The delay in packet transmission is decreased with the increasing number of rounds. It has been seen that suggested method mitigates the packet transmission delay to 0.015 s at 100 rounds. The count of alive nodes for the growing rounds is depicted in Fig. 5. By analyzing Fig. 5, it has been seen that the proposition suggested improves alive nodes for increasing number of round by examining malicious activity by earlier.

Fig. 4 Analysis of delay in terms of rounds

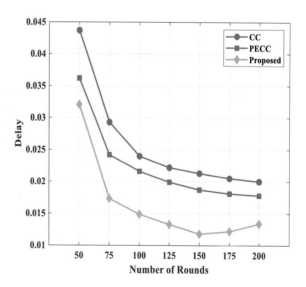

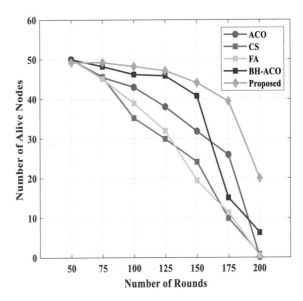

Fig. 5 Analysis of alive nodes

The average residual energy of the nodes for the increasing number of rounds is depicted in Fig. 6. Thus, the available energy is reduced with the increasing number of rounds. At 200th round, the average value of residual energy obtained by using proposed method is 20 J that is greater than the prevailing methods. Higher packets received at the destination for increasing rounds are represented in Fig. 7. The number of packets carried to the destination escalates as the number of rounds increases, according to the obtained results.

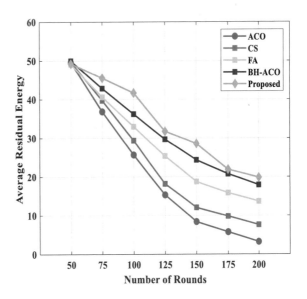

Fig. 6 Analysis of average residual energy of the nodes

Fig. 7 Analysis of packets reached in the destination

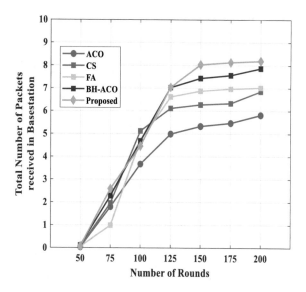

Figure 8 shows the network's life duration in terms of number of rounds played. The lifetime of the network may be increased to 180 s utilizing the suggested technique during round because the proposed method stores more energy in nodes. Figure 9 shows the network's throughput for various numbers of rounds. At 200th round, the throughput of CC, PECC, and proposed method is 3.7×10^4, 3.9×10^4, and 4.2×10^4, and the throughput of the network is increased with the increasing rounds.

Fig. 8 Analysis of network life time

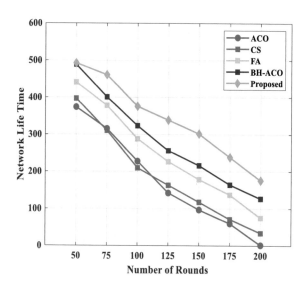

Fig. 9 Analysis of throughput

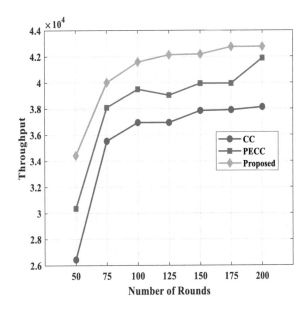

4.1 Discussion

A hybrid method of optimization using cat swarm optimization called M-CSO (monarch-cat swarm optimization) is used in this [15]. The delay and throughput offer by M-CSO at the completion of 100th round which are 13 ms and 65.24%, respectively, which are lower than the proposed method. Similarly, a trustable and secure routing algorithm is employed by Rajkumar et al. [16]. Moreover, the trust and secure route is examined by the CS algorithm. The latency of the method is 16000 ms, the lifespan of the network is 620 s, and the throughput is 68 at the network size of 8. The method used in [17] offers packet delivery of 0.93, throughput is 126.33 kbps, and delay is 22.03 ms by detecting the BH. From these analyzes, it can been determined that our proposed strategy is performs well than the current ways.

5 Conclusion

Several attacks in WSN will worsen network behavior by interrupting communication among nodes. The leading cause for any type of attack in WSN is malicious nodes. Hence to overcome this, the proposed method used DRNN along with the Bayesian approach in which DRNN is adopted for investigating malicious nodes in the WSN based on the Bayesian probabilistic approach. Furthermore, the DRNN examines the nodes' harmful activities using route information that has been tampered with. The GOA determines the best course to take after identifying a rogue node in the network.

These combined approaches of both DRNN and GOA will improve the communication through its better identification ability. For the purpose of examination of the efficacy of suggested model, comparative analysis with the existing approaches in regards alive node, throughput, energy, lifespan, and destination reached packets is made. From the result analysis, it is concluded that the projected method enhances the network performance by offering higher throughput, whereby reducing delay and increased residual energy along with increased network lifetime.

References

1. Rekik S et al (2017) Wireless sensor network based smart grid communications: challenges, protocol optimizations, and validation platforms. Wirel Pers Commun 95(4):4025–4047
2. Burunkaya M, Pars T (2017) A smart meter design and implementation using ZigBee based wireless sensor network in smart grid. In: 2017 4th international conference on electrical and electronic engineering (ICEEE). IEEE
3. Kong L et al (2018) An energy-aware routing protocol for wireless sensor network based on genetic algorithm. Telecommun Syst 67(3):451–463
4. Rathee A, Singh R, Nandini A (2016) Wireless sensor network-challenges and possibilities. Int J Comput Appl 140(2):1–15
5. Ghugar U, Pradhan J, Biswal M (2016) A novel intrusion detection system for detecting black hole attacks in wireless sensor network using AODV Protocol. IJCSN-Int J Comput Sci Netw 5(4)
6. Kalkha H, Satori H, Satori K (2019) Preventing black hole attack in wireless sensor network using HMM. Procedia Comput Sci 148:552–561
7. Ahmad B et al (2019) Hybrid anomaly detection by using clustering for wireless sensor network. Wireless Pers Commun 106(4):1841–1853
8. Kumar SP, Chiang M-J, Wu S-L (2016) An efficient distributed coverage hole detection protocol for wireless sensor networks. Sensors 16(3):386
9. Bisen D et al (2019) Detection and prevention of black hole attack using trusted and secure routing in wireless sensor network. In: International conference on hybrid intelligent systems. Springer, Cham
10. Umarani C, Kannan S (2020) Intrusion detection system using hybrid tissue growing algorithm for wireless sensor network. Peer-to-Peer Netw Appl 13(3):752–761
11. Alqahtani M et al (2019) A genetic-based extreme gradient boosting model for detecting intrusions in wireless sensor networks. Sensors 19(20):4383
12. Rezvi MA et al (2021) Data mining approach to analyzing intrusion detection of wireless sensor network. Indonesian J Electr Eng Comput Sci 21(1):516–523
13. Vimalarani C, Subramanian R, Sivanandam SN (2016) An enhanced PSO-based clustering energy optimization algorithm for wireless sensor network. Sci World J 2016
14. Sefati S, Abdi M, Ghaffari A (2021) Cluster-based data transmission scheme in wireless sensor networks using black hole and ant colony algorithms. Int J Commun Syst 34(9):e4768
15. Patil PA, Deshpande RS, Mane PB (2020) Trust and opportunity based routing framework in wireless sensor network using hybrid optimization algorithm. Wirel Pers Commun 115(1):415–437
16. Mehetre DC, Roslin SE, Wagh SJ (2019) Detection and prevention of black hole and selective forwarding attack in clustered WSN with active trust. Cluster Comput 22(1):1313–1328
17. Wazid M, Das AK (2017) A secure group-based blackhole node detection scheme for hierarchical wireless sensor networks. Wirel Pers Commun 94(3):1165–1191

Video Forgery Detection and Localization with Deep Learning Using W-NET Architecture

Bhanu Tokas, Venkata Rohit Jakkinapalli, and Neetu Singla

Abstract We propose a W-Net architecture-based approach for detecting and localizing regions of video forged using copy-move forgery technique. The proposed methodology can be utilized for the detection of forged videos with a high degree of efficiency. The model is capable of detecting forgery even when manipulations are done in complex settings, such as those with a dynamic background or complex movement of the object in the video files. The portion of the video frame that has been tampered with using temporal copy and paste (TCP) video in painting techniques can also be localized by proposed model. With lossless video clips, we were able to achieve the best video results. Finally, we were able to develop an algorithm capable of performing a task that simply required a video clip as input.

Keywords Video forensics · Deep learning · W-Net

1 Introduction

In today's modern world, the amount of information being shared and circulated locally and internationally has risen to unprecedented levels. Various social media channels use to communicate information regarding events that are taking place across the world in real time. This level of data transmission has brought everyone across the globe closer together; people can now discuss political events and share news across the globe in seconds. With this level of information interchange comes its drawbacks; because information and news are sent in real time, it is difficult to check whether the information is authentic or manipulated. Deepfakes, forged videos, manipulated photographs, and other forms of misinformation have recently surfaced the Internet.

B. Tokas · V. R. Jakkinapalli · N. Singla (✉)
Netaji Subhas University of Technology, New Delhi, India
e-mail: neetu.co19@nsut.ac.in

© The Author(s), under exclusive license to Springer Nature Singapore Pte Ltd. 2023
A. Shukla et al. (eds.), *Computational Intelligence*, Lecture Notes in Electrical
Engineering 968, https://doi.org/10.1007/978-981-19-7346-8_3

Copy-move form of manipulation is difficult to detect since the area that has been copied belongs to the same image and no traceable changes can be observed in noise, texture, and compression features. Most of research works in this area have focused on the analysis of still images. However, detecting copy-move forgery operation in videos is more challenging because it can occur in both the spatial and temporal domains.

In this paper, we propose an algorithm that is able to identify if a video file is manipulated or not and highlight the region in frames that have been manipulated. Recently, CNN-based methods have shown excellent superiority.

We propose a W-Net-based deep neural network to classify forged videos and localize the regions that have been forged.

The upcoming sections will discuss the following. In Sect. 2, we explore the previous works in the detection of forgeries in video files. The framework of the proposed W-Net-based forgery detection scheme is described in Sect. 3. In Sect. 4, experiments and results are demonstrated. Finally, in Sect. 5, we discuss the conclusions.

2 Related Work

For detecting different types of modification in digital video, a variety of algorithms have been presented in literature. Here, the review for copy-move forgery is discussed. Authors in [1, 2] extracted block-wise statistical features from frames and compared for similarity. An algorithm is presented in [1] for detecting copy-move forgery in videos by utilizing spatial and temporal correlation as a characteristic. However, when duplication is done in static background videos or shots with moving camera, these become ineffective. The authors of [3] propose a methodology for identifying region duplication based on the ghost shadow artifact as a feature. The authors of [4, 5] proposed an approach for detecting forgery based on noise and quantization residue. But, the algorithms [3–5] have only been tested in static background videos. Subramanyam et al. [6] proposed a copy-move tampering detection technique that uses the video compression features and block-wise histogram of oriented gradients (HOGs) descriptors. Because of its significantly large computational complexity, this technique is not effective for long videos. Pandey et al. [7] proposed a copy-move forgery detection approach based on (SIFT) and KNN matching in video frames. In Singh et al. [8] they have proposed a scheme to detect region duplication and frame duplication. To uncover the forgery, this technique calculates the level of similarity between regions of the frames. Both stationary and dynamic background videos are used to test the algorithms. Su et al. [9] proposed a methodology for identifying duplicated regions by exploiting exponential Fourier moments (EFMs) features. EFM features are retrieved from each block in the current frame and compared to see if there is a match. The post-verification scheme (PVS) is then used to find the forged area in the video frame. The type of forgery where certain regions were mirrored the provided solution performed well. MantraNet [10] is a solution proposed for image forgery detection and localization method using deep neural networks. MantraNet is capable of distinguishing 385 types of different image manipulation techniques. These manipulations are traced using an image manipulation classifier, which is based on VGG [11], ResNet [12], and DnCNN [13]. These CNN models are customized using similar number of filters and hyperparameters, and the best results are reported.

3   Methodology Proposed

In this section, we present an effective approach to detect and localize video forgery. In literature, we observe that deep neural networks have been proven to be useful for detecting and localizing static and moving objects. We propose a W-Net-based deep neural network that produces good forgery detection results and provides the ability to localize region where the forgery operation was performed. Our approach uses W-Net architecture as it along with its predecessor U-NET has proven to be very effective in the case image segmentation with the later providing excellent results in case of biomedical image segmentation. The proposed modified W-Net architecture takes the video frame by frame and produces two branches as outputs. One of the branches will be used for detection of forged frames and the other for localization task.

Proposed Network Architecture While W-Net-based structures have been used in various fields related to image segmentation including object segmentation [14] and density map estimation [15], it has yet to be utilized for image anomaly detection. One important difference that exists between this implementation and the previously cited implementations is that there exist connections not only between the flanking and the middle branches but also between the flanking branches themselves as well. These connections between the flanking branches enhance the capability of the network to recreate the original lower convolutional features during the upsampling by the left flanking branch. These lower convolutional features may not be sent forward by the middle branch as convolution operation may reveal some higher convolutional features. Also, even if the lower convolutional features are passed to be passed by the middle branch, it might take several epochs to train a one-to-one function instead of its close approximation. Further, after the W-NET architecture, one may notice the model diverges or branches into two wings, such as the localization wing and the classification wing. Figure 1 shows the proposed model architecture.

The localization wing consists of a simple convolution layer followed by output. The convolution layer herein uses the output of the W-Net to represent the extracted features in a single channel image of the same dimension as the original input. We have used mean squared error (MSE) loss here. The classification wing consists of a convolutional network followed by a dense network. The convolutional layers extract the spatial information encoded by the W-Net. Once extracted, these features are reshaped or "flattened", which are consequently fed to a dense network which

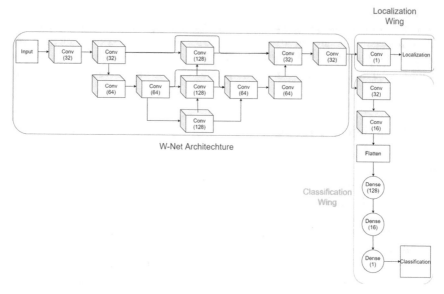

Fig. 1 Proposed W-Net-based forgery detection architecture

makes a classification of forged or normal on basis of these features. Herein, we have simply used binary cross entropy as the loss function.

Conventionally, the model would have been much deeper and had many more filters to avoid information loss in the bottleneck regions, but the added skip connections are able to somewhat overcome these limitations as early in training skip connections provide a more direct path through a shorter sub-network which leads to more meaningful gradient steps in that early and crucial phase of training. Thus, they are able to provide satisfactory performance even from a relatively miniature model. This allows training useful models even with limited computational resources.

4 Experiment Results and Discussion

Dataset To conduct experiments, we used the REWIND dataset [16, 17]. The dataset contains 20 video clips, 10 of which are authentic video clips, and 10 are tampered clips. Each clip has a size of 320×240 pixels and rate of frames set to 30 frames per second. Because the original sequences were recorded on low-end devices, they were all compressed using MJPEG/H264 video codecs. Further, uncompressed files were used to save forged sequences. Each forged clip is accompanied with a MAT file that contains the ground truth information. For our experimentation, we took a 80:20 split on the dataset for the training and testing of network. Experiments were performed on Google Colab Platform.

Localization Results As you may note from the above Tables 1,2 and Fig. 2. our model significantly outperforms MantraNet for various localization metrics.

Classification Results Now, as MantraNet concerns itself only with localization, for classification, we shall compare our model against the works of Mathai [18]. So, as one may observe from Fig. 3 and Table 3, our W-Net model performs significantly better than the Mathai model in all metrics except precision wherein in the difference ¡0.01.

As our aim for this problem is to detect forged frames in a video, our primary aim is to reduce true negatives at minimal cost from false positives. This can only be achieved when we have both high precision and high recall. Low recall can lead

Table 1 SSIM and PSNR metrics

Metric	W-Net	MantraNet
SSIM	0.464 ± 0.0001	0.353 ± 0.0001
PSNR	18.659 ± 0.004	18.564 ± 0.004

Table 2 Localization classification report (W-Net)

Class	(W-Net)			(MantraNet)		
	$F1$-score	Recall	Precision	$F1$-score	Recall	Precision
Original	0.91	0.91	0.91	0.84	0.99	0.73
Forged	0.80	0.79	0.81	0.74		
Weighted avg	0.87	0.88	0.87	0.79	0.74	0.67
Macro avg	0.86	0.85	0.85	0.83	0.58	0.56
Accuracy			0.88			0.74

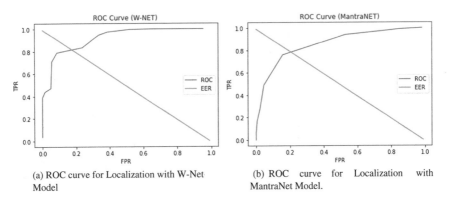

(a) ROC curve for Localization with W-Net Model

(b) ROC curve for Localization with MantraNet Model.

Fig. 2 Localization ROC

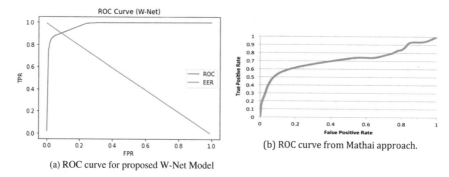

(a) ROC curve for proposed W-Net Model

(b) ROC curve from Mathai approach.

Fig. 3 Classification results

Table 3 Classification metrics

Model	$F1$-score	Recall	Precision	Accuracy
W-Net	0.91	0.92	0.91	0.9178
Mathai [18]	0.87	0.82	0.928	0.835

to missing out actual forgeries while low precision would overwhelm a system with false positives; hence, a model with good balanced performance in both these metrics is much preferable to a model that is perfect in one, but lacking in another. Thus, our model would have still been preferable even if the difference in precision was much more than presently seen.

5 Conclusion and Future Works

In this paper, we introduce a modified version of a W-NET architecture as a solution to video forgery detection and localization. The W-NET model has obtained: 92% precision, 92% recall, 92% accuracy for detection of forged frames in a video. These results have shown that it is effectively able to isolate the forged frames and also the forged regions in the video frames.

One of the primary limitations that we faced during the development of this model was lack of computational resources. One could create much more robust models by adding more layers to extract further down sampled features along with more filters. Further, this model may be retrained and utilized for various segmentation-related tasks including but not limited to, object segmentation, tumor segmentation, etc. (Fig. 4).

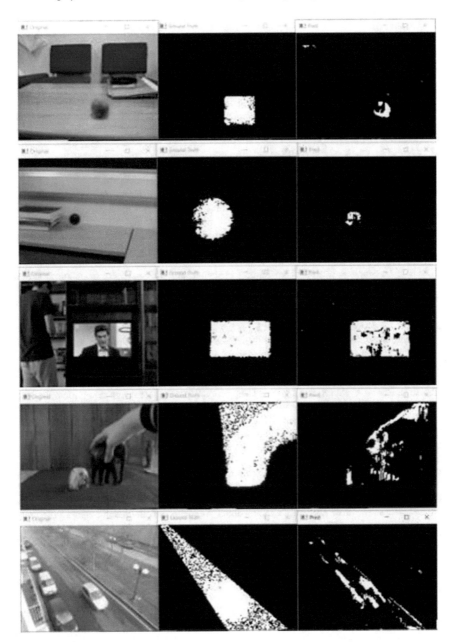

Fig. 4 Some sample results from our proposed model

References

1. Wang W, Farid H (2007) Exposing digital forgeries in interlaced and deinterlaced video. IEEE Trans Inf Forensics Secur 2(3):438–449
2. Wang W, Farid H (2007) Exposing digital forgeries in video by detecting duplication, pp 35–42
3. Zhang J, Su Y, Zhang M (2009) Exposing digital video forgery by ghost shadow artifact
4. Chetty G, Biswas M, Singh R (2010) Digital video tamper detection based on multimodal fusion of residue features. In: 2010 4th international conference on network and system security, pp 606–613
5. Goodwin J, Chetty G (2011) Blind video tamper detection based on fusion of source features. In: 2011 international conference on digital image computing: techniques and applications, pp 608–613
6. Subramanyam A, Emmanuel S (2012) Video forgery detection using hog features and compression properties. In: 2012 IEEE 14th international workshop on multimedia signal processing (MMSP), pp 89–94
7. Pandey RC, Singh SK, Shukla KK (2014) Passive copy-move forgery detection in videos. In: 2014 international conference on computer and communication technology (ICCCT), pp 301–306
8. Singh G, Singh K (2019) Video frame and region duplication forgery detection based on correlation coefficient and coefficient of variation. Multimedia Tools Appl 78:05
9. Su L, Li C, Lai Y, Yang J (2018) A fast forgery detection algorithm based on exponential-Fourier moments for video region duplication. IEEE Trans Multimedia 20(4):825–840
10. Wu Y, AbdAlmageed W, Natarajan P (2019) Mantra-net: manipulation tracing network for detection and localization of image forgeries with anomalous features. In: 2019 IEEE/CVF conference on computer vision and pattern recognition (CVPR), pp 9535–9544
11. Simonyan K, Zisserman A (2015) Very deep convolutional networks for large-scale image recognition. In: CoRR, vol abs/1409.1556
12. He K, Zhang X, Ren S, Sun J (2016) Deep residual learning for image recognition. In: 2016 IEEE conference on computer vision and pattern recognition (CVPR), pp 770–778
13. Zhang K, Zuo W, Chen Y, Meng D, Zhang L (2017) Beyond a Gaussian denoiser: residual learning of deep CNN for image denoising. IEEE Trans Image Process 26(7):3142–3155
14. Xia X, Kulis B (2017) W-net: a deep model for fully unsupervised image segmentation. In: ArXiv, vol abs/1711.08506
15. Valloli VK, Mehta K (2019) W-net: reinforced u-net for density map estimation. In: CoRR, vol abs/1903.11249
16. Bestagini P, Milani S, Tagliasacchi M, Tubaro S (2013) Local tampering detection in video sequences. In: 2013 IEEE 15th international workshop on multimedia signal processing (MMSP), pp 488–493
17. Qadir G, Yahaya S, Ho ATS (2012) Surrey university library for forensic analysis (sulfa) of video content. In: IET conference on image processing (IPR 2012), pp 1–6
18. Mathai M, Rajan D, Emmanuel S (2016) Video forgery detection and localization using normalized cross-correlation of moment features. In: 2016 IEEE southwest symposium on image analysis and interpretation (SSIAI), pp 149–152

Learning to Transfer Knowledge Between Datasets to Enhance Intrusion Detection Systems

Quang-Vinh Dang

Abstract Software-defined network (SDN) is a technology that is being used widely to reduce the time and effort required for programming network functions. However, by splitting the control layer and data layer, the SDN architecture also attracts numerous types of attacks such as spoofing or information disclosure. In the recent years, a few research articles coped with the security problem by introducing open datasets and classification techniques to detect the attacks to SDN. The state-of-the-art techniques perform very well in a single cross-validation dataset, i.e., in the situation, the training and the evaluation datasets are being withdrawn from the same source. However, their performance reduces significantly in the presence of concept drift, i.e., if the testing dataset is collected from a different source than the observed dataset. In this research study, we address this cross-dataset predictive issue by several concept drift detection techniques. The experimental results let us claim that our presented models can improve the performance in the cross-dataset scenario.

Keywords Intrusion detection system · Machine learning · Classification software-defined network

1 Introduction

Intrusion detection system (IDS) plays a crucial role in modern cyber-security systems [2]. In the recent years, there are a lot of researchers have utilized the power of modern machine learning algorithms [5] to establish a classifier that can do the classification task of an IDS. Seemingly, the most popular approaches are to build a classifier using some supervised or other methods such as unsupervised or reinforcement machine learning algorithms [4, 10] using a published dataset [17] then evaluate the algorithm using an unseen part of the same dataset.

The approaches achieve a lot of success in the recent years. The predictive performance metric such as accuracy or AUC of the state-of-the-art models is extremely

Q.-V. Dang (✉)
Industrial University of Ho Chi Minh City, Ho Chi Minh City, Vietnam
e-mail: dangquangvinh@iuh.edu.vn

© The Author(s), under exclusive license to Springer Nature Singapore Pte Ltd. 2023   39
A. Shukla et al. (eds.), *Computational Intelligence*, Lecture Notes in Electrical
Engineering 968, https://doi.org/10.1007/978-981-19-7346-8_4

high and usually be close to 1.0 [2]. However, the recent studies [6, 8] suggested that this methodology might not tell us the whole story. The researchers in the work of [6] reported that the predictive performance of supervised models will drop significantly if it has to predict the unseen data from a different dataset. Unfortunately, the researchers do not consider the phenomenon seriously and still use a same dataset in the entire process.

In this research paper, we deal with the problem of intrusion detection when the training set and the testing set come from different sources.

2 Related Works

We quickly review some state-of-the-art research studies in the topic of our paper in this section.

One of the recent works [2] studied extensively diversified machine learning models that is possibly be used for enhancing the intrusion detection systems. The authors evaluated a great deal of supervised machine learning models such as Naïve Bayes, SVM, random forest and XGBoost using a modern intrusion dataset [16]. The predictive metric such as accuracy and AUC of the algorithms is collected, evaluated, compared and analyzed. The work then be extended by following research studies [3, 7, 14]. Feature selection based on the previous works is presented in the study of [12].

The IDS problem for software-defined networks attends a lot of attention in the recent years [11]. The authors of [6, 15] address the problem by multiple machine learning algorithms.

Reinforcement learning algorithms that do not require labeled datasets have been studied recently in the work of [1, 9]. The algorithms might work well given enough time to learn.

Mostly, the published research works focusing on single-dataset evaluation, i.e., the researchers evaluate their presented algorithms using the observed training dataset and the unseen evaluating dataset that are subsets of an original dataset, such as CICIDS-2017. However, as pointed out by [6], the predictive metrics such as AUC of a well-known machine learning algorithm will drop dramatically if we switch the datasets.

The problem is partly addressed in the work of [8]. The authors built a predictive model to predict the data source. The idea of the authors is presented in Fig. 1. The authors assumed a given enough data are withdrawn from different sources so we can learn the classifier. It is not always the case; hence, we aim to solve in this study.

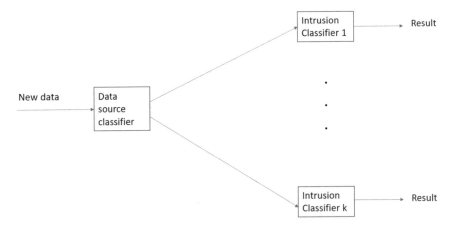

Fig. 1 Data source classification [8]

3 The InSDN Dataset

We use the dataset InSDN [11], matching with the previous work of [6].

The dataset contains three subsets: normal data that contains benign traffic only, and OVS and mealsplotable data that contain attacking traffic only. The distribution number of the traffic class is presented below:

- The normal traffic includes 68,424 instances.
- The attack traffics targeting the mealsplotable 2 servers that include 136,743 instances with the following attacks:

  - DoS: 1145 instances
  - DDoS: 73,529 instances
  - Probe: 61,757 instances
  - BFA: 295 instances
  - U2R: 17 instances

- The attack traffics targeting the Open vSwitch (OVS) machine that includes 138,722 instances with the following attacks.

  - DoS: 52,471 instances
  - DDoS: 48,413 instances
  - Probe: 36,372 instances
  - BFA: 1110 instances
  - Web Attack: 192 instances
  - BOTNET: 164 instances (Table 1).

There is in total 83 features in the dataset. They are presented in the numerical form. We consider two main scenarios: one is when we try to perform the binary

Table 1 A snapshot of the normal traffic data

Flow ID	Src IP	Src port	Dst IP	Dst port	Protocol	Timestamp	Flow duration	Tot Fwd Pkts	Tot Bwd Pkts	TotLen Fwd Pkts	TotLen Bwd Pkts	Fwd Pkt Len Max
0	185.127.17.56–192.168.20.133–443-53,648-6	185.127.17.56	443	192.168.20.133	53,648	6	5/2/2020 13:58	245,230	44	40	124,937.0	1071.0	9100
1	185.127.17.56–192.168.20.133–443-53,650-6	192.168.20.133	53,650	185.127.17.56	443	6	5/2/2020 13:58	1,605,449	107	149	1071.0 4	139537.0	517
2	192.168.20.133–192.168.20.2–35,108-53–6	192.168.20.133	35,108	192.168.20.2	53	6	5/2/2020 13:58	53,078	5	5	66.0	758.0	66
3	192.168.20.133–192.168.20.2–35,108-53–6	192.168.20.2	53	192.168.20.133	35,108	6	5/2/2020 13:58	6975	1	1	0.0	0.0	0
4	154.59.122.74–192.168.20.133–443-60,900–6	192.168.20.133	60,900	154.59.122.74	443	6	5/2/2020 13:58	190,141	13	16	780.0	11,085.0	427

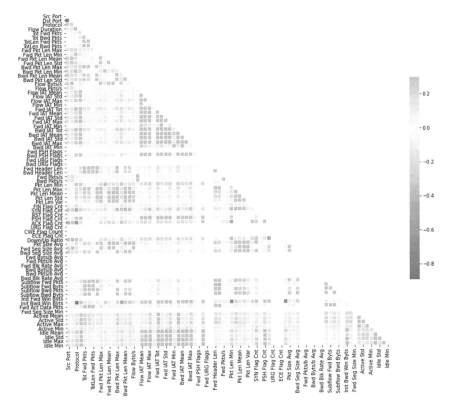

Fig. 2 Correlation matrix of the normal traffic data

classification, i.e., we need to detect a traffic is benign or attack, i.e., binary classification only; and the second one is when we try to detect not only if a traffic is attack or not but also what attack type of this traffic is.

We visualize the correlation matrix of normal data in Fig. 2. The correlation is low enough to be ignored in analysis.

4 Methods and Experimental Results

4.1 Methods

We assume the following scenario:

- We start with a single data source.
- We trained a supervised model with the first coming data.
- We start to predict the incoming data.

Fig. 3 Isolation forest

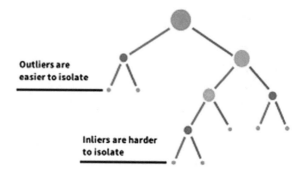

- We monitor the incoming data and collect the "strange" data from the observed data.
- We built a classifier to classify the data source when we have enough "strange" data.

In order to determine the strangeness of a new data point, we utilize the algorithm Isolation Forest [13]. The core idea of Isolation Forest is that an inlier is difficult to be distinguished from normal data points because they are just similar to others. Alternatively, an outlier is very easy to be spotted because it is different from benign data points at a certain level. Hence, the Isolation Forest algorithm builds a random forest to classify a data point from all other data points. The algorithm is visualized in Fig. 3.

We selected data points with the outlier scores larger than a threshold, then we consider this data point belongs to a different dataset.

4.2 Experimental Results

In this experiment presented with this paper, we use the InSDN dataset. We start with the mealsplotable 2 dataset, then slowly inject the OVS data points. We use the CatBoost algorithm, similar with the work of [8]. The important metric of the classifier is displayed in Table 2.

Table 2 Metrics of intrusion detection on the mealsplotable dataset

Metric	Value
Accuracy	0.9996
$F1$-score	0.9997
Precision	0.998
Recall	0.9991

Table 3 Accuracy of the data source classification by number of strange data points collected

Number of data points	Accuracy
150	0.78
500	0.83
1000	0.91
5000	0.997

We present the dependent of the accuracy score of the model on the size of the data source classification by number of data point collected in Table 3. We see that by collecting around 5000 data points or more, we can achieve a very high accuracy prediction.

5 Conclusions

In this research work, we studied the problem of intrusion detection in multiple data source scenario. Multiple data source is a critical issue that has a lot of impact both in research and practice. We resolve the issue of multiple data source by recognizing if a network flow is suitable to be classified by the established supervised classifier. The experimental results showed the validity of our proposed algorithm.

References

1. Alavizadeh H, Jang-Jaccard J, Alavizadeh H (2021) Deep q-learning based reinforcement learning approach for network intrusion detection. arXiv preprint arXiv:2111.13978
2. Dang QV (2019) Studying machine learning techniques for intrusion detection systems. In: FDSE. Lecture notes in computer science, vol 11814. Springer, pp 411–426
3. Dang QV (2020) Active learning for intrusion detection systems. In: IEEE RIVF
4. Dang QV (2020) Understanding the decision of machine learning based intrusion detection systems. In: FDSE. Lecture notes in computer science, vol 12466. Springer, pp 379–396
5. Dang QV (2021) Improving the performance of the intrusion detection systems by the machine learning explain ability. Int J Web Inf Syst
6. Dang QV (2021) Intrusion detection in software-defined networks. In: FDSE. Lecture notes in computer science. Springer
7. Dang QV (2021) Studying the fuzzy clustering algorithm for intrusion detection on the attacks to the domain name system. In: WorldS4. IEEE
8. Dang QV (2022) Detecting intrusion using multiple datasets in software-defined networks. In: International conference on research in computational intelligence and communication networks (ICRCICN). Springer
9. Dang QV, Vo TH (2021) Reinforcement learning for the problem of detecting intrusion in a computer system. In: Proceedings of ICICT
10. Data M, Aritsugi M (2021) T-dfnn: An incremental learning algorithm for intrusion detection systems. IEEE Access
11. Elsayed MS, Le-Khac NA, Jurcut AD (2020) InSDN: a novel SDN intrusion dataset. IEEE Access 8:165263–165284

12. Jaw E, Wang X (2021) Feature selection and ensemble-based intrusion detection system: an efficient and comprehensive approach. Symmetry 13(10):1764
13. Liu FT, Ting KM, Zhou ZH (2008) Isolation forest. In: 2008 8th IEEE international conference on data mining. IEEE, pp 413–422
14. Maseer ZK, Yusof R, Bahaman N, Mostafa SA, Foozy CFM (2021) Benchmarking of machine learning for anomaly based intrusion detection systems in the cicids2017 dataset. IEEE Access 9:22351–22370
15. Prabakaran S, Ramar R, Hussain I, Kavin BP, Alshamrani SS, AlGhamdi AS, Alshehri A (2022) Predicting attack pattern via machine learning by exploiting stateful firewall as virtual network function in an SDN network. Sensors 22(3):709
16. Sharafaldin I, Lashkari AH, Ghorbani AA (2018) Toward generating a new intrusion detection dataset and intrusion traffic characterization. In: ICISSP, pp 108–116
17. Thakkar A, Lohiya R (2020) A review of the advancement in intrusion detection datasets. Procedia Comput Sci 167:636–645

Retrospective Study of Convolutional Neural Network for Medical Image Analysis and a Deep Insight Through Histopathological Dataset

Shallu Sharma, Eelandula Kumaraswamy, and Sumit Kumar

Abstract Convolutional neural network (CNN) has become a prominent technology of choice in medical image analysis as CNN is easy to train and requires less pre-processing of data. The paper elaborates the concept of CNN with architectural details and their application in medical imaging to resolve the problems of detection, segmentation and classification. Limited dataset, scarcity of annotated data, non-standardized evaluation metric and model biasing are being observed as major challenges associated with the implementation of CNNs in medical image analysis. The emerging and advanced techniques to address the confronted challenges are also explored to improve the CNN performance. A brief study on BreakHis dataset classification using the fine-tuning approach of transfer learning with AlexNet is also demonstrated to select an optimal layer for fine-tuning for both binary and multi-classification task. The experimental results validate the superiority of network fine-tuning at a moderate level for the task of classifying the magnification independent histology images.

Keywords CNN · Medical image · Segmentation · Detection · Classification · Transfer learning

1 Introduction

From the past decade, CNN has been shown a great success in computer vision, viz. object recognition, face identification, visual tracking, traffic flow prediction and

S. Sharma
Neuroimaging and Neurospectroscopy Laboratory, National Brain Research Centre, Manesar 122052, India

E. Kumaraswamy
School of Electronics and Electrical Engineering, Lovely Professional University, Phagwara, Punjab 144411, India

S. Kumar (✉)
Department of Research and Innovation, Division of Research and Development, Lovely Professional University, Phagwara, Punjab 144411, India
e-mail: sumit.24786@lpu.co.in

© The Author(s), under exclusive license to Springer Nature Singapore Pte Ltd. 2023
A. Shukla et al. (eds.), *Computational Intelligence*, Lecture Notes in Electrical Engineering 968, https://doi.org/10.1007/978-981-19-7346-8_5

medical image analysis [1–5]. The deployment of CNN in medical image analysis is gaining a lot of attention because the interpretation of images is limited by humans due to large variation across expositors. Before the 2000s, many computer-aided diagnosis (CAD) systems brought in clinical practice but it has been found that these CAD systems give more false prediction than a radiologist, which results in additional assessment time [1]. With the emergence of deep learning models, it is expected that the limitation of the previously introduced CAD system will be overcome. The expectation has fulfilled to a great extent. A CAD system provides more accuracy in prediction which became possible due to new learning algorithms, availability of big datasets and enhanced computation ability with graphic processing unit (GPU).

2   Convolutional Neural Network

The CNN architecture contains four types of layers, namely convolutional, activation, pooling and the fully connected layer which are stacked on each other [2]. A typical example of CNN architecture from a technical perspective is depicted in Fig. 1.

2.1   Input Layer

First layer possesses the raw pel or pixel of the image. The dimensions of an input layer play the most important role and should be divisible by a factor of two multiple times. Let us consider an image with size [28 × 28 × 3] which implies that the height of image is 28; width is 28 and have 3 channels of red, green and blue color.

2.2   CONV Layer

The main function of this layer is to compute the output of nodes which are closely associated with the local regions of the first layer. In order to obtain the output of every node, a multiplication is carried out between the node weights and the local

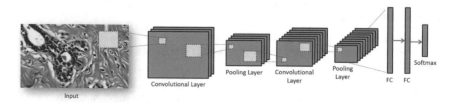

Fig. 1 Convolutional neural network architecture

area with which they are associated. For instance: if 20 filters are used; then this process may result in nodes output volume of size [28 × 28 × 20].

- Local Connectivity

Full connectivity in nodes that are present in layers raises the number of parameters in huge amount whenever high-dimensional inputs are treated and lead to the problem of overfitting. Therefore, local connectivity in an input volume is allowed. The spatial dimension (i.e., width and height) up to which this connectivity extends is a hyper-parameter named as the receptive field of the node. It is equivalent to the size of the filter. The extent of the connectivity is always full along the depth axis of the input volume.

In Fig. 2a, it is clearly depicted that each node in the hidden layer connected to a small area of the input volume. A receptive field of 5 × 5 spatial dimensions has chosen which correspond to 25 pels of the input volume. This local receptive field is then slid throughout the input volume. The sliding of the local receptive field is started from the top-left corner and shifted by one node in the right direction every time to make a connection with the next hidden node. The hidden nodes learn a weight for each connection and an overall bias. Whereas, in Fig. 2b, an input volume of dimension 28 × 28 after convolving with a receptive field of size 5 × 5 resulted in a hidden layer of dimension 24 × 24. In actual, depth, stride and zero-padding are the three hyper-parameters which control the dimension of the output volume. These hyper-parameters are explained in details below:

- Depth: This represents the number of filters used in the network to seek something different in the same area of the input volume.
- Stride: It is basically the step size with which filter is slide over the entire input volume. In the earlier example, stride value is 1 so the filters are moved over one pixel at a time as shown in Fig. 2a, b. If stride value is 2, then filters would skip 2 pels at a time.
- Zero-padding: In this feature, zeros are padded around the border of input volume to control the spatial extent of the output volume in such a way that spatial size of input volume stays preserved. The size of zero-padding helps in determining the spatial extent of output volume which is represented by three equations [6] as:

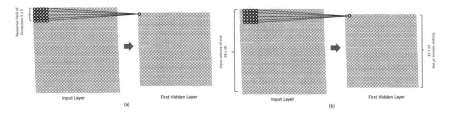

Fig. 2 **a** Convolution of the receptive field of size 5 × 5 with stride value 1 over input volume of size 28 × 28 **b** Resulted in hidden layer with size 24 × 24 after convolution

$$W' = \frac{(W - F + 2P)}{S} + 1 \qquad (1)$$

$$H' = \frac{(H - F + 2P)}{S} + 1 \qquad (2)$$

$$D' = K \qquad (3)$$

where F is the size of the receptive field, K is the number of filters, $W \times H$ represents the size of the input volume, S is the stride value and P is the number of zeros padded around the border. However, W', H' and D' represent the width, height and depth of the output volume, respectively. It is worth mentioning that the same weights and bias are used for 24×24 neurons in the first hidden layers of CNN to decrease the number of training parameters in the network. This new strategy is termed as parameter sharing. For better understanding about the preference of local connectivity and parameter sharing strategy, one more example is explained below:

Consider an image of size 28×28, without considering RGB components otherwise it would be labeled as $28 \times 28x3$. The receptive field of size 5×5 is chosen. If the number of filters used is 20, then there would be 20 feature maps corresponding to each filter, as shown in Fig. 3a. Each feature map would require a single shared bias with $5 \times 5 = 25$ shared weights. Hence, the total number of parameters required to define the convolution layer is $20 \times 26 = 520$. On the other hand, if a fully connected network is considered for this input size with relatively small 30 hidden nodes, then the total amount of parameters needed are $28 \times 28 \times 30 = 23,520 + 30$ bias $= 23,550$. Hence, the full connectivity is an unwise decision because it results in approximately 45 times additional parameters in the training of the network.

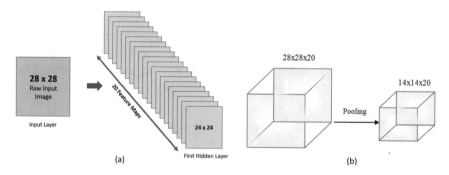

Fig. 3 **a** First hidden layer with 20 feature maps as the number of filters used is 20 and each feature map corresponds to the convolution of one filter with the input layer, **b** Reduction in dimensionality due to pooling

2.3 Rectified Linear Unit (ReLU) Layer

ReLU is an activation function [7] that defines the output of a node for the given set of inputs. It adds nonlinearity to the network and makes it capable to compute linear as well as nonlinear function for solving the considered problem. ReLU is a non-saturating activation function, which improves the performance of decision function by enhancing its nonlinear properties. It comes up with a benefit of acceleration in neural network training speed. ReLU function is also known as a max function or hard max. The size of the resulting volume remains unaltered, i.e., [28 × 28 × 20] after the application of this layer.

2.4 Pooling Layer

Pooling is a nonlinear transformation, which permits to summarize the receptive field with a single value and makes the feature descriptions more robust [8]. The pooling layer progressively reduces the dimension and controls the over fitting problem. The max pooling of size 2 × 2 on the input volume of size 28 × 28 provides an output of reduced volume having size [14 × 14 × 20], see Fig. 3b.

2.5 Fully Connected Layer

Fully connected layer like regular 'neural networks' has full connections of neurons to all activations in the former layer. This layer computes the class scores for each category to classify the input into a label. In addition, this layer is much prone to overfitting due to the occurrence of most of the parametric quantity, which trims down the efficacy of the network. Here, it leads to a volume of size [1 × 1 × 12] where 12 represent the numbers of categories (corresponds to a category score).

3 Contribution of CNN in Medical Image Analysis

3.1 Detection

Anatomical object localization and detection of lesions are the key aspects in the clinical practice employed for therapy planning and treatment. Lo et al. [9] proposed the first object detection system with four CNN layers in 1995 to detect nodules in X-ray images. Chen et al. [10] used a pre-trained CNN architecture and RBM for the same purpose to learn better feature representations even when big data is not available. CNNs have also been employed for the localization of scan planes in temporal data. Baumgartner et al. [11] proposed a way to detect twelve standardized scan planes (defined by the UK fetal abnormality-screening program) in mid-pregnancy fetal ultrasound using CNNs that were trained on 1003 video frame data. On the basis of the above discussion, it can be concluded that CNN performs considerably in detection and localization of landmarks, lesions, regions and organs for a variety of biomedical application in the diagnosis of diverse epidemics.

3.2 Segmentation

Segmentation simplifies the image analysis in a more meaningful manner [12]. Moreover, it plays a pivotal role in the CAD system. According to Ronneberger et al. [13] the task of segmentation can be significantly improved by utilizing CNN. In this context, CNN architecture has been proposed, also known as U-net. The architecture of U-net is composed of up-sampling and down-sampling layers along with skip connections between the convolution and deconvolution layers, apart from other segmentation frameworks. In another work by Cicek et al. [14] extended the U-net architecture by substituting all 2D operation with 3D. While Drozdzal et al. [15] extended the fully convolutional network (FCN) by adding short skip connections in conjunction with the long skip connections in a regular U-net architecture and further facilitates the architecture in achieving state-of-the-art results in the segmentation of biomedical images. On the other hand, Milletari et al. [16] proposed a new 3D version of U-net architecture that is also known as V-net. V-net is composed of a novel objective function that is based on dice coefficient and utilized for the segmentation of 3D MRI images of prostate using 3D convolutional layers. In addition to all, the researchers are also trying to combine CNN with graphical models to refine the segmentation output.

3.3 Classification

Classification is the most imperative application of CNN and machine learning algorithms in medical images analysis for early detection of diseases [17–22]. Despite the inadequate size of the medical dataset, a huge influx of CNN has been seen in solving the problem of medical data classification. In this context, transfer learning is growing as a potential technique in the development of a robust classification system for assessing various diseases with limited data input. In transfer learning, the knowledge gained from solving one problem is stored and applied to a different but related problem. Two strategies of transfer learning, (1) pre-trained network as a feature extractor and (2) pre-trained network with fine-tuning, have been identified as the most viable solutions in the classification of medical data. According to Antony et al. [23], fine-tuning provides better accuracy as compared to a feature extracting approach of transfer learning in multi-class grade classification of knee osteoarthritis. However, Kim et al. [24] argued that the use of CNN as a feature extractor outperforms fine-tuning in cytopathology image classification. The above statements are contradictory. However, according to Gulshan et al. [25] and Esteva et al. [26], a claim has been made that the results obtained from their pre-trained CNN model along with fine-tuning approach are as good as obtained by any human expert.

4 Challenges and Solutions

The advancement in any technology always gives rise to some complex difficulties and hurdles. Similarly, various challenges are associated with the implementation of CNN which are discussed as follows:

- The requirement of a large amount of labeled dataset is the foremost limitation in the implementation of CNN as it requires large dataset for training.
- Incomprehensive annotation of the datasets results in higher cost and scarceness of expert annotation in the medical field.
- Model biasing in the prediction of a healthy label from disease label due to underrepresentation of rare diseases is of main concern in medical image analysis.
- High data dimensionality is a big paramount in the implementation of CNN. It happens due to the presence of uninformative and redundant training samples.
- The requirement of metadata along with medical image data is highly desirable in early diagnosis but it is very hard to extract conclusive outcomes from different data modalities with a single-type CNN.
- Non-standardized evaluation metrics are of big concern in healthcare domain for comparing and determining the advancements in the field.
- Diversity in the datasets, different methods of data collection and data preprocessing are few important facts which also need to be addressed.

The requirement of large and well-annotated data can be overcome by utilizing a variety of data augmentation techniques. However, the data augmentation arises high data dimensionality problem because of the redundancy in data samples. Low variance filter, high correlation filter, ensemble trees, backward feature elimination, and forward feature construction, etc., are emerging as state-of-the-art data dimensionality reduction techniques. The problem of non-standardized evaluation metrics can be resolved by imposing common standards for statistical testing [27, 28].

5 A Case Study on Histology Digital Image Analysis

Whole slide imaging (WSI) is the recent advancement in pathology where the scanner captures images of tissues from histopathological slides and digitize to generate a digital image. All types of disease affect individual cells in various manners. There-fore, it is essential to have a sound knowledge of underlying disease processes to interpret the changes in the cells and tissues. Due to the fact, it is very difficult and challenging to extract discerning information related to tissue and cell characteristics for analyzing the digital histopathological images. In the pursuit above, we determine the potential of CNN in solving the problem of relevant feature extraction from the histopathological images of breast cancer. A well-known pre-existing architecture the 'AlexNet' has been employed for feature extraction as well as classification of breast cancer images [29–34].

In the present study, the histopathological image samples are obtained from BreakHis dataset which is composed of 7909 images of various magnification factors such as $40 \times$, $100 \times$, $200 \times$ and $400 \times$. The size of the dataset is not very large so the layer-wise fine-tuning approach of transfer learning has been utilized in the training of the AlexNet. We have followed the same training strategy that was utilized by Sharma et al. [32] in the classification of histopathological images with considera-tion of the magnification factor of images. However, the novelty of this case study lies in the fact that here we have executed the layer-wise fine-tuning approach for the classification of BreakHis dataset which is independent to a magnification factor of images for both cases, i.e., binary and multi-classification. This is considered a more complex task in contrast to a magnification-dependent case especially for any histopathological dataset. Moreover, we have reported for the first time magni-fication independent classification by considering layer-wise fine-tuning approach on histopathological images as best of authors knowledge concern. Therefore, we cannot compare the results of two entirely different cases, whereas the same transfer learning approach was adopted for demonstrating the effectiveness of CNN approach for magnification-dependent case.

5.1 Binary Classification

In this section, the performance of layer-wise fine-tuning approach for the classifica-tion of BreakHis dataset in two classes: benign and malignant has been demonstrated. Receiver operating characteristic (ROC) curve analysis has been performed to verify the potential of CNN in the magnification independent classification of breast cancer images. The whole dataset is partitioned into two sets: a training set and testing set by a ratio of 80 and 20, respectively. By comparing the performances obtained with the fine-tuning, it has been observed that the tuning of last two layers 'FC7-FC8' is providing the highest AUC (0.60) and APS (0.56), see Table 1. While the insertion of more network layers in fine-tuning degrades the classification performance of the network. A deeply and moderately fine-tuned network is outperformed by the shallow fine-tuning in case of binary classification. Therefore, the layers from 'FC7-FC8' are the most appropriate for tuning because of enough transfer of knowledge.

Table 1 Layer-wise fine-tuning of AlexNet model for magnification independent binary classification of BreakHis dataset

Layers	Weighted Avg			APS	AUC
	Pre	Rec	F_1		
FC8	0.53	0.53	0.53	0.51	0.52
FC7-FC8	0.61	0.60	0.60	0.56	0.60
FC6-FC8	0.17	0.13	0.14	0.13	0.52
Conv5-FC8	0.24	0.20	0.21	0.14	0.46
Conv4-FC8	0.22	0.17	0.18	0.13	0.53
Conv3-FC8	0.28	0.19	0.21	0.14	0.44
Conv2-FC8	0.24	0.16	0.17	0.13	0.50
Conv1-FC8	0.22	0.14	0.15	0.13	0.46

5.2 Multi-classification

It has been analyzed from ROC curve analysis for magnification independent multi-classification of breast cancer that the fine-tuning of the last four layers 'Conv5-FC8' of the AlexNet network provides the highest sensitivity (0.27), AUC (0.58) and APS (0.17). While the inclusion as well as exclusion of more layers of the pre-trained network in fine-tuning lowers the performance of the network, successively, see Table 2. A moderately fine-tuned network outperforms the deep as well as shallow fine-tuning in case of multi-classification. In magnification independent multi-classification, more layers of the network are required for fine-tuning as compared to binary classification due to eight sub-classes of breast cancer which give rise to more intra-correlation and inter-correlation in the extracted features. Therefore, more convolutional layers of the pre-trained network are fine-tuned to extract the most relevant features from the images to develop a robust model.

Table 2 Layer-wise fine-tuning of AlexNet model for magnification independent multi-classification of BreakHis dataset

Layers	Weighted Avg			APS	AUC
	Pre	Rec	F_1		
FC8	0.53	0.53	0.53	0.51	0.52
FC7-FC8	0.61	0.60	0.60	0.56	0.60
FC6-FC8	0.17	0.13	0.14	0.13	0.52
Conv5-FC8	0.24	0.20	0.21	0.14	0.46
Conv4-FC8	0.22	0.17	0.18	0.13	0.53
Conv3-FC8	0.28	0.19	0.21	0.14	0.44
Conv2-FC8	0.24	0.16	0.17	0.13	0.50
Conv1-FC8	0.22	0.14	0.15	0.13	0.46

6 Conclusion

This rigorous study signifies the importance of CNN in almost entire aspects of medical image analysis. Transfer learning is emerging as a new throwback in technologies to deploy CNN even when a huge and well-annotated data is not available. The employment of CNNs with training from scratch is growing as the standard approach for medical image interpretation. Moreover, it has been observed that the CNNs and its variants are capable in achieving human level performance in lot of computer-aided applications. Therefore, CNN has an impressive impact on medical image analysis. Despite their success, some aspects like image reconstruction in medical imaging are still unexplored. Furthermore, from a comprehensive set of experimentation, it has been observed that the shallow fine-tuning yields the most optimal solution for the magnification independent binary classification. However, more layers of the network are needed to be fine-tuned for the multi-classification of the breast cancer histopathological images with the highest accuracy. Deep fine-tuning significantly lowers the performance of the network and becomes a sub-optimal choice for the breast cancer classification.

References

1. Kumar S, Sharma S (2021) Sub-classification of invasive and non-invasive cancer from magnification independent histopathological images using hybrid neural networks. Evol Intell 1–13
2. Sharma S Mehra R (2018) Automatic magnification independent classification of breast cancer tissue in histological images using deep convolutional neural network. In: International conference on advanced informatics for computing research. Springer, pp 772–781
3. Sharma S, Mehra R, Kumar S (2020) Optimised CNN in conjunction with efficient pooling strategy for the multi-classification of breast cancer. IET Image Process 1–12
4. Sharma S, Mandal PK (2022) A comprehensive report on machine learning-based detection of AD using multi-modal neuroimaging data. ACM Comput Surv 55(02):1–43
5. Sharma S, Kumar S (2021) The Xception model: a potential feature extractor in breast cancer histology images classification. ICT Express 1–7
6. https://cs231n.github.io/convolutional-networks/. Last Accessed 22 Feb 2022
7. Xu B, Wang N, Chen T, Li M (2015) Empirical evaluation of rectified activations in convolutional network. arXiv preprint arXiv:150500853
8. Druzhkov P, Kustikova V (2016) A survey of deep learning methods and software tools for image classification and object detection. Pattern Recognit Image Anal 26(1):9–15
9. Lo S-C, Lou S-L, Lin J-S, Freedman MT et al (1995) Artificial convolution neural network techniques and applications for lung nodule detection. IEEE Trans Med Imaging 14(4):711–718
10. Chen H, Ni D, Qin J, Li S, Yang X, Wang T, Heng PA (2015) Standard plane localization in fetal ultrasound via domain transferred deep neural networks. IEEE J Health Inform 19(5):1627–1636
11. Baumgartner CF, Kamnitsas K, Matthew J, Smith S, Kainz B, Rueckert D (2016) Real-time standard scan plane detection and localisation in fetal ultrasound using fully convolutional neural networks. In: International conference on medical image computing and computer-assisted intervention, Springer, pp 203–211
12. Gupta S, Kumar S (2012) Variational level set formulation and filtering techniques on CT images. Int J Eng Sci Technol (IJEST) 4(07):3509–3513

13. Ronneberger O, Fischer P, Brox T (2015) U-net: Convolutional networks for biomedical image segmentation. In: International conference on medical image computing and computer-assisted intervention. Springer, pp 234–241

14. Çiçek Ö, Abdulkadir A, Lienkamp SS, Brox T, Ronneberger O (2016) 3D U-net: learning dense volumetric segmentation from sparse annotation. In: International conference on medical image computing and computer-assisted intervention. Springer, pp 424–432

15. Drozdzal M, Vorontsov E, Chartrand G, Kadoury S, Pal C (2016) The importance of skip connections in biomedical image segmentation. In: Deep learning and data labeling for medical applications. Springer, pp 179–187

16. Milletari F, Navab N, Ahmadi S-A (2016) V-net: Fully convolutional neural networks for volumetric medical image segmentation. In: 2016 4th international conference on 3D vision (3DV). pp 565–571

17. Nanglia P, Kumar S, Mahajan AN, Singh P, Rathee D (2021) A hybrid algorithm for lung cancer classification using SVM and neural networks. ICT Express 7(3):335–341

18. Nanglia P, Mahajan AN, Rathee DS, Kumar S (2020) Lung cancer classification using feed forward back propagation neural network for CT images. Int J Med Eng Inform 12(5):447–456

19. Narayan Y, Kumar D, Kumar S (2020) Comparative analysis of sEMG signal classification using different K-NN algorithms. Int J Adv Sci Technol 29(10):2257–2266

20. Narayan Y, Ahlawat V, Kumar S (2020) Pattern recognition of sEMG signals using DWT based feature and SVM Classifier. Int J Adv Sci Technol 29(10S):2243–2256

21. Nanglia P, Kumar S, Rathi D, Singh P (2018) Comparative investigation of different feature extraction techniques for lung cancer detection system. In: International conference on advanced informatics for computing research. Springer, pp 296–307

22. Kumaraswamy E, Sharma S, Kumar S (2021) A review on cancer detection strategies with help of biomedical images using machine learning techniques. In: ICRSET-2021 AIP conference @SRITW -Warangal, Telangana

23. Antony J, McGuinness K, O'Connor NE, Moran K (2016) Quantifying radiographic knee osteoarthritis severity using deep convolutional neural networks. In: 2016 23rd international conference on pattern recognition (ICPR), 2016. IEEE, pp 1195–1200

24. Kim E, Corte-Real M, Baloch Z (2016) A deep semantic mobile application for thyroid cytopathology. In: Medical imaging 2016: PACS and imaging informatics: next generation and innovations. International Society for Optics and Photonics, p 97890A

25. Gulshan V, Peng L, Coram M, Stumpe MC, Wu D, Narayanaswamy A, Venugopalan S, Widner K, Madams T, Cuadros J (2016) Development and validation of a deep learning algorithm for detection of diabetic retinopathy in retinal fundus photographs. 316 (22):2402–2410

26. Esteva A, Kuprel B, Novoa RA, Ko J, Swetter SM, Blau HM, Thrun S (2017) Dermatologist-level classification of skin cancer with deep neural networks. Nature 542(7639):115–118

27. Tatli S, Gerbaudo VH, Mamede M, Tuncali K, Shyn PB, Silverman SG (2010) Abdominal masses sampled at PET/CT-guided percutaneous biopsy: initial experience with registration of prior PET/CT images. Radiology 256(1):305–311

28. Begley C, Ellis L (2012) Drug development: Raise standards for preclinical cancer research. Nature 483(7391)

29. Sharma S, Mehra R (2018) Breast cancer histology images classification: training from scratch or transfer learning? ICT Express 4(4):247–254

30. Voets M, Møllersen K, Bongo LA (2018) Replication study: development and validation of deep learning algorithm for detection of diabetic retinopathy in retinal fundus photographs. arXiv preprint arXiv:180304337

31. Nanglia P, Kumar S, Luhach AK (2019) Detection and analysis of lung cancer using radiomic approach. In: Smart computational strategies: theoretical and practical aspects. Springer, pp 13–24

32. Sharma S, Mehra R (2019) Effect of layer-wise fine-tuning in magnification-dependent classification of breast cancer histopathological image. Vis Comput 1–15

33. Sharma S, Mehra R (2019) Implications of pooling strategies in convolutional neural networks: a deep insight. Found Comput Decis Sci 44(3):303–330
34. Sharma S, Mehra R (2020) Conventional machine learning and deep learning approach for multi-classification of breast cancer histopathology images—a comparative insight. J Dig Imag 1–23

Building Web-Based Subject-Specific Corpora on the Desktop: Evaluation of Search Metrics

Jean-Paul Van Belle

Abstract Building subject-specific or domain corpora from Web data is well-researched. However, most approaches start by using seed articles as inputs to Web crawlers and take document similarity algorithms for selection. We take a different lean resource approach by applying traditional search metrics with a relatively large (more than 100 search terms) 'bag of domain words' approach on the colossal clean crawled corpus. This approach enables one to build rich domain corpora of text documents quickly in a resource-poor environment (e.g., a few CPU cores). This paper tests several metrics using three different subject domains—language, Colossal Clean Crawled Corpus basic mathematics, and information science—and finds that there are significant performance differences between the various metrics. Surprisingly, a naïve, simple metric, outperforms TD-IDF and performs almost as well as our top ranked algorithm, Okapi BM25. This demonstrates that the performance of search metrics using a relatively larger number of search key words (> 100) is different than when a small set of search key words is used. We also demonstrate how to optimize the free parameters for Okapi BM25.

Keywords Search metrics · Subject-specific domain corpora · Information retrieval (IR) · Colossal clean crawled corpus (C4) · TD-IDF · Okapi BM25 · Large search keyword set

1 Introduction

Natural language processing and artificial intelligence (e.g., chatbots) researchers and practitioners often require domain- or subject-specific corpora [1]. For advanced scientific domains, academic databases of literature can be used, e.g., ArXiv, PubMed, CiteSeer, etc. However, in many cases, the intellectual level of the documents should not be advanced research publications but be of a more general knowledge-level nature. In that case, the World Wide Web is normally used as a

J.-P. Van Belle (✉)
University of Cape Town, Rondebosch, South Africa
e-mail: jean-paul.vanbelle@uct.ac.za

© The Author(s), under exclusive license to Springer Nature Singapore Pte Ltd. 2023
A. Shukla et al. (eds.), *Computational Intelligence*, Lecture Notes in Electrical Engineering 968, https://doi.org/10.1007/978-981-19-7346-8_6

source for candidate text or documents [2]. One traditional approach to building these types on domain corpora is to use a search engine and download the highest-ranked documents [3]. The limitation of getting many (i.e., tens or hundreds of) thousands of results can be circumvented by using APIs or specialist search engines. However, search engines usually allow only a small number of search terms, and their ranking algorithms is often opaque. In addition, the documents returned must still be validated. Another approach to building domain-specific corpora entails starting with a suitable set of seed documents (as specified through URLs), using these to crawl the Web oneself, and using document similarity indices to vet the documents [4]. This is usually a resource-intensive exercise, even if done using cloud-based machine instances.

In this paper, we describe a light-weight approach that can be used to build multiple high-quality custom domain corpora consisting of millions of relevant text documents quickly and efficiently (typically creating a corpus in less than a day on a single desktop computer). This addresses the gaps identified above, i.e., being able to bypass the use of standard search engines and not to have to rely on heavy computing processing power or network capacity. Our approach uses a recently released high-quality terabyte corpus, the colossal clean crawled corpus aka C4 [5]. C4 is a cleaned and de-duplicated version of the April 2019 shard of the multi-terabyte common crawl corpus; the latter was also used to train the GPT-3 machine learning task [6]. We apply a customized selection process using a relatively extensive set of domain-specific keywords (well over 100) covering all the base concepts of the domain. In this paper, we evaluate a number of different search metrics which select the relevant documents using large keyword sets. Thus, this paper has the following two research objectives:

- How to quickly and efficiently generate large (million document) subject-specific corpora, i.e., a 'light-weight' instead of a resource-intensive approach.
- Determine which search metrics return the highest quality (most specific) corpora when given a large set of search keywords.

The findings of this research will be of interest to researchers who regularly need to create different domain corpora—e.g., for building subject-specific resources such as chatbots or training documents for ML applications. It may also be used in educational and research settings where you want to have many different document datasets for evaluation purposes. Finally, this will also be of interest to less technically-minded researchers and practitioners who need, or want to explore, quality corpora without needing cloud computing or crawling skills.

This paper proceeds with a quick overview of relevant concepts and literature. We then describe the specific method used in preparing and building these corpora. Finally, we present an evaluation of different search criteria (algorithms) across three sample domains: language, basic mathematics, and information/knowledge science.

2 Definitions and Prior Work

Web-based corpora has proven useful in many domains, including linguistic analysis, natural language processing, machine learning, computer-based lexicography, chatbot construction, and many other domains [7]. The most commonly discussed method is by means of a crawlers seeded with suitably selected candidate documents (URLs) as e.g., in the WebBootCat approach [8]. This is resource-intensive because documents first have to be crawled, cleaned, and then inspected for suitability. Architectures for 'focussed' Web-crawling that increase the proportion of domain-specific documents (i.e., targeting URLs that are both in the language and within the domain of interest) have been explored but add other complexities [4]. Other approaches use keywords as inputs to search engines so that only more suitable candidates have to be downloaded (Shäfer 2013), [9]. However, commercial search engines do not provide insight in their search ranking algorithms. Additionally, they usually only allow the specification of a very limited number of keywords (32 or less), and only some allow the returning of several thousands of candidate results. Finally, these results still have to be retrieved, cleaned, and checked [3].

In 2020, the colossal clean crawled corpus (C4) was released to the public. It takes a snapshot of the 2019 common crawl (https://commoncrawl.org/), but the clean version applies a number of filters to dramatically increase the quality of the documents: leaving out short sentences or lines that do not terminate in punctuation marks, remove short documents, remove markup tags and programming code, and finally only retain English documents (99%) and is de-duplicated [10]. The C4 is available in three different versions as shown in Table 1 [5] and can be downloaded from https://c4-search.apps.allenai.org/. This corpus offers an amazing resource and opportunity for quickly and relatively effortlessly building more specialized subcorpora. Even if a sub-selection contains, say, only 0.5% of the documents, it can still be considered as a large corpus, especially considering the quality of the documents. In this project, we use C4 for building subject-specific (sub) corpora and show that this is useful in low-resource contexts.

Information retrieval research has explored and suggests a number of different search metrics or algorithms. The 'classic' search algorithm is the TF-IDF metric (also referred to as TF-IDF, TF*IDF) which is short for term frequency-inverse document frequency and is still used widely, e.g., [11]. The basic version of TF-IDF uses the number of times a given search term is found in a document (TF) but adjusts this for the relatively frequency of its occurrence to prevent frequently used or common words from skewing search results. Thus, it multiplies TF with

Table 1 Three available versions of the colossal clean crawled corpus (C4)

Dataset	Nr of documents	Nr of tokens	Compressed size (.gz)
C4.EN.NOCLEAN	1.1 billion	1.4 trillion	2.3 Terabyte
C4.EN.NOBLOCKLIST	395 million	198 billion	280 Gigabyte
C4.EN	365 million	156 billion	305 Gigabyte

inverse document frequency (IDF), i.e., the number of documents in the entire search collection divided by the number of times the term is found in the collection. Usually, the logarithm of the IDF is used [12]. Many variants and weighting schemes for the TF-IDF exist, especially when multiple search terms are used [13].

However, despite the TF-IDF variants, a persisting problem remains that it does not suitably account for the overall document length in which the terms are found. Thus, an enhanced metric, the Okapi BM25, was suggested (sometimes referred to as the Okapi model or BM25). This formula uses two free parameters, usually denoted as k_1 and b, and refines the use of TF and IDF as follows.

$$\text{BM25} = \text{score}(D, Q) = \sum_{i=1}^{n} \text{IDF}(q_i) \cdot \frac{f(q_i, D) \cdot (k_1 + 1)}{f(q_i, D) + k_1 \cdot \left(1 - b + b \cdot \frac{|D|}{\text{avgdl}}\right)} \quad (1)$$

Okapi BM25 has generally been found to outperform TF-IDF [14], although this is still the subject of ongoing research in various contexts, e.g., [15] found no significant performance difference between the two metrics for Twitter analysis. Interestingly, [16] used Okapi BM25 for plagiarism detection and found no accuracy difference with the Rabin–Karp algorithm. This paper hopes to contribute further to the body of knowledge since we have found no prior literature on using these algorithms when using big sets of search terms (> 100 words).

3   Method

The source corpus used for this research is the C4 colossal clean crawled corpus described above. This corpus is pre-cleaned so consists of English language text documents only without embedded tags or java code. (During analysis, it was found that a very few documents were *not* in English). In addition, the corpus is de-duplicated so almost all text is unique. The most time-consuming task was to tokenize the entire corpus and create inverted indices, but this only has to be done once after which the selection of domain-specific documents is relatively quick. A handcrafted lemmatization tool was used, but the standard Python NLTK tools were used to remove stopwords and tokenize the entire corpus. Finally, inverted indices were created. This process took, on average, between 20 and 30 min on a single i5 core for each of the 1024 C4 text data blocks but can easily be run in parallel on multiple CPUs or machine instances. For this research, the entire C4 corpus was tokenized and inverted on three standard Windows i5 PCs, using two threads on each machine, running in the background of normal PC operations, during the course of one week, with the data processed on an external hard drive.

For purposes of this research, we selected a number of Naïve metrics (e.g., number of hits, number of unique hits, average distance between hits), several variants of the TF-IDF algorithm and Okapi BM25 with 4 different parameter sets (Table 2).

Table 2 List of metrics which were tested (with brief description)

Metric	Brief description
Hits	Total hits = total number of search terms found in document
Hits/Doclen	Total hits adjusted for (divided by) document length (in tokens)
UniqHits	How many of the search terms occur in the document, i.e., the number of unique hits
UnHits/Doclen	Unique hits divided by the document length (in tokens)
Median collocation distance	The median distance between hits (i.e., the median number of non-matched tokens between hits). Lower is better
Median closest neighbour	As above, but only measured between the closest neighbour pairs (lower is better)
TF-IDF	Original formulation of the TF-IDF metric, i.e., TF*IDF
BoolTF-IDF	Boolean version of TF-IDF, i.e., (1 if search term appears in the document, 0 if it does not)*IDF
TF-IDF/dl	TF-IDF adjusted for (i.e., just divided by the) document length (in tokens)
LogTF-IDF	Log-scaled TF-IDF, i.e., $(1 + \log(TF))*IDF$
NormTF-IDF	Normalized TF-IDF here $(1-3/(3 + TF))*IDF$. This ensures the result is in range [0, 1]
NormLog TF-DF	Normalized version of the logarithmically-scaled TF-IDF here: $(1-(1/\ln2(TF + 1) + 1))*IDF$
Okapi BM25a	Okapi BM25 with 'defaults' $k_1 = 1.2$ and $b = 0.8$
Okapi BM25b	Okapi BM with $k_1 = 1.4$ and $b = 0.85$
Okapi BM25c	Okapi BM with $k_1 = 1.4$ and $b = 0.70$
Okapi BM25d	Okapi BM with $k_1 = 1.6$ and $b = 0.85$

To test the metrics for particular domains, we created search keyword lists for three fairly general domains involving abstract concepts: language, basic mathematics, and information/knowledge. A key consideration was that some basic search terms would be polysemous but the domain would also have several more specific words. Additionally, the domains would have 'sub-domains' using quite distinct word/terminology sets. Given that the corpus was lemmatized in the earlier step, only the (word) lemmas were required in the search list. Column 2 in Table 3 shows the number of search terms, i.e., between 100 and 200 keywords for each domain. Note that this would not be acceptable for submission to a standard search engine (API).

4 Analysis

This section presents the evaluation of the various metrics tested. First, we present the raw evaluation of the various metrics (or algorithms) based on their classification

Table 3 Accuracy levels for selected search metrics for top-n ranking documents

Top n	Hits (%)	Hits/Dlen (%)	Uniq Hits (%)	UHits/Dlen (%)	Med Coll Dist (%)	Bool TF-IDF (%)	TF-IDF (%)	TF-IDF/dl (%)	NLog TF-IDF (%)	Okapi BM25 a (%)	Okapi BM25 c (%)
100	45	81	68	84	66	77	58	89	77	99	99
200	48	83	65	78	64	71	51	86	71	96	98
300	45	80	59	73	60	65	52	85	65	91	93
400	46	75	59	68	58	59	52	80	60	89	90
500	43	72	50	65	58	56	51	76	56	83	83
600	42	68	52	61	56	52	48	73	52	75	77

accuracy. We explore the optimal parameter setting for the best-performing metric, namely the Okapi BM25.

4.1 Relative Performance of the Different Metrics

We ran the 16 metrics using the 'language' domain search words on C4's first block of documents (consisting of 356 K documents, i.e., 1/1024th of the entire corpus). The results were ordered, for each of the metric, by matching score (normally from high to low). The better the matching score, the higher ranked the document, i.e., the probability of the document matching the keyword profile.

The base 'truth' was defined by manually inspecting and classifying documents to see if they actually fitted the domain. Due to resource constraints, the documents were classified by the researcher only, thus inevitably introducing some bias. However, the higher-ranked documents (for each of the metrics) were given a particularly close inspection and careful consideration, with only documents clearly *not* in the domain being classified as non-matches. For the language domain, only 2384 documents (i.e., less than 1% of the total documents) had to be manually classified of which 639 documents (< 0.2% of the total number of documents) were classified as being 'language' domain documents.

Table 3 shows the accuracy levels for selected metrics' highest-ranked documents at various cut-off levels, i.e., the proportion of the n-top ranked documents (according to that metric) that is indeed a relevant hit, i.e., a document that should be selected for the domain corpus. Figure 1 visualizes the absolute number of correct documents found for all the metrics for the first 1278 documents (twice the actual number of relevant documents).

It can be seen that Okapi BM25 clearly outranks any other search algorithm. The second-best algorithm is the TF-IDF normalized for (divided by) document length. Indeed, this is what the Okapi BM25 metric set out to do, although the latter is more complicated to calculate and understand. However, the TF-IDF is a parameter-free metric, whereas Okapi BM25 requires two parameters, although the default value (Okapi BM25a) does very well. The next section will look at how the parameters affect the performance and can be fine-tuned to optimize the search even further.

Perhaps the biggest surprise is the fact that two 'Naïve' search metrics: the total and the unique number of hits, both divided by document length outscore any of the alternative and substantially more sophisticated TF-IDF variants (standard, Boolean, normalized, logarithmic, or normalized-logarithmic) by a fair margin. It must be born in mind that the IDF calculation requires pre-calculation of all term frequencies across the entire corpus, whereas the naïve metrics just require a quick and simple local document calculation. They can therefore be calculated if one does not have access to, or knowledge about the entire corpus.

Similar results were found for the other search domains (these required separate determinations of 'ground truth' documents), and the Okapi BM was found to be superior to all other metrics, closely followed by TF-IDF/DocLength and NrHits/DocLength.

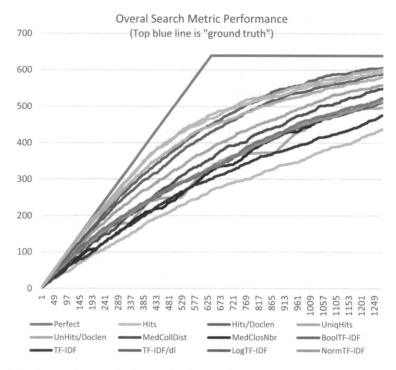

Fig. 1 Performance (accuracy) of 15 metrics for top-1278 ranked documents

However, before determining document selection cut-off values for Okapi BM25 metric, its parameter values were first optimized for accuracy as explained below.

4.2 Optimizing the Okapi BM25 Parameters

The Okapi BM25 has two free parameters, the so-called k_1 and b. The metric's authors suggest that these default to 0.8 and 1.2, respectively (referred to as Okapi BM25a in the above analysis), but little else is suggested in the literature about these parameters. Figure 2 shows how the accuracy of Okapi BM25 changes across a range of b (which has to be between 0 and 1) and various values of k_1 (ranging here from 1 to 3) for the language domain.

It is evident that, in this case, the accuracy is not at all sensitive to k_1 but quite sensitive to the value of b. In general, the optimal parameters will differ for different domains, keyword sets, and source corpus. In our case, we found the following optimal parameters for the three domains (Table 4).

From this admittedly limited sample, it appears that the optimal b remains very close to 0.80, but k_1 increases with the number of search terms in the search word sets, although further research could examine this further. In all three domains, the

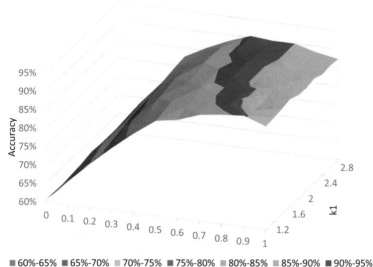

Fig. 2 Accuracy of Okapi BM25 as a function of b (0–1) and k_1 (1–3)

Table 4 Optimal parameters for Okapi BM25	Domain	Nr of search terms	k_1	b
	Language	131	1.2	0.8
	Basic mathematics	148	2.2	0.75
	Information/knowledge	194	2.6	0.75

accuracy was very sensitive to changes in b and relatively insensitive to changes in k_1, i.e., a similar accuracy profile as in Fig. 2 was found throughout.

4.3 Using the Metrics for Creating Subject-Specific Corpora

For the actual selection of documents, the user has to select a cut-off value for a metric, balancing the accuracy with the number of documents required. Accuracy generally falls off quite quickly, and the marginal number of valid documents retrieved decreases quite quickly. By choosing an appropriate metric cut-off, the research can control for the number of false positives—since accuracy generally declines as the cut-off values are reduced (see Table 3). However, false negatives can be reduced by combining multiple diverse metrics. Analysis (omitted due to space limitations) revealed that metrics from different models ('families') 'catch' some of the false negatives from other metrics so a combination approach is recommended. For the

Table 5 Metric cut-off values for 90% and 80% accuracy, respectively

Domain	Nr of search terms	Okapi BM25 90%	Nr docs selected (90 + %)	Okapi BM25 80%	#Hits/DocLen	TF-IDF/DocLen	Nr docs selected (80 + %)
Language	131	31.3053	**396,838**	30.2365	0.07447	0.27083	**569,826**
Basic maths	148	40.4290	**479,906**	38.0809	0.10171	0.39445	**805,818**
Information	194	57.0468	**160,191**	53.9160	0.26000	0.72000	**270,296**

selected domains, we combined three search metrics and set the cut-off value after analysis of the first block of (365,000) documents, i.e., by determining the 'ground truth' by inspecting all selected documents. The metric cut-off values in Table 5 were chosen to generate a set of '90% accuracy' (Okapi BM25 only) and a second set of 80–90% lower accuracy corpora (last three columns/metrics). Clearly, and logically, the cut-off values for each domain increase as the number of search terms increases.

Table 5 also shows how many documents were selected, i.e., classified into three respective sub-corpora using a 90% (column 4) and 80% (last column) accuracy criterion, respectively. Each of the 3 sets of corpora, consisting of hundreds of thousands of relevant documents, were generated in less than 1 day using a single i5 CPU core.

5 Conclusion

This paper showed how the recently released colossal clean crawled corpus can be used to create subject-specific corpora quickly in a lean resource environment. In particular, the research looked at the performance, i.e., accuracy of common information retrieval models/algorithms when relatively large set of search terms (> 100) are used to define or delineate the domain of interest. The various variants of TF-IDF and Okapi BM25 were tested, and Okapi BM25 was found to outperform TF-IDF. This is consistent with the literature. This research also confirmed that the suggested default values for Okapi BM25 ($k_1 = 0.8$; $b = 1.2$) perform reasonably well, but further optimizations are possible, depending on the domain. In particular, the value of k_1 can be increased further. But the optimal parameter space tends to be a plateau that is sensitive to b, less so to k_1 (Fig. 2). This finding is also a contribution to the literature. More surprisingly, the simple metric of (unique) hits/document length—which requires no information or preprocessing outside the document itself—performed almost as well as the more complicated information retrieval metrics such as TD-IDF. This unexpected finding may be a unique attribute for a search with a large number (> 100) of search terms, and it is suggested that future research seeks to confirm this in order to generalize this particular finding.

This research only probed three sample subjects, and even though the findings were consistent across the three radically different domains, they may not be generalizable to other domains. Also, invariably, there is some subjectivity in determining both the extensive sets of search terms as well as establishing the 'base truth', i.e., determining which of the high-ranked documents do not belong to the domain. Further research could test other domains or search term sets of different lengths as well as apply the approach to other types of base corpora.

References

1. Schäfer R, and Bildhauer F (2013) Web corpus construction. In synthesis lectures on human language technologies 6(4):1–145.
2. Barbaresi A (2014) Finding viable seed URLs for web corpora: a scouting approach and comparative study of available sources. In 9th Web as Corpus Workshop (WaC-9), 14th conference of the european chapter of the association for computational linguistics, pp. 1–8.
3. Kilgarriff A, Baisa V, Bušta J, Jakubíček M, Kovář V, Michelfeit J, Rychlý P, Suchomel V (2014) The sketch engine: ten years on. Lexicography 1(1):7–36
4. Remus S, Biemann C (2016) Domain-specific corpus expansion with focused webcrawling. In: Proceedings of the 10th international conference on language resources and evaluation (LREC'16), pp 3607–3611
5. Dodge J, Sap M, Marasović A, Agnew W, Ilharco G, Groeneveld D, Mitchell M, Gardner M (2021) Documenting large Webtext corpora: a case study on the colossal clean crawled corpus. arXiv preprint arXiv:2104.08758
6. Brown TB, Mann B, Ryder N, Subbiah M, Kaplan J, Dhariwal P, Neelakantan A, Shyam P, Sastry G, Askell A, Agarwal S (2020) Language models are few-shot learners. arXiv preprint arXiv:2005.14165
7. Jakubícek M, Kovár V, Rychlý P, Suchomel V (2020) Current challenges in web corpus building. In: 12th web as corpus workshop (LREC). ELRA, Marseille, pp 1–4
8. Baroni M, Kilgarriff A, Pomikálek J, Rychlý P (2006) WebBootCaT. Instant domain-specific corpora to support human translators. In: Proceedings of the 11th annual conference of the European association for machine translation
9. Barbaresi A (2014) Finding viable seed URLs for web corpora: a scouting approach and comparative study of available sources. In: 9th web as corpus workshop (WaC-9), 14th conference of European chapter of association for computational linguistics, pp 1–8
10. Raffel C, Shazeer N, Roberts A, Lee K, Narang S, Matena M, Zhou Y, Li W, Liu PJ (2020) Exploring the limits of transfer learning with a unified text-to-text transformer. J Mach Learn Res
11. Qaiser S, Ali R (2018) Text mining: use of TF-IDF to examine the relevance of words to documents. Int J Comput Appl 181(1):25–29
12. Jabri S, Dahbi A, Gadi T, Bassir A (2018) Ranking of text documents using TF-IDF weighting and association rules mining. In: 2018 4th international conference on optimization and applications (ICOA). IEEE (2018), pp 1–6
13. Mishra A, Vishwakarma S (2015) Analysis of TF-IDF model and its variant for document retrieval. In International conference on computational intelligence and communication networks (CICN). IEEE (2015), pp 772–776
14. Fautsch C, Savoy J (2010) Adapting the TF-IDF vector-space model to domain specific information retrieval. In: Proceedings of 2010 ACM symposium on applied computing, pp 1708–1712

15. Kadhim AI (2019) Term weighting for feature extraction on Twitter: a comparison between BM25 and TF-IDF. In: International conference on advanced science and engineering (ICOASE). IEEE (2019), pp 124–128
16. Wijaya I, Seputra K, Parwita W (2021) Comparison of the BM25 and Rabinkarp algorithm for plagiarism detection. J Phys Conf Ser 1810:012032

Studying Effectiveness of Transformers Over FastText

Jitendra Singh Malik

Abstract In this project, we are using various embedding namely FastText, Bert Expert-MEDLINE/PubMed + CNN, Bert Expert-Talking Head along with the presence of convolution neural network (CNN) and multi-layer perceptron (MLP) for detection of the fake new in the dataset. In order to remove biasness towards a particular model, we have chosen a balanced dataset which is obtained from Kaggle. The models have been trained with different number of epochs varying from 10 to 20 epochs with accuracy measure been relied upon the usage of weighted average $F1$-score, to find which method is outperforming other methods. We have also try to give an overview as to how which model is able to complete each epochs faster. All these models have the same architecture of CNN and MLP's to analyze which model is performing better.

Keywords Machine learning · Natural language processing · Deep learning · Transformers

1 Introduction

There is huge amount of content available for users on web but not all of it is true and contains the mix of fake, real, hatred, offensive, etc. This fake content or fake news causes the spreading of false information among users which eventually lead to certain disturbance in the community.

For example, it was noted that the 62% of all United States adults received the news from some kind of the social media in the year 2016, where in the year 2012, there were approximate 49 percent who acknowledged the seeing of news of the social media.

It was also found that the social media now has a better reach than television for the source of news. There are plenty of pros of social media, despite that the authentication of the news which is present online on the social media is very far

J. S. Malik (✉)
Lucid Insights Pty Ltd., Adelaide, South Australia 5000, Australia
e-mail: jmjmalik22@gmail.com

© The Author(s), under exclusive license to Springer Nature Singapore Pte Ltd. 2023
A. Shukla et al. (eds.), *Computational Intelligence*, Lecture Notes in Electrical
Engineering 968, https://doi.org/10.1007/978-981-19-7346-8_7

less t compared to other sources such as newspaper, news channel, radio, etc. In compared to other sources of news circulation, it is also noted that it is easier to circulate the news through the means of social media to target a large audience, i.e., there are different purpose which are behind the creation false information on the Internet which ranges from the advantage such as financial or political.

An effective measure is thus necessary to stop the spread of such misinformation in the society. This gives us the motivation to develop such steps or measure which can be effective against such problems. Through this study we are developing different models which can help us analyze and figure out the piece of news sentence is real or fake.

2   Related Work

Detection of fake news has been in the world for a long time now with many authors and researchers taking the step to remove such content. Huge varieties of machine learning models have been employed to remove such problems.

Use of machine learning methods such as KNN, SVM, XGB and AdaBoost have been widely adopted to separate the fake content [1]. Use of SVM, network analysis approaches and linguistic cue approach were also adopted to figure out the fake content by authors [5]. Researchers also employed many machine learning methods such as Naive Bayes, SVM, deep learning and random forest to create the model to detect the fake content. The dataset was collected with the help of the API from the Twitter [7]. We have also used deep learning in our approach to detect the fake content.

Researchers have also taken advantage of kernels available to us in SVM and used linear and sigmoid kernel along with use the various other methods like decision trees, KNN, Naive Bayes and logistic regression along with combination of embedding like TF-IDF and bag of words. The research was done on dataset "Fake and True News" collected from online available news websites and was labeled manually [9]. We have instead relied on taking the dataset from the website Kaggle and have not used any API to get the tweets or to manually annotate.

A study also showed the use of n-gram and other machine learning models. An accuracy of 92% was obtained by using linear support vector machine in combination with TF-IDF [2]. A study also took into consideration the writers profile such as title and location of the writer which resulted in the increased accuracy. The model was trained with the help of low short-term memory (LSTM) [11].

The authors of [18] suggested that detection of the deception based on the dataset which was named "LIAR". This dataset was evidently increased the accuracy in finding out the fake content. The authors of study used text to predict different scenarios such as rumors, political, etc. The main focus of the authors was on using the CNN which was used to find the relation between the vectors of metadata and Bi-LSTM layers. In our study, we have also tried to implement the use of CNN to figure the fake news detection with the presence of transformers embedding. Our study does not consider the use of LSTM's.

The authors in [15] found out that nearly 14 million messages were on web nearly again around 400 k times. This instances happened after the presidential election of 2016 in United States.

Some researchers used the fake news as multi-label classification problem and used the ensemble machine learning framework (gradient boosting) to experiment on the fake news challenge-challenge (FNC) [8], whereas our case is most of binary classification.

One of the best solution was found in the study [14] that is a detection system which works in detecting the stance and then categorize into different types of classes such as agree, disagree, discuss and unrelated.

A study also expanded the instance detection system which consisted of four classes which depended upon the text. The study used the TF-IDF for the feature set along with combination of MLP. There were different accuracy which were noted in the dataset. The system was able to obtain the accuracy of 88.46%. We have also used MLP in our approach where we have directly used it with the FastText and transformers embedding.

A study [13] conducted few experiments with FastText embeddings which also used the pre-trained models in the study like ours. The embedding dimension in both the studies is kept same with 300.

3 Dataset

One of the important feature for machine/deep learning is the presence of data which is properly labeled. Thereby, it is utterly important to focus on how the data we are going to use for the study is produced. There are different ways to collect dataset such as by using the tweeter API to get the tweets and string of characters. Whereas it has been observed that in respect to privacy concerns, the data only constitute of the text sentence and the category it belong to. In order to save the time and get the dataset using the help of API and manually annotate it later, we have relied upon the dataset which is taken from the Kaggle website. The data has been made by public and can be used in any project free of copyrights. The details of the users who have tweeted are kept hidden according to policy of different websites like Twitter. Our dataset contains different classes as explained in Table 1

Table 1 Dataset description

Dataset	Description
Kaggle fake-news dataset	Fake—3164 (49.94%)
	Real—3171 (50.06%)
	Total—6335

Our dataset is balanced dataset and contains two category that are fake and real. Both the classes are almost equal in number to train the model better. The total length of the dataset is 6335 which is ranging with *fake* classes containing 3164 tweets and 3171 belonging to *real* class.

4 Implementation Details

In this section, we have described step-by-step implementation of our experiments. Section 4.1 has explained with the first phase in our modeling of generation of embedding through the help of different embeddings. The next section describes our approach to the modeling architecture, and we are using in our dataset with Sect. 4.2 for the CNN and MLP. Section 4.3 provides outline of loss function that has been recycled by us in our approach. Section 4.4 defines the evaluation metrics or the accuracy parameters which we are using in our approach.

4.1 Embedding

In this study, we are using three embeddings in their combination with CNN and MLP's as described in Sect. 4.1 and Table 2.

Each of the sentence is passed through these word embeddings layer followed by the model with a classification layer in the end to use for the prediction. To get an overview for FastText embedding, we have passed the data to generate the embeddings and connected the layer to respected CNN/MLP architecture. In case of transformer embeddings, we have relied on generating the embedding by taking into consideration the sequence layer which is then fed into CNN/MLP. Instead of directly predicting through the transformer by using the pooled layer, we have used the deep learning to do the job of prediction.

FastText FastText is a free to use available to public lightweight library. It makes users to understand the representation of the text and classifying the text. In our study, we are using wiki-news-300d- 1 M.vec.zip [12]. FastText embeddings defines

Table 2 Embedding and classifiers

Dataset	Description
FastText	FastText + CNN
	FastText + MLP
Talking Head	Talking Head + CNN
	Talking Head + MLP
PubMed/MEDLINE	PubMed/MEDLINE + CNN
	PubMed/MEDLINE + MLP

a word vector as the sum of the vectors of its n-grams. FastText aims to find word representations that do the best job of predicting which words in the vocabulary appear in context. Let's assume, we have word w_t, ..., w_T in the context and C_t be the index of words. FastText aims to maximize the log-likelihood as given in 1

$$\sum_{i=1}^{T} \sum_{c \in Ct} \log \ p(\omega_c / \omega_t) \tag{1}$$

where $p \ (w_c \ w_t)$ is the probability of a word w_c given the current word w_t. FastText does binary-prediction for every word present in our vocabulary, if that the given word present in the sample of the word against the negative sample of words which are not in the context, given as N_t, c. So for the word at position c considers log-likelihood as given in 2 as described the authors of study [3]

$$\sum_{i=1}^{T} \left[\sum_{c \in Ct} l(s(\omega_t, \omega_{tc})) + \sum_{n \in Nt,c} l(-s(\omega_t, n)) \right] \tag{2}$$

Talking Head BERT [6] with Talking-Heads attention and gated GELU position-wise feed-forward networks is a transformer encoder which modifies BERT architecture by using talking-heads attention along with gated linear unit with GELU activation as the first layer of the position-wise feed-forward networks. It has 12 hidden layers, a hidden size of 768 between the transformer blocks and 4H inside the position-wise feed-forward networks. The number of attention heads is 12, identically for the separate "heads dimensions" h, h_k and h_v. [16, 17]

PubMed/MEDLINE This model has been trained from scratch on MEDLINE/PubMed. The model was trained using a joint self-supervised masked language model and next sentence prediction task. The language modeling task uses whole word masking up to three contiguous masked words with a maximum of 76 predictions. The model has been trained for 500 K steps, with a batch size of 2048, and a max sequence length of 512. The training is done with an Adam optimizer using a learning rate schedule of linear warm up for 2500 steps to a 4e-4 learning rate, followed by polynomial decay. After pretraining, the pooling layer is replaced with an identity matrix before the model is exported which we have observed to be more stable during downstream tasks [19].

4.2 Model Architecture

We have used two different model type with each of the embeddings as already explained in Table 2. In this section, we are going to explore the detailed architecture of the models we have used to classify the classes for our dataset. Usually CNN or

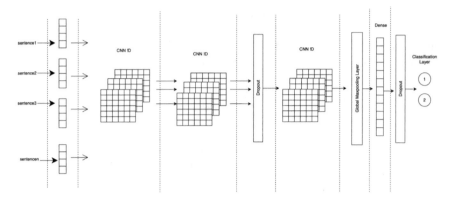

Fig. 1 CNN network architecture used in this study

MLP layers are usually lie secondly to embeddings layers in our network. All the layout with the probability for each layers are described in the below sections.

CNN—Architecture CNN models are used after embedding in first layer as the second layer. For our CNN network, we have used convolution 1D layer followed by another 1D convolution layer with is joint to dropout later with probability of 0.1 going to another 1D convolution layer feeding to global maxpooling layer followed by a dense layer and dropout layer with 256 neurons and 0.1 probability of dropout joining finally at dense layer with two neurons. The architecture can be viewed in Fig. 1.

The 1D convolution works for each vector of the word, the one-dimension filter is reading the vector and transforming it using the convolution layer. This explains how there is reduction in the length of vector on passing of the 1D filter through the input. The filter can be used by us of any size depending on our requirement.

MLP—Architecture MLP model on other hand consists of embedding layer feeding the input to flatten layer which is joining to a dense layer of 64 neurons followed by dropout layer of 0.1 probability joining to another dense layer with 64 neurons This is connected to a dense layer with 128 neurons again followed by a dropout layer which is joining to dropout layer with probability of 0.2 and finally going to classification layer with two neurons for prediction. The architecture can be viewed in Fig. 2.

4.3 Loss Function and Optimizer

This study is dependent upon the sparse categorical cross entropy along with the Adam optimizer for training of our models. It is due to the fact the it has been noted

Fig. 2 MLP network architecture used in this study

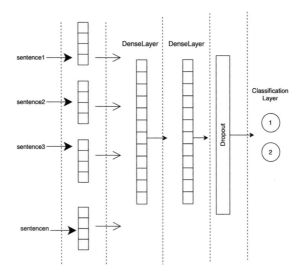

that, for the classification tasks the former is very effective. Adam optimizer here is used to give improvements in the stochastic gradient descent [10].

4.4 Evaluation Metrics

The problem in our case is binary classification problem, and we will use the $F1$-score [4] as described in Eq. (3). The one with higher $F1$-score will be given superiority over other methods.

$$F_1 = 2 \times \frac{\text{precision} \times \text{recall}}{\text{precision} + \text{recall}} \tag{3}$$

In order to measure the accuracy of all the models, we have taken into consideration the weighted average F_1-score to carry out the result and provide the best models.

5 Result

In terms of result, we will take into consideration the accuracy and computation speed which is measured in ms for each epoch which are described in Sects. 5.1 and 5.2.

Table 3 F_1-weighted average score for our experiment

Embedding	Model	Weighted average		
		P	R	F1
FastText	FastText + CNN	0.94	0.94	0.94
	FastText + MLP	0.93	0.92	0.92
Transformer	Talking Head + CNN	0.97	0.97	0.97
	Talking Head + MLP	0.97	0.97	0.97
	MEDLINE/PubMed + CNN	0.96	0.96	0.96
	MEDLINE/PubMed + MLP	0.97	0.97	0.97

5.1 Accuracy

We have defined our accuracy in terms of F_1-score to measure each model. The model with highest F_1-score will considered best of all for the prediction.

In our study, we have observed that in terms of best accuracy, Talking Head in combination of CNN and MLP both provided the F_1-score of 0.97 which was equal to the F_1-score provided by MEDLINE/PubMed model in combination with MLP, which is trained on medical data. The number of epochs though varied in Talking Head and MEDLINE/PubMed, the former was trained for 25 epochs whereas the later consisted of only 10 epochs. CNN on the other hand in combination with MEDLINE/PubMed provided the just a percent less score of 0.96 compared to other transformers embedding (Table 3).

FastText embedding was also quite successful in providing good results. The CNN combination of it provided the F_1-score of 0.92 which is the lowest recorded for any model whereas its combination with MLP provided the F_1-score of 0.94 which is still low compared to transformer embedding.

We can say that despite being giving good results, it can be note that transformers embedding are quite better than FastText which may be due to large number of dataset it is being trained upon compared to other traditional embeddings which are trained on much less data.

5.2 Computational Time

In terms of computational time, we are measuring the time for data to completely pass, i.e., the number of ms it took to complete one epoch for each model (Fig. 3).

It is observed that Talking Head takes the lowest time to complete one epoch which is 124 ms which is higher compared to FastText which took approximately 155 ms to complete one pass of data. It can be noted that MEDLINE/PubMed is taking the highest time to complete an epoch and thus comes as a slowest model even though its accuracy is quite similar to Talking Head.

Fig. 3 Computational time
for each model

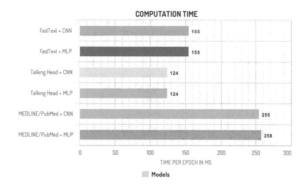

6 Conclusion

It can be observed that transformer embeddings comes as quite successful in our
case to deliver higher result that FastText in terms of weighted average F_1-score.
MEDLINE/PubMed in combination with both MLP and CNN provided the very best
result along with the Talking Head in combination with CNN and MLP. The performance of the models were almost equal but better than the FastText embedding which
provided the good results but still lacked to Talking Head and MEDLINE/PubMed. In
terms of speed, it can be noted that Talking Head came as the fastest model followed
by FastText and then MEDLINE/PubMed. It can also be observed that Talking Head
is faster than MEDLINE/PubMed and provided the same accuracy but it took more
number of epochs, i.e., 25 epochs to reach that accuracy which MEDLINE/PubMed
achieved in 10 epochs.

References

1. Ahmad I, Yousaf M, Yousaf S, Ahmad MO (2020) Fake news detection using machine learning
   ensemble methods. Complexity 2020
2. Ahmed H, Traore I, Saad S (2017) Detection of online fake news using n-gram analysis and
   machine learning techniques. In: International conference on intelligent, secure, and dependable
   systems in distributed and cloud environments. Springer, pp 127–138
3. Bojanowski P, Grave E, Joulin A, Mikolov T (2017) Enriching word vectors with subword
   information. Trans Assoc Comput Linguist 5:135–146
4. Chicco D, Jurman G (2020) The advantages of the Matthews correlation coefficient (mcc) over
   f1 score and accuracy in binary classification evaluation. BMC Genomics 21(1):1–13
5. Conroy NK, Rubin VL, Chen Y (2015) Automatic deception detection: Methods for finding
   fake news. Proc Assoc Inf Sci Technol 52(1):1–4
6. Devlin J, Chang MW, Lee K, Toutanova K (2018) Bert: pre-training of deep bidirectional
   transformers for language understanding. arXiv preprint arXiv:1810.04805
7. Helmstetter S, Paulheim H (2018) Weakly supervised learning for fake news detection on
   twitter. In: 2018 IEEE/ACM international conference on advances in social networks analysis
   and mining (ASONAM). IEEE, pp 274–277

8. Kaliyar RK, Goswami A, Narang P (2019) Multiclass fake news detection using ensemble machine learning. In: 2019 IEEE 9th international conference on advanced computing (IACC), pp 103–107. https://doi.org/10.1109/IACC48062.2019.8971579

9. Kareem I, Awan SM (2019) Pakistani media fake news classification using machine learning classifiers. In: 2019 international conference on innovative computing (ICIC), pp 1–6. https://doi.org/10.1109/ICIC48496.2019.8966734

10. Kingma DP, Ba J (2014) Adam: a method for stochastic optimization. arXiv preprint arXiv: 1412.6980

11. Long Y (2017) Fake news detection through multi-perspective speaker profiles. Assoc Comput Linguist

12. Mikolov T, Grave E, Bojanowski P, Puhrsch C, Joulin A (2018) Advances in pre-training distributed word representations. In: Proceedings of the international conference on language resources and evaluation (LREC 2018)

13. Pamungkas EW, Basile V, Patti V (2020) Misogyny detection in twitter: a multilingual and cross-domain study. Inf Process Manage 57(6):102360

14. Riedel B, Augenstein I, Spithourakis GP, Riedel S (2017) A simple but tough-to-beat baseline for the fake news challenge stance detection task. arXiv preprint arXiv:1707.03264

15. Shao C, Ciampaglia GL, Varol O, Yang KC, Flammini A, Menczer F (2018) The spread of low-credibility content by social bots. Nat Commun 9(1):1–9

16. Shazeer N (2020) Glu variants improve transformer. arXiv preprint arXiv:2002.05202

17. Shazeer N (2020) Zhenzhong lan, youlong cheng, nan ding, and le hou. talking-heads attention. arXiv preprint arXiv:2003.02436

18. Wang WY (2017) liar, liar pants on fire: a new benchmark dataset for fake news detection. arXiv preprint arXiv:1705.00648

19. Yu H, Chen C, Du X, Li Y, Rashwan A, Hou L, Jin P, Yang F, Liu F, Kim J, Li J (2020) TensorFlow model garden. https://github.com/tensorflow/models

Aerial Object Detection Using Deep Learning: A Review

Vinat Goyal, Rishu Singh, Mrudul Dhawley, Aveekal Kumar, and Sanjeev Sharma

Abstract Aerial object detection is a key to many functionalities like animal population estimation, pedestrian counting, security systems and many more. Traditional methods made use of machine learning algorithms that involved hand-made features. At present, deep learning models have surpassed the conventional machine learning models. Some of the major problems faced in aerial object detection are angle shift and the small sizes of objects in aerial images. This paper discusses the developments in aerial object detection. It discusses the various approaches presented by researchers, the different datasets available today for the task, and some of the standard evaluation and metrics followed to evaluate models.

Keywords Aerial object detection · Supersampling · Denoising · Deep learning · Machine learning · Computer vision

1 Introduction

Object detection is a subdomain of the computer vision field. It refers to the capability of computer and software systems to locate objects in an image/scene and identify each object. It has been widely used for face detection, vehicle detection, pedestrian counting, web images, security systems and autonomous cars. The motive of object detection [1] is to recognise and locate (localise) all known objects in a scene. The

V. Goyal (✉) · R. Singh · M. Dhawley · A. Kumar · S. Sharma
Indian Institute of Information Technology Pune, Pune, India
e-mail: vinatgoyal19@cse.iiitp.ac.in

R. Singh
e-mail: rishusingh19@ece.iiitp.ac.in

M. Dhawley
e-mail: mruduldhawley19@cse.iiitp.ac.in

A. Kumar
e-mail: aveekalkumar19@ece.iiitp.ac.in

S. Sharma
e-mail: sanjeevsharma@iiitp.ac.in

© The Author(s), under exclusive license to Springer Nature Singapore Pte Ltd. 2023
A. Shukla et al. (eds.), *Computational Intelligence*, Lecture Notes in Electrical
Engineering 968, https://doi.org/10.1007/978-981-19-7346-8_8

primary goal behind object detection is to determine whether there are any instances of an object, such as a human, car, bike, etc., in an image. If present, return a specific instance of that object with some labelling in the image. Object detection can be performed on static as well as dynamic images.

A more specific area of research in object detection is aerial object detection. It detects various objects in images taken from an altitude like satellites. Aerial object detection can be accommodating in many aspects like visible and thermal infrared remote sensing for the detection of white-tailed deer using an unmanned aerial system [2]. It can help in improving the precision and accuracy of animal population [3] estimates with aerial image object detection. It aid via robust detection of small and dense objects in images from autonomous aerial vehicles. In the field of geography, it assists by accurate landslide detection leveraging UAV-based aerial remote sensing. One of the critical uses of aerial object detection is to detect unmanned aerial vehicles (UAVs) like a drone or the US Boeing Eagle Eye. Other uses include detecting vehicles from satellite images.

This paper is a survey on the various methodologies and past work done in aerial object detection. This paper discusses the past methods and the ones used in the present for the task. Model training requires a lot of data. Later, this paper also discusses the various open datasets that are available online to train a model. This survey reviews the existing literature on aerial object detection with deep learning architecture. The significant contribution of this survey can be summarised as:

1. Discuss the previous methods successfully applied for aerial object detection.
2. Identify the popular datasets for aerial object detection.
3. Identify famous metrics of evaluation for the aerial objects detection.

The rest of the paper is organised as follows: Sect. 2 discusses the literature survey of the related work. Section 3 covers the various datasets available for aerial object detection. At last, we are concluding work in Sect. 4.

2 Survey Related to Aerial Object Detection

A total of 200 published research papers were collected from reputed publishers. The papers were then read, and the ones related to aerial object detection were only kept and used for the survey. A total of 40 papers out of the 100 were found to be relevant to the study of the paper. The publications used for the methodology are IEEE Xplore, MDPI, Springer, Science Direct, ACM Digital Library, Wiley Online Library and Google Scholar.

Object detection has been a hot topic of research owing to its benefits. One of the early models created for the task is the Viola-Jones detectors. They were made 20 years ago. It was the first real-time face detection model. They were followed by the HOG detector [4] in 2005, which had an essential improvement of the scale-invariant feature. Then came the deformable part-based model (DPM) [5] in 2008, which was an extension of HOG. Neural Networks have been there for a long time, but it was until recently that neural network architectures found a place in object detection research. One of the neural network models that have been extensively used is the CNN model. CNN model is the go-to model for researchers in the computer vision domain and has inspired a lot of computer vision architectures. Today, the majority of the state-of-the-art object detection architectures are based on CNN. The critical difference between the old traditional methods and neural network architectures is that the feature representation is learned in neural networks instead of pre-determined by the researcher as in the old methods. A CCN-based two-stage detector was proposed in 2010, after which a more complex version of the same called the R-CNN was built. Some more examples of such architectures are the single-shot MultiBox detector (SSD), feature pyramid networks (FPN) and deep neural networks like DeepID-Net, Fast R-CNN, etc. From the above discussion, it can be seen that object detection has come a way long way. Today, many models are present that give excellent performance for object detection. However, these models tend to underperform on aerial images.

According to a study, when the size of an image is enormous or quite extensive, object detection becomes more challenging as the object we are trying to find would be small and will acquire only a few pixels of the entire image. The model would easily confuse it with some random background features.

In general, object detection has high performance on large objects [6] and gets strenuous as the object's size decreases as it gets densely packed. Thus, the accuracy of the model will rapidly fall due to the degradation of image resolution. One specific characteristic of an aerial image is the object can align in any direction. Therefore, orientation plays a vital role in the extraction of essential features. If orientation is not taken care of, it can lead to rotation bounding box problems.

There has been a lot of research dedicated to object detection owing to the value it brings. Today, object detection is widely used in several applications like image retrieval, security, surveillance, automated vehicle systems, machine inspection, etc. Generally, two methods are used for this task (1) machine learning-based and (2) deep learning-based, a summary of which can be seen in Fig. 1.

2.1 Traditional Approach

Previously, the traditional manual methods were used for this task. Some of the traditional object detection algorithms used were Haar and AdaBoost [7], histogram of oriented gradient (HOG) and support vector machine (SVM) [8] and deformable parts model (DPM) [9]. These algorithms are difficult to adapt to the diversity of the object, and the robustness of these detection models are not strong. As computer vision is developing, these traditional algorithms are losing application.

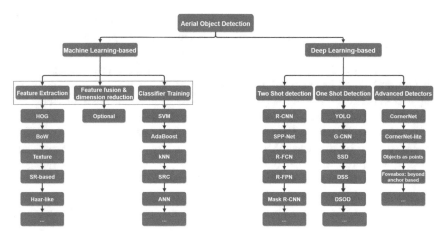

Fig. 1 Summary of approaches for aerial object detection

2.2 Deep Learning-Based Approach

Today deep learning models are primarily used for such purposes. Deep learning models have been there for decades but did not get popular then because of lack of data and computational power. Today there is much data available, the computational power of computers has drastically increased, and there has been a lot of development in developing better optimisation algorithms. Algorithms like stochastic gradient descent with momentum (SGDM) and RMSprop have emerged as the favourites for optimisation. All these factors have contributed to the success of deep learning models today.

Generic object detection is done by two important steps, classifying existing objects in an image and labelling them with rectangular bounding boxes. CNN's have been the go-to architecture for the task of image classification, but object detection also requires the task of object localisation. This is achieved by proposing different models developed on the foundational idea of a CNN. There are still some difficulties that make this task difficult, like foreground–background class imbalance [10]. There are generally two frameworks of generic object detection. The first framework, called the two-step detector, proposes two different pipelines for classification and localisation. Algorithms belonging to this category are R-CNN [11], SPP-net [12], R-FCN [13], FPN [14] and Mask R-CNN [15]. On the other hand, the second framework, called the one-step detector, proposes a single pipeline and regards object detection as a regression or classification problem and achieves final results in one go. Examples of the second model framework include MultiBox, AttentionNet, YOLO [16], G-CNN, SSD [17], DSS and DSOD. R-CNN uses selective search to select bounding box regions that contain object instances. Later Fast R-CNN [18] unified different models used in R-CNN by sharing features. Faster R-CNN [19] uses a single unified model composed of region proposal network (RPN) and Fast R-CNN with shared convolutional layers. Mask R-CNN extends Faster R-CNN to pixel-level image segmentation. YOLO was the first to introduce a single-stage detector, which can infer quickly compared to two-stage detectors. SSD was the first to use feature pyramids for multi- scale object detection. RetinaNet [20] is based on a feature pyramid network and uses focal loss, which tends to leverage the complex examples by assigning more weight to complex examples and less weight to easy examples.

R-CNN selects bounding box regions that are likely to contain object instances via selective search. Different, unified models are afterwards employed in R-CNN by sharing characteristics in Fast R-CNN. Faster R-CNN combines region proposal network (RPN) and Fast R-CNN with shared convolutional layers into a single model. Faster R-CNN is extended to pixel-level image segmentation with Mask R-CNN. Compared to two-stage detectors, YOLO was the first to introduce a single-stage detector, which can infer quickly. SSD was the first algorithm to use feature pyramids to detect multi-scale objects. RetinaNet is based on a feature pyramid network and employs focus loss, which tends to leverage complex cases by giving them more weight and giving easy ones less.

Object aerial detection is a subdomain of object detection which has also attracted a lot of attention. Aerial object detection is the task of detecting objects from aerial images. It has many applications like animal population estimation, security surveillance, crop yield analysis, traffic flow analysis, videography for multiple purposes, etc. The algorithms mentioned above for generic object detection have performed well on scenery images. Still, they fail to deliver similar accuracy for aerial images. This is major because of the shift in the object's angle and the uncertainty in the object's size.

A lot of other methods are also used to tackle the problems mentioned. Some of these methods are briefly explained below:

Rotated Bounding Boxes As discussed, one of the problems faced in aerial object detection is the shift in the angle of bounding boxes.

Lei et al. [21] proposes a model, You Only Look Twice: Rapid multi-scale object detection in satellite imagery, to tackle the problem of the orientation of bounding boxes. The model extends DarkNet neural network and got competitive results on standard datasets.

Han et al. [22] proposes a single-shot alignment network (S2A-Net) consisting of two modules: a feature alignment module (FAM) and an oriented detection module (ODM). The FAM generates high-quality anchors with an anchor refinement. Network and adaptively align the convolutional features according to the anchor boxes with a novel alignment convolution. The ODM first adopts active rotating filters to encode the orientation information and then produces orientation-sensitive and orientation-invariant features to alleviate the inconsistency between classification score and localisation accuracy. The model achieves state-of-the-art performance on the DOTA and HRSC2016 datasets.

Ding et al. [23] proposes a region of interest (RoI) transformer to model the geometry transformation, avoiding misalignment. The core idea of the RoI transformer is to apply spatial changes on ROIs and learn the transformation parameters under the supervision of oriented bounding box (OBB) annotations.

Han et al. [24] proposes ReDet: A rotation-equivariant detector for aerial object detection to address the orientation variation problem, which explicitly encodes rotation equivariance and rotation invariance. They present rotation-invariant RoI align (RiRoI align), which adaptively extracts rotation-invariant features from equivariant features according to the orientation of RoI. Compared with previous best results, the ReDet gains 1.2, 3.5 and 2.6 mAP on DOTA-v1.0, DOTA-v1.5 and HRSC2016, respectively, while reducing the number of parameters by 60 [25] has tackled the problem of the shift in angle by proposing the RepVGG-YOLO model, which uses an improved RepVGG module as the backbone feature extraction network (backbone) of the model, and uses SPP, feature pyramid network (FPN) and path aggregation network (PANet) as the enhanced feature extraction networks. The model combines context information on multiple scales, accumulates multi-layer features and strengthens feature information extraction. They also use four target detection scales to enhance the feature extraction of remote sensing small target pixels and the CSL method to increase the detection accuracy of objects at any angle.

Segmentation Image segmentation is a process used to create partitions in an image according to its features and properties. The partitioned regions must relate to the targeted part of interest. This process is also called pixel-level classification. One of the most well-known architectures used for this task is the U-Net architecture.

U-Net is a well-known network for the fast and precise segmentation of images. It was developed by Olaf Ronneberger et al. for biomedical image segmentation. It consists of an encoder–decoder path scheme. The encoder decreases the spatial dimensions in every layer and increases the channels. At the same time, the decoder increases the spatial dimensions while reducing the channels.

Ulmas and Liiv [26] proposes a modified U-Net model for land cover classification. The revised model consisted of two neural networks. The pre-trained ResNet50 classification model was used as the encoder in the U-Net model to solve the segmentation task. The analysis was done in three levels on two datasets: A custom CORINE Land dataset for training the segmentation model and a BigEarthNet dataset for classification. The classification model showed good results overall.

The segmentation model showed a high 91.4

Heidler et al. [27] proposes a combined model HED-U-Net for monitoring the Antarctic coastline. This approach makes use of both segmentation and edge detection. This model also uses an encoder–decoder architecture but with an addition of skip connections to predict segmentation masks and edges at multiple resolutions. With the addition of the edge detection task as an output for a segmentation model, the segmentation results seemed to improve. The model performed much better than its traditional counterparts.

Iglovikov and Shvets [28] proposes a modification of U-Net using pre-trained weights to improve performance in image segmentation tasks. VGG11was used as the encoder for the U-Net model. VGG is a simple CNN with small convolution filters to increase the depth of the neural network. VGG11 contains seven convolutional layers, each followed by a ReLU activation function and five max polling operations, decreasing the feature map by 2. The model was tested on different initialisation weights. This architecture ranked first in the Kaggle: Carvana image masking challenge.

Super-resolution Super-resolution is an attempt to improve the resolution of images, helping to improve the quality of small objects in images. This technique has been widely used in the pre-processing stage. Joint-SRVDNet: joint super-resolution and vehicle detection network [29] is one such model that uses super-resolution for the task of improving small objects resolution. The model generates discriminative, high-resolution images of vehicles from low-resolution aerial images. First, the aerial images are up-scaled by a factor of $4 \times$ using a multi-scale generative adversarial network (MsGAN), which has multiple intermediate outputs with increasing resolutions. Second, a detector is trained on super-resolved images up-scaled by factor $4 \times$ using MsGAN architecture. Finally, the detection loss is minimised jointly with the super-resolution loss to encourage the target detector to be sensitive to the subsequent super-resolution training. The network learns hierarchical and discriminative features of targets and produces optimal super-resolution results. This model has attained state-of-the-art results on the VEDAI, xView and DOTA datasets.

Courtrai et al. [30] uses super-resolution along with standard object detection algorithms for the task of aerial object detection. The super-resolution network is performed using deep residual blocks integrated with a Wasserstein generative adversarial network. Then, the detection task is achieved by using two state-of-the-art detectors, Faster R-CNN and YOLOv3. The model was tested on the VEDAI and the xView datasets. The model performed better than the state-of-the-art detectors.

Pest bird detection, classification and recognition in vineyard environments are challenging because the motion of birds causes the images captured to be a blur. Bhusal et al. [31] has solved this problem by using the super-resolution mechanism in the pre-processing stage.

Courtrai et al., and Schubert et al. [32, 33] further discuss the impact of using supersampling and denoising on aerial object detection.

Many more approaches have been proposed for the task of aerial object detection. Ahmad et al. [34] uses RetinaNet for the task and tackles the problem of uncertainty in object size using the crow search algorithm to search for optimal ratios and scales of anchors on the training set. For search space reduction, they followed the work of [35].

Lin et al. [36] uses a modified version of the YOLOv3 model. To tackle the small object detection problem, they used k-means clustering to optimise the anchor size and adjust the YoloV3 model based on the object characteristics. According to the experimental data, compared with the original YoloV3 network, their improved YoloV3 model's mAP is improved by 5.86%, and model convergence is faster. Liu and Li [37] uses the Faster R-CNN model for the task of aerial object detection and uses R-NMS in the post-processing stage. They achieved mAP, which was 16.31% greater than the standard baseline.

Additionally, aerial object detection on drone images tends to face more problems. Drones tend to navigate from different altitudes because of which the object scale varies a lot. Also, the images captured by the drone travelling at high-speed and low altitudes tend to be a blur. Zhu et al. [38] proposes TPH-YOLOv5 to overcome these problems. It is a YOLOv5-based model in which one more prediction head is added to detect different scale objects. Then, the original prediction head is replaced with transformer prediction heads (TPH) to explore the prediction potential with the self-attention mechanism. Convolutional block attention model (CBAM) is also integrated to find attention regions on scenarios with dense objects. The model has gained competitive results on the DET-test-challenge and VisDrone2021 datasets.

Autoencoders are also used for a wide range of computer vision tasks. Prystavka et al. [39] uses autoencoders to enhance the accuracy of aerial object detection. Another mechanism that has gained a lot of attention is the ensemble learning mechanism, wherein multiple models are individually trained and then stacked together, and then the stacked model is trained. Walambe et al. [40] makes use of the ensemble learning approach using standard architectures.

3   Dataset Related to Aerial Object Detection

Aerial object detection tasks are subject to a variety of issues when compared to general object detection tasks as follows:

- The target size is smaller
- The target is affected by external changes (weather)
- The target might be in large number and crowded
- Range of variations direction
- Limited number of categories.

We have made use of three standard benchmark datasets of the aerial object detection problem. These are the DOTA-v 2.0 dataset, iSAID dataset and the VEDAI dataset. Below is the summary of these datasets (Table 1).

4   Performance Metrics

Performance metrics are a vital part of any research. Models proposed need to be evaluated based on standard metrics to compare their performance to previous models. This section discusses some of the widely used evaluation metrics to assess models for aerial object detection. Precision-recall curve (PRC), mean average precision (mAP) and F-measure ($F1$) are three well-known and widely applied standard measures approach for comparisons. PRC is obtained from four well-established evaluation components in information retrieval, true positive (TP), false positive (FP), false negative (FN) and true negative (TN). TP and FP indicate the number of correct

Table 1 Summary of the datasets

Dataset	Number of images	Number of classes	Instances
DOTA-v 2.0	11,268	18	1.8 million
iSAID	2806	15	655,451
VEDAI	12,009	9	3700
UCAS-AOD	1510	2	14,596
NWPU VHR-10	800	10	3755
RSOD	2326	4	6950
DIOR	23,463	20	192,472
HRSC2016	1061	1	2976
COWC	53	1	32,716
DLR 3 K	20	2	14,235
VisDrone	10,209	10	540 K
XVIEW	7400	60	1 million

predictions and error predictions. FN is the sum of regions not proposed. Based on these four components, we define precision and recall rate as

$$f(\text{precision}) = \frac{\text{TP}}{\text{TP} + \text{FP}} \tag{1}$$

$$f(\text{recall}) = \frac{\text{TP}}{\text{TP} + \text{FN}} \tag{2}$$

$F1$ is a statistic that is commonly used in the field of object detection. The higher the $F1$ value, the better the performance. The definition is as follows:

$$f(f_1) = \frac{2 * \text{recall} * \text{precision}}{\text{recall} + \text{precision}} \tag{3}$$

mAP is to solve the single-point value limitations of P, R, $F1$, it can get an indicator that reflects global performance. The definition is as follows:

$$mAP = \int_0^1 P(R)dR \tag{4}$$

For deeper insight on the result of a model, we can derive other metrics like AP50, AP75 and AP [0.5:0.5:0.95].

In object detection problems, IoU evaluates the overlap between ground-truth mask (gt) and the predicted mask (pd). It is calculated as the area of intersection between gt and pd divided by the area of the union of the two, that is,

$$IoU = \frac{area(gt \cap pd)}{area(gt \cup pd)} \tag{5}$$

IoU metric ranges from 0 and 1, with 0 signifying no overlap and 1 implying perfect overlap between gt and pd. With the IoU metric, we need to define a threshold (α, say) that is used to distinguish a valid detection from the one which is not. AP50, AP75 and AP [0.5:0.5:0.95] helps get a deeper understanding of results from a model where AP50 AP75 mean average precision at 0.5 and 0.75 IoU, respectively.

5   Conclusion and Future Trends

It can be concluded that there has been a lot of development in natural scene object detection. Problems arise when it is done on aerial images. The shift in the angle of anchor boxes and the small sizes of the objects are the primary concerns in aerial object detection.

While various researchers have tried to approach the issues, they have tackled only a single problem at a time, and only a few considered both of the problems. In future, we would like to make a hybrid version of a generic object detection algorithm that will tackle both issues. To tackle the angle shift problem, we would like to use the You Only Look Twice: rapid multi-scale object detection [21] model as our base model. We will use supersampling and denoising models for the small object concern in the pre-processing stage. We would also like to use the search crow optimisation in the post-processing stage for the angle shift.

References

1. Papageorgiou CP, Oren M, Poggio T (1998) A general framework for object detection. In: 6th International conference on computer vision (IEEE Cat. No.98CH36271), pp 555–562
2. Chrétien LP, Théau J, Ménard P (2016) Visible and thermal infrared remote sensing for the detection of white-tailed deer using an unmanned aerial system. Wildlife Soc Bull 40(1):181–191
3. Eikelboom JA, Wind J, van de Ven E, Kenana LM, Schroder B, de Knegt HJ, van Langevelde F, Prins HH (2019) Improving the precision and accuracy of animal population estimates with aerial image object detection. Methods Ecol Evol 10(11):1875–1887
4. Creusen IM, Wijnhoven RG, Herbschleb E, de With PH (2010) Color exploitation in hog-based traffic sign detection. In: 2010 IEEE international conference on image processing, pp 2669–2672
5. Li J, Wong H-C, Lo S-L, Xin Y (2018) Multiple object detection by a deformable part-based model and an r-cnn. IEEE Signal Process Lett 25(2):288–292
6. Yue L, Shen H, Li J, Yuan Q, Zhang H, Zhang L (2016) Image super-resolution: the techniques, applications, and future. Sign Process 128:389–408
7. Lienhart R, Maydt J (2002) An extended set of haar-like features for rapid object detection. In: Proceedings of the international conference on image processing, vol 1, p I

8. Dalal N, Triggs B. (2005) Histograms of oriented gradients for human detection. In: Schmid C, Soatto S, Tomasi C (eds) International conference on computer vision pattern recognition (CVPR '05), vol 1. IEEE Computer Society San Diego, pp 886–893

9. Felzenszwalb PF, Girshick RB, McAllester D (2010) Cascade object detection with deformable part models. In: 2010 IEEE computer society conference on computer vision and pattern recognition, pp 2241–2248

10. Lin TY, Goyal P, Girshick R, He K, Dollár P (2020) Focal loss for dense object detection. IEEE Trans Pattern Anal Mach Intell 42(2):318–327

11. Girshick R, Donahue J, Darrell T, Malik J (2014) Rich feature hierarchies for accurate object detection and semantic segmentation

12. He K, Zhang X, Ren S, Sun J (2014) Spatial pyramid pooling in deep convolutional networks for visual recognition. Lecture Notes Comput Sci 346–361

13. Dai J, Li Y, He K, Sun J, Fcn R (2016) Object detection via region-based fully convolutional networks. In: Lee D, Sugiyama M, Luxburg U, Guyon I, Garnett R (eds) Advances in neural information processing systems, vol 29. Curran Associates, Inc

14. Lin TY, Dollár P, Girshick R, He K, Hariharan B, Belongie S (2017) Feature pyramid networks for object detection

15. He K, Gkioxari G, Dollár P, Girshick R (2018) Mask r-cnn

16. Redmon J, Divvala S, Girshick R, Farhadi A (2016) You only look once: unified, real-time object detection

17. Liu W, Anguelov D, Erhan D, Szegedy C, Reed S, Fu CY, Berg AC (2016) Ssd: single shot multibox detector. In: Lecture notes in computer science, pp 21–37

18. Girshick R (2015) Fast r-cnn

19. Ren S, He K, Girshick R, Sun J (2016) Faster r-cnn: towards real-time object detection with region proposal networks

20. Lin TY, Goyal P, Girshick R, He K, Dollár P (2018) Focal loss for dense object detection

21. Lei J, Gao C, Hu J, Gao C, Sang N. Orientation adaptive yolov3 for object detection in remote sensing images. In: Lin Z, Wang L, Yang J, Shi G, Tan T, Zheng N, Chen X, Zhang Y (eds) Pattern recognition and computer vision. Springer, Cham, pp 586–597

22. Han J, Ding J, Li J, Xia GS (2021) Align deep features for oriented object detection

23. Ding J, Xue N, Long Y, Xia GS, Lu Q. (2018) Learning roi transformer for detecting oriented objects in aerial images

24. Han J, Ding J, Xue N, Xia GS (2021) Redet: a rotation-equivariant detector for aerial object detection

25. Qing Y, Liu W, Feng L, Gao W (2021) Improved Yolo network for free-angle remote sensing target detection. Remote Sens 13(11)

26. Ulmas P, Liiv I (2020) Segmentation of satellite imagery using u-net models for land cover classification. ArXiv, abs/2003.02899

27. Heidler K, Mou L, Baumhoer C, Dietz A, Zhu XX (2021) HED-UNet: combined segmentation and edge detection for monitoring the antarctic coastline. IEEE Trans Geosci Remote Sens 03:1–14

28. Iglovikov VI, Shvets AA (2018) Ternausnet: U-net with vgg11 encoder pre-trained on imagenet for image segmentation. ArXiv, abs/1801.05746

29. Mostofa M, Ferdous SN, Riggan BS, Nasrabadi NM (2020) Joint-srvdnet: joint super resolution and vehicle detection network. IEEE Access 8:82306–82319

30. Courtrai L, Pham MT, Friguet C, Lefèvre S (2020) Small object detection from remote sensing images with the help of object-focused super- resolution using wasserstein GANs. In: IGARSS 2020 IEEE international geoscience and remote sensing symposium, pp 260–263

31. Bhusal S, Bhattarai U, Karkee M (2019) Improving pest bird detection in a vineyard environment using super-resolution and deep learning. IFAC- PapersOnLine 52(30):18–23; 6th IFAC conference on sensing, control and automation technologies for agriculture AGRICONTROL 2019

32. Courtrai L, Pham MT, Lefèvre S (2020) Small object detection in remote sensing images based on super-resolution with auxiliary generative adver sarial networks. Remote Sensing 12(19)

33. Schubert M Chowdhury S Chao D Rabbi J, Ray N (2020) Small-object detection in remote sensing images with end-to-end edge-enhanced GAN and object detector network. Remote Sens 12:1432.

34. Ahmad M, Abdullah M, Han D (2020) Small object detection in aerial imagery using RetinaNet with anchor optimization. In: 2020 International conference on electronics, information, and communication (ICEIC), pp 1–3

35. Zlocha M, Dou Q, Glocker B (2019) Improving RetinaNet for CT lesion detection with dense masks from weak RECIST labels. In Shen D, Liu T, Peters TM, Staib LH, Essert C, Zhou S, Yap PT, Khan A (eds) Medical image computing and computer assisted intervention—MICCAI 2019. Springer, Cham, pp 402–410

36. Lin F, Zheng X, Wu Q. Small object detection in aerial view based on improved yolov3 neural network. In *2020 IEEE International Conference on Advances in Electrical Engineering and Computer Applications(AEECA)*, pages 522–525, 2020.

37. Liu QQ, Li JB (2019) Orientation robust object detection in aerial images based on r-nms. Procedia Comput Sci 154:650–656; Proceedings of the 9th international conference of information and communication technology [ICICT-2019] Nanning, Guangxi, China January 11–13, 2019

38. Zhu X, Lyu S, Wang X, Zhao Q (2021) Tph-yolov5: improved yolov5 based on transformer prediction head for object detection on drone-captured scenarios

39. Prystavka P, Cholyshkina O, Dolgikh S, Karpenko D (2020) Automated object recognition system based on convolutional autoencoder. In: 2020 10th international conference on advanced computer information technologies (ACIT), pp 830–833

40. Walambe R, Marathe A, Kotecha K (2021) Multiscale object detection from drone imagery using ensemble transfer learning. Drones 5(3)

Precise Temperature Control Scheme for Nonlinear CSTR Using Equilibrium Optimizer Tuned 2-DOF FOPID Controller

Riya Shivhare, Nandini Rastogi, Muskan Bhardwaj, Ekta Kumari, Nitin Agrawal, and Mohit Jain

Abstract Precise temperature control of an unstable nonlinear continuous stirred tank reactor (CSTR) is a significant problem of process industries and to serve this purpose, a fractional two degree of freedom PID (2-DOF FOPID) control algorithm is designed in this research study. Incorporation of additional degree of freedom and fractional order operators to the fundamental PID algorithm provides more flexible control structure at the expenditure of a massive number of design parameters. Therefore, a parametric tuning scheme is suggested based on the equilibrium optimizer (EO) to tackle the problem of controller design. The convergence rate analysis of EO proves its efficiency as a tuning technique in contrast to several cutting-edge techniques such as dragonfly algorithm, genetic algorithm and multi-verse optimizer. Conducted investigations show that the EO optimized 2-DOF FOPID (E2FPID) provides more accurate and quick temperature control of nonlinear CSTR as compared to EO tuned classical PID as well as its fractional order variant.

Keywords Equilibrium optimizer · Nonlinear continuous stirred tank reactor · Controller parametric tuning · 2-DOF FOPID

1 Introduction

The core functional element of numerous process industries such as petroleum, polymers, beverages, biochemicals and pharmaceuticals are chemical reactors as they process and alter the raw materials into chemically useful substances. They are the subject of research because of their extremely nonlinear behaviour as well as open loop unstable nature [1]. In chemical reactors, the temperature is a critical characteristic that must be precisely regulated to achieve the needed product quality

R. Shivhare (✉) · N. Rastogi · M. Bhardwaj · E. Kumari · M. Jain
School of Automation, Banasthali Vidyapith, Aliyabad, Rajasthan 304022, India
e-mail: riyashivhare35@gmail.com

N. Agrawal
Indusmic Private Limited, Lucknow 226101, India

© The Author(s), under exclusive license to Springer Nature Singapore Pte Ltd. 2023
A. Shukla et al. (eds.), *Computational Intelligence*, Lecture Notes in Electrical
Engineering 968, https://doi.org/10.1007/978-981-19-7346-8_9

while ensuring a consistent, safe and cost-effective operation of the plant. The situation "thermal runaway" creates a serious difficulty with temperature management that often arises when exothermic and irreversible reactions occur within a chemical reactor [2]. In such a circumstance, these reactors are unable to self-regulate, resulting in greater temperatures which enhances the rate of reaction and this elevates the temperature even higher. Therefore, chemical reactors require more reliable, proficient and precise temperature control scheme to operate them safely and efficiently. Therefore, a diverse range of control solutions is suggested and developed in literature for temperature and concentration control of CSTR.

Recently, model reference neural fuzzy adaptive control scheme is suggested to control the concentration of CSTR [3]. In [4], neural networks inspired adaptive predictive control scheme is proposed for CSTR. In [5], fuzzy (Takagi–Sugeno) model dependent control structure is proposed for precise concentration control of the CSTR plant. The authors claimed excellent servo-regulatory control from the developed controller. In [6], an asynchronous sliding mode control structure is recommended for the CSTR. In spite of several advance control schemes, the fundamental PID controller is most commonly used in a diverse range of control applications because of its straightforward but robust architecture and ease of implementation. Nevertheless, it lacks in flexibility of parametric settings. The issue of flexibility motivated the researchers to employ fractional calculus in basic PID control algorithm which leads to fractional order PID controller (FOPID). The fundamental PID controllers and their fractional order variants have only one degree of freedom (1-DOF) which refers that they comprised of only one closed loop in their control architecture. Thus, they are not highly effective in rejecting disturbances and set point tracking, simultaneously. This drawback led the researchers to expand this control algorithm by further incorporating another loop, i.e. an additional degree of freedom which is commonly called as two degree of freedom PID controller (2-DOF PID) [7]. Moreover, the amalgamation of fractional order operators with this basic structure of 2-DOF PID results in a more flexible but precise control architecture referred as 2-DOF FOPID controller. However, the precision and flexibility are attained at the price of enhanced number of parametric settings in controller design which makes it more challenging optimization problem.

This problem of parametric tuning of 2-DOF PID/FOPID control algorithms may be proficiently tackled via metaheuristics algorithms such as dragonfly algorithm (DA) [8], nondominated sorting genetic algorithm (NSGA-II) [9], spider monkey optimization (SMO) [10], stochastic fractal search (SFS) [11] and water cycle algorithm (WCA) [7]. Recently, equilibrium optimizer is claimed highly efficient [12] than the state-of-the-art metaheuristic optimizers. Therefore, in this study, the potential of EO is explored as a controller tuning technique for the designed 2-DOF FOPID control structure. The key contribution of this research study is that the EO tuned 2-DOF FOPID control algorithm is first time validated on the nonlinear CSTR model as per the current understanding of authors.

The remaining article is structured as follows: Nonlinear CSTR and its mathematical modelling are provided in Sect. 2. The control architecture of 2-DOF FOPID is introduced in Sect. 3. A brief description of EO algorithm and its controller tuning procedure is illustrated in Sect. 4. Section 5 explores about the numerous simulation experiments conducted in this study. Lastly, the conclusion of the current research work is provided in Sect. 6.

2 Mathematical Modelling of Nonlinear CSTR

The majority of nonlinear processes depict highly complex system dynamics which makes them challenging to analyse as well as control. A typical example of such a sophisticated nonlinear system unit is most commonly found in process industries called as CSTR. It is known as the heart of any chemical process industry because it plays a key role in processing and altering the raw materials into useful substances via chemical reactions. CSTR is a benchmark in nonlinearity and exhibits multiple equilibrium points (two stable and one unstable) which makes the task of control system design more challenging [1]. Further, the system dynamics is highly susceptible to parametric variations. Therefore, this system is an ideal choice for control engineers to validate a new control scheme and thus considered in the present work. The governing modelling equations and parameters (Table 1) of nonlinear CSTR are defined as follows [1, 13]

$$\dot{x}_c = -x_c + D_a(1 - x_c) \exp\left(\frac{x_T}{1 + x_T/\Psi_A}\right); \tag{1}$$

$$\dot{x}_T = -x_T + B_{Hr} D_a(1 - x_c) \exp\left(\frac{x_T}{1 + x_T/\Psi_A}\right) + \alpha_h(\mu_{Tj} - x_T); \tag{2}$$

$$y_o = x_T \tag{3}$$

where the output of plant is represented by $y_o = x_T$ and the control input is cooling jacket temperature denoted as μ_{Tj}.

There are several equilibrium states in this system such as $(x_c, x_T)_a = (0.144, 0.886)$, $(x_c, x_T)_b = (0.4472, 2.7517)$ and $(x_c, x_T)_c = (0.7646, 4.705)$. The points

Table 1 Dimensionless modelling parameters of nonlinear CSTR

Variables	Description	Parameters
x_c	Reactor concentration	0.445
x_T	Reactor temperature	2.75
D_a	Damökhler number	0.072
Ψ_A	Activated energy	20
B_{Hr}	Heat of reaction	8
μ_{Tj}	Cooling jacket temperature	0
α_h	Heat transfer coefficient	0.3

$(x_c, x_T)_a$ and $(x_c, x_T)_c$ are stable equilibrium points while point $(x_c, x_T)_b$ is referred to as middle unstable equilibrium state. However, the CSTR is generally operated on the middle unstable equilibrium state to achieve proper production rate within safe operating temperature limits. Therefore, the objective of this research work is to come up with an appropriate control mechanism which makes sure that the output of CSTR must follow the desired set point changes while suppressing all the external random disturbances.

3   2-DOF FOPID Control Algorithm

The central idea of proposed control scheme (Fig. 1) is to deliver an effortless and robust control algorithm which offers commercially efficient results to the chemical industry. The output $U(s)$ of 2-DOF FOPID control algorithm is provided as [14]

$$U(s) = K_p\big[R(s)P_{spw} - Y(s)\big] + K_i\frac{[R(s) - Y(s)]}{s^\lambda} + K_d s^\mu\big[R(s)D_{spw} - Y(s)\big]$$

(4)

where $Y(s)$ is output of the CSTR and $R(s)$ is the reference input. K_p, K_i and K_d are proportional, integral and derivative gain factors, respectively, while P_{spw} is proportional weighing factor and D_{spw} is derivative weighing factor applied on set point (Fig. 1). λ and μ are fractional order operators for integrator and differentiator, respectively. These fractional order operators are approximated and implemented using Oustaloup's filter method [15]. In this control scheme, proportional and derivative set point weights provide quick disturbance rejection ability without creating notable amount of overshoot while following the desired set point [7]. These weighing factors also reduce the impact of external disturbances on the execution of CSTR. Further, the fractional operators help in achieving fine tuning in the output. However, several unknown but interdependent parameters are involved in the controller design and

Fig. 1 Internal architecture of 2-DOF FOPID controller

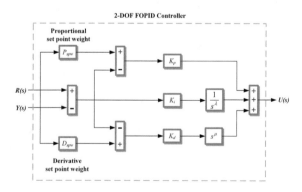

optimal value of them may deliver efficient, stable and safe operation of the plant. Thus, highly acclaimed optimization algorithms are employed to serve the purpose.

4 EO-Based Parametric Tuning Technique of 2-DOF FOPID Control Algorithm

Due to the significant increment in controller design parameters of 2-DOF FOPID control scheme, an efficient optimization algorithm is needed to solve this issue of tuning. Thus, EO is utilized for optimal parametric selection of advised control algorithm for CSTR. EO algorithm is formulated on the fundamental law of dynamic mass balance within a controlled volume. According to this mass balance principle of physics, mass entered, left out and formed in a control volume is always conserved. Similar to other population-based optimizers, EO also relies on search agents known as particles or solutions of the considered optimization problem [16]. Each particle is associated with its concentration level which literally signifies about the current locality of the search agent. Particles in existing population randomly update their level of concentration with reference to current equilibrium candidate (best solution obtained till now) and try to reach in equilibrium state (optimal solution). EO involves following phases to achieve the optimal solution:

4.1 Phase1: Random Initialization

At the beginning, N_p number of particles (population) are formed with random commencing level of concentrations based on uniform distribution as follows:

$$\text{Co}^{k,\text{in}} = \text{Co}^{\text{Min}} + r^k \left(\text{Co}^{\text{Max}} - \text{Co}^{\text{Min}} \right) \quad k = 1, 2, \ldots N_p \qquad (5)$$

where $\text{Co}^{k,\text{in}}$ is a vector and represents the initial concentrations of kth particle, r^k is a real number randomly selected out of the uniform distribution, while Co^{Max} and Co^{Min} are maximal and minimal value of the variable under optimization, respectively. Based on the user defined cost function, fitness of each particle is computed and stored to find the equilibrium candidate solution.

4.2 Phase2: Equilibrium Pool and Candidates

EO is said to be in equilibrium state (optimal state) after final convergence of the algorithm. It utilizes the idea of equilibrium pool (Co^{EqPl}) to ensure the equilibrium state of the algorithm without trapping in local optimal regions. It is a vector formed by best four particles found in existing population along with the average of these best four particles

$$Co^{EqPl} = \left[Co^{Eq(1)}, Co^{Eq(2)}, Co^{Eq(3)}, Co^{Eq(4)}, Co^{EqAvg} \right] \tag{6}$$

where

$$Co^{EqAvg} = \frac{Co^{Eq(1)} + Co^{Eq(2)} + Co^{Eq(3)} + Co^{Eq(4)}}{4} \tag{7}$$

4.3   Phase3: Concentration Level Updating Mechanism:

The concentration level (Co) updating mechanism for each particle is defined as

$$Co = Co^{Eq} + \left(Co - Co^{Eq} \right) \times T^{Exp} + \frac{g^R}{\lambda^{rand} \times V^0} \left(1 - T^{Exp} \right) \tag{8}$$

where g^R is generation rate [16], Co^{Eq} is a randomly selected vector from Co^{EqPl}, λ^{rand} is a number picked at random from a uniform distribution and V^0 is taken as unity. T^{Exp} is called as exponential term and described as

$$T^{Exp} = e^{-\lambda^{rand}} \left(t - t^0 \right) \tag{9}$$

where time (t) is defined as follows:

$$t = \left[1 - \frac{i}{i^{max}} \right]^{A_2 \times \frac{i}{i^{max}}} \tag{10}$$

where A_2 is considered as 1, i is present iteration and i^{max} is the total iterations defined by user. Further

$$t^0 = \frac{1}{\lambda^{rand}} \ln \left(-A_1 \times sign(R - 0.5) \left[1 - e^{\lambda^{rand}_t} \right] \right) + t \tag{11}$$

where R is a number randomly considered from the uniform distribution, A_1 is taken as 2.

Almost every optimizer demands a suitable objective function and by minimizing it (for minimization problems) finds the optimal set of the parameters under optimization. Several primitive cost functions are defined in the literature [7] and after rigorous analysis integral absolute error IAE is judged to be the most appropriate for the present purpose. Apart from EO, other existing optimizers such as MVO, DA and GA are also deployed for the optimal selection of design parameters of 2-DOF

Fig. 2 Convergence rate curve of various metaheuristics while optimizing the parameters of 2-DOF FOPID controller

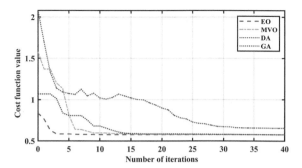

FOPID control scheme and the assessment of their convergence rate is presented in Fig. 2. It is examined here that the EO quickly discovers optimal solution and shows a fast convergence rate in comparison with other optimizers. This makes EO more adequate solution for industrial applications in which online tuning of controller is required and thus preferred for the parametric tuning of suggested control algorithm. The EO-based tuning technique for 2-DOF FOPID is depicted via flowchart (Fig. 3).

5   Results and Discussion

Various standard experimental studies such as step response analysis, disturbance rejection and set point tracking are performed to assess the efficacy of proposed EO tuned 2-DOF FOPID (E2FPID) controller. Further, its performance is analysed against conventional FOPID and PID controllers. However, for fair comparison both FOPID and PID are tuned via EO and called as EFPID and EPID, respectively. All intended simulation studies are performed on the PC having Intel(R) Core (TM) i7-6600U CPU operated on 2.60 GHz, 8.00 GB RAM installed with MATLAB R2019a.

5.1   Step Response Analysis

The aim of this analysis is to analyse the quickness of the designed controllers while dealing with the inputs having abrupt changes. Therefore, this simulation study is carried out by elevating the target temperature of CSTR to 2.75 at time $t = 0$ which means that the operating point is considered at the middle unstable equilibrium point. The responses generated by the designed controllers during step change in the set point temperature are recorded and depicted in Fig. 4 along with their corresponding IAE (Fig. 5). It is revealed from the obtained responses (Fig. 4) that the E2FPID controller quickly achieves the target temperature with the least deviation from set point and thus demonstrates superior performance in comparison with the EFPID

Fig. 3 EO-based parametric
selection scheme for 2-DOF
FOPID controller

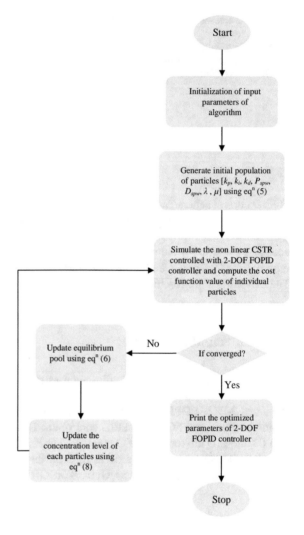

and EPID controllers. The quantitative examination (Fig. 5) also confirms that the
proposed E2FPID controller notably reduces IAE.

5.2 Disturbance Rejection Study

External disturbances are random in nature and cause the plant output to deviate from
the present operating point, and this situation is called as a regulatory problem often
encountered in chemical process industries. Therefore, the designed controllers are
also tested under similar situation to assess their disturbance rejection capability. An

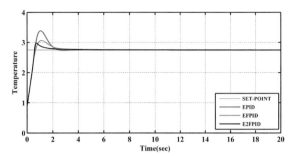

Fig. 4 Performance comparison of implemented control algorithms during step response analysis

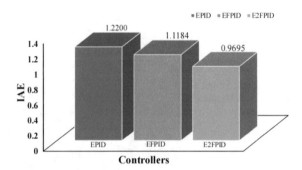

Fig. 5 IAE offered by implemented control algorithms during step response study

external disturbance $d = 0.55$ (approximately 20% of the nominal value) is assumed [17] and introduced into the right-hand side of the modelling Eq. (2) of CSTR for $11 \leq t \leq 13$ to perform this analysis. The recorded responses of reactor controlled by the developed controllers when exposed to high intensity external disturbance are presented in Fig. 6. It may be deduced from the findings that the proposed E2FPID controller rapidly suppresses the disturbance in comparison with the other conventional controllers and maintains the reactor temperature at the desired operating point with minimal deviation during the period of disturbance. The quantitative analysis in terms of IAE (Fig. 7) also validates the dominance of the E2FPID controller as it significantly reduces the error.

5.3 Set Point Tracking

Set point tracking is commonly referred as a servo problem often encountered in process industries and thus performed to examine the performance of the implemented control schemes. Disturbances and parametric variations are also incorporated to analyse the sturdiness of the implemented control algorithms. In this simulation study, the operating temperature of reactor is initially elevated to 2.75 and maintained for the next 25 s, afterwards an elevation of 1.75 is introduced for next 25 s. Further, at time instant $t = 50$ s, another elevation of same intensity (i.e.

Fig. 6 Disturbance rejection analysis of the designed controllers

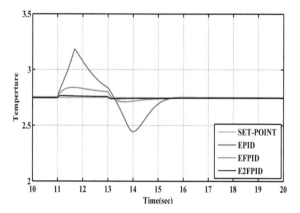

Fig. 7 IAE comparison of the designed controllers during the occurrence of disturbance

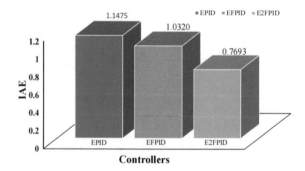

1.75) is incorporated. However, during $11 \leq t \leq 13$ and $60 \leq t \leq 62$ parametric variations of about 15% is taken into account in the nominal values of B_{Hr} and $\alpha_{h,}$ respectively, while for $35 \leq t \leq 36$ an external disturbance of value 0.23 is introduced. The recorded responses of reactor controlled by the implemented controllers under the set point tracking are depicted in Fig. 8. The findings reveal that the proposed control configuration outperforms the conventional controllers during servo problem. The quantitative analysis (Table 2) also confirms the efficaciousness of the proposed E2FPID controller in explicitly dealing with uncertainty. It is also evident here that the enhanced performance of proposed control system is achieved via an additional degree of freedom and fractional order operators due to which the proposed controller efficiently handles servo and regulatory problem at the same time.

6   Conclusion

In this research work, a precise, effective and robust control scheme is designed for temperature control of nonlinear CSTR. The parametric settings of suggested 2-DOF FOPID are obtained and optimized via EO algorithm. The algorithm is first

Fig. 8 Comparative analysis
of implemented controllers
during servo problem with
external disturbances and
parametric variations

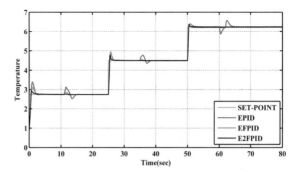

Table 2 IAE offered by the
designed control techniques
during servo problem

Controllers	EPID	EFPID	E2FPID
IAE	4.5711	3.2977	**2.3440**

time validated on nonlinear CSTR and employed to tune the designed controller. The
results reveal that EO offers a quick and precise solution for the controller tuning
problem in comparison with GA, DA and MVO as it shows fast convergence rate. The
EO is also employed for the tuning of FOPID and traditional PID controllers which are
intended to be used for comparative analysis. In all the conducted simulation studies
such as step response analysis, disturbance rejection study and set point tracking,
the designed E2FPID is found superior in comparison with its basic variants such as
EFPID and EPID. IAE-based quantitative investigation of results also validates the
pre-eminence of the proposed E2FPID controller. The findings lead to the conclusion
that EO-based optimized 2-DOF FOPID controller provides effective, precise and
robust temperature control of nonlinear CSTR. Thus, the EO algorithm must also
be employed and tested for other control studies. In future, the adaptive version of
the proposed E2FPID controller may be designed and validated on nonlinear CSTR
with time varying modelling parameters.

Acknowledgements The authors are grateful to Indusmic Private Limited for providing continuous
guidance and unwavering support throughout the research work. This study is the result of their
tenacious and resolute attitude, which enabled us to meet our aims.

References

1. Chang WD (2013) Nonlinear CSTR control system design using an artificial bee colony
   algorithm. Simul Model Pract Theory 31:1–9
2. Luyben WL (2007) Chemical reactor design and control. Wiley
3. Bahita M, Belarbi K (2016) Model reference neural-fuzzy adaptive control of the concentration
   in a chemical reactor (CSTR). IFAC-PapersOnLine 49(29):158–162
4. Li S, Gong M, Liu Y (2016) Neural network-based adaptive control for a class of chemical
   reactor systems with non-symmetric dead-zone. Neurocomputing 174:597–604

5. Arasu SK, Prakash J (2020) Design and implementation of Takagi-Sugeno fuzzy model based control scheme for the continuous stirred tank reactor. IFAC-PapersOnLine 53(1):447–452
6. Li F, Cao X, Zhou C, Yang C (2021) Event-triggered asynchronous sliding mode control of CSTR based on Markov model. J Franklin Inst 358(9):4687–4704
7. Jain M, Rani A, Pachauri N, Singh V, Mittal AP (2019) Design of fractional order 2-DOF PI controller for real-time control of heat flow experiment. Eng Sci Technol Int J 22(1):215–228
8. Mishra S, Prusty RC, Panda S (2020) Design and analysis of 2dof-PID controller for frequency regulation of multi-microgrid using hybrid dragonfly and pattern Search algorithm. J Control Autom Electr Syst 1–15
9. Gaidhane PJ, Kumar A, Nigam MJ (2017) Tuning of two-DOF-FOPID controller for magnetic levitation system: a multi-objective optimization approach. In: 2017 6th international conference on computer applications in electrical engineering-recent advances (CERA). IEEE, pp 479–484
10. Tripathy D, Sahu BK, Patnaik B, Choudhury ND (2018) Spider monkey optimization based fuzzy-2D-PID controller for load frequency control in two-area multi source interconnected power system. In: 2018 technologies for smart-city energy security and power (ICSESP). IEEE, pp 1–6
11. Mandala II, Nazaruddin YY (2019) Optimization of two degree of freedom PID controller for quadrotor with stochastic fractal search algorithm. In: 2019 IEEE conference on control technology and applications (CCTA). IEEE, pp 1062–1067
12. Guha D, Roy PK, Banerjee S (2021) Equilibrium optimizer-tuned cascade fractional-order 3DOF-PID controller in load frequency control of power system having renewable energy resource integrated. Int Trans Electr Energy Sys 31(1):e12702
13. Chen CT, Peng ST (1999) Intelligent process control using neural fuzzy techniques. J Process Control 9(6):493–503
14. Mohapatra TK, Dey AK, Sahu BK (2020) Employment of quasi oppositional SSA-based two-degree-of-freedom fractional order PID controller for AGC of assorted source of generations. IET Gener Transm Distrib 14(17):3365–3376
15. Pachauri N, Rani A, Singh V (2017) Bioreactor temperature control using modified fractional order IMC-PID for ethanol production. Chem Eng Res Des 122:97–112
16. Faramarzi A, Heidarinejad M, Stephens B, Mirjalili S (2020) Equilibrium optimizer: a novel optimization algorithm. Knowl-Based Syst 191:105190
17. Colantonio MC, Desages AC, Romagnoli JA, Palazoglu A (1995) Nonlinear control of a CSTR: disturbance rejection using sliding mode control. Ind Eng Chem Res 34(7):2383–2392

Descriptive Predictive Model for Parkinson's Disease Analysis

Akbar Ali, Ranjeet Kumar Rout, and Saiyed Umer

Abstract The popularity of machine learning applications is increasing day by day. Uses of prediction model are in so many fields, for example weather forecast, industry, medical, technology, and data science also. In this paper, we are going to analyze a very common disease in human known as Parkinson disease. It is mostly found in older people but some time also reflects its presence in younger people. Model objective is to predict according to the past information. It is better to have high accuracy of the proposed model. The goal of Parkinson disease predicting model is detecting Parkinson's disease in humans according to the data we will provide. We have used two types of audio dataset in this model, namely dysphonia and dysarthria. We have used four machine learning algorithms namely **support vector machine**, **logistic regression**, **K_nearest neighbor,** and **random forest**. First dataset has 24 feature-set and other have 756 feature-set, so we have used PCA also to provide feature-set according to need. We divided the dataset into four modules in first module we have used only five feature-set, and then in second module, we have used ten feature-set and then 15 and 20 feature-set, respectively, and then all features at a time. We have feeded these module to all algorithms and noted which one is best for which case. So we have four algorithm and we have tested these module one by one to these algorithm carefully and noted down there result, and then we analyzed that for dysphonia if we use random forest, then it gives better result for ten feature-set while for dysarthria it is support vector machine that gives better performance for ten feature-set.

Keywords Parkinson's disease · Dysphonia · Dysarthria

A. Ali (✉) · S. Umer
Computer Science and Engineering, Aliah University, Kolkata, India
e-mail: aamaanakbarali@gmail.com

S. Umer
e-mail: saiyed.umer@aliah.ac.in

R. K. Rout
Computer Science and Engineering, National Institute of Technology, Srinagar, Jammu and Kashmir, India
e-mail: ranjeetkumarrout@nitsri.net

© The Author(s), under exclusive license to Springer Nature Singapore Pte Ltd. 2023   105
A. Shukla et al. (eds.), *Computational Intelligence*, Lecture Notes in Electrical
Engineering 968, https://doi.org/10.1007/978-981-19-7346-8_10

1 Introduction

Parkinson's disease [17] is a fundamental disease. It affects us very slowly. We cannot be sure about its starting period. It affects the human body for a long period. After a long period, when its frequency increases, then it is noticed. After investigation, it comes to the knowledge that it does not come instantly. It was present for a long period. Initially, its frequency was too low to be noticed. It is impossible to say its origin point. Sometimes we linked it for certain accidents like fever, pain, and similar problems [26]. After that it comes to notice that oh my god, I have this disease too. Most of the time Parkinson's disease starts with tremble and dodder. Parkinson's disease starts with the damage of a specific type of shell of the brain. It starts with tremble and dodder of finger and hand. Initially, it was low, but suddenly it increased in the patient. Often, it is observed that patients feel it right side most of the time, but some patients also feel it on the left side and both sides. You can feel it when you closely observe your hand, elbow, and wrist on the table at rest. At peak, it makes a serious problem like one cannot write, shave, cut mustache, thread the needle, and sew on a button, similar to a problem. It is also observed that walking steps are very short or slow walk. One with this disease cannot feel anger, pain, or pleasure, i.e., expression cannot be seen on the patient's face. In older people, sometimes it is so difficult to say that the problem is due to age or Parkinson's disease, and the person suffers from laziness, slowness, acrimony, ugliness, tetchiness, sleeplessness, weakness, sex problem, weight problem, dizziness, and some other side effects. The name Parkinson's disease comes after researcher Dr. James Parkinson. His work in this field was too perfect that it cannot be replaced, so it is known as Parkinson's disease to honor him. Parkinson's is present worldwide in all human species, breeds, and gender. It is mainly present in older people. The developed country has older adults so it is often found there. In America total population is 300 million, and almost six lakes have this disease. In India, total population is 1300 million. Data is not present, but it is speculated that it may be 2.4 million. Parkinson's is still a mystery, although many research papers have proposed. Research on this field is still going, and doctors and scientists are trying to find its reason. The cause of Parkinson's disease is unknown, then how can we cure it? Parkinson's disease cannot be cured, but its effect can be slowed down with early diagnosis and treatment. Sort visit of the clinic is not sufficient for this problem we must have patience for it, so remote monitoring of this is needed. The Parkinson's disease symptoms are divided into two parts, namely motor and non-motor. Motor symptoms include micrographic, dysphonia, bradykinesia, fog, trimmer, and giant. Non-motor symptoms include cognitive parameters that can be measured with MRI, DAT-SCAN, and F-MRI, etc. Parkinson's disease does not cause a big problem, but sometimes it causes pain. [19] Four different types of pain is described as follows (i) musculoskeletal, (ii) radicular neuropathic, (iii) dystonic pain, and (iv) central and primary pain. Scientists are finding the cure to this dramatic problem. Accident or tension is no reason behind it. Sometimes it is linked with the accidental patient, and it is said that this disease comes after the accident, but this claim is false. It may be present in a person having no such incident. Sometimes it is observed that all accidental patients do not have this disease. A very few symptoms like Parkinson's disease come from other diseases like (a) some medicinal uses for the human brain, (b) blocking of brain bloodline, and (c) infection in the brain, till now the original reason behind it not known. But it is known that it happens due to some internal brain tissue damage.no of some specific neuron decreases. Due to it, dopamine is not formed as per the required amount.

Why it happens is not known till now, but some theories are present. Some may find it possible due to genes or genetic, but not all. It is unknown why it attacks some specific people; it may be due to the environment or pollution, I do not know. This disease is not contagious; eating, sleeping, or living with Parkinson's patients is perfectly safe. In early 60, it is come to notice that some people taking brown sugar and similar like intoxicant and an injection having m.p.t.p mixture is spreading Parkinson disease fast, so it is said that m.p.t.p like chemical is from the environment or any source is damaging brain shell. So due to brain tissue problem this disease comes. According to recent work on Parkinson's disease (PD), it is the most common disease in Asia and especially in China among older people. It is also present in other Continent. Research says it is not so powerful to cause death but causes many problems, as mentioned above. This work proposes a PD prediction system based on identifying the dysphonia and dysarthria symptoms. The organization of this paper is as follows: Sect. 2 describes the related work; Sect. 3 demonstrates the proposed methodology. The experimental results and discussion have been described in Sect. 4. This paper is concluded at Sect. 5.

2 Related Work

Hu et al. [8] had proposed a method for the identification of the Parkinson disease problem, diagnosis, and treatment by the use of multi-omics joint analysis. In-guanzo et al. [9] had proposed cluster analysis based on cortical thickness for the detection of Parkinson's disease problem. Solana-Lavalle et al. [21], detection of Parkinson's disease based on voice-based analysis provides physician a decision tool to help why Parkinson's disease happens. Simplifying diagnosis with classification model for the patient who is suffering from Parkinson disease had been developed by Gottapu et al. [6]. Landers et al. [11] had explored the validity of the theoretical model for the Parkinson's disease problem. Diaz et al. [3] had identified Parkinson's symptoms by common characteristics like rigidity, akinesia, and tremor for analyzing patient motor control over handwriting with the help of an online tool to support

Parkinson disease assessment. Jankovic et al. [10] had described the characteristic of Parkinson's disease with emphasis on motor and non-motor characteristics.

Advancement in technology with different collected data with the help of health professionals had been used for providing diagnosis with more accurate and cheaper by Wroge et al. [27]. Massano et al. [13] had derived an instance of Parkinson disease make the practical test for Parkinson disease problem. Tsoulos et al. [22] had provided evidence that with the help of artificial intelligence system allow discriminating the patient from the further treatment of Parkinson disease problem. The different aspects had been detected through the computer-assisted habitation technique by Goyal et al. [7]. Rana et al. [19] had identified the intensity of pain in different patients to realize the improvement in the health of a person. Zhang et al. [28] had performed a histogram-based analysis system for the investigation of Parkinson's disease patients. Kadar.

N. Prasad et al. [18] had identified the types of treatment that can be provided to a Parkinson's disease patient. He also classified the treatment and gave a detailed analysis. Nussbaum et al. [16] identifying the genetic contribution to complex diseases like Alzheimer's and Parkinson's for genomic medicine. Davie et al. [2] had described the detailed reviews of such a vast neurological disease. Forno et al. [4] had described the neuropathological treatment of Parkinson's through various methods and then summarized the most important point about it. Moore et al. [14] had defined the contribution of genes to the pathogenesis of Parkinson's disease. He also described the heritability relation of the gene to cause Parkinson's disease. Lotharius et al. [12] had given the importance of dopamine and their effect on the patient mind also described the relation between them. Damier et al. [1] had given the accuracy in studying the pattern of dopamine in midbrain dopamine-containing neurons in Parkinson's disease. The order of dopamine loss in the brain is causing the genetic problem.

3 Methodology

In this work, we have implemented a system to identify Parkinson's disease problems in patients. From the above literature, it has been studied that this problem identification has been performed based on text, image, and video data. In our work, we have considered the secondary data distributed publicly for the Parkinson's disease identification problem research areas. The block diagram for our proposed methodology has been shown in Fig. 1.

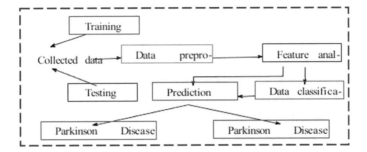

Fig. 1 Block diagram of the proposed methodology

The figure is all about the methodology. It says that first, we have used the secondary data. Then, we prepossess data, and then, we will do feature analysis. Then, divide the data into two parts, training and testing. After that, we use different data classifications like k-nearest neighbor, random forest, logistic regression, and SVM. Now our prediction model is ready to predict whether a person has Parkinson's disease or not.

3.1 Data Pre-processing

Data pre-processing is nothing but making the data suitable for the machine. The data we are getting from any source may always not come with that format we need. It is essential to clean the data and make it suitable. There are several ways to make it do it. In machine learning, pre-processing is all about the objective and our vision to make the best model. We follow some steps for pre-processing: (i) get the data, (ii) import libraries, (iii) import dataset, (iv) find missing data, (v) encode data, (vi) split data into training and testing, and (vii) scale the feature. This is the complete process of how we are processing our data into the format we need. If the data are corrupted or missing, training, and testing will become wrong. Before understanding it, we must know which type of data will be there. Data may be in the form of text, image, video, audio, etc. The pre-processing data [15] technique involves the process of converting the raw (original) data to its meaningful format. The data obtained from any source may be incomplete, inconsistent, missing, noisy, and duplicate values. So we need to convert those data into meaningful data which have the following properties: (i) accuracy, (ii) completeness, (iii) consistency, (iv) believably, and (v) interpretability. Hence, there are some major steps in the data pre-processing techniques:

- Data cleaning is the step for filling missing values, correcting inconsistent data and smooth noise during processing.

- Data integration is a technique to merge multiple sources and combine them for further processing.
- Data reduction is the technique to reduce the data size by aggregating and eliminating redundant features.
- Data discretization is the technique to transform numeric data by mapping values to interval or concept labels.
- Data transformation is the method to normalize the data from one scale to another scale such that the values in each feature must lie within the range.

In our model, the data we got is also not clean. The information we have is taken from an external source. Our model is the prediction model, so it must take good data so to predict right.

3.2 Feature Analysis

It is an important component of any data analysis task. In feature analysis, mainly the more discriminant and discriminating feature or parameter has been analyzed from the capture data [23, 24]. The performance of a prediction system is based on methods of feature analysis. There are several methods of feature analysis that are being used to extract more discriminate features. This method is divided into two parts (i) feature engineering and (ii) feature selection. Feature engineering means extracting features from the giver information, while the feature selection task is refining features from the given feature information. Feature engineering needs both theoretical and mathematical approaches for feature computation from that research domain for which the data belongs [23]. The feature selection methods are used widely in several domains of research areas. It needs techniques from unsupervised machine learning approaches. The technique under this category is principal component analysis, linear discriminate analysis, and other dimensionality reduction methods [5, 15]. In this work, we have employed a feature selection technique to analyze each feature for the Parkinson's disease problem. The most important features after feature analysis derives a good predictive model (f) that establishes better relationship between input features ($X \in 1 \times n$) and target feature (y) such that $y = f (X)$. Principle component analysis (PCA) [15] is one of the best dimensionality reduction techniques. PCA is a technique for reducing the dimensionality of dataset, increasing interpretability while minimizing information loss.

Let us say you have a massive table of data that it is hard to process and you want to make it as small as possible while keeping information as much as possible. So dimensionality reduction refers to the technique that reduces the number of input variables in a dataset. The importance of dimensionality reduction are: (i) it reduces dimension for a dataset due to less computation or training time required; (ii) similar entries from a dataset redundancy is also removed; (iii) space is reduced to store data; (iv) 2D and 3D plotting is easy; (v) it is helping to find out the most significant feature and skipping the rest; and (vi) better human interpretations. PCA performs the following operation to evaluate the PCA for a given dataset (i) standardization (ii) covariance matrix computation (iii) eigenvectors, eigenvalues, and (iv) feature vector standardization is the step to standardize the range of attribute, so each of them falls within similar boundaries. The $z =$ (variable values-mean)/standard deviation/the covariance matrix shows the correlation between parameters in the matrix form. Now eigenvalues and eigenvectors are extracted from the covariance matrix that computes the principle component.

3.3 Classification

Classification [23] is a technique in which you are predicting something from labeled data. It says whether something is falling to a definite group or not. Here, we want to know whether the algorithm can separate an object into categories, i.e., classification is something upcoming input and says this data is falling to this category. Classification is a technique of supervised learning which means it uses labeled data to train the algorithm. It is the process of categorizing a given set of data into class. Classification can perform on both structured and unstructured data points. The types are often referred to as target, label, or categories.

- Logistic regression It is a classification algorithm in machine learning that uses one or more independent variables to determine an outcome [23]. The outcome is measured with a dichotomous variable, meaning it will have only two possible outcomes. Like all regression, analysis logistic is a predictive analysis. Logistic regression is used to establish the relationship between one dependent binary variable to one or more nominal, ordinal, and interval variables. It is used when the data is conditional, i.e., output data is either categorized or binary. We have the following four types of logistic regression binary. When there are two categories. Two level characteristic e.g., on or off, 1 or 0 ordinal. Three or more categories. Natural ordering characteristic e.g., cricket match out, not out, bold, etc., nominal. Three or more than three categories. Characteristics are not according to nature. E.g., color (blue, gray, and red) passion. Three or more than three categories. No of occupancy of an event e.g., 0, 1, 2, 3.
- SVM The main purpose of SVM is to divide the data into two parts. It is used to classify image, recognize handwriting, and categorize text. It is one type of non-binary linear classifier that divides labeled data into two parts.
- KNN It is used for regression and classification problems. It uses data to classify new data points according to the similarity measure. It is also called a lazy algorithm. It is simple, but it gives a high competitive result. With the help of The KNN algorithm, one can classify a customer that will buy a shirt or not, will buy which color shirt so on. It is also used in many fields like image recognition and speech recognition, etc.
- Random forest It creates a decision tree from a subset of the training set. It is a bagging technique. Create many decision trees; combine them to get a good prediction. It is also used in both classification and regression.

4 Experiments

In this section, we have discussed the experimental setting and results. The experiment of this work has been performed using Python in windows10 OS with an 8 GB RAM core i5 processor. Here, both dysphonia and dysarthria symptoms [20] have been considered for Parkinson's disease analysis. In this project, we are making a model with the help of Python to detect Parkinson's disease in someone 's body. We will use Python libraries like NumPy, panda, and xgboost to build our model. We will load data, get features and labels, scale the feature, and split the dataset to calculate accuracy. Our project will deal with two types of Parkinson's disease problems known as dysphonia and dysarthria. Both the problem is the speech-related problem. Dysphonia is all about abnormal voice. It may come suddenly or over the time. The voice may suffer from hoarse, raspy, rough, weak, strained, etc. While on the other hand, dysarthria comes when your mussel used for speech is weak or has some problem controlling them. It can cause voice slurred or slow that can be difficult to understand.

4.1 Dataset Used

In this section, we have discussed the dataset used for this experiment. We have used two dataset taken from kaggle.com. The dataset we used is vocal and known as dysphonia, and the other is known as dysarthria. First, dataset [20] has 24 features, and 196 patients were the part of that survey while on the other hand, the second dataset [21, 25] having 752 features and 756 patients were the part of that dataset. The number of male patients in the dataset is 390 out of 756, while female patients are 366 out of 756.

4.2 Results and Discussion

We have two dataset so; first, we start our experiment using dysphonia. We have divided into training and testing using 50–50, 60–40, 70–30, 80–20 and 90–10 (%) and observer which one gives the best result and note down. We start an experiment using dysarthria. We have done the same procedure with this dataset divided training and testing into several formats as given above. We pick the best result from both experiments and then apply PCA to that selected one to observe its behavior to make the best model.

4.3 Experiment for Dysphonia

First case where we divided our data into 50–50% for training and testing of Parkinson's dataset. At first, we took all attributes and found the result, then we took five attributes and got the result, then ten attributes, 15 attributes, and 20 attributes were selected for finding the individual result. From the following classification, we have found that random forest gives the best result among all. So we can say that random forest is best for 50–50% cases. Table 1 gives the information about it.

This is the case where we divided our data into 60–40% for training and testing of Parkinson's dataset. At first, we took all attributes and found the result, then we took five attributes and got result, then ten attributes, 15 attributes, and 20 attributes were selected for finding the respective result. From the following classification, we found that random forest gives the best result among all. So we can say that random forest is best for 60–40% cases. Table 2 gives the information about it.

This is the case where we divided our data into 70–30% for training and testing of Parkinson's dataset. At first, we have taken all attributes and found the result, then we took five attributes and got result, then ten attributes, 15 attributes and 20 attributes selected for finding the respective result. From the following classification, we found that random forest and KNN give the best result, so we can say that random forest and KNN are best for70–30% case. Table 3 gives the information about it.

This is the case where we divided our data into 80–20% for training and testing of Parkinson's dataset. At first, we took all attributes and found the result, then we took five attributes and got result, then ten attributes, 15 attributes, and 20 attributes selected for finding the respective result. From the following classification, we found

Table 1 Performance of dysphonia symptom for the proposed system using 50–50% training–testing protocol

Classifier	F	G_1	G_2	G_3	G_4
LR	70.00	81.63	83.67	82.65	83.67
SVM	85.00	84.69	88.77	84.69	87.75
KNN	95.00	87.75	91.83	84.69	88.77
RF	90.00	85.71	92.85	82.65	85.71

Table 2 Performance of dysphonia symptom for the proposed system using 60–40% training–testing protocol

Classifier	F	G_1	G_2	G_3	G_4
LR	84.61	85.71	89.79	88.77	84.69
SVM	89.74	81.63	92.85	87.75	90.81
KNN	87.17	84.69	87.75	89.79	88.77
RF	89.74	84.69	94.89	86.73	88.77

Table 3 Performance of dysphonia symptom for the proposed system using 70–30% training–testing protocol

Classifier	F	G_1	G_2	G_3	G_4
LR	88.77	81.63	83.05	82.65	87.75
SVM	85.71	86.73	86.44	89.79	89.79
KNN	88.77	88.77	89.83	91.83	86.73
RF	91.83	86.73	89.83	85.71	90.81

that random forest gives the best result among all. So we can say that random forest is best for 80–20% case. Table 4 gives the information about it.

This is the case where we divided our data into 90–10% for training and testing of Parkinson's dataset. At first, we took all attributes and found the result, then we took five attributes and got result, then ten attributes, 15 attributes, and 20 attributes selected for finding the respective result. From the following classification, we found that KNN gives the best result among all. So we can say that KNN is best for 90–10% case. The following Table 5 gives the information about it.

Finally, from Tables 1, 2, 3, 4 and 5, it has been observed that for dysphonia, ten-dimensional PCA has obtained better performance for the random forest. These ten dimension features are sufficient for the prediction of dysphonia problems in a person.

Table 4 Performance of dysphonia symptom for the proposed system using 80–20% training–testing protocol

Classifier	F	G_1	G_2	G_3	G_4
LR	82.05	85.71	85.71	84.69	80.61
SVM	82.05	88.77	87.75	87.75	86.73
KNN	87.17	87.75	94.89	87.75	91.83
RF	87.17	92.85	91.83	90.81	88.77

Table 5 Performance of dysphonia symptom for the proposed system using 90–10% training–testing protocol

Classifier	F	G_1	G_2	G_3	G_4
LR	80.61	80.61	78.57	86.73	86.73
SVM	87.75	80.61	83.67	92.85	90.81
KNN	89.79	83.67	85.71	96.93	88.77
RF	85.71	85.71	84.69	91.83	79.59

Table 6 Performance of dysarthria symptom for the proposed system using 50–50% training–testing protocol

Classifier	H	L_1	L_2	L_3	L_4
LR	82.53	80.42	82.80	82.01	79.36
SVM	83.33	80.15	81.48	84.12	83.06
KNN	84.12	79.89	83.86	83.33	81.21
RF	85.44	80.95	84.65	82.80	82.53

4.4 Experiment for Dysarthria

In this experiment, the dataset for dysarthria has been fragmented into 50–50, 60–40, 70–30, 80–20, and 90–10 (%) as training–testing datasets. The individual of these experiments have been shown in Tables 6, 7, 8, 9 and 10, respectively. Here, in each Table, H represents the actual feature-set, L_1 is the 5% PCA feature-set, L_2 be the 10% PCA feature-set, L_3 is the 15% PCA feature-set, L_4 is the 20%

Table 7 Performance of dysarthria symptom for the proposed system using 60–40% training–testing protocol

Classifier	H	L_1	L_2	L_3	L_4
LR	88.44	80.68	82.80	82.27	84.39
SVM	86.13	79.10	82.53	83.33	84.92
KNN	85.80	81.21	85.71	83.33	84.65
RF	88.77	81.21	85.97	82.80	85.44

Table 8 Performance of dysarthria symptom for the proposed system using 70–30% training–testing protocol

Classifier	H	L_1	L_2	L_3	L_4
LR	84.14	80.15	80.82	82.27	80.42
SVM	85.90	81.21	81.74	84.39	85.18
KNN	83.70	79.10	78.83	83.86	84.39
RF	87.66	79.89	83.59	85.44	84.65

Table 9 Performance of dysarthria symptom for the proposed system using 80–20% training–testing protocol

Classifier	H	L_1	L_2	L_3	L_4
LR	84.21	81.48	83.59	85.44	81.48
SVM	85.52	83.33	83.59	85.44	81.21
KNN	88.15	82.80	81.21	86.50	81.48
RF	88.15	84.39	87.56	84.39	80.95

Table 10 Performance of dysarthria symptom for the proposed system using 90–10% training–testing protocol

Classifier	H	L_1	L_2	L_3	L_4
LR	82.89	80.42	79.62	83.33	80.15
SVM	86.84	81.48	79.62	82.53	85.71
KNN	84.21	79.62	80.15	79.89	83.33
RF	85.52	78.57	82.27	84.65	85.18

PCA feature-set of the total PCA feature-set. So, from the Tables 6, 7, 8, 9, it has been observed that for RF classifier the performance is good while in Table 10, it is seen that for SVM classifier the performance is better.

5 Conclusion

A Parkinson's disease prediction system has been proposed in this paper. The implementation of the system has been divided into three components (i) data preprocessing, (iii) feature analysis, and (iii) data classification. Here, two different symptoms, dysphonia and dysarthria of Parkinson's disease, have been considered. Dysphonia problem is related to the vocal sound disturbance in a patient, while dysarthria is related to speech analysis where the patient faces when talking with someone; hence, two different datasets for both dysphonia and dysarthria are two different datasets have been considered. Then, data pre-processing followed by feature analysis using the principal component method has been adopted in both these databases. Finally, logistic regression, support vector machine, KNN, and random forest classifiers are used to obtain the prediction model individually for these systems. The accepted prediction model of dysphonia symptoms is eligible to predict the symptoms within a patient based on some discriminant features. The prediction model of dysarthria symptoms is also suitable for Parkinson's disease in a patient.

References

1. Damier P, Hirsch E, Agid Y, Graybiel A (1999) The substantia nigra of the human brain: Ii. Patterns of loss of dopamine-containing neurons in Parkinson's disease. Brain 122(8):1437–1448
2. Davie CA (2008) A review of Parkinson's disease. Br Med Bull 86(1):109–127
3. Diaz M, Moetesum M, Siddiqi I, Vessio G (2021) Sequence-based dynamic handwriting analysis for Parkinson's disease detection with one-dimensional convolutions and bigrus. Expert Syst Appl 168:114405
4. Forno LS (1988) The neuropathology of Parkinson's disease. Prog Parkinson Res 11–21

5. Ghosh A, Umer S, Khan MK, Rout RK, Dhara BC (2022) Smart sentiment analysissystem for pain detection using cutting edge techniques in a smart healthcare framework. Cluster Comput 1–17

6. Gottapu RD, Dagli CH (2018) Analysis of Parkinson's disease data. Procedia Comput Sci 140:334–341

7. Goyal J, Khandnor, P, Aseri TC (2020) Classification, prediction, and monitoring of Parkinson's disease using computer assisted technologies: a comparative analysis. Eng Appl Artif Intell 96:103955

8. Hu C, Ke CJ, Wu C (2020) Identification of biomarkers for early diagnosis of Parkinson's disease by multi-omics joint analysis. Saudi J Biol Sci 27(8):2082–2088

9. Inguanzo A, Sala-Llonch R, Segura B, Erostarbe H, Abós A, Campabadal A, Uribe C, Baggio HC, Compta Y, Marti MJ, Valldeoriola F (2021) Hierarchical cluster analysis of multimodal imaging data identifies brain atrophy and cognitive patterns in Parkinson's disease. Parkinsonism Relat Disord 82:16–23

10. Jankovic J (2008) Parkinson's disease: clinical features and diagnosis. J Neurol Neurosurg Psychiatry 79(4):368–376

11. Landers MR, Jacobson KM, Matsunami NE, McCarl HE, Regis MT, Long-hurst JK (2021) A vicious cycle of fear of falling avoidance behavior in Parkinson's disease: a path analysis. Clin Parkinsonism Relat Disord 4:100089

12. Lotharius J, Brundin P (2002) Pathogenesis of Parkinson's disease: dopamine, vesicles and α-syncline. Nat Rev Neurosci 3(12):932–942

13. Massano J, Bhatia KP (2012) Clinical approach to Parkinson's disease: features, diagnosis, and principles of management. Cold Spring Harb Perspect Med 2(6):a008870

14. Moore DJ, West AB, Dawson VL, Dawson TM (2005) Molecular pathophysiology of Parkinson's disease. Annu Rev Neurosci 28:57–87

15. Nasar N, Ray S, Umer S, Mohan Pandey H (2020) Design and data analytics of electronic human resource management activities through internet of things in an organization. Softw Pract Experience

16. Nussbaum RL, Ellis CE (2003) Alzheimer's disease and Parkinson's disease. N Engl J Med 348(14):1356–1364

17. Poewe W, Seppi K, Tanner CM, Halliday GM, Brundin P, Volkmann J, Schrag AE, Lang AE (2017) Parkinson disease. Nat Rev Dis Primers 3(1):1–21

18. Prasad KN, Cole WC, Kumar B (1999) Multiple antioxidants in the prevention and treatment of Parkinson's disease. J Am Coll Nutr 18(5):413–423

19. Rana AQ, Kabir A, Jesudasan M, Siddiqui I, Khondker S (2013) Pain in Parkinson's disease: analysis and literature review. Clin Neurol Neurosurg 115(11):2313–2317

20. Rosen KM, Kent RD, Delaney AL, Duffy JR (2006) Parametric quantitative acoustic analysis of conversation produced by speakers with dysarthria and healthy speakers

21. Solana-Lavalle G, Rosas-Romero R (2021) Analysis of voice as an assisting tool for detection of Parkinson's disease and its subsequent clinical interpretation. Biomed Sign Process Control 66:102415

22. Tsoulos IG, Mitsi G, Stavrakoudis A, Papapetropoulos S (2019) Application of machine learning in a Parkinson's disease digital biomarker dataset using neural network construction (NNC) methodology discriminates patient motor status. Frontiers in ICT 6:10

23. Umer S, Mohanta PP, Rout RK, Pandey HM (2020) Machine learning method for cosmetic product recognition: a visual searching approach. Multimedia Tools Appl 1–27

24. Umer S, Mondal R, Pandey HM, Rout RK (2021) Deep features based convolutional neural network model for text and non-text region segmentation from document images. Appl Soft Comput 113:107917

25. Umer S, Rout RK, Pero C, Nappi M (2021) Facial expression recognition with tradeoffs between data augmentation and deep learning features. J Ambient Intell Hum Comput 1–15

26. Verbaan D, Marinus J, Visser M, van Rooden SM, Stiggelbout AM, van Hilten JJ (2007) Patient-reported autonomic symptoms in Parkinson disease. Neurology 69(4):333–341

27. Wroge TJ, Özkanca Y, Demiroglu C, Si D, Atkins DC, Ghomi RH (2018) Parkinson's disease diagnosis using machine learning and voice data. In: 2018 IEEE signal processing in medicine and biology symposium (SPMB). IEEE, pp 1–7
28. Zhang Y, Yang M, Wang F, Chen Y, Liu R, Zhang Z, Jiang Z (2020) Histogram analysis of quantitative susceptibility mapping for the diagnosis of Parkinson's disease. Acad Radiol

Recommender System Based on Network Structure Link Analysis Technique Through Community Detection in Social Network to Handle the Cold-Start Problem

Honey Pasricha, Shano Solanki, and Sumit Kumar

Abstract The recommendation system plays a pivotal role in electronic commerce sites to trade their products and services to consumers. The system further helps in identifying the items for the user and might be worthful for creating buying options that are purely rely on recommender. Especially, in the case of a book recommender system, it is necessary to recommend the books to learn in conformity with the learning environment of the learner. The prime objective is to propose an efficient book recommender system for social networking communities. Moreover, the paper addresses all the important parameters required to compute the efficiency of the system. The authors also outlined existing approaches utilize in book recommendation system along with major findings. The author took one of the fundamental challenges of cold-start problem and proposed a new book recommender system's framework using the "link analysis technique" of network structure. The proposed structure utilizes a hybrid technique for a book recommender system. Besides, the paper has tremendous potential to appraise the reader to explore a new book recommender system which further suggests the books to a novice on the social network as well as in-network communities.

Keywords Recommendation techniques · Book recommender system · Opinion leader · Social network

H. Pasricha · S. Solanki
Department of Computer Science and Engineering, National Institute of Technical Teachers Training and Research, Chandigarh, Chandigarh, India
e-mail: honey.cse@nitttrchd.ac.in

S. Kumar (✉)
Department of Research and Innovation, Division of Research and Development, Lovely Professional University, Phagwara, Punjab, India
e-mail: sumit.24786@lpu.co.in

1 Introduction

The enormous growth in information technology enhances the availability of items on the Internet for users. Thus, selection becomes uneasy for the user to prefer the items having almost similar features and leads to the creation of a positive dilemma in the user's perceiving. This has increased the interest of recommender systems like never before previously [1]. Recommendation systems (RS) are utilized by various online business destinations like Amazon eBay and so on for proposing relevant items dependent on clients' inclinations just as things bought by individuals of comparable interests [2]. The RS encourages the users to recognize the items which might justify the purchasing by them. Rather, recommender systems can also learn the user behaviour from his past purchasing history and items bought in the past by other users. According to the possibility that recommender system is probably going to utilize divergent sources of data (communitarian, content, information-based, labels, understood and express information procurement and so forth) to anticipate the things [3]. The RS's techniques address the number of issues which can sprung up while the traditional recommendation researches such as personalization from the abundance of retrieved products [4]. The RS utilizes an assortment of sifting strategies and appeared in Fig. 1.

Collaborative filtering (CF) is a standard recommendation method used to prescribe things to clients of comparative interests. Generally, CF is used for missing value analysis [6]. The idea behind the CF is similarity measures between the users. The advantage of CF is easy to implement, but difficult to actualize the suggested things without knowing the thing. The CF has various representative techniques such as neighbour-based CF and Bayesian belief nets CF. A content-based recommender system considers the historical backdrop of the user's preferences to suggest the items and products. This method relies on the item description and users' preference profiles. Knowledge-based recommender systems depend upon explicit information about users' tastes, item classification, and recommendation measures [7]. However, hybrid recommender systems depend on the fusion of content-based and collaborative recommender procedures [8]. In the view of the foregoing, a hybrid system can

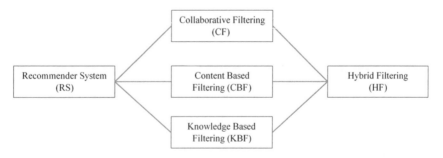

Fig. 1 Classification of recommender system [5]

take lead of techniques (collaborative and content-based techniques) to overcome the limitation used for making an efficient recommender engine.

2 Literature Review

In this context, the number of different studies had been done in the direction to know the effective use of a recommender system in commercial systems, such as Amazon [9]. However, such challenges which were sprung up during the recommendation are tackled by authors in their research, as different challenges exist at various levels of recommendation. Some challenges apart from traditional recommendation challenges are explained following [10]. To embark on the challenge of sparsity and curse of dimensionality, when the data render to the recommender system, the recommender generates recommendations based on given data for items that are available to the system. This can be defined with coverage of items by the recommender system to recommend. Hence, the recommender system will be incompetent to generate the recommendation for such items, which rated less by users, rather having fewer similarities between users tastes. One of the significant difficulties is the cold-start situation that infers the recommender system does not have adequate information of a new user who has quite recently shown up for looking for a suggestion, cold-start problem likewise alludes as item cold-start problem when a certain thing is simply included and none of the users has rated like (no explicit as well as implicit ratings) it till now [3, 11, 12]. Another challenge is scalability, once several consumers, users, and items increase in huge amounts in that the situation of recommender system becomes ineffective to cope with this colossal data in that case computational resources beat practical limitation [13, 14]. The prediction is measured by offline analysis. In general, it is to predict the behaviour of users and items without recommendations. Thus, accuracy is a concern for hybrid recommender systems.

According to Kim et al., the use of collaborative filtering after locating the neighbours for online communities using their feature information [15]. Furthermore, Xin et al. had analysed three methods for community detection while considering the influential nodes for recommendation further which implies the invalidate results on real dataset [16]. However, Liu had successfully proposed the blend of a collaborative filtering algorithm to enhance the result of the recommender system [17]. In a study by Yada proposed the book recommender system for infrequent readers which had used the unlike recommender methods for recommender system but not able to provide any robust way to evaluate the system [18]. On the contrary, Devika et al. had proposed the novel book recommender system which enhanced the searching time for recommendations [19]. Similarly, Nunez-Valdez et al. purported a book recommender system based on explicit along with implicit ratings [20]. In terms of the dataset, as practiced by the authors in their research has either collected from library book databases or fetched through social networking APIs like Tweepy, etc.

In this work, we are going to demonstrate the novelty of paper as work constitutes in different sections as in Sect. 2 related to relevant approaches for book recommender system followed by methods and dataset used in paper's novel work in Sect. 3. After that, the outcomes and discourse have been mentioned before the conclusion part. The proposed recommender alleviated the meta-cognitive activities by perceiving the learner's preferences [21]. Correspondingly, the creator assessed the trust of the target client on companions of users and companions of companions utilize nearby and worldwide trust measures to outline a trusted network. The author used collaborative filtering along with association mining rules for the recommendation of a book to the target user. The outcomes of the technique, used by the author, gave better results than traditional individual collaborative algorithms [17]. Mathew et al. suggested a hybrid algorithm that has utilized content-based filtering along with item-based CF techniques for a book recommendation [22]. Most of the papers used standard extended multi-criteria approaches which illustrated incorporation of traditional recommendation framework for better output of recommendation. Kumari et al. proposed the system based on classification and opinion mining techniques for recommending books [23]. The recommendations are generated based on some weighted features such as language, publisher, content, price, etc. Extracted features from reviews uniquely classified as documents.

3   Material and Methods

3.1   Dataset Used

In this study, we have generated a real-time dataset in which the data is retrieved from the Twitter platform. The dataset comprises information of followers of such online platform and the number of identifiers like several friends, followers of accounts, followers ID, followers list, friend show, etc. The collected data consists various Twitter users' accounts and their followers' information. The downloaded data is in JavaScript Object Notation (JSON) format. The JSON format is based on named attributes, key-value pairs, and their associated values. The process of data accumulation has been executed for two months. Twitter serves many objects as in JSON formats such as users, and tweets. Each tweet comprises of author ID, a message, user screen name, a timestamp when it gets published, and seldom geo metadata shared by the user. Every account holder of Twitter also comprises of a Twitter name, an ID, user followers, user location, user language, etc.

3.2   Proposed Methodology

There is an assorted variety of opinions in the learning and information field. Each individual has its feelings about any item, and it might be specific when it comes for books. The proposed framework is enforced with the help of Python language and gephi analysis tool. Initially, the Tweepy package with authentication handler of python has been used to generate a dataset from the Twitter platform. The section elaborates the creation of a framework for an efficient recommendation system and extends the work of author Pasricha [24]. The proposed methodology for recommender system takes place in five major steps as which are illustrated in Fig. 2 and explained below:

Step 1. Retrieve the user's data from Twitter, an online networking system where people make a connection with others rather than having supporters and following other people.

Step 2. Target people group can be distinguished and bunched, which rely on the online accessibility of the profile's inclinations. In any case, in the wake of choosing the most reasonable relevant parameters.

Step 3. From target networks, locate the persuasive users, this is to say vital nodes by utilizing different centrality estimates measures.

Step 4. Maps the comparability of each opinion leader for a given book by utilizing the tags and phrases of books from every community.

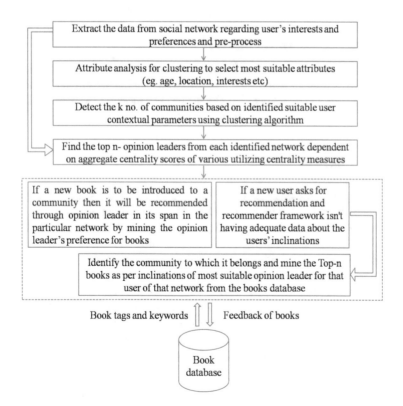

Fig. 2 Methodology for a book recommendation

Step 5. At last, based on the inclinations and interests of opinion leader, then books are prescribed to users.

The recommendations are performed for a new user who is looking for a recommendation. As the recommender system can have the inputs related to users such as users' interests as in the form of posted tweets, hashtags, location, a language that can be mined from the online social networks using sort of data extraction techniques. After extracting the interests, we find the opinion leader based on link analysis techniques namely betweenness, closeness, and eigenvector centrality. After determining the interests of opinion leaders, we can recommend the Top-n books in the community that can correspond to opinion leaders. The recommendation is evaluated basis on the quality metrics that includes precision, recall, and F-measure.

3.3   Pre-processing

To perform the first step of the methodology, the sub-steps are performed as follows:

Creating a Twitter Application: The whole process of creating an application is performed using the standard that is called open authorization. The authentication process involves three things a developer, application, and Twitter platform as a data source. In this process, various steps are involved.

Streaming of Data from Twitter: Twitter's developer platform is comprised of a number of APIs which is helping to build the Twitter application and further includes different features and endpoints. Developers can use these API's to stream the data from platform. Twitter API delivers data in JSON format.

3.3.1   User Interests

Frequency-inverse document frequency (TF-IDF) technique is utilized to get the users interest from their timeline as this highlights the word's importance in document collection which is also known as corpus-based on its frequency. TF-IDF value increases directly when the same word occurs in a document. It also filters the stop words like is, am, are and other helping verbs used commonly. State, when a 100-word archive contains the expression "bird" multiple times, the TF for the word "bird" can be determined as TF bird $= 12/100$ for example 0.12. IDF [25] of a word is the proportion of how noteworthy or uniqueness is the term in the entire corpus. We can say that it can scale up the significance of one of a kind words from the corpus. For instance, say the expression "bird" shows up x measure of times in a 10,000,000 million record estimated corpus, i.e. web. How about we accept that there are 0.3 million records that contain the expression "bird". For example, log (DF) is given by the all-out number of documents (10,000,000) divided by the quantity of documents containing the expression "bird" (300,000). IDF (bird) $=$ log

$(10,000,000/300,000) = 1.52$. This implies that TF $\times$ IDF (bird) $= 0.12 \times 1.52 = 0.182$.

Therefore, users' interests can be obtained by analysing the timeline tweets and hashtags with TF-IDF techniques. We retrieved the Top-n words from the tweets, re-tweets, and hashtags which are used by the user during tweets.

3.3.2 Users Interests-Labelling

All interests are labelled using the coding technique which means every interest has been encoded with a numerical value apart from the users' locations and users' languages. As labelled data helped to implement the clustering algorithm in a better way because most of the attributes are fetched as categorical shapes which cannot be implemented by basic clustering techniques. In the same way, we have to exercise to get the correct information shape before feeding into the algorithm.

3.3.3 Normalization and Missing Value Treatment

As the dataset contains null values, that is to say, while fetching the information some attributes contain the no value. This is because of the privacy enabled the feature of a user profile. Therefore, some records possessed null values. Using the missing value analyses all those null values can be replaced with zero. Correspondingly, various algorithms may be utilized to treat the collected data to organize it in such form thus further analytical processing is possible by an algorithm. In this way, normalization may assist with getting the fitting outcomes and even be required in some artificial intelligence calculations in which information feed as input values of various scales.

Such values can be standardized by following plan $y = (x - min)/(max - min)$, where y is the yield estimation of one record for specific property which is being a part of the analysis, min means the base estimation of an attribute, max speaks to the greatest estimation of the trait, and x is a present estimation of the record in an attribute.

3.3.4 Clustering of Networking Data

The clustering technique is utilized to discover the network that comparable tastes among individuals by improving the recommendation's prediction. By exploiting informal organization information, we can improve the prediction results. Here, K-means clustering calculation is utilized for the network discovery in which the K number of networks could be discovered. Consequently, target networks are considered to locate the influential nodes utilizing centrality measures. Although various existing methods are accessible for opinion leader mining, however, we have opted the link analysis procedure of graph theory for finding the nodes which interacting to the key nodes. This method has investigated in the framework.

3.3.5 Identification of Opinion Leader

According to Bavelas, a researcher, at Massachusetts Institute of Technology introduced the centrality technique around the years 1948 and 1950, in which the central node could play the role of a leader in the network. An opinion leader is the persuasive individual in the network. To discover the leader in social network communities, the eigenvector centrality metric is utilized [25]. At that point, Top-N books are mined by leader inclinations that consider to address the cold-start problem of recommendation. Even though there are additionally other centrality estimates such as closeness centrality and betweenness centrality. These schemes can be utilized for recognizing the important nodes from the online informal community. These centrality techniques may further help to generate a better recommendation system. We have computed the most influential node using different approaches like centrality scores namely betweenness, closeness, eigenvector and compare the recommendation system for corresponding influential finding techniques in recommender system. During finding it has observed that which technique is the best in terms of precision, recall of recommendation for the community. The data files have been collected and used to generate the centrality scores rely on the number of the associations among users and their followers.

4 Results and Discussion

In the present work, the performance of the entire model was based on different detection of opinion leaders at the cluster level. The findings imply that the recommendation based on the betweenness performed better results than the other two centrality measures which were also used for the detection of opinion leader for further recommendation. The calibre of the proposed work would be evaluated through two measurements. The performance evaluation of effectuating through at community level by exploring the opinion leaders is average precision and average recall. The F-score is taken into account to check the overall performance of the model by comparison with the use of different centrality techniques at the community level and the precision is determined. It provides how much proportion of recommended books is collected by users to the total number of recommended books. In other words, number of relevant retrieved documents divided by the total number of retrieved, however, recall points that how many books are relevant to the total number of actual relevant books.

The different cases of recommendation systems are analysed at the community level which is identified on the network that is grounded on the user's locations, interests, and language. In this context, six communities are identified and recommendation is performed and evaluated using precision and recall. However, the other cases are also evaluated for different communities that are identified to perform the recommendation.

All the data is extracted and processed using different data mining techniques. As we performed the recommendation through the opinion leader and corresponding found its precision that is up to what extent the recommendation is relevant to the novice or new user in a particular community. The recall rate is also measured for every community to check whether the recommended books to cold-start users in a community have been accepted. Recall rate shows the ratio of the number of acknowledged books to the total number of relevant items found during recommendation.

The performance of the entire model is based on different detection of opinion leader's measures at the cluster level and tabularized in Tables 1 and 2. The advantage of using a comparative approach in the methodology enhanced our understanding of link analysis technique along with employment of data mining techniques for the recommendation. In this context, Fig. 3 shows the overall performance of the proposed work in the percentage. Hence, the proposed book recommender system is one of the convincing and important recommender systems which may help in college and online web businesses to endorse relevant books to clients in a customized and modified manner.

From the previous studies, we have resonated that the majority of the work has been executed based on opinion mining and association rules to build the book

Table 1 Average precision based on different centrality measures

Number of clusters for network G	Average precision rate based on eigenvector	Average precision rate based on closeness	Average precision rate based on betweenness
1	32.75	32.75	32.75
2	28.62	7.725	53.125
3	27.8	53.2544333	31.5929333
4	31.72	25.65673	56.50285
5	39.74	39.56998	50.93312
6	56.11	49.41022	56.45327
7	47.3	50.05654	64.3026

Table 2 Average recall based on different centrality

Number of clusters for network G	Average recall rate based on eigenvector	Average recall rate based on closeness	Average precision rate based on betweenness
1	37.53	34.59	34.59
2	32.37	32.578	64.25
3	37.88	23.9517333	28.47746667
4	40.205	28.08043	40.96445
5	41.52	25.70073	35.86948
6	33.95	27.0767	40.92532
7	33.94	30.44733	34.71004

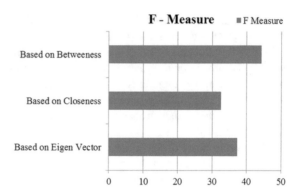

Fig. 3 Performance of proposed methodology

recommender system. Also, we determined that the existing techniques are not sufficient to solve the fundamental problem of recommender systems such as cold-start problem, and data sparsity. In this context, the proposed work has used the link analysis techniques along with data mining techniques and specifically focused in order to resolve the cold-start problems. Moreover, the link analysis techniques such as centrality measure could help to find out the opinion leader from the network, where the network is considered at the cluster level. The results of the recommendation were not anticipated. Despite this, results can still state that different measures which may help us to choose the best opinion leader, must affect the recommendation in the community. Since the present study used a limited extent of data but our results are encouraging and may be validated by a large sample size.

5   Conclusion and Future Work

The importance of this study to present the novel way of recommendation for users in social network communities in which link analysis techniques incorporate with traditional techniques. This framework is also signified to tackle the cold-start problem, rather, able to recommend the product and items to new users who have just been introduced to the community. The accuracy of the model is measured through precision, recall, and F-measure. These parameters separately measured for different link analysis techniques as comparison made for all in which eigenvector performed well. However, the result is not exactly as we were anticipated. The dataset which is being used, extracted from Twitter Inc. through API. In the same way, the present results may need to be validated on a large dataset in future. The proposed recommender platform represents the hybrid recommender system and having the ability to recommend the books to users on the social network. Efforts would also be made to key out the items for users and use it as a reliable buying option.

In future, the experimentation could be performed with other diverse datasets from other platforms like Facebook, Instagram, and YouTube also, to check the effectiveness of the recommender system using the diverse nature of the information.

Further studies, which may take account of the sentimental analysis approach with current methodology, add weightage to an additional user. The framework could be extended also by collecting data related to the user's feedback about the product, book item that will help other users to generate the relevant recommendation.

References

1. Isinkaye F, Folajimi Y, Ojokoh B (2015) Recommendation systems: principles, methods and evaluation. Egypt Inform J 16(3):261–273
2. Eirinaki M, Gao J, Varlamis I, Tserpes K (2018) Recommender systems for large-scale social networks: a review of challenges and solutions. Elsevier
3. Bobadilla J, Ortega F, Hernando A, Gutiérrez A (2013) Recommender systems survey. Knowl-Based Syst 46:109–132
4. Pradhan T, Pal S (2020) A hybrid personalized scholarly venue recommender system integrating social network analysis and contextual similarity. Futur Gener Comput Syst 110:1139–1166
5. Seyednezhad S, Cozart KN, Bowllan JA, Smith AOJAPA (2018) A review on recommendation systems: context-aware to social-based
6. Wang F, Zhu H, Srivastava G, Li S, Khosravi M, Qi L (2021) Robust collaborative filtering recommendation with user-item-trust records. IEEE Trans Comput Soc Syst 1–11. https://doi.org/10.1109/tcss.2021.3064213
7. Aggarwal C (2016) Knowledge-based recommender systems. In: Recommender systems, pp 167–197
8. Cheng LC, Ming-Chan L (2020) A hybrid recommender system for the mining of consumer preferences from their reviews. J Inf Sci 46(5):664–682; Kim KJ, Joukov N (2016) Information science and applications (ICISA) 2016. Springer.
9. Sharma L, Gera A (2013) A survey of recommendation system: research challenges. Int J Eng Trends Technol (IJETT) V4(5):1989–1992. ISSN: 2231-5381
10. Khusro S, Ali Z, Ullah I (2016) Recommender systems: issues, challenges, and research opportunities. In: Lecture notes in electrical engineering, pp 1179–1189
11. Gonzalez Camacho L, Alves-Souza S (2018) Social network data to alleviate cold-start in recommender system: a systematic review. Inf Process Manag 54(4):529–544
12. Su X, Khoshgoftaar T (2009) A survey of collaborative filtering techniques. Adv Artif Intell 2009:1–19
13. Sarwar BM, Karypis G, Konstan J, Riedl J (2002) Recommender systems for large-scale e-commerce: Scalable neighborhood formation using clustering. In: Proceedings of the fifth international conference on computer and information technology, vol 1, pp 291–324
14. Kim H, Oh H, Gu J, Kim J (2011) Commenders: a recommendation procedure for online book communities. Electron Commer Res Appl 10(5):501–509
15. Cheng V, Li C-H (2007) Combining supervised and semi-supervised classifier for personalized spam filtering. In: Pacific-Asia conference on knowledge discovery and data mining. Springer, pp 449–456
16. Xin L, Haihong E, Junde S, Meina S, Junjie T (2013) Collaborative book recommendation based on readers' borrowing records. In: 2013 international conference on advanced cloud and big data. IEEE, pp 159–163
17. Yada S (2014) Development of a book recommendation system to inspire "infrequent readers". In: International conference on Asian digital libraries. Springer, pp 399–404
18. Devika P, Jisha R, Sajeev G (2016) A novel approach for book recommendation systems. In: 2016 IEEE international conference on computational intelligence and computing research (ICCIC). IEEE, pp 1–6

19. Núñez-Valdéz ER, Lovelle JMC, Martínez OS, García-Díaz V, De Pablos PO, Marín CEM (2012) Implicit feedback techniques on recommender systems applied to electronic books. Computers in Human Behavior 28(4):1186–1193

20. Tewari S, Kumar A, Barman AG (2014) Book recommendation system based on combine features of content based filtering, collaborative filtering and association rule mining. In: 2014 IEEE international advance computing conference (IACC). IEEE, pp 500–503

21. Zheng X-L, Chen C-C, Hung J-L, He W, Hong F-X, Lin Z (2015) A hybrid trust-based recommender system for online communities of practice. IEEE Trans Learn Technol 8(4):345–356

22. Mathew P, Kuriakose B, Hegde V (2016) Book recommendation system through content based and collaborative filtering method. In: 2016 international conference on data mining and advanced computing (SAPIENCE). IEEE, pp 47–52

23. Priyanka K, Tewari AS, Barman AG (2015) Personalised book recommendation system based on opinion mining technique. In: 2015 global conference on communication technologies (GCCT). IEEE, pp 285–289

24. Pasricha H, Solanki S (2019) A new approach for book recommendation using opinion leader mining. In: Lecture notes in electrical engineering, pp 501–515

25. The TF*IDF algorithm explained | Onely (2021) Onely. [Online]. Available: https://www.onely.com/blog/what-is-tf-idf/. Accessed 20 July 2021

Performance Analysis of NOMA Over Hybrid Satellite Terrestrial Communication Systems

Priyanka Prasad, M. K. Arti, and Aarti Jain

Abstract The Non-Orthogonal Multiple Access (NOMA) is one of the most promising technology for upcoming future communication system due to its ability to offer services to multiple users with better connectivity and low latency. The NOMA with Cooperative Relaying System (NOMA-CRS) is used to improve the spectral efficiency as compared to conventional Cooperative Relay System (CRS). In our paper, the performance of NOMA over hybrid satellite terrestrial communication system is analyzed over two different fading channels—Shadowed Rician fading channel and Nakagami-m fading channel in two phases of downlink transmission and also derived the expression for end to end Bit Error Probability of proposed model assisted with cooperative relays. The significant performance with variations in the derived parameters is shown in results and validated in extensive simulations.

Keywords Bit Error Probability (BEP) · Bit Error Rate (BER) · Non-Orthogonal Multiple Access (NOMA) · Cooperative Relaying System (CRS) · Shadowed Rician (SR) fading · Nakagami-m fading · Successive Interference Cancellation (SIC)

1 Introduction

Non-Orthogonal Multiple Access (NOMA) is one of the most promising and interesting technology for future wireless communication networks, 5G cellular communication, etc., because of its capability to serve more users with low latency, massive connectivity and high spectral efficiency. Various features of NOMA make it a potential tool for cellular radio access discussed in [1]. NOMA uses the concept of superposition coding at the transmitting unit and Successive Interference Cancellation (SIC) at receiving unit making it more robust for future communication technologies. NOMA is basically classified into two categories—code domain NOMA (CD-NOMA) and power domain NOMA (PD-NOMA). Code domain NOMA multiplexes

P. Prasad (✉) · M. K. Arti · A. Jain
Department of Electronics and Communication Engineering, NSUT, New Delhi, India
e-mail: manavpriyanka1992@gmail.com

© The Author(s), under exclusive license to Springer Nature Singapore Pte Ltd. 2023
A. Shukla et al. (eds.), *Computational Intelligence*, Lecture Notes in Electrical Engineering 968, https://doi.org/10.1007/978-981-19-7346-8_12

the signal in code domain and can further be classified into various multiple access techniques that rely on sparse code multiple access and low-density spreading. Power domain NOMA multiplexes the signal from different users in power domain. There are numerous potentials and challenges associated with power domain NOMA when used in 5G systems, which are given in [1]. NOMA with Cooperative Relay System (CRS) is one of the most attracted topics because of the spectral inefficiency in the case of conventional CRS. The performance statistics of NOMA-based CRS using machine learning algorithms is derived over fading channels in [2] where the power optimization for minimum error is performed over Nakagami-m fading channel. The power has been optimized by selecting the appropriate values of parameters involved keeping the computation complexity lesser using machine learning algorithms. Various research works are done on PD-NOMA-based CRS schemes to analyze and evaluate different performance parameters in [3]. The PD-NOMA-based CRS in [4] used relaying system to analyze and proved that the efficacy is much better than conventional CRS. The capacity analysis of NOMA-based CRS is done in [5], and their expressions are derived. A novel method for receiver design is proposed for NOMA-based CRS in [6] suitable to be used for better transmission of symbols with reduced error. There are various types of relays and strategies which helps in proper transmission of the signal. The relay can be half duplex or full duplex with different operating characteristics, as discussed in [7]. The selection of appropriate relay for transmission is necessary to avoid interference and signal loss. The selection of Relay for Cooperative NOMA is discussed in [8]. Besides selection of relay, there are three types of relay strategies which are adopted according to the mode of transmission. The strategies are—amplify and forward (AF) relay, decode and forward (DF) relay and compress and forward (CF) relay. Various researches and investigations are done using different types of strategies. An AF relay in NOMA inspired CRS is used to analyze the performance in [9]. Diamond relaying network is a kind of network in which all the nodes are arranged forming diamond network, used in NOMA for error performance analysis as given in [10]. It is also used to investigate the sum rate in [11]. Another analysis with a comparative study on cooperative NOMA with spatial modulation in discussed [12]. The efficient resource allocation is much needed in transmission using NOMA as presented in [13]. The proper allocation of power ensures fair reception of data at the receiver.

The transmission channel is often gets affected by two mechanisms—fading [14] and shadowing [15]. The fading occurs due to multipath propagation of signal with different statistics as discussed in [16], and shadowing is the process occurring due to presence of obstacles in the path of signal such as buildings, trees and, mountains. There are various distributions proposed to model small-scale fading such as Nakagami-m [17], Rayleigh, Hoyt and, Weibull. Some other generalized statistical distributions are cascaded generalized K fading, double generalized Gamma distribution, etc. The log-normal and Gamma distribution are used to model shadowing phenomenon. Different channel models for communication system like land mobile satellite network, mobile to mobile networks, land mobile network, etc., with their statistical behavior are discussed in [18–22]. The digital transmission through land mobile and satellite channel is given in [23], and a new simple model on land mobile and satellite channel with first- and second-order statistics and model are presented in [24–27]. These model presents an insight to the land mobile satellite network, their mathematical analysis and adaptability to the future models. The modeling and estimation of various wireless fading channels have been proposed in [28], and the computer models of fading channels with their applications are proposed in [29]. In addition to it, a modified model of fading signals has been presented in [30] where the effect of fading signal is investigated over proposed model.

To model the channel experiencing shadowing and fading, a new fading model such as Shadowed Rician (SR) model [31] can be used. The SR model is very effective in analyzing the performance. The characteristics of double shadowed Rician fading model is given in [32] to consider the effects of fading with double-sided distribution. Various approaches and studies are done to evaluate the performance statistics such as error probability, outage probability, etc., over Rician fading channels using NOMA-based CRS as discussed in [33, 34]. Rician fading is also used with different relay strategies to improve the parameters of performance and obtain closed form performance in [35–37]. The performance analysis using one of the most commonly used fading model, i.e., Nakagami-m fading channels are investigated in [38]. Various other approaches for evaluation of error rates over different fading channels are discussed in [39, 40]. The investigation and derivation in modeling a channel requires various mathematical functions, algorithms and properties, presented in [41–43].

This paper has shown the performance analysis of NOMA over hybrid satellite terrestrial communication system over the Shadowed Rician fading channel in first phase and Nakagami-m fading in second phase of downlink transmission. The paper is organized into five sections. An introduction and related literature review are discussed in Sect. 1. Section 2 contains system model and its description. Numerical results are discussed in Sect. 3 and useful conclusions of the paper are given in Sect. 4 followed by references.

2 System Model

In this paper, the system model is a Non-Orthogonal Multiple Access (NOMA) assisted with Cooperative Relay system (CRS) consisting of a source (S), destination (D) and a half duplex relay (R) for downlink transmission. The source is a satellite and the destination is an earth station. The relay is also present at the earth. There are two type of channel links—the satellite channel link, present between source and destination and source and relay, and the terrestrial channel link, present between destination and relay. Here, the satellite channel link is assumed to have followed Shadowed Rician distribution [21] and the terrestrial channel link is assumed to have followed Nakagami-m distribution [17, 27].

Figure 1 shows the block diagram of system model for downlink transmission. In this, a single antenna is assumed to be connected with all the nodes for downlink transmission with flat fading channel coefficient [2] between each node, i.e., h_λ where

Fig. 1 Block diagram of
system model for downlink
transmission

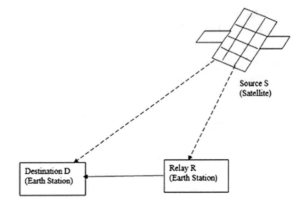

$\lambda = $ sr, sd and, rd and sr, sd and, rd are source to relay, source to destination and relay to destination, respectively. The fading distribution has shape parameter m_λ and spread parameter Ω_λ. For better efficiency, NOMA is applied for two consecutive symbols x_1 and x_2 to be transmitted simultaneously with an assumption that symbols x_2 contains more information than symbol x_1. . NOMA uses the concept of using superposition coding at transmitter unit and the Successive Interference Cancellation (SIC) at receiver unit. Then, using this superposition coded symbol is conveyed to destination and relay, and the received signal in the first phase of transmission can be given in [2] as

$$y = \sqrt{P_s}\left(\sqrt{\alpha}x_1 + \sqrt{(1-\alpha)}x_2\right)h_\lambda + W_\lambda \tag{1}$$

where $\lambda = $ sr and, sd are source to relay and source to destination, respectively, and α is the power allocation coefficient and assumed to be less than 0.5. P_s is the transmit power of source. W_λ denotes additive white Gaussian noise with variance N_0. x_1 and x_2 are the two symbols that are transmitted simultaneously. In the first phase of transmission, symbol x_2 is detected by relay and destination while symbol x_1 is treated as noise. To detect the x_1 symbol, relay uses SIC to estimate $\hat{x}_1$ and forward it to destination in second phase and is given in [2] as

$$y_{rd} = \sqrt{P_r}\hat{x}_1 h_{srd} + W_{rd} \tag{2}$$

where P_r denotes the transmit power of relay. The above equation shows that symbol x_1 is detected by the destination based on y_{rd}. . We want to obtain the total Bit Error Probability (BEP) of the NOMA-CRS; therefore, the average Bit Error Probability (ABEP) of the NOMA-CRS with symbols x_1 and x_2 are given as

$$P_{e2e}(e) = \frac{P_{x_1}(e) + P_{x_2}(e)}{2} \tag{3}$$

where P_{x_1} and P_{x_2} are the BEPs of symbols x_1 and x_2, respectively. Since only symbol x_2 is transmitted to destination in first phase, and symbol x_1 is detected as noise; therefore, the BEP for symbol x_2 appears same as the BEP of far end user in downlink NOMA transmission which is dependent on chosen constellation pairs. The conditional BEP of far end user in NOMA given in [2] as

$$P_{x_2}(e|\gamma_{sd}) = \sum_{i=1}^{N} P_i Q\left(\sqrt{2\upsilon_i \rho_s \gamma_{sd}}\right) \tag{4}$$

where $\gamma_\lambda = |h_\lambda|^2$ and $\rho_s = P_s/N_0$. The coefficients P_i and υ_i vary according to the chosen constellation pair for symbols x_1 and x_2. If the chosen constellation pair is BPSK, then as given in [2], $\upsilon_i = 1 \mp 2\left(\sqrt{(\alpha)} - (\alpha)^2\right)$ and $N = 2$ with $P_i = 0.5$ for all i. Since the satellite channel link is assumed to have followed Shadowed Rician fading in first phase; therefore, with help of Eq. (4), the SR fading distribution for symbol x_2 discussed in [21] can be expressed as given below

$$P_{x_2}^{sd}(e) = \int_0^\infty \sum_{i=1}^{N} P_i Q\sqrt{2\upsilon_i \rho_s \gamma_{sd}}\alpha_i e^{-\beta_i x} {}_1F_1(m_i; 1; \delta_i x)dx; \quad x > 0 \tag{5}$$

where ${}_1F_1$ denotes confluent hypergeometric function given in [23], and m_i is the shaping parameter of the Shadowed Rician (SR) fading distribution with $i = 0, 1$, and the other parameters are given in [21] as follows

$$\alpha_i = 0.5\left(\frac{2b_i m_i}{2b_i m_i + \Omega_i}\right)^{m_i}/b_i; \quad \beta_i = \frac{0.5}{b_i}; \quad \delta_i = \frac{0.5b_i}{2b_i^2 m_i + b_i \Omega_i}$$

The parameter Ω_i is the average power of LOS component and $2b_i$ is average power of multipath component. The range of shaping parameter m_i varies from 0 to ∞. Now for evaluation of BEP, the approximated Q function is used and given as

$$Q(x) = \frac{1}{12}e^{\frac{-x^2}{2}} + \frac{1}{4}e^{\frac{-2x^2}{3}} \tag{6}$$

Now, Eq. (5) can be rewritten using Eq. (6) and properties of hypergeometric function [23] to obtain the BEP of symbol x_2 as follows

$$P_{x_2}^{sd}(e) = \frac{1}{12}\sum_{i=1}^{N} \frac{P_i \alpha_i}{\upsilon_i \rho_s + \beta_i} {}_2F_1\left(1; m_i; 1; \frac{\delta_i}{\upsilon_i \rho_s + \beta_i}\right)$$

$$+ \frac{1}{4}\sum_{i=1}^{N} \frac{P_i \alpha_i}{\frac{4}{3}\upsilon_i \rho_s + \beta_i} {}_2F_1\left(1; m_i; 1; \frac{\delta_i}{\frac{4}{3}\upsilon_i \rho_s + \beta_i}\right) \tag{7}$$

while, the symbols x_1 are detected in the first phase of transmission at relay and forwarded to destination in very next phase. The detection is erroneous in both the phases but are totally independent. So, with the help of law of total probability, the BEP of symbol x_1 is given as

$$P_{x_1}(e) = P_{x_1}^{sr}(e)\left[1 - P_{x_1}^{rd}(e)\right] + \left[1 - P_{x_1}^{sr}(e)\right]P_{x_1}^{rd}(e) \tag{8}$$

where $P_{x_1}^{sr}$ and $P_{x_1}^{rd}$ are BEPs of symbol x_1 from source to relay in first phase and relay to destination in second phase, respectively. The second phase transmission is not encountered with interference resulting in transmission from R-D which makes it easier to obtain the BEP of symbol x_1 in the second phase. For BPSK, the conditional probability of x_1 is given with the help of [2] as

$$P_{x_1}^{rd}(e|\gamma_{rd}) = Q\sqrt{2\rho_s\gamma_{rd}} \tag{9}$$

where $\rho_s = P_s/N_0$. The use of SIC resulted in detection of symbols x_1 which followed Nakagami-m fading distribution. Therefore, the BEP of symbol x_1 from [2] can be given as

$$P_{x_1}^{rd}(e) = \frac{1}{2\sqrt{\pi}}\frac{\sqrt{P}}{(1+P)^{m_{rd}+0.5}}\frac{\Gamma(m_{rd}+0.5)}{\Gamma(m_{rd}+1)}\,_2F_1\left(1; m_{rd}+0.5; m_{rd}+1; \frac{1}{1+P}\right) \tag{10}$$

where $P = \frac{\rho_r\Omega_{rd}}{m_{rd}}$ and $\rho_r = P_r/N_0$. The equation given above is applicable when m_{rd} is non-integer. The error propagation to SIC during transmission should be taken into consideration in order to analyze BEP of symbol x_1 in first phase, i.e., $P_{x_1}^{sr}(e)$, and it adds more effort for further detection. To detect symbol x_1, the symbol x_2 should be detected first and the detected symbol $\hat{x}_2$ should be subtracted from the received signal. Since the satellite channel link is assumed to have followed Shadowed Rician fading, therefore, the conditional BEP of symbols x_1 in first phase in downlink transmission from [2] as

$$P_{x_1}(e|\gamma_{sr}) = \sum_{i=1}^{L} \eta_i Q\sqrt{2V_i\rho_s\gamma_{sr}} \tag{11}$$

where L, η_i and V_i varies with modulation pairs and for BPSK from [2], $L = 5$ and $\eta_i = 0.5[2, -1, 1, 1, -1]$ and the roots values are $V_i = [\alpha_i, 1 \pm 2\sqrt{\alpha - \alpha^2}, 4 - 3\alpha \pm 4\sqrt{\alpha - \alpha^2}]$. Now, by substituting the approximated Q function from Eq. (6) in Eq. (11), the BEP for symbol x_1 is evaluated as follows

$$P_{x_1}^{sr}(e) = \frac{1}{12}\sum_{i=1}^{L}\frac{\eta_i\alpha_{isr}}{v_i\rho_s + \beta_{isr}}\,_2F_1\left(1; m_{sr}; 1; \frac{\delta_i}{v_i\rho_s + \beta_{isr}}\right)$$

$$+\frac{1}{4}\sum_{i=1}^{L}\frac{\eta_i\alpha_{isr}}{\frac{4}{3}\upsilon_i\rho_s+\beta_{isr}}\,{}_2F_1\left(1;m_{sr};1;\frac{\delta_i}{\frac{4}{3}\upsilon_i\rho_s+\beta_{isr}}\right)\quad(12)$$

where the parameters α_{isr} and β_{isr} can be defined in the similar way as α_i and β_i. m_{sr} is the shaping parameter. Now the total end to end Bit Error Probability (BEP) of the system can be evaluated by substituting Eqs. (10) and (12) in (8) and the obtained result and Eq. (7) in Eq. (3), we get final equation (A) as follows

$$P_{e2e}(e)=\frac{1}{2}\Bigg[\Bigg[\frac{1}{12}\sum_{i=1}^{L}\frac{\eta_i\alpha_{isr}}{\upsilon_i\rho_s+\beta_{isr}}\,{}_2F_1\left(1;m_{sr};1;\frac{\delta_i}{\upsilon_i\rho_s+\beta_{isr}}\right)$$

$$+\frac{1}{4}\sum_{i=1}^{L}\frac{\eta_i\alpha_{isr}}{\frac{4}{3}\upsilon_i\rho_s+\beta_{isr}}\,{}_2F_1\left(1;m_{sr};1;\frac{\delta_i}{\frac{4}{3}\upsilon_i\rho_s+\beta_{isr}}\right)\Bigg]$$

$$\Bigg[1-\frac{1}{2\sqrt{\pi}}\frac{\sqrt{P}}{(1+P)^{m_{rd}+0.5}}\frac{\Gamma(m_{rd}+0.5)}{\Gamma(m_{rd}+1)}$$

$$_2F_1\left(1;m_{rd}+0.5;m_{rd}+1;\frac{1}{1+P}\right)\Bigg]$$

$$+\Bigg[\frac{1}{2\sqrt{\pi}}\frac{\sqrt{P}}{(1+P)^{m_{rd}+0.5}}\frac{\Gamma(m_{rd}+0.5)}{\Gamma(m_{rd}+1)}$$

$$_2F_1\left(1;m_{rd}+0.5;m_{rd}+1;\frac{1}{1+P}\right)\Bigg]$$

$$\Bigg[1-\frac{1}{12}\sum_{i=1}^{L}\frac{\eta_i\alpha_{isr}}{\upsilon_i\rho_s+\beta_{isr}}\,{}_2F_1\left(1;m_{sr};1;\frac{\delta_i}{\upsilon_i\rho_s+\beta_{isr}}\right)$$

$$-\frac{1}{4}\sum_{i=1}^{L}\frac{\eta_i\alpha_{isr}}{\frac{4}{3}\upsilon_i\rho_s+\beta_{isr}}\,{}_2F_1\left(1;m_{sr};1;\frac{\delta_i}{\frac{4}{3}\upsilon_i\rho_s+\beta_{isr}}\right)\Bigg]$$

$$+\Bigg[\frac{1}{12}\sum_{i=1}^{N}\frac{P_i\alpha_i}{\upsilon_i\rho_s+\beta_i}\,{}_2F_1\left(1;m_i;1;\frac{\delta_i}{\upsilon_i\rho_s+\beta_i}\right)$$

$$+\frac{1}{4}\sum_{i=1}^{N}\frac{P_i\alpha_i}{\frac{4}{3}\upsilon_i\rho_s+\beta_i}\,{}_2F_1\left(1;m_i;1;\frac{\delta_i}{\frac{4}{3}\upsilon_i\rho_s+\beta_i}\right)\Bigg]\Bigg]$$

3 Numerical Results

The results of analysis and simulations show the bit error performance of NOMA over hybrid satellite terrestrial communication system over Shadowed Rician fading channel and Nakagami-m fading in two phases of downlink transmission by varying the parameters, i.e., power allocation coefficient α, power sharing coefficient β, shaping parameters m_λ and spread parameters Ω_λ in derived expressions. The shaping parameters denote the diversity order, and spread parameters show the gain in error performance of the system. It can be apparently observed from the plots that with increase in values of shaping parameter for symbol x_1 and symbol x_2, the fall becomes more steeper which indicates better channel statistics for transmission of symbols. The value of power allocation coefficient is kept less than equal to 0.5 for fair reception of the symbols at the destination. Since, the concept of superposition coding is prevalent is NOMA; therefore, one of the symbol has been allotted power coefficient greater than the other symbol. The plot in Fig. 2 shows error performance of proposed model with variation in parameters for symbol x_2 in first phase from source to destination (Shadow Rician fading). The BEP plot for second phase of transmission of symbol x_1 from relay to destination (Nakagami fading distribution) and source to relay (Shadowed Rician fading) are shown in Figs. 3 and 4, respectively with varying values of α, β, m_λ and Ω_λ. The plot of end to end BER of proposed system is shown in Fig. 5. All the simulations of this paper are performed in MATLAB.

Figure 2 represents the plot of BEP of symbol x_2 with different values over SR fading in first phase. Figure 3 shows the plot of symbol x_1 with different values over Nakagami-m fading when shaping parameter m_{rd} is non integer. Figure 4 represents the plot of BEP of symbol x_1 with different values over SR fading in second phase. Figure 5 shows the plot of total end to end BEP of our proposed system with various values in downlink transmission.

Fig. 2 The plot of BEP of symbol x_2 with different values over SR fading in first phase

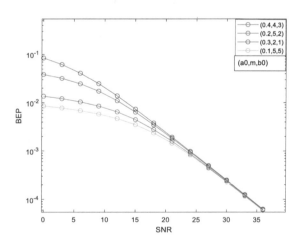

Fig. 3 The plot of symbol x_1 with different values over Nakagami-m fading when shaping parameter m_{rd} is non integer

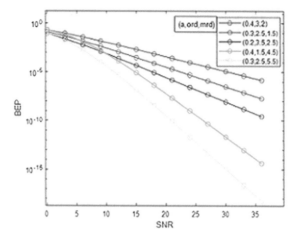

Fig. 4 The plot of BEP of symbol x_1 with different values over SR fading in second phase

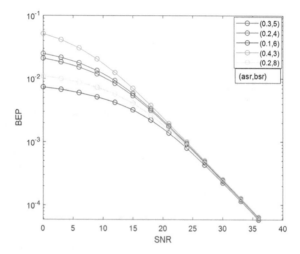

4 Conclusions

In this paper, the performance of NOMA over hybrid satellite terrestrial communication system is analyzed and the expression of end to end Bit Error Probability of proposed model has been derived over Shadowed Rician fading channel and Nakagami-m fading channel in two phases of downlink transmission. Based on simulations in MATLAB, remarkable improvement in error performance is achieved by varying the derived parameters. The study is done to obtain the minimum bit error rate. It has been concluded from analysis and simulations that the performance of proposed model improves with increasing values of shaping parameter to a finite range and smaller values of power allocation coefficient. Furthermore, the study can be extended to achieve optimization in error performance using artificial intelligence.

Fig. 5 The plot of total end to end BEP of our proposed system with various values in downlink transmission

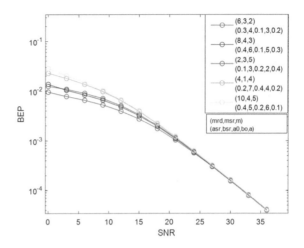

References

1. Saito Y, Kishiyama Y, Benjebbour A, Nakamura T, Li A, Higuchi K (2013) Non-orthogonal multiple access (NOMA) for cellular future radio access. In: Proceedings of IEEE vehicular technology conference, Dresden, Germany, June 2013
2. Kara F, Kaya H, Yanikomeroglu H (2021) A light weight machine assisted power optimization for minimum error in NOMA-CRS over Nakagami-m channels. IEEE Trans Veh Technol
3. Avazov N, Dobre OA, Kwak KS (2016) Power domain non orthogonal multiple access (NOMA) in 5G systems: potentials and challenges. IEEE Commun Surv Tutor 9(2)
4. Zhang Y, Yang Z, Feng Y, Yan S (2018) Performance analysis of cooperative relaying systems with power-domain non-orthogonal multiple access. IEEE Access 6:39839–39848
5. Kim J-B, Lee IH (2015) Capacity analysis of cooperative relaying systems using non-orthogonal multiple access. IEEE Commun Lett 19(11):1949–1952
6. Xu M, Ji F, Wen M, Duan W (2016) Novel receiver design for the cooperative relaying system with non-orthogonal multiple access. IEEE Commun Lett 20(8):1679–1682
7. Shinde N, Gurbuz O, Erkip E (2013) Half duplex or full duplex relaying: a capacity analysis under self interference. arXiv:1303.0088v2 [cs.IT], 11 Mar 2013
8. Ding Z, Dai H, Poor HV (2016) Relay selection for cooperative NOMA. IEEE Wireless Commun Lett 2162–2337
9. Abbasi O, Ebrahimi A, Mokari N (2019) NOMA inspired cooperative relaying system using an AF relay. IEEE Wireless Commun Lett 8(1):261–264
10. Kara F, Kaya H (2020) Error probability analysis of NOMA-based diamond relaying network. IEEE Trans Veh Technol 69(2):2280–2285
11. Wan D, Wen M, Ji F, Yu H, Chen F (2019) On the achievable sum-rate of NOMA-based diamond relay networks. IEEE Trans Veh Technol 68(2):1472–1486
12. Li Q, Wen M, Basar E, Poor HV, Chen F (2019) Spatial modulation aided cooperative NOMA: performance analysis and comparative study. IEEE J Sel Top Signal Process 13(3):715–728
13. Fang F, Zhang H, Cheng J, Leung V (2016) Energy-efficient resource allocation for downlink non-orthogonal multiple access network. IEEE Trans Commun 64(9):3722–3732
14. Simon MK, Alouini MS (2000) Digital communication over fading channels: a unified approach to performance analysis. Wiley, New York
15. Abdi A, Kaveh M (1999) On the utility of gamma pdf in modeling shadow fading (slow fading). In: Proceedings of IEEE vehicular technology conference, 1999, Houston, TX, USA, May 1999, pp 2308–2312

16. Lin SH (1971) Statistical behavior of a fading signal. Bell Syst Tech J 50(10):3211–3269
17. Youssef N, Munakata T, Takeda M (1996) Fade statistics in Nakagami fading environments. In: Proceedings of IEEE international symposium on spread spectrum techniques and applications, Mainz, Germany, pp 1244–1247
18. Loo C, Butterworth JS (1998) Land mobile satellite channel measurements and modelling. Proc IEEE 44:1442–1463
19. Lutz E, Cygan D, Dipplold M, Dolainsky F, Papke W (1991) The land mobile satellite communication channel-recording, statistics and channel model. IEEE Trans Veh Technol 40:375–386
20. Xie Y, Fang Y (2000) A general statistical channel model for mobile satellite systems. IEEE Trans Veh Technol 49:744–752
21. Mehrnia A, Hashemi H (1999) Mobile satellite propagation channel: part I—a comparative evaluation of current models. In: Proceedings of IEEE vehicular technology conference, Amsterdam, The Netherlands, pp 2775–2779
22. Talha B, Ptzold M (2011) Channel models for mobile to mobile cooperative communication systems: a state of the art review. IEEE Trans Veh Technol 6(2):33–43
23. Loo C (1990) Digital transmission through a land mobile satellite channel. IEEE Trans Commun 27:693–697
24. Abdi A, Lau WC, Alouini MS, Kaveh M (2001) A new simple model for land mobile satellite channels. In: Proceedings of the international conference on communications, Helsinki, Finland, pp 2630–2634
25. Abdi A, Lau WC, Alouini MS, Kaveh M (2001) On the second order statistics of a new simple model for land mobile satellite channels. In: Proceedings of the international conference on communications, Atlantic City, NJ, pp 301–304
26. Loo C (1985) A statistical model for a land mobile satellite link. IEEE Trans Veh Technol 34(3)
27. Abdi A, Lau WC, Alouini MS, Kaveh M (2003) A new simple model for land mobile satellite channels: first and second order statistics. IEEE Trans Wireless Commun 2(3)
28. Abdi A (2001) Modeling and estimation of wireless fading channels with applications to array based communications. Ph.D. dissertation, Department of Electrical and Computer Engineering, University of Minnesota, Minneapolis, MN
29. Loo C, Secord N (1991) Computer models for fading channels with applications to digital transmission. IEEE Trans Veh Technol 40(2):700–707
30. Aulin T (1979) A modified model for the fading signal at a mobile radio channel. IEEE Trans Veh Technol 28(3):182–203
31. Stefanovic H, Stefanovic V, Cvetkovic A (2008) Some statistical characteristics of a new shadowed Rician fading channel mode. In: The fourth international conference on wireless and mobile communications. IEEE
32. Browning JW, Cotton SL, Jimenez DM, Sofotasios PC, Yacoub MD (2019) A double shadowed Rician fading model: a useful characterization. In: WPMC 2019, the 22nd international symposium on wireless personal multimedia communications
33. Jiao R, Dai L, Zhang J, Mackenzie R, Hao M (2017) On the performance of NOMA-based cooperative relaying systems over Rician fading channels. IEEE Trans Veh Technol 66(12):11409–11413
34. Yue X, Liu Y, Yao Y, Li T, Li X, Liu R, Nallanathan A (2020) Outage behaviors of NOMA based satellite network over shadowed Rician fading channels. IEEE Trans Veh Technol 69(6)
35. Bhatnagar MR, Arti MK (2014) On the closed-form performance analysis of maximal ratio combining in shadowed-Rician fading LMS channels. IEEE Commun Lett 18(1)
36. Bhatnagar MR (2013) On the capacity of decode-and-forward relaying over Rician fading channels. IEEE Commun Lett 17(6):1100–1103
37. Jiao R, Dai L, Zhang J, Mackenzie R, Hao M (2017) On the performance of NOMA-based cooperative relaying systems over Rician fading channels. IEEE Trans Veh Technol 66(12):11409–11413
38. Yue X, Liu Y, Kang S, Nallanathan A (2017) Performance analysis of NOMA with fixed gain relaying over Nakagami-m fading channels. IEEE Access 5:5445–5454

39. Alouini MS (1999) A unified approach for calculating error rates of linearly modulated signals over generalized fading channels. IEEE Trans Commun 47(9):1324–1334
40. Bhatnagar MR, Arti MK (2013) Performance analysis of AF based hybrid satellite-terrestrial cooperative network over generalized fading channels. IEEE Commun Lett 17(10)
41. Adamchik VS, Marichev OI (1990) The algorithm for calculating integrals of hypergeometric type functions and its realizations in reduce systems. ACM 089791-401-5/90/0008/0212
42. Gradshteyn IS, Ryzhik IM (2014) Table of integrals, series, and products. Academic Press
43. Magnus W, Oberhettinger F, Soni RP (1966) Formulas and theorems for the special functions of mathematical physics, 3rd edn. Springer, New York

Multivariate Partially Blind Signature Scheme

Satyam Omar⬤, Sahadeo Padhye⬤, and Dhananjoy Dey⬤

Abstract In e-cash, blind signature provides perfect anonymity, which causes the threat of managing an infinite database to prevent double-spending of e-coins. Partial blind signature is used to tackle this problem using a piece of common information shared between the signer and the user. In the family of multivariate public key cryptography, there is no partially blind signature till now. We here design the first multivariate partially blind signature deploying the Rainbow signature scheme and MQDSS signature scheme. Moreover, we discuss partial blindness and one more unforgeability of the proposed scheme, along with its efficiency.

Keywords Partially blind signature · Multivariate public key cryptography · Identification scheme

1 Introduction

Chaum [1] introduced the model of blind signature in 1983. Blind signature setup contains three parties; a user, a signer, and a verifier. It provides the service of obtaining the signature on a message from a signer to the user without disclosing the content of the message to the signer. Blind signature provides perfect anonymity, and hence, it is used in e-cash, e-voting, and sensor networks. Specifically, it is very handy to use electronic cash in a rapidly growing digital payment infrastructure. If the applied signature scheme provides perfect anonymity, the problem of managing an ever-mounting database for the banks, comes into the picture.

In 1996, Abe and Fujisaki [2] proposed a partially blind signature scheme in which the signer and the user consort on a piece of common information ($info$). This $info$ may be the date of signature, the amount of coins to be spent, or any other information. At first, the customer computes the blinded message using $info$,

S. Omar (✉) · S. Padhye
Motilal Nehru National Institute of Technology Allahabad, Prayagraj 211004, India
e-mail: satyamomar@mnnit.ac.in

D. Dey
Indian Institute of Information Technology, Lucknow 226002, India

© The Author(s), under exclusive license to Springer Nature Singapore Pte Ltd. 2023   143
A. Shukla et al. (eds.), *Computational Intelligence*, Lecture Notes in Electrical
Engineering 968, https://doi.org/10.1007/978-981-19-7346-8_13

a random number and the amount e-coins, and sends it to the bank. The bank issues a signature to the customer, and the customer recovers a ticket from the signature, through which the customer can exchange e-coins. Before the ticket's expiry date, the customer sends the amount of e-coins along with the ticket to the bank. The bank issues the e-coins to the customer after validating the ticket with $info$. To spend e-coins, the customer sends the issued e-coins, amount of transaction, and $info$ to the merchant, who verifies e-coins at the bank. It then traces its database corresponding to $info$. If the bank does not get the data corresponding to the e-coins, it rejects the transaction; otherwise, it verifies the transaction. Thereafter, the bank removes the data of spent e-coins corresponding to $info$ from the bank's database. It can be considered as the generalization of the blind signature scheme if null information is used in place of $info$.

Various partially blind signature schemes [2–5] are present in the literature, but all of these are either factoring-based or DLP-based. In 1997, Shor [6] gave the quantum algorithm for solving the factoring problem and DLP problem within polynomial time. Multivariate public key cryptography (MPKC) is an efficient alternative to construct small-size signatures against quantum adversaries. Due to modest computational expenses, these schemes can be used on RFID chips and smart cards [7, 8]. The early digital signature constructions [9–11] were not secure enough to implement, but the evolution of unbalanced oil and vinegar construction brought a revolutionary enhancement in the efficiency and security of multivariate digital signature schemes. Some of the efficient and secure signature schemes in MPKC are UOV signature scheme [12] and Rainbow signature scheme [13]. The only problem with these schemes is their large key size. MQDSS [14] is another efficient and secure digital signature scheme in MPKC, which has small key sizes, unlike other MPKC's. These basic signature schemes can be used to construct the signature scheme having additional properties like *Ring*, *Group*, *Blind signature*, etc. We use Rainbow signature scheme [13] and MQDSS [14] to design our scheme. There is already a blind signature scheme [15] in MPKC. We have modified the scheme [15] to design our multivariate partially blind signature scheme (MPBS). The outline of the remaining paper is given below.

The preliminary concepts having notation table, partially blind signature model, multivariate digital signature construction, and some special tools to construct the scheme are discussed in Sect. 2. In Sect. 3, the proposed scheme and its properties are discussed. In Sect. 4, the parameters and efficiency of the proposed scheme are discussed. At last, Sect. 5 concludes the paper.

2 Preliminaries

In this section, we discuss the notations that we use in the construction of the proposed signature scheme. Further, we discuss the model of the general partially blind signature scheme along with commitment function [16], identification scheme [17], and

Table 1 Notation table

Notation	Quantity
λ	Security parameter
M	Original message
M^*	Blinded message
$info$	Agreed information
μ, D, ν	Secret key
$\bar{D}$	Public key
$\mathbb{E}$	Finite field
$\mathbb{E}^*$	Arbitrary length string of the elements of finite field $\mathbb{E}$
$\{0, 1\}^*$	Arbitrary length binary string
F_{PR}	Pseudo random function
$\parallel$	Concatenation
$\cdot$	Component wise product
$\overline{G}$	Polar form of $\bar{D}$
x_1, x_2, x, y	Unknown variables
$H, \overline{H}, H_0, H_1, H_2$	Hash functions

Fiat–Shamir transformation [18]. Then we discuss multivariate digital signature construction and the underlying basic signature schemes Rainbow [13] and MQDSS [14] as well.

2.1 Notations

We use the notations according to Table 1.

2.2 Partially Blind Signature (PBS)

PBS comprises the algorithms, setup (*STP*), Key Generation (*KG*), Partially Blind Signature Issue (*PBSI*), Partially Blind Signature Verification (*PBSV*), which are as follows:

- **Setup (*STP*)**: It takes 1^λ as input and outputs the parameters to design MPBS. Here, λ is the security parameter.
- **Key Generation (*KG*)**: It takes parameters as input and provides the private key *sk* and public key *pk* for the signer as output.
- **Partially Blind Signature Issue (*PBSI*)**: It has four sub-algorithms, namely, *Agree*, *Blind*, *Sign*, and *Unblind*, which are as follows:

- *Agree*: In this algorithm, a common information $info$ is consorted between the user and the signer.
- *Blind*: In this algorithm, the user computes a blinded message M^* using original message M, common agreed information $info$, and a random number u^*. Then the user delivers M^* to the signer.
- *Sign*: On receiving M^*, the signer signs on the blinded message M^* and sends the blinded signature u to the user.
- *Unblind*: In this algorithm, the user unblinds the blinded signature u to obtain the signature σ. The user sends $(\sigma, M, info)$ to the verifier.

Partially Blind Signature Verification (*PBSV*): In this algorithm, the verifier verifies the signature σ using the signer's public key pk, message M, and the agreed information $info$.

A PBS has security notions in the form of *completeness, partial blindness*, and *universal one more unforgeability*. *Completeness* shows the precision of the partially blind signature scheme for honest involving parties. *Partial blindness* gives the assurance to the signer about the non-removable embedding of $info$ in the blinded message; and for the same $info$, the signer could not link a signature to the signer's view that produces the corresponding partially blind signature. For *universal one more unforgeability* [19], the challenger C and the adversary A participate in an interactive game. The adversary is provided the public key pk of the signer, and the advantage of obtaining l message-signature pairs for a common information $info$ from the challenger C who contains the public key pk and the private key sk of the signer. At last, a message d^* different from the obtained message-signature pairs is given to the adversary A. The adversary A is successful in forging the signature if he outputs the valid partially blind signature σ^* on d^* with non-negligible probability.

2.3   Cryptographic Tools

We use three cryptographic tools, namely commitment scheme [16], identification scheme [17], and Fiat–Shamir transformation [18].

Commitment Scheme: A commitment scheme allows to commit a value in cryptographic schemes. It keeps the committed value hidden from others and committing party can disclose the committed value later on. It has two important properties, namely hiding and binding. *Hiding* property hides the committed value from the verifier, and *binding* property hampers the prover to vary the committed value.

Identification Scheme: The identification scheme [17] is used for a prover, so that he could convince a verifier of its identity without leaking any secret information possessed by him. Zero knowledge proof is deployed to achieve this goal. An identification scheme must satisfy two important properties, namely soundness and honest verifier zero knowledge. Soundness means a cheating prover cannot prove someone else's identity, while honest verifier zero knowledge indicates that a cheating verifier cannot get any secret information of the prover.

Fiat–Shamir Transformation: Fiat and Shamir [18] proposed the way of transforming a $(2n + 1)$-pass identification scheme into a digital signature scheme using hash functions called Fiat–Shamir (FS) transformation. If the underlying identification scheme is unforgeable, it implies the unforgeability of the resulting signature scheme. Here, we have implemented the same idea to design our scheme with the help of the scheme [17].

2.4 MQ Problem

Let $\mathbb{E}$ be a finite field, and a system of equations of degree two having l variables and v equations is given as

$$f_1(z_1, z_2, \ldots, z_l) = e_1$$
$$f_2(z_1, z_2, \ldots, z_l) = e_2$$
$$\vdots$$
$$f_v(z_1, z_2, \ldots, z_l) = e_v$$

where $f_1, f_2, \ldots, f_v \in \mathbb{E}[z_1, z_2, \ldots, z_l]$ and $e_1, e_2, \ldots, e_v \in \mathbb{E}$.

MQ Problem is to solve the above quadratic system, i.e., to find the solution $(c_1, c_2, \ldots, c_l)$ of above quadratic system such that $f_j(c_1, c_2, \ldots, c_l) = e_j$ is satisfied $\forall j = 1, 2, \ldots, v$. MQ Problem has been stated and proven NP-complete [20].

2.5 Multivariate Digital Signature Construction

Multivariate digital signature construction [9] requires the following procedure to be followed.

Key Generation: The signer chooses two invertible affine transformations $\mu : \mathbb{E}^v \to \mathbb{E}^v$ and $v : \mathbb{E}^l \to \mathbb{E}^l$ and a random quadratic system of equations $D : \mathbb{E}^l \to \mathbb{E}^v$ which is easy to solve. The signer computes $\overline{D} = \mu \circ D \circ v$.

The quadratic system of equations $\overline{D}$ works as the public key and the tuple (μ, D, v) works as the private key of the signer. One thing is noticeable here that after hiding the quadratic system of equations D by μ and v, the obtained quadratic system of equations $\overline{D}$ must be intractable to solve.

Signature Generation: Let M be the document to sign. To sign on M, the signer works as follows.

1. Uses a hash function $H : \{0, 1\}^* \to \mathbb{E}^v$ and computes $MSG = H(M)$.
2. Computes $\sigma = \overline{D}^{-1}(MSG) = (\mu \circ D \circ v)^{-1}(MSG) = v^{-1} \circ D^{-1} \circ \mu^{-1}(MSG)$.
3. The signature on the message M is σ.

Signature Verification: The verifier checks whether the equality $\overline{D}(\sigma) = MSG = H(M)$ holds to verify the signature σ on the document M. If the equality holds, the signature is successfully verified, otherwise rejected.

2.6 Rainbow Signature Scheme

Ding and Schmidt proposed Rainbow signature scheme [13] as the multi-layer construction of the UOV signature scheme [12]. It optimizes the efficiency and security of the UOV signature scheme. Here, the central map contains a special structure in which all the variables of a layer become the vinegar variables of the next layer. Due to this, the central map is very easy to invert at the time of signing a message. Moreover, after hiding this central map using two invertible affine transformations, it looses its special structure and becomes very tough to invert. The two layer Rainbow signature construction is denoted by $(\hat{v}_1, o_1, o_2)$, where $\hat{v}_1$ denotes the number of vinegar variables used in the first layer, and o_1 and, o_2 denote the number of oil variables used in the first layer and the second layer, respectively. The number of vinegar variables in a subsequent layer is equal to the summation of the number of vinegar variables and oil variables of the previous layer.

2.7 MQDSS

MQDSS [14] is a multivariate digital signature scheme constructed by applying the FS transformation [18] on the Sakumoto et al.'s identification scheme [17]. They gave two variants, namely 3-pass and 5-pass. The success probability of the 3-pass variant is $\frac{1}{3}$, on the other hand, the success probability of the 5-pass variant is $\frac{1}{2} - \frac{1}{2q}$, where q denotes the number of elements in the underlying finite field. Moreover, the size of the transcript in the 3-pass variant is also large. So, the 5-pass variant is used in the construction of MQDSS [14]. This scheme is constructed using an identification scheme, so multiple rounds are required to be performed for a significant security level.

3 Multivariate Partial Blind Signature Scheme (MPBS)

Here, we propose the specific construction of MPBS. According to the definition of partially blind signature, our scheme works as follows.

- **Setup (*STP*)**: A finite field $\mathbb{E}$, the positive integers l, v, ω. Here, l, v, and ω are generated using the desired security level λ. The publicly known pseudo random function $F_{PR} : \mathbb{E}^{\frac{l(v+1)(v+2)}{2}} \rightarrow \mathbb{E}^{\frac{v(v+1)(v+2)}{2}}$, hash functions $H : \{0, 1\}^* \rightarrow \mathbb{E}^v$,

$\overline{H} : \mathbb{E}^* \times \{0, 1\}^* \to \{0, 1\}^{2\lambda}$, $H_0 : \{0, 1\}^* \to \{0, 1\}^{2\lambda}$, $H_1 : \{0, 1\}^* \to \mathbb{E}^\omega$, $H_2 : \{0, 1\}^* \times \mathbb{E}^* \to \{0, 1\}^\omega$, and the commitment functions $Com_0 : \mathbb{E}^* \to \{0, 1\}^{2\lambda}$, $Com_1 : \mathbb{E}^* \to \{0, 1\}^{2\lambda}$ are used.

- **Key Generation (*KG*):** The signer chooses two invertible affine transformations $\mu : \mathbb{E}^v \to \mathbb{E}^v$, $v : \mathbb{E}^l \to \mathbb{E}^l$ and Rainbow central map $D : \mathbb{E}^l \to \mathbb{E}^v$, and computes $\overline{D} = \mu \circ D \circ v$. The private key of the signer is (μ, D, v) and the corresponding public key is $\overline{D}$. A quadratic system of equation $\mathcal{R} : \mathbb{E}^v \to \mathbb{E}^v$ which is generated as $\mathcal{R} = F_{PR}(\overline{D})$ is also used in *PBSI* and *PBSV*.

- **Partially Blind Signature Issue (*PBSI*):** It has four sub-algorithms, which work as follows:

  - *Agree:* At first, a piece of common information $info \in \{0, 1\}^*$ is negotiated between the user and the signer.
  - *Blind:* Let the user has a message $M \in \{0, 1\}^*$ on which he wants the signature of the signer. The user proceeds in the following way.
    i. Computes $\varrho = H(M)$.
    ii. Chooses $u^* \in \mathbb{E}^v$ uniformly at random and computes $\tilde{\varrho} = \varrho - H(info) \cdot \mathcal{R}(u^*)$; where $(\cdot)$ represents the component wise multiplication in $\mathbb{E}^v$.
    iii. Sends the blinded message $\tilde{\varrho}$ to the signer.
  - *Sign:* After receiving $\tilde{\varrho}$, the signer computes $u = \overline{D}^{-1}(\tilde{\varrho}) = (\mu \circ D \circ v)^{-1}(\tilde{\varrho}) = v^{-1} \circ D^{-1} \circ \mu^{-1}(\tilde{\varrho})$ using his secret key. The signer sends u to the user.
  - *Unblind:* Unblinding of the signature u is performed using the technique of constructing the *Signature of knowledge*. After the algorithm *Sign*, the user has the solution $(u\|u^*)$ of the quadratic system of equations $\overline{D}'(x) = \overline{D}(x_1) + H(info) \cdot \mathcal{R}(x_2) = H(M)$ for the variables x, x_1, x_2. The user wants to prove this knowledge of the solution $(u\|u^*)$ of the system $\overline{D}'$ in a zero knowledge way. The concept of MQDSS [14] signature scheme for the message M is used to do this. According to MQDSS setup, $\overline{D}'(x) = \overline{D}(x_1) + H(info) \cdot \mathcal{R}(x_2)$ is used as the public parameter, $H(M)$ is used as the public key, and $u\|u^*$ is used as the secret key to construct the *Signature of knowledge*. The polar form [21] of the system $\overline{D}'(x)$ is $\overline{G}'(x)$, such that $\overline{G}'(x + y) = \overline{D}'(x + y) - \overline{D}'(x) - \overline{D}'(y)$. Now, the user proceeds as follows:
    i. Computes the random value $R = \overline{H}(\overline{D}\|M)$.
    ii. Using R and M, computes message digest $MD = \overline{H}(R\|M)$.
    iii. Chooses $u_{0_1}, u_{0_2}, \ldots, u_{0_\omega}, v_{0_1}, v_{0_2}, \ldots, v_{0_\omega} \in \mathbb{E}^{l+v}$, $w_{0_2}, \ldots, w_{0_\omega} \in \mathbb{E}^v$ randomly, and computes the commitments $c_{0_j} = Com_0(u_{0_j}, v_{0_j}, w_{0_j})$, $c_{1_j} = Com_1(u_{1_j}, \overline{G}'(v_{0_j}, u_{1_j}) + w_{0_j})$, where

    $$u_{1_j} = u\|u^* - u_{0_j}, \quad \forall\ j = 1, 2, \ldots \omega.$$

    iv. Computes $Sign_0 = H_0(c_{0_1}\|c_{1_1}\|c_{0_2}\|c_{1_2} \ldots \|c_{0_\omega}\|c_{1_\omega})$.

v. Extracts the challenges α_j from $h_1 = H_1(MD, Sign_0)$ as its components, and computes $v_{1_j} = \alpha_j u_{0_j} - v_{0_j}$ and $w_{1_j} = \alpha_j \overline{D}'(u_{0_j}) - w_{0_j}$, $\forall j = 1, 2, \ldots \omega$.

vi. Constructs $Sign_1 = (v_{1_1} \| w_{1_1} \| v_{1_2} \| w_{1_2} \ldots \| v_{1_\omega} \| w_{1_\omega})$.

vii. Produces the final challenges $ch_1, ch_2, \ldots, ch_\omega$ as the digits of $h_2 = H_2(MD, Sign_0, h_1, Sign_1)$ and constructs $Sign_2 = (u_{ch_1}, u_{ch_2}, \ldots, u_{ch_\omega}, c_{1-ch_1}, c_{1-ch_2}, \ldots, c_{1-ch_\omega})$.

viii. $Sign = (Sign_0, Sign_1, Sign_2)$ is the *Signature of knowledge* of blind signature u on the document M. The user sends $(Sign, M, info)$ to the verifier.

- **Partially Blind Signature Verification (PBSV)**: On receiving $(Sign, M, info)$, the verifier proceeds as follows.

(a) Computes $\varrho = H(M)$.

(b) Computes the random value $R = \overline{H}(\overline{D} \| M)$.

(c) Computes message digest $MD = \overline{H}(R \| M)$.

(d) Computes $h_1 = H_1(MD, Sign_0)$, and parses h_1 to obtain $\alpha_1, \alpha_2, \ldots, \alpha_\omega$.

(e) Computes $h_2 = H_2(MD, Sign_0, h_1, Sign_1)$, and parses h_2 to obtain $ch_1, ch_2, \ldots, ch_\omega$.

(f) Parses $Sign_1$ and $Sign_2$ to obtain $v_{1_1}, w_{1_1}, v_{1_2}, w_{1_2}, \ldots, v_{1_\omega}, w_{1_\omega}$, and $u_{ch_1}, u_{ch_2}, \ldots, u_{ch_\omega}$.

(g) The component $Sign_2$ provides some of the commitments and the remaining are computed according to the challenge values $ch_1, ch_2, \ldots, ch_\omega$ as

$$\text{If } ch_j = 0, c_{0_j} = Com_0\left(u_{0_j}, \alpha_j u_{0_j} - v_1, \alpha_j \overline{D}'(u_{0_j}) - u_{1_j}\right).$$

$$\text{If } ch_j = 1, c_{1_j} = Com_1\left(u_{1_j}, \alpha_j\left(\varrho - \overline{D}'(u_{1_j})\right) - \overline{G}'(v_{1_j}, u_{1_j}) - w_{1_j}\right).$$

(h) Computes $Sign_0' = H_0(c_{0_1} \| c_{1_1} \| c_{0_2} \| c_{1_2} \ldots \| c_{0_\omega} \| c_{1_\omega})$, and checks whether the equation $Sign_0' \stackrel{?}{=} Sign_0$ is valid or not. If it is valid, accepts the signature; otherwise rejects.

Completeness: Using the polar form $\overline{G}'$ of the system $\overline{D}'$ and its bilinear nature, the validity of the following equality can be checked

$$\overline{G}'(v_{0_j}, u_{1_j}) + w_{0_j} = \alpha_j\left(\varrho - \overline{D}'(u_{1_j})\right) - \overline{G}'(v_{1_j}, u_{1_j}) - w_{1_j}; \quad \forall j = 1, 2, \ldots \omega.$$

In *PBSV*, this equality is used to verify the commitments, which shows the precision of the scheme.

Partial Blindness: In *PBSI*, the use of random number u^* in *Blind* phase assures the randomization of the blinded message $\tilde{\varrho}$. So, this blinded message $\tilde{\varrho} = \varrho - H(info) \cdot \mathcal{R}(u^*)$ does not leak any information about the signer and original message. The signer is assured about the inclusion of $info$ because of the use of $info$ in *PBSV*. Moreover, the user obtains the solution of the equation $\overline{D}(x_1) + H(info) \cdot \mathcal{R}(x_2) = H(M)$ from the signer. Now, if the user having $info'$ instead of $info$ wants to forge the signature, the user would have the solution of $\overline{D}(x_1) + H(info') \cdot \mathcal{R}(x_2) = H(M)$. Let $u \| u^*$ be the obtained solution from the signer, then for the successful execution of $PBSV$, the following equation is necessary to be satisfied

$$\overline{D}(x_1) + H(info) \cdot \mathcal{R}(x_2) = \overline{D}(x_1) + H(info') \cdot \mathcal{R}(x_2).$$

It implies that

$$H(info) = H(info').$$

It shows that we find a collision on a cryptographically secure hash function H, which is not possible. So any user not having $info$ cannot forge the signature. Hence, the signer is assured that the common information $info$ cannot be removed from the signature.

Given a valid signature $Sign = (Sign_0, Sign_1, Sign_2)$, the signer cannot link any of his view $(\tilde{\varrho}, u)$ due to the random numbers used in *Unblind* phase. Even if the same common information is embedded in two signatures, the signer cannot link the signatures due to randomization in *Unblind* phase.

Hence, the proposed PBS preserves the partial blindness.

Universal One More Unforgeability: If an adversary can break the universal one more unforgeability of the proposed *MPBS* with non-negligible probability, then one can solve *MQ problem* which is NP-complete [20].

Initial: Given the security level λ and the setup parameters $l, v, \omega \in \mathbb{N}$, the challenger $\mathcal{C}$ generates the private keys $sk = (\mu, D, v)$, and the corresponding public key $pk = \overline{D} = \mu \circ D \circ v$ of the signer. The challenger $\mathcal{C}$ sends the public key pk of the signer to the polynomially bounded adversary $\mathcal{A}$.

Attack: The adversary $\mathcal{A}$ interacts with the challenger $\mathcal{C}$ to perform polynomially bounded number of queries adaptively. It means $\mathcal{A}$ may query according to the outcome of the previous query.

Signature Issuing Queries: At first, $\mathcal{A}$ and $\mathcal{C}$ agree on a common information $info$. Now, $\mathcal{A}$ computes the blinded message $\tilde{\varrho}$ for the message m to which it wants to query and sends $\tilde{\varrho}$ to $\mathcal{C}$. The challenger $\mathcal{C}$ computes the partially blind signature u on $\tilde{\varrho}$ corresponding to $info$, and sends it to $\mathcal{A}$. Then $\mathcal{A}$ unblinds u to obtain the signature $Sign = (Sign_0, Sign_1, Sign_2)$.

Forgery: The adversary $\mathcal{A}$ outputs a valid signature $Sign' = (Sign'_0, Sign'_1, Sign'_2)$ on a message m' different for the queried messages for the same $info$. It means $\mathcal{A}$ would have forged either of the problems, the first is $\mathcal{A}$ would have invert $\overline{D}$ without having the private key of the signer, and the second is $\mathcal{A}$ would have solved MQ problem for the successful verification of the commitments without inverting $\overline{D}$.

The signer's public key $\overline{D}$ is the Rainbow public key [13] which is considered secure, and hence, it cannot be inverted by anyone except the legitimate signer. The previous queries on the different messages do not help in computing the targeted inverse, otherwise, it would be the attack on Rainbow signature scheme [13]. It means that there is just one way left to successfully design the signature by satisfying the commitments as follows.

$$\overline{G}'(v_0, u_1) + w_0 = \alpha\left(\varrho - \overline{D}'(u_1)\right) - \overline{G}'(v_1, u_1) - w_1$$

$$\overline{G}'(v_0, u_1) + \overline{G}'(v_1, u_1) + w_0 + w_1 = \alpha\left(\varrho - \overline{D}'(u_1)\right)$$

$$\overline{G}'(v_0 + v_1, u_1) + \alpha\overline{D}'(u_0) = \alpha\left(\varrho - \overline{D}'(u_1)\right)$$

$$\overline{G}'(\alpha u_0, u_1) + \alpha\overline{D}'(u_0) = \alpha\left(\varrho - \overline{D}'(u_1)\right)$$

$$\alpha\overline{G}'(u_0, u_1) + \alpha\overline{D}'(u_0) = \alpha\left(\varrho - \overline{D}'(u_1)\right)$$

The probability of α being a zero vector is $\frac{1}{|\mathbb{E}|}$ which is very less, and hence,

$$\overline{G}'(u_0, u_1) + \overline{D}'(u_0) = \varrho - \overline{D}'(u_1)$$

$$\overline{G}'(u_0, u_1) + \overline{D}'(u_0) + \overline{D}'(u_1) = \varrho \tag{1}$$

$$\overline{D}'(u_0 + u_1) = \varrho$$

$$\overline{D}'(u \| u^*) = \varrho$$

Equation (1) indicates that the adversary $\mathcal{A}$ has solved MQ problem which is NP-complete [20]. Thus, both the ways are not possible to forge the signature.

4 Parameters and Efficiency

4.1 Parameters

The *Sign* phase in the proposed scheme is just the Rainbow signature generation [13], so the size of the PBS is the same as in Rainbow signature scheme [13]. The signature in Rainbow construction is just an element of $\mathbb{E}^l$, and hence, the size of the PBS is just the size of l field elements. The *Unblind* phase in our PBS follows the MQDSS [14] signature construction in which the concatenation of the blind signature u obtained from the signer and random number u^* used in *Blind* phase works as the private key, and the message works as the public key. The unblinded signature $Sign = (Sign_0, Sign_1, Sign_2)$ has three components, whose individual sizes for $\mathbb{E} = GF(31)$ are equal to $2 \cdot \lambda, 5 \cdot l \cdot \omega + 5 \cdot v \cdot \omega, 5 \cdot l \cdot \omega + 2 \cdot \lambda \cdot \omega$ bits, respectively. Thus, the size of $Sign$ is $2 \cdot \lambda \cdot (\omega + 1) + 5 \cdot \omega \cdot (2 \cdot l + v)$ bits. According to updated MQDSS [14], the formula to find the number of repetitions of Sakumoto et al.'s identification [17] is $\omega = \frac{1.4\lambda}{\log_2\left(\frac{1}{2} + \frac{1}{|\mathbb{E}|}\right)}$, where 1.4 has been included

Table 2 Key and signature sizes

Security	Parameters	Pub	Pri	PBS	US
(bit)	$(\hat{v}_1, o_1, o_1)$	(kilobyte)	(kilobyte)	(kilobyte)	(kilobyte)
80	(19, 16, 17)	30.7	20.2	0.031	12.19
100	(23, 20, 21)	51.7	40.2	0.039	18.77
128	(27, 26, 26)	101.3	74.0	0.048	30.00

Table 3 Comparison table

Scheme	Public key	Private key	Signature size
	(kilobyte)	(kilobyte)	(kilo byte)
Scheme [22]	852	5865	868
Our scheme	51.7	40.2	18.77

due to improvement in MQDSS [14] in the second round of NIST post-quantum standardized project. The sizes of the public key (Pub) and the private key (Pri) of the signer, partially blind signature (PBS), and unblinded signature (US) for different security levels are given in Table 2.

Table 3 gives the comparison between our scheme and the recently proposed lattice-based partially blind signatures [22] in terms of the public key size, private key size, and signature size for 100-bit security. According to Table 3, the size of public key is nearly 16 times, private key is nearly 145 times, and signature size is nearly 46 times shorter in our scheme.

4.2 Efficiency

- In *PBSI*, one hash computation in computing ϱ, and one hash computation and one public key evaluation in computing $\tilde{\varrho}$ are performed in the *Blind* phase.
- In *Sign* phase, a secret key evaluation is performed to compute the partially blind signature u, which is exactly the Rainbow signature generation [13].
- In *Unblind* phase, two hash computations are performed in computing R and MD, respectively. To compute the commitments for all the rounds 2ω times commitment functions which are also hash computations are performed. Moreover, ω times public key evaluations are also performed to compute the commitments for all the rounds. Then, three times hash computations are performed in computing $Sign_0$, h_1, and h_2. In computing w_1 for all the rounds, ω times public key evaluations are performed. So, overall $2\omega + 7$ times hash computations, $2\omega + 1$ times public key evaluations, and one secret key evaluation are performed in *PBSI*.
- In *PBSV*, $2\omega + 6$ times hash computations and 2ω times public key evaluations are performed.

5 Conclusion

In this paper, we have introduced the first multivariate partially blind signature scheme, which can be handy in managing the size of the database in the banks to prevent double-spending of e-coins. Being a multivariate signature, the proposed scheme is a good suite against the post-quantum adversaries. The signature sizes in the proposed scheme are quite small to communicate for the signer and the user both.

References

1. Chaum D (1983) Blind signatures for untraceable payments. In: Advances in cryptology. Springer, Boston, MA, pp 199–203
2. Abe M, Fujisaki E (1996) How to date blind signatures. In: Advances in cryptology—ASIACRYPT 1996. LNCS, vol 1163. Springer-Verlag, pp 244–251
3. Chow SSM, Hui LCK, Yiu SM, Chow KP (2005) Two improved partially blind signature schemes from bilinear pairings. In: Information security and privacy—ACISP 2005. LNCS, vol 3574. Springer, Berlin, Heidelberg, pp 316–328
4. Hu X, Huang S (2007) An efficient ID-based partially blind signature scheme. In: 8th ACIS international conference on software engineering, artificial intelligence, networking and parallel/distributed computing—SNPD, 2007. IEEE Computer Society, Qingdao, China, pp 291–296
5. Li F, Zhang M, Takagi T (2013) Identity-based partially blind signature in the standard model for electronic cash. Math Comput Model 58(1–2):196–203
6. Shor P (1997) Polynomial-time algorithms for prime factorization and discrete logarithms on a quantum computer. SIAM J Comput 26(5):1484–1509
7. Bogdanov A, Eisenbarth T, Rupp A, Wolf C (2008) Time-area optimized public-key engines: MQ-cryptosystems as replacement for elliptic curves? In: CHES 2008. LNCS, vol 5154. Springer, Berlin, Heidelberg, pp 45–61
8. Chen AIT, Chen M-S, Chen T-R et al (2009) SSE implementation of multivariate PKCs on modern x86 CPUs. In: CHES 2009. LNCS, vol 5747. Springer, Berlin, Heidelberg, pp 33–48
9. Matsumoto T, Imai H (1988) Public quadratic polynomial-tuples for efficient signature-verification and message-encryption. In: Barstow D et al (eds) Advances in cryptology—EUROCRYPT'88. EUROCRYPT 1988. LNCS, vol 330. Springer, Berlin, Heidelberg, pp 419–453
10. Patarin J (1996) Hidden fields equations (HFE) and isomorphisms of polynomials (IP): two new families of asymmetric algorithms. In: International conference on the theory and applications of cryptographic techniques, vol 1070. Springer, Berlin, Heidelberg, pp 33–48
11. Patarin J (1997) The oil and vinegar signature scheme. In: Dagstuhl workshop on cryptography, Sept 1997
12. Kipnis A, Patarin J, Goubin L (1999) Unbalanced oil and vinegar schemes. In: EUROCRYPT 1999. LNCS, vol 1592. Springer, Berlin, Heidelberg, pp 206–222
13. Ding J, Schmidt DS (2005) Rainbow, a new multivariate polynomial signature scheme. In: ACNS 2005. LNCS, vol 3531. Springer, Berlin, Heidelberg, pp 164–175
14. Chen MS, Hülsing A, Rijneveld J, Samardjiska S, Schwabe P (2016) From 5-pass MQ-based identification to MQ-based signatures. In: Advances in cryptology—ASIACRYPT 2016—22nd international conference on the theory and application of cryptology and information security. LNCS, vol 10032. Springer, Berlin, Heidelberg, pp 135–165
15. Petzoldt A, Szepieniec A, Mohamed MSE (2017) A practical multivariate blind signature scheme. In: Kiayias A (ed) Financial cryptography and data security. FC 2017. LNCS, vol 10322. Springer, Cham, pp 437–454

16. Goldreich O (2001) Foundations of cryptography, volume 1, basic tools. Cambridge University Press
17. Sakumoto K, Shirai T, Hiwatari H (2011) Public-key identification schemes based on multivariate quadratic polynomials. In: Rogaway P (ed) Advances in cryptology—CRYPTO 2011. CRYPTO 2011. LNCS, vol 6841. Springer, Berlin, Heidelberg, pp 706–723
18. Fiat A, Shamir A (1987) How to prove yourself: practical solutions to identification and signature problems. In: Odlyzko AM (ed) Advances in cryptology—CRYPTO'86. CRYPTO 1986. LNCS, vol 263. Springer, Berlin, Heidelberg, pp 186–194
19. Pointcheval D, Stern J (2000) Security arguments for digital signatures and blind signatures. J Cryptol 13(3):361–396
20. Garey MR, Johnson DS (1991) Computers and intractability: a guide to the theory of NP-completeness. W. H. Freeman
21. Fouque PA, Granboulan L, Stern J (2005) Differential cryptanalysis for multivariate schemes. In: Cramer R (ed) Advances in cryptology—EUROCRYPT 2005. EUROCRYPT 2005. LNCS, vol 3494. Springer, Berlin, Heidelberg, pp 341–353
22. Bouaziz-Ermann S, Canard S, Eberhart G, Kaim G, Roux-Langlois A, Traore J (2020) Lattice-based (partially) blind signature without restart. IACR Cryptol ePrint Arch 2020/260
23. Courtois NT, Goubin L, Patarin J (2003) SFLASHv3, a fast asymmetric signature scheme. IACR Cryptol ePrint Arch Rep 2003/211. Citeseer
24. Petzoldt S, Bulygin J, Buchmann A (2010) Selecting parameters for the rainbow signature scheme-extended version. IACR Cryptol ePrint Arch 2010:435
25. Petzoldt A, Chen MS, Yang BY, Tao C, Ding J (2015) Design principles for HFEv-based signature schemes. In: ASIACRYPT 2015—part 1. LNCS, vol 9452. Springer, Berlin, Heidelberg, pp 311–334

Topic Analysis and Visualisation of Peer-to-Peer Platform Data: An Airbnb Case Study

Juanita Subroyen⬤, Marita Turpin⬤, Alta de Waal⬤, and Jean-Paul Van Belle⬤

Abstract Peer-to-peer (P2P) platforms play an important economic role as they bring together buyers and sellers and allow them to directly interact with each other. People who sell goods or services on P2P platforms are often ordinary citizens who do not have sophisticated marketing knowledge or skills. It is important that service providers are empowered to know how to market themselves since they are competing globally. In a review of accommodation P2P platforms, previous studies have shown the importance of marketer-generated content (MGC), and how it relates to aspects such as pricing, demand, and customer experience. The same holds for user-generated content (UGC), which typically takes the form of customer reviews. However, there was a lack of studies considering both host and guest generated data. This study addresses the identified gap by using topic modelling to analyse both host and guest data obtained from the Airbnb platform. A Latent Dirichlet Allocation algorithm is used to discover latent topics in the Airbnb MGC and UGC, respectively. The discovered topics are first ranked and then analysed using Tableau's data visualisation tool, by using various dimensions such a geographic location, review scores, or number of reviews. The analysis, even among the top-ranked properties, shows that there are still many mismatches between host and guest topics. The paper contributes by illustrating the value of topic modelling and analysis to gain practical insights into the data accumulated on an accommodation booking platform.

Keywords Topic modelling · Digital economy · Peer-to-peer platforms · Latent Dirichlet Allocation · Data visualisation · Airbnb

J. Subroyen
Department of Computer Science, University of Pretoria, Pretoria 0001, South Africa

M. Turpin (✉)
Department of Informatics, University of Pretoria, Pretoria 0001, South Africa
e-mail: marita.turpin@up.ac.za

A. de Waal
Department of Statistics, University of Pretoria, Pretoria 0001, South Africa

Centre for Artificial Intelligence Research (CAIR), Pretoria 0001, South Africa

J.-P. Van Belle
University of Cape Town, Cape Town 7700, South Africa

© The Author(s), under exclusive license to Springer Nature Singapore Pte Ltd. 2023
A. Shukla et al. (eds.), *Computational Intelligence*, Lecture Notes in Electrical Engineering 968, https://doi.org/10.1007/978-981-19-7346-8_14

1 Introduction

Peer-to-peer (P2P) platforms allow buyers and sellers to directly interact with each other by facilitating communication between the parties that exchange goods and services [1]. Content presented by buyers (user-generated content or UGC) and sellers (marketer-generated content or MGC) has been shown to provide insight into aspects such as pricing, demand, and customer experience [2–4]. This study focused on applying topic modelling, analysis, and visualisation to content from one of the largest P2P platforms, Airbnb. A literature review on accommodation P2P platforms showed that previous studies have focused mainly on UGC (the user/guest perspective) when studying customer experience attributes and have not looked at MGC (the marketer/host perspective) or how these perspectives may relate to each other. This study contributes to research that leverages both MGC and UGC to derive customer experience attributes, as described by hosts and guests, through the application of topic modelling, analysis, and visualisation. The study is based on Airbnb listings in Cape Town, South Africa.

This paper reports on work related to the following research question: *How can the customer experience attributes derived from topic models be analysed and visualised to understand the difference between host and guest perspectives across listing and host dimensions in Cape Town, South Africa?*

Topic modelling was performed using Latent Dirichlet Allocation to discover latent topics in Airbnb listing descriptions (MGC) and reviews (UGC). These topics represented the key customer experience attributes referred to by hosts and guests. The topics where then used as input into a topic analysis and visualisation process in which an interactive visualisation tool was developed. The tool facilitates the study of customer experience attributes across various dimensions including geographic location, host and property attributes, guest ratings, and pricing and demand.

The paper is structured as follows. Section 2 provides background on Airbnb and an overview of previous studies analysing accommodation platform data. Section 3 presents the topic modelling and analysis method. Section 4 presents the analysis with a focus on conclusions drawn from the data visualisation. Section 5 concludes the study.

2 Background

2.1 The Airbnb Platform

The Airbnb accommodation platform was founded in 2008 and currently has more than five million listings in over a hundred thousand cities across the world [5]. Airbnb hosts (service providers) do not necessarily have knowledge or skills in the tourism and hospitality industry. Hosts promote their listings by providing attributes of their properties, adding photographs, and further describing the space, location, and host. This MGC gives the guest an understanding of the property and host. Guests can review their stay and experience. Guest reviews (UGC) provides potential new guests with additional information about the property to assist in their decision-making. MGC and UGC are crucial components of the overall experience for both the host and guests who use the digital service offered on Airbnb. For researchers, this content can be a valuable source of information to study pricing, demand, and customer experience.

2.2 Previous Studies Analysing Accommodation Platform Data

Several researchers have investigated how textual content relates to price [4, 6, 7], or booking demand [2, 8, 9]. These studies tend to focus on using only one type of content—MGC or UGC—to study the impact of content on price or demand. For example, review sentiment [7] and various word attributes in the listing title (such as "luxury", "penthouse", or "duplex") [4] were shown to be associated with higher pricing. Hosts who had more text fields completed, with more words, tended to have more booking activity for their listings [2]. Chen and Chang [8] found that information quality on Airbnb listings positively affected purchase intention on the platform.

Textual content on accommodation P2P platforms has also been used to study customer experience and satisfaction [3, 10–13]. Some authors used review sentiment as a measure of customer satisfaction when studying how accommodation attributes impact guest satisfaction [12, 13]. Other authors employed text-mining techniques to discover experience attributes that are outlined by guests through their reviews [10, 11, 13]. Since these studies have used UGC exclusively, they have focused on the accommodation experience from the guest perspective, and none of the studies have studied the intended customer experience from a host perspective. This study aims to fill this research gap by studying the perspectives of both the guest and host through considering both UGC and MGC.

2.3 Techniques for Analysing Textual UGC and MGC

The unstructured textual data of UGC and MGC needs to be transformed into a format in which it can be analysed. In previous studies, researchers have utilised a variety of techniques to accomplish this. For example, lexical features [2, 14] and semantic features [4, 10] have been derived from MGC and UGC text data. Semantic features allow for the understanding of the meaning of text and is hence suitable for the study of customer experience attributes. A common approach is to identify themes in the text using a machine learning process known as topic modelling. In previous studies, some authors have used topic modelling as the technique for analysis [10, 11, 13]; however, there has not been significant focus on studying how these attributes differ across various listing dimensions such as regions (neighbourhoods) and property features. By analysing how the host and guest perspectives relate to listing features, hosts can identify gaps and opportunities for improving their listing to suit guest preferences. Therefore, this project also aims to make a research contribution by using data visualisation techniques to present guest and host perspectives in a format that can be used for analysis and insights generation.

3   Methods

Latent Dirichlet Allocation (LDA) has been selected as the algorithm to perform topic modelling on UGC and MGC, as it is a commonly used topic modelling algorithm [15]. The method as summarised in Fig. 1 has been derived from the work of Metaxas [16] and Prabhakaran [17].

In Phase A, topic modelling was performed on datasets extracted on 17 July 2019 from Cape Town Airbnb listings; this comprises 22,230 host listings and 117,522 user reviews (after cleaning) (Step 1). Python's NLTK package was used for the data pre-processing and tokenisation (Step 2). The topic modelling was done using LDA in Gensim using genism.models.ldamodel with 10 topics extracted, using a chunk size of 500 documents (reviews) per training chunk. Ten passes were made through

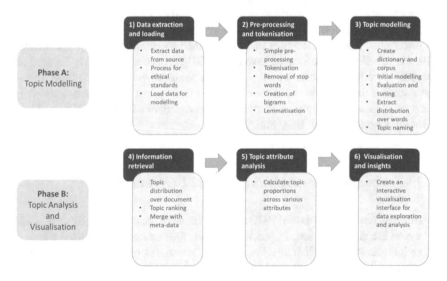

Fig. 1 Method overview

the training with alpha (a-priori belief for topic probability) and eta (a-priori belief on word probability) set to "automatic", and a random state seed of 100 was used. The final topic coherence score was 0.59 (as opposed to 0.49 using default values) (Step 3). Details of the above steps can be found in [18].

In the end, ten topics (themes) were defined for Airbnb listings, i.e. the host perspective using the word clusters generated by the LDA analysis: lifestyle, property rooms, outdoor features, surrounding views, privacy and autonomy, comfort, facilities and amenities, lifestyle, Décor and Design, and attractions. Note that the theme labels (names) were generated by the researchers to most accurately but succinctly describe the actual words within the word clusters. Seven themes (word clusters) were identified for Airbnb user reviews, i.e. the guest perspective: general positive experience, additional touches, home-like experience, distance to points of interest, property and features, miscellaneous, and Non-English.

4 Analysis

This section describes topic analysis and visualisation in order to study trends in customer experience attributes from a guest and host perspective (Phase B in Fig. 1). This was achieved by retrieval of the relevant information from the topic models as described in a prior paper (Step 4), developing computations for attribute analysis (Step 5), and analysing the data using an interactive visualisation tool (Step 6). The analyses and visualisations were conducted across various dimensions which were categorised into five main groups: overall host and guest perspectives, geo-spatial trends, trends across property and host dimensions, ratings, and pricing and demand.

4.1 Topic Prominence

To retrieve relevant information from the topic models and identify the most important topic, the topic distribution over documents for both topic models were extracted. These distributions provide the probability of each topic per item (listing or review). The topics for each review and listing were then ranked by the probabilities in order to identify the most important topics per listing and review. Table 1 gives the final topic ranking based on the topic probability distribution.

As can be seen, not only is there a mismatch between the *nature* of the topics between host and guest perspectives, but the *relative importance* of those topics where there is at least a partial overlap is also quite different in relative priority. For instance, the highly ranked user topics such as additional touches and home-like experience are congruent with the host perspectives of comfort, décor and design, and other lifestyle aspects which rank them much lower. Distances to points of interest

Table 1 Topic probabilities and ranking for host and guest perspectives

Listing description topic (host perspective)	Prob	Rank	Review topic (guest perspective)	Prob	Rank
Lifestyle	0.233	1	General positive experience	0.321	1
Outdoor features	0.180	2	Additional touches	0.273	2
Property rooms	0.147	3	Distance to points of interest	0.190	3
Surrounding views	0.116	4	Home-like experience	0.113	4
Comfort	0.114	5	Miscellaneous	0.095	5
Décor and design	0.105	6	Property and features	0.006	6
Lifestyle2	0.095	7	Non-English	0.002	7
Facilities and amenities	0.006	8			
Privacy and autonomy	0.003	9			
Attractions	0.001	10			

are very (third-most) important to guests (with a topic probability of 0.190), but the hosts put attractions very low down on their property descriptions (topic probability 0.001). On the other hand, hosts feature outdoor features (#2) and property rooms (#3) prominently in their listing descriptions (probabilities of 0.180 and 0.147) but in reviews, these rank near the bottom (second-lowest) in the review topic importance (probability is only 0.006). This mismatch will be discussed in more detail and nuance below. In order to clearly differentiate between the owner (host listing) and customer/guest (user reviews) perspectives, the term "attributes" will be used to refer to the guest themes/topics and the term "dimensions" refers to the listing or host topics. "Property and host attributes" refer to listing attributes such as listing price, property type, number of rooms, "superhost" status, or how long a property has been listed.

4.2 Topic Attribute Analysis and Visualisation

Using the full feature set derived above, a topic attribute analysis was performed (Step 5 in Fig. 1). This comprised the computation of proportions, averages, or distributions across the various dimensions and customer experience attributes. The computations were determined for both the host and guest perspectives, using Airbnb listing and review topics, respectively. The intention of these computations was to allow for the analysis of trends relating to customer experience attributes across the various dimensions.

To further facilitate the study and analysis of customer experience attributes across the various dimensions and to support data exploration, an interactive visualisation tool was developed using the output of the topic analysis (Step 6 in Fig. 1). The tool was developed in Tableau, a business intelligence platform used for data visualisation and analytics, and employs various visualisation techniques such as geo-spatial mapping, bar charts, bubble charts, heat maps, box-and-whisker plots, and cross-tabulations to aid in the exploration of customer experience attributes across various dimensions. The visualisation tool provides the user with the opportunity to view the customer experience attributes from both the host and guest perspectives.

The analyses and visualisations were categorised into five key groups: overall host and guest perspectives (baseline analysis with no dimensions), geo-spatial trends, property and host dimensions, ratings, and pricing and demand. In each category, computations across the dimensions were specified. These computations were combined with the topic ranking (i.e. topic importance) and served as input to the interactive visualisation tool. Given space limitations, we only report on a few salient observations but also demonstrate the variety of analysis and visualisations of the tool. The following investigations aimed to focus on the top-ranked hosts, i.e. listings with rank 1.

Average review scores across host and guest perspectives. Figure 2 presents a cross-tabulation of the review scores across customer experience attributes. The maximum score possible for these metrics is 10. Hosts who mention Décor and Design as a primary theme in their listings appear to perform well across all metrics, while hosts who mention facilities and amenities or lifestyle and activities have lower scores overall. From the guest perspective, those who refer to property and features appear to give lower scores overall. Overall, this suggests that hosts who focus on details beyond the basic property features and associated amenities can possibly have properties that perform better. It can be noted that the scores are generally quite high.

Host Perspective: Review Scores Across Customer Experience Attributes

	Accuracy	Check-in	Cleanliness	Communication	Location	Value
Attractions	10.00	10.00	10.00	9.67	9.67	9.67
Comfort	9.66	9.77	9.49	9.73	9.63	9.54
Commute	9.58	9.77	9.48	9.76	9.64	9.42
Decor and Design	9.93	10.00	9.81	9.85	9.93	9.81
Facilities and Amen..	9.33	9.58	9.56	9.52	9.56	9.23
Lifestyle and Activi..	9.29	9.69	9.33	9.64	9.38	9.31
Outdoor Features	9.62	9.82	9.57	9.83	9.69	9.53
Privacy and Autono..	9.66	9.87	9.73	9.83	9.64	9.64
Property Rooms	9.63	9.72	9.53	9.71	9.68	9.42
Surrounding Views	9.68	9.82	9.65	9.77	9.81	9.48

Host Rank
1 2 3 4 5 6 7 8 9 10

Guest Perspective: Review Scores Across Customer Experience Attributes

	Accuracy	Check-in	Cleanliness	Communication	Location	Value
General Positive Ex..	9.83	9.91	9.74	9.90	9.83	9.68
Additional Touches	9.79	9.86	9.71	9.86	9.76	9.65
Property and Featu..	9.55	9.77	9.43	9.71	9.71	9.36
Non-English	9.85	9.92	9.77	9.92	9.84	9.71
Distance to Points ..	9.76	9.91	9.69	9.89	9.78	9.65
Home-Like Experien..	9.88	9.90	9.80	9.90	9.78	9.84

Guest Rank
1 2 3 4 5 6 7

Avg. Review Score
9.00   10.00

Fig. 2 Average review scores (heatmap) across host and guest perspectives

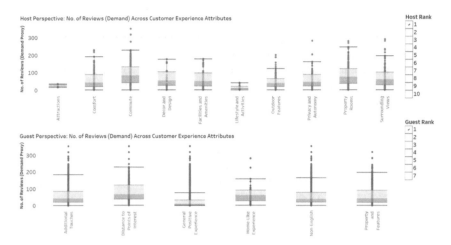

Fig. 3 Review distributions (box-and-whisker) across host and guest perspectives

Number of user reviews across host and guest perspectives. In Fig. 3, the demand distribution across customer experience attributes from a host and guest perspective is presented, using box-and-whisker plots to better highlight the distribution of the measures. The number of reviews per listing was used as a proxy for demand, as has been done in previous studies [2, 9, 14].

Properties that have commute as a primary theme in their listing have higher demand as the median (84 reviews) and upper hinge (133 reviews) for this group is higher than other customer experience attributes. A similar trend is seen from the guest perspective, where distance to points of interest has the highest demand compared to other topics mentioned in reviews. Hosts who mention attractions and lifestyle and activities have the lowest demand. These findings differ from that of Zhang [9], who found that reviews that mention home-like experiences, surrounding views and gardens, and comfortable accommodation were related to the highest demand for Airbnb listing in various USA cities. This suggests that guest preferences differ across regions, as in the context of Cape Town, demand appears to be more closely related to ease of commute and proximity to main points of interest. Ease of commute and proximity to points of interest possibly have higher demand in Cape Town due to limited public transport.

Listings which have a focus on commute (host perspective) or distance to point of interest (guest perspective) have the highest demand, but these listings do not necessarily have the highest prices or guest review scores (performance). There are also fewer "superhosts" who focus on attributes like commute than normal hosts. Properties that stand out have a more targeted focus on other aspects of the guest experience. Hosts who have Décor and Design are more likely to perform better in terms of guest ratings and "superhost" status and are able to charge higher prices than other properties.

5 Conclusion

In this paper, topic modelling is applied to Airbnb data, to answer the following research question: *How can the customer experience attributes derived from topic models be analysed and visualised to understand the difference between host and guest perspectives across listing and host dimensions in Cape Town, South Africa?*

The differences in content and relative rankings (4.1) show that there appears to be a substantial mismatch across the board between owners (host) and customer (guest) perspectives. A data visualisation tool investigated this in more detail and shows that these differences persist, even for the top-ranked properties, across a number of analysis dimensions. These insights demonstrate the value of using topic attribute analysis and visualisation techniques to observe trends in guest and host perspectives. These trends can potentially help current hosts understand the dynamics at play for listings in Cape Town, and they can use the insights to adapt their offering on Airbnb. The findings can also assist potential Airbnb hosts in identifying what is important from a guest perspective, and what they can consider when establishing their own Airbnb listing.

This analysis is based on unstructured natural language text provided by hosts and guests. While most hosts provide a well-thought description of their own listing, the same is not necessarily true for guests. One should keep in mind that guest comments might be biased in terms of (1) the guest profile most likely to provide feedback and (2) the circumstances under which guests are likely to provide feedback. Still, topic models do not elucidate sentiment, but rather topics and this case study illustrates the benefits of this technique for this purpose.

Future work includes alignment of topics between guest reviews and host listings. This technique is often used in multilingual corpora and statistical machine translation where a mapping between documents in two corpora is possible [18]. This might lead to more accurate comparison of host and guest perspectives. Lastly, the correlation between review scores and topics could be further investigated by using supervised LDA.

References

1. Henama US (2018) Disruptive entrepreneurship using Airbnb: the South African experience. Tour Leis 7:1
2. Liang S, Schuckert M, Law R, Chen C (2020) The importance of marketer-generated content to peer-to-peer property rental platforms: evidence from Airbnb. Int J Hosp Manag 84:102329
3. Khotimah DAK, Sarno R (2018) Sentiment detection of comment titles in booking.com using probabilistic latent semantic analysis. In: 2018 6th international conference on information and communication technology (ICoICT). Bandung, p 514
4. Falk M, Larpin B, Scaglione M (2019) The role of specific attributes in determining prices of Airbnb listings in rural and urban locations. Int J Hosp Manag 83:132
5. Airbnb. https://news.airbnb.com/about-us/
6. Kalehbasti PR, Nikolenko L, Rezaei H (2019) Airbnb price prediction using machine learning and sentiment analysis. arXiv preprint arXiv:1907.12665

7. Lawani A, Reed MR, Mark T, Zheng Y (2019) Reviews and price on online platforms: evidence from sentiment analysis of Airbnb reviews in Boston. Reg Sci Urban Econ 75:22

8. Chen C, Chang Y (2018) What drives purchase intention on Airbnb? Perspectives of consumer reviews, information quality, and media richness. Telemat Inform 35:1512

9. Zhang J (2019) Listening to the consumer: exploring review topics on Airbnb and their impact on listing performance. J Mark Theory Pract 27:371

10. Ju Y, Back KJ, Choi Y, Lee JS (2019) Exploring Airbnb service quality attributes and their asymmetric effects on customer satisfaction. Int J Hosp Manag 77:342

11. Luo Y, Tang R (2019) Understanding hidden dimensions in textual reviews on Airbnb: an application of modified latent aspect rating analysis (LARA). Int J Hosp Manag 80:144

12. Moro S, Rita P, Esmerado J, Oliveira C (2019) Unfolding the drivers for sentiments generated by Airbnb experiences. Int J Cult Tour Hosp Res 13:430

13. Situmorang KM, Hidayanto AN, Wicaksono AF, Yuliawati A (2018) Analysis on customer satisfaction dimensions in peer-to-peer accommodation using Latent Dirichlet Allocation: a case study of Airbnb. In: 2018 5th international conference on electrical engineering, computer science and informatics (EECSI). Malang, p 542

14. Zhang L, Yan Q, Zhang L (2020) A text analytics framework for understanding the relationships among host self-description, trust perception and purchase behavior on Airbnb. Decis Support Syst 133:113288

15. Boyd-Graber J, Hu Y, Mimno D (2017) Applications of topic models. Found Trends Inf Retr 11:143

16. Metaxas O (2018) Data4Impact—topic modelling workflow report. Project report. European Commission

17. Prabhakaran S (2018) Topic modeling in Python with Gensim. Machine Learning Plus

18. Subroyen J, Turpin M, De Waal A (2021) Empowering peer-to-peer platform role-players by means of topic modelling: a case study of Airbnb in Cape Town, South Africa. In: Proceedings of the second southern African conference for artificial intelligence research. South Africa, pp 107–120

Pandemic-Induced Behavioral Change in Mobile Banking Adoption: An Opportunity for Indian Banking Industry for Embracing Artificial Intelligence

Nitin Shankar, Sana Moid, Fatima Beena, and Vinod Kumar Shukla

Abstract The present study aims at understanding and analyzing the COVID-19-induced behavioral change spurting artificial intelligence (AI) adoption in Indian banking industry. The study has further identified and analyzed the usage pattern of Indian customers for mobile banking/online banking services in the pre-pandemic phase and progression of Indian customers for mobile banking/online banking services during the pandemic. Secondary data has been used for deep understanding of the AI adoption in Indian banking industry, with reports from McKinsey, PWC, RBI, NPCI, BIS, etc., to form the base. The period of study was taken from 2016 to 20, and this was taken keeping in mind the timing of another unprecedented event of demonetization. Behavioral change of Indian banking industry customer was assessed on three broad parameters change in value and volume of mobile banking transactions on year on year basis. COVID-19-induced behavioral change translating in massive jump of 178% in volume of mobile transactions between March 2019 and 2021. The increase in number of smart phone users and access to connectivity and desired technology has helped the cause. With 2020–21 punctuated by several nationwide as well as localized lockdowns adoption of AI for customer engagement has been crucial for Indian banking industry, which has further translated in to designing and customizing products and risk profiling of customers further resulting in increased operational efficiency and intuitive decision making. The behavioral change induced by COVID-19 in the Indian baking industry achieves competitive advantage by truly responding to huge customer data base which has been utilized by other financial industries as now it can have systems which understand and are responsive to behavior of varied customers. From responses feeded chatbots to intuitively responsive AI bots, the customer engagement is going to be a whole new experience which will help in customer acquisition and retention. Further, with falling data

N. Shankar · S. Moid
Amity Business School, Amity University Uttar Pradesh, Lucknow Campus, Lucknow, India

F. Beena
American College of Dubai, Dubai, United Arab Emirates

V. K. Shukla (✉)
Department of Engineering and Architecture, Amity University, Dubai, United Arab Emirates
e-mail: vinodkumarshukla@gmail.com

storage costs, increasing processing speeds and capabilities and improved connectivity and access for all has helped the rapid automation and AI adoption. Enterprise level adoption of AI has led to revenue generation and optimization of functional resources this reducing the cost at functional level. The AI adoption has been continuous from the banks over the years though banks have started to harness its potential in the recent years with customer's adoption of smart hand-held devices.

Keywords Artificial intelligence · Banking sector · Banking services · Mobile payments · Pandemic · Behavioral transition/progression/change

1 Introduction

Lockdowns and distancing protocols threw a gamut of challenge in front of businesses from safeguarding their resources to safeguarding their business interests. Businesses have been circumspect in embracing technology in the past though COVID-19 has changed this mindset both at business and client end. Lot of industries are reconfiguring their offering keeping technology as the delivery platform and those already using technology are rethinking on ways to connect with their clients in an effective manner. Banking is one industry where 'Trust' is the corner block of the relationship between various stakeholders. Banking industry in India has been trying for more than decade to migrate banking services and its users from branch to online platforms like Internet/online and mobile banking. The progress has long been hampered by doubt in the minds of users over the security concerns over the usage of these platforms. Advancement in telecom technologies from 2 to 4G and subsequently 5G coupled with rapid advancements in the sphere of hand-held devices opened flood gates for user friendly banking application user interface. COVID-19 has brought massive shift in customer's mind of Indian banking industry. The behavioral change has also spurted the industry to look at ways for effective customer's engagement using AI in their strategy.

   With this backdrop, the present study has been undertaken with the primary aim of identifying and analyzing the usage pattern of Indian customers for mobile banking/online banking services in the pre-pandemic phase and understanding the transition/progression of Indian customers toward mobile banking/online banking services during the pandemic.

2 Review of Literature

2.1 The Initial Designs

The earlier mobile banking interface was perceived to be difficult to use which effected the user's intention to embrace mobile banking service [1]. This also arose from the not so aesthetically designed user interface with small texts and icons specially when compared to laptop/desktop versions of Internet banking [2]. Initial versions mobile banking applications had complex paths and navigation to services proved to be deterrent for some customers toward its adoption [3].

2.2 Risks and Trust

Moreover, some researches have emphasized the need to have simple though robust authorization processes instead of complex and inconvenient process like of code card [3]. Apart from the ease of usage which deters customers from mobile banking adoption, history indicates that innovation come with their set of risks and this has an effect on the customer's intention to adopt mobile banking [4] or online banking [5]. Further, the earlier models of cell phones having battery backup issues and a non-reliable Internet connectivity also deterred the customers usage of mobile banking as these two challenges can lead to interruption during a banking transaction [3]. Moreover customer feared making errors while doing banking transaction using the computer [3] or a mobile [6]. Also earlier banks generated portable codes were used which increased a potential threat of fraud in case were lost [3]. Authors [7] show an explicit linkage between risk perception and adoption of online banking services.

3 Human Engagement Versus User (Technological) Interface

The long standing brick and mortar structures which gives lot of people a routine and a social need to interact with a human is also barrier in adoption of the self-service technologies [8], and this social need also moves balance against the technology adoption [9]. Further, there is resistance to change and changing ones ways and learning new technology heavily influences their adoption toward online banking [10]. Commercial banks in India for a long time offering product and services through their trust worthy branch networks. The customer segment they focus on, activities they get involved in and channel they choose have effect on their customer relationships [11]. With advent of technology how we pay and receive money, how we invest and where we invest is fast changing. It is important to observe the steady rise in non-cash transactions which is helping banks in better understanding of their customer basis the huge data which come along with these transactions [12]. Demonetization gave the impetus to cash less transactions in India and it has been on the rise since then.

Author's [13] explained the impact of new information systems on financial industry has been immense especially in the area of mathematical finance. With the availability of big data and availability of affordable computing power which has pushed the envelope of AI and more complex problems can be addressed using AI

[4]. Though the technology is present and banking industry has developed mobile banking application to migrate its existing customers to a different channel and also allure new customer with this convenience of location free 24 × 7 banking.

4 Research Gap and Problem Statement

India has been working on taking its digital imprints across width and depth of the country through its digital India campaign, and people have also embraced technology though when it comes adopting online and banking transactions financial institutions have find it hard cover the entire contours of the nation. In the current study, we have studied the impact of digitization and COVID-19 on behavioral change toward mobile banking adoption Urban and Rural Areas of India. Following are the objective to conduct this research.

- To identify and analyze the usage pattern of Indian customers for mobile banking/online banking services in the pre-pandemic phase.
- To identify and analyze the transition/progression of Indian customers for mobile banking/online banking services during the pandemic.
- To identify the impact of behavioral change toward usage of banking services during pandemic on AI adoption by Indian banking sector.

The pre-pandemic phase was dotted with significant events and that laid the basis for period of current study (2016–20). As mentioned Fig. 1 illustrates the pre-pandemic sluggish growth in number of mobile banking transactions from 2016 to 2018. The gear shift from brick mortar to online and mobile banking can be attributed to events in Indian economy demonetization and introduction of Goods and Services Tax (GST). With economy maturing and banking industry investing AI, refer to Table 1 since 2017, there is clear spike in increase in growth rate of mobile banking transactions.

Behavioral intention (BI) in adoption of online technology has been closely related with performance expectancy (PE) [15, 16]. Further, authors [17] found mobile phones as a top choice for financial inclusion concluded mobile phones as the best alternatives for financial inclusion. Taking Internet connectivity as an important part of performance expectancy (PE), secondary data from TRAI, RBI and world data was collated and structured, in Table 1, to reflect the wireless connectivity in urban as well as rural setup. Data collated included the period from 2016 to 2020, which included the three important events demonetization, launch and implementation of GST and the pandemic, COVID-19. The current study is looking at these events as enabler toward adoption of online and mobile banking. Table 1 also collated from various sources like World Bank and RBI. The settlement and payment volume and value, as per transactions settled by both RBI as well as NPCI, give us trend of behavioral intention (BI).

Figure 2 exhibits strong percentage growth in both value and volume mobile banking transactions; in pre-pandemic phase, there is distinct difference in growth

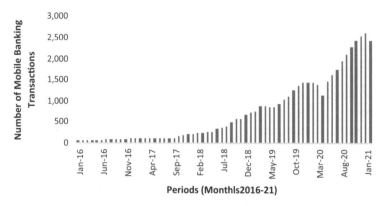

Fig. 1 Clearly exhibits the substantial increase in number of mobile banking transactions from 1000 to 1787 million transactions post COVID-19 registering nearly 79% growth

rate in terms of volume and value, giving a clear indication that in pre-pandemic phase, major transactions in terms of value were still transacted from branch channels. This has significantly changed post pandemic with big transactions in terms of value also moving through online and mobile banking giving banks an opportunity to leverage this scenario.

Table 1 throws important numbers which gives us the insight in terms of India's growing infrastructure of wireless telecommunication services on various parameters at both urban and rural sectors, and the mobile banking transaction was collated from bank settlement data and was placed in the same time line (2016–2020). The table is structured to understand the impact of telecommunication services (PE) on behavioral intention (BI), trend in mobile banking transactions. The degree of user willingness to adopt any new technology is defined as behavioral intention (BI), and it has been mostly used in technology framework as dependent variables [18, 19].

Regression analysis was run to test the hypothesis on following equation for the period 2016–2020. Y (BI—volume of mobile banking transactions) $= \square + \square X$ (PE—wireless Internet connectivity). Following hypothesis was tested using the linear regression model separately for urban and rural centers, and results were collated in Table 2. The results in Table 2 clearly reflect significant impact of improved wireless telecommunications and Internet services in adoption mobile banking at both rural and urban centers.

- Hypothesis: PE about mobile banking service technology has a positive impact on user's intention to adopt it.

Hypothesis:

- H01: There is no significant impact of (PE), wireless Internet services on BI, mobile banking adoption of urban population.
- H02: There is no significant impact of (PE), wireless Internet services on BI, mobile banking adoption of rural population.

Table 1 India's growing infrastructure of wireless telecommunication

	2016	2017	2018	2019	2020
Total telecommunication users	1151.78	1190.67	1197.87	1172.44	1173.83
Urban telecommunication users	683.14	688.25	666.28	662.45	647.91
Rural telecommunication users	468.64	502.42	531.59	509.99	525.92
Tele-density	89.9	91.9	91.45	88.56	86.38%
Wireless telecommunication users					
Total wireless telecommunication users	1127.37	1167.44	1176.00	1151.44	1153.77
% change over the previous year	11.52	3.55	0.57	−2.09	2.34
Urban telecommunication users	662.6	668.44	647.52	643.97	629.67
Rural telecommunication users	464.78	499	528.48	507.46	524.11
Wireline telecommunication users					
Total wireline telecommunication users	24.4	23.23	21.87	21	20.05
% change over the previous year	−4.37	−4.79	−1.11	−3.95	−4.53
Urban telecommunication users	20.55	19.81	18.76	18.47	18.24
Rural telecommunication users	3.86	3.42	3.11	2.53	1.81
Internet/broadband telecommunication users					
Total Internet telecommunication users	391.5	445.96	604.21	718.74	795.18
% change over previous year	18.04	13.91	7.89	18.95	10.64
Narrowband telecommunication users	155.41	83.09	78.86	56.806	4777.00%
Broadband telecommunication users	236.09	362.87	525.36	661.938	747.41
Wired Internet telecommunication users	21.51	21.28	21.42	22.386	25.54
Wireless Internet telecommunication users	370	424.67	582.79	696.36	769.64
Urban Internet telecommunication users	276.44	313.92	390.91	450.31	487.01
Rural Internet telecommunication users	115.06	132.03	213.3	268.43	308.17
Total Internet telecommunication users per 100 population	30.56	34.42	46.13	54.29	58.51 M
Urban Internet telecommunication users per 100 population	68.86	76.76	93.86	106.22	103.98
Rural Internet telecommunication users per 100 population	13.08	14.89	23.87	29.83	34.6

(continued)

Table 1 (continued)

	2016	2017	2018	2019	2020
Volume of transactions via mobile banking (in millions)	67	126	380	1000	1787
Value of transactions via mobile banking (in millions)	7,857,809	4,786,694	2,942,974	1,177,460	1,309,054

Source Author compilation from TRAI, RBI, World Bank reports [20–26]

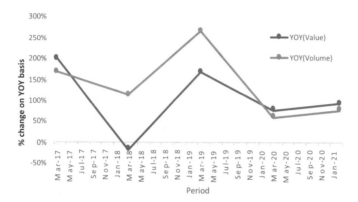

Fig. 2 Percent change in value versus volume of mobile transaction (2016–2021)

Table 2 Results obtained based on linear regression model

		Coefficients	Standard error	*P* value
Urban	Intercept	−2223.826164	703.926304	0.05090352
	Slope	7.546078645	1.796416378	0.02462231
Rural	Intercept	−1007.669269	385.0352093	0.07920005
	Slope	8.097519314	1.745869077	0.01888677

The result shows significant impact of wireless Internet connectivity on mobile banking adoption for both rural and urban sets with *P* value being 0.0246 and 0.01888, respectively. The other aspect which needs to kept in mind that though both banks and Government of India have been investing heavily on mobile banking application technology and on wireless Internet connectivity, respectively, the behavioral intention (BI) toward mobile banking has seen shift during the post COVID-19 periods which is illustrated through Fig. 3 takes it forward with explicit view of both rural and urban centers.

- The null hypothesis H01 is rejected. With *P* value found to less than 0.05; 0.0246.
- The null hypothesis H02 is rejected. With *P* value found to less than 0.05; 0.01889.

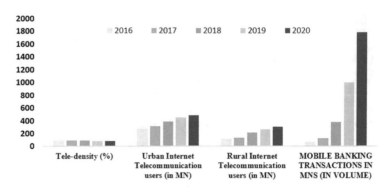

Fig. 3 Growth in Internet subscribers versus growth in mobile banking transactions

The results in Table 2 clearly reflect significant impact of improved wireless telecommunications and Internet services in adoption mobile banking at both rural and urban centers. Figure 3 gives a clear picture how pandemic has rocketed the adoption of mobile banking with other aspects like tele-density, urban and rural telecommunication wireless connection remaining of a similar number. This is a welcome picture for banking industry which has been trying to utilize benefits of technology to reduce cost and increase margins and also tap in to huge unbanked segments of the nation where servicing via brick mortar channel is a costly affair.

India has been a savings economy for a long time like any other developing nation it grinds hard to save and save for future generations and this where trust in banks comes, however, the distrust in technology has hindered the financial services penetration which transpires in to business loss for financial institutions as well. Over the past decade, the improvement and affordability of technology in smartphones and laptops coupled with ever improving affordable Internet services have created conducive environment taking host of financial services on line armed with technologies are artificial intelligence. The COVID-19's scale has led the organizations to rejig their strategy by prioritizing health of its stakeholders, employees and customers and tries to balance it with business goals. Though AI has been around since 1950, its capability has increased multi-fold. Technology, especially AI, is proving to be a key enabler for businesses in their endeavor to rebuild, restructure customer engagement programs and processes.

5   AI-Enabled Resilience

Banking industry can encash on the COVID-19-induced mobile banking adoption, and the banks can work on two fronts:

Table 3 Bank and their chatbot

Bank name	Name of chatbot	Launch date
State Bank of India (SBI)	SBI intelligent assistant (SIA)	2017
HDFC Bank	Electronic virtual assistant (EVA)	2017
ICICI Bank	iPal	2017
Axis Bank	Axis Aha	2018
Kotak Mahindra Bank	Keya	2019
Andhra Bank	ABHi	2019

5.1 Transforming Customer Engagement Methods

Customer service, sales, personalized experience through 'Human interface,' bank branches were severely impacted; in spite of being very few selected sectors which were functioning during the lockdown. These fats tracked the Indian banking sector automation plans.

5.2 Chatbots

The concept of chatbot is getting very popular in banking [27]. Chatbots have been able to cover-up 46% of the support function during pandemic. The chatbots have been a pre-pandemic investment by banks, and Table 3 is snap shot of bank and their chatbots.

5.3 Automating Decision Making

Around 44% of the organization has been able automate decision making using AI to delegate front end office protocols.

5.4 Redesigning Processes Around AI

Industries which have been severely impacted by pandemic have been adapted AI in a concrete manner, as now, it is not an add on feature its business requirement. Around 82% of organizations involved financial services have embraced technology. The major focus has been utilizing the cutting edge technology in the domain of customer engagement, fraud prevention and detection and adherence to regulations. The dearth of useful data has been solved to a great extent with customers getting online and technology capturing patterns and requirements. The variables of external environment as well at micro customers such as credit history, spending patterns and lifestyle maps are synthesized from the data, and an individual customized product is offered without escalating the customization cost [28, 29].

6 Conclusion

COVID-19 has fast tracked the country's financial inclusion mission by inducing technology adoption among masses. There has been significant impact of pandemic as well as Internet connectivity on the growth of mobile banking transactions between the periods of 2016 and 2021. The water shed year being the 2018—demonetization followed by GST and COVID-19. With strong impetus on improving tele connectivity and coverage, the entire environment is a perfect foil for Indian banking industry to embrace AI and ensure engaging customer hooks for sustainable growth and financial inclusion. The latest RBI regulations and guidelines for Account Aggregator (AA) licenses, the control of data now lies with customer. With AA being an intermediary between the Financial Information Users (FIU) and Financial Information Providers (FIP). This can be a game changer as the financial institutions specially the new age technology powered Neo banks will be able to leverage the client approved data fully. Artificial intelligence, machine learning, blockchain are truly changing the landscape of financial services, raising the bar of delivery and increasing the accuracy and reducing the various human induced errors.

For a developing nation like India which has spent a lot on digital infrastructure, the current study highlights how significant this has proved to be in mobile banking adoption with P value for urban and rural center coming less than 0.05, i.e., 0.0246 and 0.01888, respectively, the pandemic has given Indian banks a wonderful opportunity to serve the last mile without losing on operations cost by infusing and improving online banking and mobile banking technology where people have migrated flying above the self-doubts, social stigma and digital literacy, without losing focus on risk mitigating and protection measures. A similar growth rate in terms of value and volume of mobile banking transactions since March 2020 also echoes the aforesaid sentiments.

References

1. Lee Y-K, Park J-H, Chung N, Blakeney A (2012) A unified perspective on the factors influencing usage intention toward mobile financial services. J Bus Res 65(11):1590–1599
2. Bruner GC II, Kumar A (2005) Explaining consumer acceptance of handheld internet devices. J Bus Res 58(5):553–558
3. Kuisma T, Laukkanen T, Hiltunen M (2007) Mapping the reasons for resistance to internet banking: a means-end approach. Int J Inf Manage 27(2):75–85

4. Chen C (2013) Perceived risk, usage frequency of mobile banking services. Manag Serv Qual 23(5):410–436
5. Martins C, Oliveira T, Popovic A (2014) Understanding the internet banking adoption: a unified theory of acceptance and use of technology and perceived risk application. Int J Inf Manage 34(1):1–13
6. Laukkanen T, Lauronen J (2005) Consumer value creation in mobile banking services. Int J Mobile Commun 3(4):325–338
7. Yiu CS, Grant K, Edgar D (2007) Factors affecting the adoption of internet banking in Hong Kong: implications for the banking sector. Int J Inf Manage 27(5):336–351
8. Marr NE, Prendergast GP (1993) Consumer adoption of self-service technologies in retail banking: is expert opinion supported by consumer research. Int J Bank Mark 11(1):3–10
9. Dabholkar P (1996) Consumer evaluations of new technology-based self-service options: an investigation of alternative models of service quality. Int J Res Mark 13(1):29–51
10. Al-Somali SA, Gholami R, Clegg B (2009) An investigation into the acceptance of online banking in Saudi Arabia. Technovation 29(2):130–141
11. Casu B, Girardone C, Molyneux P (2016) Introduction to banking, 2nd edn. Pearson Education Limited, Harlow
12. Harasim J (2016) Europe: the shift from cash to non-cash transactions. In: Transforming payment systems in Europe. Springer, pp 28–69
13. Seese D, Weinhardt C, Schlottmann F (2008) Handbook on information technology in finance. Springer Science & Business Media
14. Halevy A, Norvig P, Pereira F (2009) The unreasonable effectiveness of data. IEEE Intell Syst 24(2):8–12
15. Koufaris M (2002) Applying the technology acceptance model and flow theory to online consumer behavior. Inf Syst Res 13(2):205–223
16. Lin JCC, Lu H (2000) Towards an understanding of the behavioural intention to use a web site. Int J Inf Manage 20(3):197–208
17. Bina M, Giaglis GM (2007) Perceived value and usage patterns of mobile data services: a cross-cultural study. Electron Mark 17:241–252
18. DeLone WH, McLean ER (2003) The DeLone and McLean model of information systems success: a ten-year update. J Manag Inf Syst 19(4):9–30
19. Venkatesh V, Morris MG, Davis GB, Davis FD (2003) User acceptance of information technology: toward a unified view. MIS Q 27(3):425–478
20. Reserve Bank of India (2021) Settlement data of payment systems. Retrieved from https://rbi.org.in/Scripts/Statistics.aspx
21. Telecom Regulatory Authority of India (2021) Performance on key quality of service parameters—comparative performance report. Retrieved from https://www.trai.gov.in/release-public ation/reports/performance-reports
22. Telecom Regulatory Authority of India (2016) Yearly performance indicators of Indian telecom sector. Retrieved from https://www.trai.gov.in/sites/default/files/Yearly_PI_Reports_2016.pdf
23. Telecom Regulatory Authority of India (2017) The Indian telecom services performance indicators. Retrieved from https://www.trai.gov.in/sites/default/files/Performance_Indicator_Rep orts_28Sep2017.pdf
24. Telecom Regulatory Authority of India (2018) Yearly performance indicators of Indian telecom sector. Retrieved from https://www.trai.gov.in/sites/default/files/PIR_25092019.pdf
25. The World Bank (2017) The Global Findex database 2017. Retrieved from https://globalfin dex.worldbank.org
26. The World Bank (2021) COVID-19 related shocks survey in rural India 2020. Retrieved from https://microdata.worldbank.org/index.php/catalog/3769
27. Suhel SF, Shukla VK, Vyas S, Mishra VP (2020) Conversation to automation in banking through chatbot using artificial machine intelligence language. In: 2020 8th international conference on reliability, infocom technologies and optimization (trends and future directions) (ICRITO), pp 611–618. https://doi.org/10.1109/ICRITO48877.2020.9197825

28. Thekkethil MS, Shukla VK, Beena F, Chopra A (2021) Robotic process automation in banking and finance sector for loan processing and fraud detection. In: 2021 9th international conference on reliability, infocom technologies and optimization (trends and future directions) (ICRITO), pp 1–6. https://doi.org/10.1109/ICRITO51393.2021.9596076
29. Beena F, Mearaj I, Shukla VK, Anwar S (2021) Mitigating financial fraud using data science—"a case study on credit card frauds". In: 2021 international conference on innovative practices in technology and management (ICIPTM), pp 38–43. https://doi.org/10.1109/ICIPTM52218.2021.9388345

On the Efficacy of Boosting-Based Ensemble Learning Techniques for Predicting Employee Absenteeism

Kusum Lata⊙

Abstract Employee absenteeism is a substantial problem faced by many organizations. It severely affects the productive operations in organizations. In the recent years, predictive modeling using ensemble learning techniques has increased the attention of researchers to develop competent models for various predictive tasks. Ensemble techniques combine the predictions of multiple models to yield a single consolidated decision. Thus, predictive modeling with the help of ensemble learning techniques can predict employee absenteeism so that the human resource department can devise intervention policies and lessen monetary losses due to absenteeism. Therefore, in this direction, this study develops models with the help of boosting-based ensemble learning aggregate with data balancing to predict employee absenteeism. The predictive accuracy of the absenteeism prediction models is evaluated using strong performance measures; area under receiver operator characteristics curve, balance and geometric mean. The results are also examined statistically with statistical analysis. The boosting-based ensemble learning techniques are effective for predicting employee absenteeism according to the results of the study.

Keywords Machine learning · Ensemble learning · Data balancing

1 Introduction

The workforce is an important asset for an organization. A happier and healthy workforce is critical for the organization's success. However, many organizations face employee-related issues that may hinder its success. Absenteeism is one such issue that is defined as a habitual and intentional absence from work [1]. It incurs high costs for an organization and loss of productivity. Absenteeism is very difficult to tackle. It is very challenging for organizations to effectively monitor, control, and reduce absenteeism. Therefore, the absenteeism prediction mechanisms are imperative for organizations relying heavily on human resources. Such a prediction mechanism

K. Lata (✉)
University School of Management and Entrepreneurship, Delhi Technological University, Delhi, India
e-mail: kusumlata@dtu.ac.in

helps the managers take preventive measures to tackle absenteeism and reduce its financial costs.

The absenteeism prediction problem is dealt as a classification problem in this work. The data for this work has been downloaded from UCI repository. The target variable to be predicted is absenteeism having two values: absent and not absent. This data is imbalanced to treat the absenteeism prediction as classification problem. The issue with machine learning (ML) techniques is that the imbalanced data cannot be learned properly.

Ensemble learning (EL) techniques, a subclass of ML techniques, have been used competently in various real-life prediction problems. The EL techniques combine the predictions of multiple ML models to provide a single and consolidated prediction outcome [2]. But with imbalanced dataset, the prediction models developed by EL may not be accurate [3]. Therefore, we have used boosting-based EL techniques that combine data balancing techniques to develop absenteeism prediction models. The boosting-based EL techniques tend to build a robust classification model by combining the prediction outcomes of weak classifiers. The weights training examples are adjusted depending upon last classification.

The boosting process in boosting-based EL techniques lowers the bias in the final classification models, but it might not hold with imbalanced data [3]. Therefore, combining data balancing within each round of boosting in boosting-based EL techniques enables the classification model to sample a more significant number of instances of the minority class. In this study, we used boosting-based EL techniques to develop absenteeism prediction models. These techniques are Synthetic Minority Oversampling with Boosting (SMOTEBoosting), Modified Synthetic Minority Oversampling with Boosting (MSMOTEBoosting), Synthetic Minority Oversampling with Boosting (SMOTEBoosting), Evolutionary Undersampling with Boosting (EUSBoosting), DataBoosting and Random Undersampling with Boosting (RUSBoosting). In this study, the training and test data ratio taken for model learning and validation is 10:1, and area under receiver operator characteristic curve (AUC), geometric mean (GMEAN), and balance (BAL) are used for evaluation of models. The organization of different sections is structured as: Sect. 2 defines the research framework, Sect. 3 states the results and analysis, and conclusions are documented in Sect. 5.

2   Related Work

Predictive modeling with the help of ML techniques has contributed to building models for diverse predictive modeling tasks like fraud detection [4–5], defect prediction in software classes [6], predicting fraud transactions [7], etc. In HR analytics, employee absenteeism prediction is the concern of the researchers that led to various prediction models with the ML techniques. The study by Oliveira et al. [8] developed models to predict absenteeism with the help of ML techniques with dataset collected from call centers in Brazil. The results of this study advocated XGBoost as a competent technique for predicting absenteeism with 72% value of AUC. The paper by Shah et al. [1] applied deep neural networks (DNN) to predict absenteeism. They developed models using DNN with ML-based models: support vector machine (SVM) and decision tree (DT). According to the study results, DNN-based models were the best predictor compared to ML-based models. Tiwari el al. [9] used linear regression to develop absenteeism prediction models. Rista et al. [10] analyzed the performance of DT, logistic regression, and SVM-based absenteeism prediction models on employee absenteeism dataset of an Oil Refinery. Lawrance et al. [11] employed cost sensitivity classification for employee absenteeism prediction. Thus, in the literature, ML techniques are investigated for predicting absenteeism. However, the EL techniques have rarely been investigated in the literature.

3   Research Framework

The experimental framework is shown in Fig. 1.

3.1   Data Collection and Preprocessing

The dataset for conducting this research work is downloaded from UCI repository. This dataset corresponds to employee absenteeism of a Brazilian Courier Company.

The dataset comprises 740 records, and each record consists of values of 21 features. These features are: Individual identification, Month of absence, Distance from Residence to Work, Seasons, Reason for absence, Day of the week, Social smoker, Transportation expense, Age, Social drinker, Workload Average/day, Disciplinary failure, Hit target, number of children, Education, Service time, Pet, Body mass index, Height, Weight, and Absenteeism time in hours. The feature, Individual identification corresponds to the identification number for an employee. This feature is removed in the data preprocessing phase as this feature will not add any predictive power to the model. Also, the feature, workload average/day has huge values ranging from 205,917 to 378,884. The feature with such large values makes the

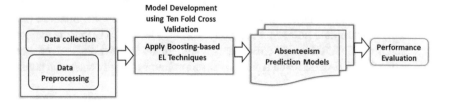

Fig. 1 Experimental framework

training process slow. So, the attribute values of this feature were standardized using min − max normalization given by Eq. 1, where x_i is the feature value from the workload average/day, min(x) and max(x) are, respectively, the minimum and maximum value of the workload average/day and x_{new} is the transformed value after min − max normalization.

$$x_{new} = \frac{x_i - \min(x)}{\max(x) - \min(x)} \tag{1}$$

The absenteeism time in hours was defined as the target variable that contains real values ranging from 0 to 120. The absenteeism prediction problem is treated as a classification in this study; the target variable is transformed to a binary variable, absenteeism. For the absenteeism time per hour zero value, absenteeism is assigned value 0 (Not Absent) otherwise 1 (Absent). We got 696 data points where the absenteeism value was 1, and in 44 data points, the absenteeism value was 0.

3.2   Model Development and Validation

The data points belonging target class (i.e., absenteeism value 0 and 1) are very uneven in the dataset. The majority of the data points are from class where the absenteeism value is 1, and for very few data points, the target class has 0 values. The data points where the absenteeism value is 1 are regarded as majority class data points, while the others for which this value is 0 are regarded as minority class data points. As discussed in Sect. 1, considering the imbalanced nature of the dataset, we opt to apply the EL techniques that include data balancing in its successive iterations. Combination of EL techniques with data balancing yields a robust classification model. For the development of the model, the cross-validation method used is ten-fold cross-validation (TenFCV). With TenFCV, the dataset is subdivided in ten equal sized partitions containing nine out of which are used model training with ML techniques. The remaining partition is reserved for validation. This process is repeated ten times; each time model is validated on a different partition. A brief explanation of boosting-based EL techniques used in the study is given below.

(i)   SMOTEBoost [12]: It hybridizes AdaBoost [13] with SMOTE [14]. AdaBoost is an ensemble based on boosting methodology that emphasizes on hard to learn data points. In the beginning, the data points are presented to the classifier by giving equal weights that are updated in the subsequent iterations so that hard to learn data points get increased weight. The SMOTEBoost technique combines SMOTE with AdaBoost; synthetic minority data points are created using SMOTE so that each AdaBoost iteration learns minority class data points better.

(ii) MSMOTEBoost [15]: This technique combines MSMOTE with AdaBoost. In respective iterations of AdaBoost, data points belonging to minority class are oversampled by employing the MSMOTE oversampling method.

(iii) DataBoost [16]: This technique aggregated data generation and boosting as one single technique to effectively train the model. It works in three stages. In the first stage, equal weights are given to each training data point. This weighted dataset is then used for training. In the second stage, the hard instances are identified called seed instances, and synthetic instances are generated corresponding to them. These synthetic instances are added to the training data in the third stage. The class distribution of both of the classes is balanced in this way.

(iv) RUSBoost [17]: It is the hybridization of RUS [18] and AdaBoost. Unlike SMOTEBoost and MSMOTEBoost, it does not oversample the minority class by creating the synthetic data points in AdaBoost iterations rather data balancing in AdaBoost iterations is performed with random undersampling by randomly removing the majority class data points from the dataset. This is a faster technique than SMOTEBoost and MSMOTEBoost.

(v) EUSBoost [19]: It extends RUSBoost by including evolutionary undersampling in each iteration of AdaBoost instead of pure random undersampling.

3.3 Performance Evaluation

In this study, AUC, GMEAN, and BAL performance measures are used to evaluate models. These are strong performance metrics particularly recommended for evaluation of models developed with imbalanced training data. The prediction model with high value of these performance measures is considered a good model.

AUC: It is the area under curve constructed by plotting sensitivity values on the y-axis and (false positive rate) fpr on the x-axis. The trade-off between the sensitivity and fpr by the curve serves as the model's performance. A good prediction model produces higher AUC.

GMEAN: It is measured as the geometric mean of sensitivity (i.e., true positive rate) and specificity (i.e., true negative rate).

$$GMEAN = \sqrt{Sensitivity * Specificity} \qquad (2)$$

BAL: It measures Euclidean distance between the pair of true positive rate: tpr and false positive rate: fpr and the optimal value of this pair. The optimal value for fpr is 0 and that of tpr is 1.

$$BAL = 1 - \sqrt{\frac{(0 - fpr)^2 + (1 - tpr)^2}{2}} \qquad (3)$$

4 Results and Analysis

The absenteeism prediction models in this study are developed using boosting-based EL techniques discussed in Sect. 3. These techniques are used to develop the absenteeism prediction models using the KEEL tool with parameter values stated in Table 1.

The predictive accuracy of the models is measured by analyzing the AUC, GMEAN, and BAL values. All the models are developed using TenFCV. In Table 2, the AUC values are in the range of 96.40–98.80. Similarly, the GMEAN and BAL range of the models are 96.25–98.71 and 95.16–98.38, respectively. As depicted in Fig. 2, models developed using DataBoost technique have given best average AUC, GMEAN, and BAL values. The second-best performer is RUSBoost. The performance of EUSBoost is also nearly same as that of RUSBoost as shown in Fig. 2. The performance of EL techniques that include SMOTE and MSMOTE oversampling is reported almost similar for all performance measures. As, the DataBoost shown the best accuracy in terms of all the three performance measures, we further analyzed the test results statistically by Wilcoxon-signed rank test. For analysis with this test level of significance, $\alpha = 0.05$ is considered. We performed a pair-wise comparison of all the techniques with DataBoost technique according to all the performance measures. In Table 3, the outcomes of the Wilcoxon test are stated.

Table 1 Parameter settings

EL technique	Parameter values
SMOTEBoost	Pruning = true, number of instances/leaf = 2, confidence = 0.25, no. of classifiers = 10
MSMOTEBoost	Pruning = true, number of instances/leaf = 2, confidence = 0.25, no. of classifiers = 10
RUSBoost	Pruning = true, number of instances/leaf = 2, confidence = 0.25, classifiers = 10, percentage majority class = 50
DataBoost	Pruning = true, number of instances/leaf = 2, confidence = 0.25, no. of classifiers = 10
EUSBoost	Pruning = true, number of instances/leaf = 2, confidence = 0.25, classifiers = 10, percentage majority class = 50

Table 2 Performance of developed models w.r.t. AUC, GMEAN, and BAL

EL technique	AUC	GMEAN	BAL
SMOTEBoost	96.90	96.64	96.44
MSMOTEBoost	96.40	96.25	95.16
RUSBoost	98.40	98.28	98.20
DataBoost	98.80	98.71	98.38
EUSBoost	98.20	98.11	98.07

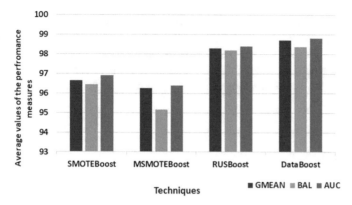

Fig. 2 Performance of boosting-based EL techniques

Table 3 Wilcoxon test

Pair examined	p-value
DataBoost and SMOTEBoost	1.09
DataBoost and MSMOTEBoost	1.09
DataBoost and RUSBoost	1.09
DataBoost and EUSBoost	1.02

The p-value of the comparison of DataBoost with all other examined techniques is reported in Table 3. In Table 3, for all four comparisons carried out with Wilcoxon test, the p-values obtained are greater than the significance level of significance, i.e., $\alpha = 0.05$. This indicated that the performance of DataBoost is best among all other investigated techniques in terms of all the three performance measures, but statistically no significant difference is noted in the performance of DataBoost, RUSBoost MSMOTEBoost, SMOTEBoost, and EUSBoost. Therefore, the Wilcoxon test results led to the conclusion that when the data is imbalanced, all the boosting-based EL techniques used in the study are competent techniques to develop effective employee absenteeism prediction models.

5 Conclusions

Punctuality of employees is a critical factor for organizations to make progress. It is very troublesome for an organization, if the employees are habitually late and absent from the work. As it is very challenging for the organizations to effectually monitor, control, and reduce absenteeism, the present study puts forward a mechanism to predict employee absenteeism using predictive modeling with EL techniques. The dataset for this research work pertains to Brazilian Courier company with different feature reflecting the human behavior. Considering the imbalanced nature of this dataset, EL techniques aggregated with data balancing are used to develop effective absenteeism prediction models. The predictive power of the investigated techniques is assessed with AUC, BAL, and GMEAN performance measures and statistical analysis is carried out. This study advocates that boosting-based EL techniques used in the study are competent techniques to develop effective absenteeism prediction models.

References

1. Ali Shah SA, Uddin I, Aziz F, Ahmad S, Al-Khasawneh MA, Sharaf M (2020) An enhanced deep neural network for predicting workplace absenteeism. Complexity 2020
2. Oza NC, Tumer K (2008) Classifier ensembles: select real-world applications. Inf Fusion 9(1):4–20
3. Galar M, Fernandez A, Barrenechea E, Bustince H, Herrera F (2011) A review on ensembles for the class imbalance problem: bagging-, boosting-, and hybrid-based approaches. IEEE Trans Syst Man Cybern Part C (Appl Rev) 42(4):463–484
4. Randhawa K, Loo CK, Seera M, Lim CP, Nandi AK (2018) Credit card fraud detection using AdaBoost and majority voting. IEEE Access 6:14277–14284
5. Patil S, Nemade V, Soni PK (2018) Predictive modelling for credit card fraud detection using data analytics. Procedia Comput Sci 132:385–395
6. Alsaeedi A, Khan MZ (2019) Software defect prediction using supervised machine learning and ensemble techniques: a comparative study. J Softw Eng Appl 12(5):85–100
7. Gao J, Zhou Z, Ai J, Xia B, Coggeshall S (2019) Predicting credit card transaction fraud using machine learning algorithms. J Intell Learn Syst Appl 11(3):33–63
8. de Oliveira EL, Torres JM, Moreira RS, de Lima RAF (2019) Absenteeism prediction in call center using machine learning algorithms. In: World conference on information systems and technologies, Apr 2019. Springer, Cham, pp 958–968
9. Tewari K, Vandita S, Jain S (2020) Predictive analysis of absenteeism in MNCs using machine learning algorithm. In: Proceedings of ICRIC 2019. Springer, Cham, pp 3–14
10. Rista A, Ajdari J, Zenuni X (2020) Predicting and analyzing absenteeism at workplace using machine learning algorithms. In: 2020 43rd international convention on information, communication and electronic technology (MIPRO). IEEE, pp 485–490
11. Lawrance N, Petrides G, Guerry MA (2021) Predicting employee absenteeism for cost effective interventions. Decis Support Syst 113539
12. Chawla NV, Lazarevic A, Hall LO, Bowyer KW (2003) SMOTEBoost: improving prediction of the minority class in boosting. In: European conference on principles of data mining and knowledge discovery, Sept 2003. Springer, Berlin, Heidelberg, pp 107–119
13. Freund Y, Schapire RE (1997) A decision-theoretic generalization of on-line learning and an application to boosting. J Comput Syst Sci 55(1):119–139
14. Chawla NV, Bowyer KW, Hall LO, Kegelmeyer WP (2002) SMOTE: synthetic minority over-sampling technique. J Artif Intell Res 16:321–357
15. Hu S, Liang Y, Ma L, He Y (2009) MSMOTE: improving classification performance when training data is imbalanced. In: 2009 second international workshop on computer science and engineering, Oct 2009, vol 2. IEEE, pp 13–17
16. Guo H, Viktor HL (2004) Learning from imbalanced data sets with boosting and data generation: the DataBoost-IM approach. ACM SIGKDD Explor Newsl 6(1):30–39

17. Seiffert C, Khoshgoftaar TM, Van Hulse J, Napolitano A (2009) RUSBoost: a hybrid approach to alleviating class imbalance. IEEE Trans Syst Man Cybern Part A Syst Hum 40(1):185–197
18. Batista GE, Prati RC, Monard MC (2004) A study of the behavior of several methods for balancing machine learning training data. ACM SIGKDD Explor Newsl 6(1):20–29
19. Galar M, Fernández A, Barrenechea E, Herrera F (2013) EUSBoost: enhancing ensembles for highly imbalanced data-sets by evolutionary undersampling. Pattern Recogn 46(12):3460–3471

Framework to Impute Missing Values in Datasets

Manoj Kumar⬦, Saiesh Kaul⬦, Sarthak Sethi⬦, and Siddhant Jain⬦

Abstract Many real-time databases are facing the problem of missing data values, which may lead to a variety of problems like improper results, less accuracy and other errors due to the absence of automatic manipulation of missing values in different Python libraries, making the imputation of these missing values of utmost priority for better results. The primary intent of our research is to create a framework that would try to give the most optimal method for the imputation of these missing data-points in datasets using the best possible methods like DataWig, K-nearest neighbor (KNN), multiple imputation by chained equations (MICE), MissForest, multivariate feature, mean, median, most frequent element and use the method which is most appropriate with the particular dataset to impute that dataset.

Keywords Missing values imputation · KNN · MICE · Null values · Mean absolute error · MissForest · Root mean square error · DataWig · Hyperparameter tuning · Symmetric mean absolute percentage error

1 Introduction

In today's time, data is everything. Data is being used in almost everything. However, it should be clean to analyze this data and not contain empty spaces. These empty spaces are called missing data. Missing data has become a big problem nowadays, and it has grown much interest in statistical analysis. Also, missing data is a significant problem in all types of research, and it introduces uncertainty in the data analysis.

M. Kumar · S. Kaul · S. Sethi (✉) · S. Jain
Department of Computer Science and Engineering, Delhi Technological University, Delhi, India
e-mail: sarthaksethi_2k18co326@dtu.ac.in

M. Kumar
e-mail: mkumarg@dce.ac.in

S. Kaul
e-mail: saieshkaul_2k18co311@dtu.ac.in

S. Jain
e-mail: siddhantjain_2k18co350@dtu.ac.in

© The Author(s), under exclusive license to Springer Nature Singapore Pte Ltd. 2023
A. Shukla et al. (eds.), *Computational Intelligence*, Lecture Notes in Electrical Engineering 968, https://doi.org/10.1007/978-981-19-7346-8_17

Missing data has become a significant problem with data quality. The most typical approach to get rid of this missing data is to impute these missing values. By imputing missing values, the complexity of analysis is decreased as a complete dataset is created, eradicating the problem of handling intricate missing patterns in the datasets [1]. Older methods for data imputation are effortless to understand and implement, but they introduce unfairness in the data.

By considering certain assumptions, present and mixed methods work better and perform better than these older methods. We have tried to implement a mixed imputation method. We have combined various single and multiple imputation techniques in this mixed method for a particular dataset [2]. It is also possible that different columns of a dataset require different imputation techniques, which also we have covered in this mixed method. To evaluate how a particular method is performing on a particular dataset, we have compared the results by different techniques like mean absolute error (MAE), root mean square error (RMSE) and symmetric mean absolute percentage error (SMAPE). Missing values of a particular attribute of a dataset were filled using the best imputation technique meant for that particular attribute/column. In the end, we have generated the final imputed dataset according to the best imputation techniques according to our framework.

Various imputation techniques behave differently on different datasets. In other words, one imputation technique can be the most refined imputation technique for a particular type of dataset. However, it cannot be the most refined imputation technique for other dataset types. E.g., it is possible that the DataWig is not the most refined imputation technique for dataset A, but DataWig can be the most refined imputation technique for another dataset B. That is why we have tried to run different imputation techniques on a particular dataset [3], to find the best techniques out of them and compare their result by various evaluation metrics.

2   Proposed Framework

Our research proposed a one-stop framework that would work with different datasets of different domains and give the desired results. Figure 1 shows the basic flow of our framework. This framework consists of four main parts, namely generation of the dataset, imputed methods, comparison of methods and imputation on the original dataset, and we discuss it in detail below.

2.1   Generation of Dataset

Every model needs a good dataset to provide the desired results. Our method revolves around comparing the outcomes of the imputations on the artificially injected null values, which requires a dataset without any null values. Following the above approach, the raw dataset is neglected as mostly all the datasets have null values,

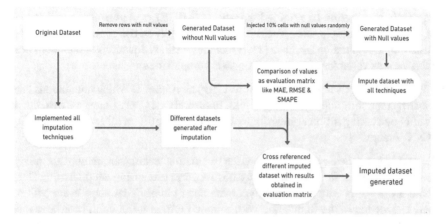

Fig. 1 Proposed framework

which we are trying to impute. So, we created another dataset by deleting all the null values already present in the raw dataset, making our dataset ready for further steps. Now we have a dataset without any null values, in which we would artificially inject null values (10%) randomly and store it as another dataset, which would complete all the datasets we require for our model: the original dataset, a dataset without null values and a dataset with injected null values. Moreover, now we can use these datasets in our model.

2.2 Imputed Methods

Various techniques exist to impute null values that perform differently for different datasets. Some approaches do not consider the correlation between different attributes, and it only considers individual attribute levels like **mean, median and most frequent** imputation in which we require a constant value for all the imputations [4]. This constant value can be either the mean, median or the most frequent value in the column. These methods are fast, easy and give satisfactory result for small numerical datasets but are sometimes not very accurate.

Multivariate imputation techniques calculate null values using the whole set of accessible feature dimensions [5], rather than ith feature dimension using just non-missing data in that dimension [6].

K-nearest neighbors (KNN) imputes values dynamically, meaning it imputes the missing values considering the nearby values in the dataset. KNN is beneficial due to the absence of a training phase in it, and also it uses all the data during training which is why it is a lazy learning algorithm [7]. It has a limitation also that this algorithm becomes ineffective or very slow if the size of the dataset increases.

Multiple imputation by chained equations (MICE) works on filling null values in the dataset multiple times, measuring the ambiguity in the missing values better than simple imputation methods [1]. It includes chained equations, which help deal with various variables, and their asymptotic complexities.

MissForest is derived from random forest (RF) algorithm, in which imputations take place by randomly adding missing values. In several cases, MissForest gave excellent results outscoring all other algorithms by more than 50% in all metrics, including KNN-impute.

DataWig is a robust approach, where the imputation techniques are used for mixed data types present in tables, including unstructured text fusing deep learning (DL) feature extractors with automatic hyperparameter tuning. That helps users with no ML knowledge in the imputation with minimal efforts in different heterogeneous data types present in different tables. Also, it offers better and more flexible modeling options.

2.3   Comparison of Methods

According to Table 1, various techniques were applied to the dataset, which was generated by artificially injected null values. We need to decide which one to use on which datasets to precisely do this. Upon the successful imputation of the artificially injected missing values and comparing the dataset generated to the dataset we had earlier without any missing values, we found which method performs better on any particular dataset [8]. We compare this by calculating how different the imputed dataset is from an original dataset with the help of different error calculating methods. Lower error in an imputation method means better accuracy and performance than the imputation method having a higher error value [9]. We also plotted a graph between the original dataset values and the imputed ones. The overlapping between original and imputed values would indicate better imputation in the plotted graph.

Table 1 List of algorithms/techniques used for imputation of datasets

S. No.	Algorithms/techniques
1	Mean
2	Median
3	Most frequent
4	Multivariate feature
5	K-nearest neighbors (KNN)
6	Multiple imputation by chained equations (MICE)
7	MissForest
8	DataWig

2.4 Imputation on the Original Dataset

Different columns in the particular dataset might present dissimilar results for distinct imputation methods. Moreover, situations may arise where any particular imputation method would perform satisfactorily for some columns but would give terrible results for some other columns, which means we cannot just use one approach for the entire dataset. So, for all different columns, we need to find the accurate imputation technique to deliver the best results. We compare all the methods on the error values generated, compare the imputed and original values in the graph, and then complete the imputation with the method, which provides the lowest error and the maximum overlap for each column.

As we cannot just use different imputation techniques for different columns separately as some methods might use features of the other columns too for better imputations, the entire dataset was imputed using all algorithms and then chose the respective columns from the dataset where that particular method has performed best. For example, in a dataset with just two columns, KNN performs best for column A [10], and MissForest performs most suitable for column B. We will impute the whole dataset with KNN and MissForest separately, forming two datasets. Then, for the resultant dataset, we will choose the first column from the dataset imputed with KNN and the latter from MissForest. We scale this approach to different datasets across multiple columns.

3 Experiments and Results

In our proposed approach, we experimented on a variety of datasets like the air quality dataset of India which contains data on air quality of different dates of different cities, data consist of approximately 30,000 rows and 16 columns, including two columns of date and city which have no missing values, but for the rest 14 columns, it contains missing values. The temperature change dataset contains temperature for the past 60 years, which was cross-matched with country and time of year. Data consists of approximately 10,000 rows and 66 columns in which area code, country, month contains no missing values but the following columns, which contain the temperature of the particular year, contains null values. Using our proposed framework, we tried to impute these datasets according to the best-suited technique to fill these null values. The proposed framework experimented on the Air Quality dataset, which initially has 30,000 rows after deleting rows containing null values. The new dataset without any null values was generated with 6000 rows. Each column that initially contained null values was forcefully imputed with null values of 10% of the values.

Both datasets, which had 6000 rows with no null values and datasets in which null values were injected, were saved to understand the nature and cross-relation between attributes. In the next step, we needed to understand the dataset by imputing the dataset in which we had injected the null values and compared the result with the dataset without null values.

The dataset we injected the null values was imputed using the algorithm or techniques listed in Table 1, we need to figure out for which column which technique is best suited. In this dataset, we can see in Table 2 the comparison of various techniques. Calculation of error between dataset without null values and dataset in which we applied the algorithm to impute the dataset can be calculated using to various techniques.

In this research, we have considered MAE, RMSE and SMAPE to show a particular column in Tables 2 and 3. Now that we have scores of all columns with all the techniques, we are at our final step, in which we have to impute the original dataset according to the score we calculated in the previous step. For that, we imputed the original dataset according to all methods given in Table 1, and now for all attributes, we compare methods for which we are getting the best results, it can be according to the evaluation technique of our choice like MAE, RMSE and SMAPE [11]. The best technique on the Air Quality dataset using RMSE as evaluation criteria is given in Table 4.

Table 2 Evaluation metric for imputation of CO attribute in air quality

S. No.	Algorithms/techniques	MAE	RMSE	SMAPE
1	Mean	0.58	1.29	55.21
2	Median	0.53	1.32	51.04
3	Most frequent	0.53	1.34	52.45
4	Multivariate feature	0.55	0.91	68.70
5	KNN	0.15	0.32	20.72
6	MICE	0.55	0.91	69.15
7	MissForest	0.12	0.24	17.25
8	DataWig	0.27	0.41	35.36

Table 3 Evaluation metric for imputation of Benzene attribute in air quality

S. No.	Algorithms/techniques	MAE	RMSE	SMAPE
1	Mean	2.86	4.58	84.05
2	Median	2.63	4.72	83.36
3	Most frequent	3.50	5.76	164.35
4	Multivariate feature	1.93	2.85	77.34
5	KNN	0.83	1.53	31.03
6	MICE	1.93	2.85	77.51
7	MissForest	2.72	2.07	24.44
8	DataWig	1.43	2.11	60.30

Table 4 Imputation results of air quality dataset considering RMSE as evaluation metric

S. No.	Attribute/column	Best imputation technique
1	PM2.5	MissForest
2	PM10	MissForest
3	NO	MissForest
4	NO_2	MissForest
5	NO_x	MissForest
6	NH_3	MissForest
7	CO	MissForest
8	SO_2	MissForest
9	O_3	MissForest
10	Benzene	KNN
11	Toluene	MissForest
12	Xylene	KNN
13	AQI	MissForest
14	AQI_Bucket	MissForest

Furthermore, using the values generated in Table 4 and comparing the graphical representation shown in Fig. 2, the entire dataset was imputed. Dataset of any type can be imputed using this framework. We also tried this framework on other datasets from which we generated the imputed dataset, which was successfully imputed using this framework.

4   Conclusion

In a real-world application, datasets are used in various machine learning/deep learning models. The dataset that consists of null/missing values can be due to equipment failure or any other reason. Imputation is necessary prior to using this dataset in any model. There are various techniques to deal with these missing values. However, while applying these various techniques to a particular dataset, firstly, we need to have comprehensive knowledge about the dataset to understand which imputation technique will work on the dataset. Further, by imputing a single dataset multiple times, we can increase imputation accuracy. One attribute of the dataset may be imputed with one of the imputation techniques. Another attribute may need some other imputation technique on the same dataset to provide better results than applying a single imputation technique on the whole dataset.

As a result, we concluded that we could develop a framework that provided the best imputation technique for all the columns/attributes (which contains null values) in a particular dataset. For the Air Quality dataset, when we considered RMSE as an evaluation metric, various attributes which contained null values were filled by MissForest and the rest by KNN. Due to our framework, we were able to identify for

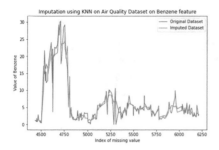

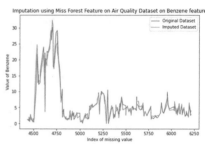

a) Imputation of Benzene attribute using KNN

b) Imputation of Benzene attribute using MissForest

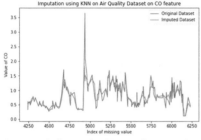

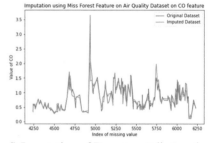

c) Imputation of CO attribute using KNN

d) Imputation of Benzene attribute using MissForest

Fig. 2 Graphical representation of KNN and MissForest on CO and Benzene attribute of air quality dataset

which column, which imputation technique was providing the best result. Figure 2 shows the graphical representation of MissForest and KNN on CO and Benzene attributes provides comparative results. However, from Table 2, MissForest provides better results than KNN in CO attribute, and from Table 3, KNN provides better results than MissForest in Benzene attribute. That is how our framework provides the best imputation technique for a particular attribute in a dataset.

References

1. Royston P, Cancer Group (2005) Multiple imputation of missing values: update of ice. Stata J 5(4):527–536
2. Soley-Bori M (2013) Dealing with missing data: key assumptions and methods for applied analysis
3. Junninen H, Niska H, Tuppurainen K, Ruuskanen J (2004) Methods for imputation of missing values in air quality data sets
4. Wongoutong C (2022) Imputation methods for missing response values in the three parts of a central composite design with two factors
5. Buck SF (1960) A method of estimation of missing values in multivariate data suitable for use with an electronic computer. J R Stat Soc B22(2):302–306

6. Royston P (2004) Multiple imputation of missing values. Stata J 4(3):227–241
7. Zakaria NA, Noor NM (2018) Imputation methods for filling missing data in urban air pollution data for Malaysia. Urban Arch Constr 9(2):159–166
8. Grzymala-Busse JW, Hu M (2000) A comparison of several approaches to missing attribute values in data mining. In: Rough sets and current trends in computing 2000, pp 340–347
9. Berkelmans GFN, Read SH, Gudbjornsdottir S (2022) Population median imputation was noninferior to complex approaches for imputing missing values in cardiovascular prediction models in clinical practice
10. Batista GEAPA, Monard MC (2002) K-nearest neighbour as imputation method: experimental results. Technical report 186. ICMCUSP
11. Lin W-C, Tsai C-F (2019) Missing value imputation: a review and analysis of the literature (2006–2017). Artif Intell Rev 53:1487–1509

Modelling of an Efficient System for Predicting Ships' Estimated Time of Arrival Using Artificial Neural Network

Md. Raqibur Rahman⑩, **Ehtashamul Haque**⑩, **Sadia Tasneem Rahman**⑩, **K. Habibul Kabir**⑩, and **Yaseen Adnan Ahmed**⑩

Abstract Ports act as a hub to the global economy. As such, port efficiency is an important factor that has to be maintained properly. A smart port system aims to increase port efficiency by integrating state-of-the-art technology with port management and predicting a ship's estimated time of arrival (ETA) is a critical step towards establishing a smart port system. This study aims to develop a data-driven model to estimate the ETA of incoming ships to Port Klang of Malaysia based on past voyage data. An artificial neural network (ANN)-based model to predict ETA has been proposed in the study. The proposed model achieves a mean absolute percentage error (MAPE) value of 36.99% with a mean absolute error (MAE) value of 4603.1367 s. The model's coefficient of determination was calculated to be 78.67% indicating a satisfactory fit to the data set.

Keywords Smart port · Estimated time of arrival · Artificial neural network · 4th Industrial Revolution

Md. R. Rahman (✉) · E. Haque · S. T. Rahman · K. Habibul Kabir
Department of Electrical and Electronic Engineering, Islamic University of Technology (IUT), Dhaka 1704, Bangladesh
e-mail: raqiburrahmana@iut-dhaka.edu

E. Haque
e-mail: ehtashamulhaqueb@iut-dhaka.edu

S. T. Rahman
e-mail: sadiatasneemc@iut-dhaka.edu

Y. A. Ahmed
Department of Naval Architecture, Ocean and Marine Engineering, University of Strathclyde, Glasgow, UK
e-mail: yaseen.ahmed@strath.ac.uke

© The Author(s), under exclusive license to Springer Nature Singapore Pte Ltd. 2023
A. Shukla et al. (eds.), *Computational Intelligence*, Lecture Notes in Electrical Engineering 968, https://doi.org/10.1007/978-981-19-7346-8_18

1 Introduction

Maritime transports are crucial to the global economy, with ships carrying out approximately 80% of the global trade logistics [1]. With the increase in container sizes and their carrying capacity, port congestion is becoming a bigger problem by the day, and it currently accounts for 93.6% of the port delays [2]. Congestion can significantly increase the ships' and the port's operation cost if not managed properly [3, 4]. Proper planning and distribution of port resources are required to prevent congestion. This study focuses on one of the factors of port planning, and the prediction of the estimated time of arrival (ETA) of ships coming to the port.

The most rudimentary form of ETA calculation, still employed by some shipping lines, is done by dividing the distance to port by the ship's speed over ground (SOG). This form of estimation does not consider external factors and is often erroneous. So, the recent literatures focusing on ETA prediction proposes methodologies based on pathfinding and machine learning (ML) algorithms trained on historic automatic identification system (AIS) data. While they do produce acceptable results, most of these data-driven methodologies are not suitable for continuous real-time predictions. Therefore, this study aims to present an accurate data-driven methodology that will be able to predict ship arrival times in real time using an artificial neural network (ANN), as shown in 1. Port planners may use the predicted ETA to generate cost-effective port schedules. Such an intricate system will assist smart ports in facing the challenges of the ongoing 4th Industrial Revolution (Fig. 1).

The next section is an overview of relevant research in this field. Section 3 expands more on AIS, Sect. 4 explains the pre-processing steps with the methodology of ETA prediction and Sect. 5 presents the results. The overall conclusion of this research is presented in Sect. 6.

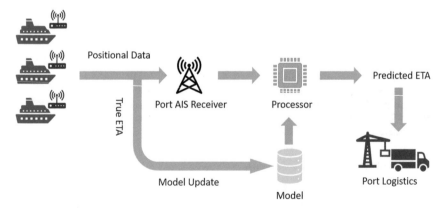

Fig. 1 Overview of the proposed system

2   Related Works

In this section, we will discuss some of the research related to ETA prediction of ships.

Pathfinding algorithms and statistical model-based solutions are common in many studies. Researchers in [5] proposed algorithms that use trajectory mining to determine the ship's route to predict the ETA from a point on the map to a port. In [6], a Markov decision process (MDP)-based reinforcement learning (RL) framework is used for optimal pathfinding and the Metropolis-Hastings algorithm to estimate the SOG. The ship's ETA was calculated by dividing the distance of the predicted trajectory by the estimated SOG. Another study proposed a pathfinder that exploits historical ship tracking data in [7]. Then, they extracted the ship's SOG from its AIS message to predict ETA. In [8], authors map the entire Mediterranean Sea in a spatial grind. A seq2seq model predicts the ship trajectory as a sequence of cells. Then, the average acceleration of ships moving from cell to cell is used to predict the arrival time.

No stand-alone ANN-based models were found in the current literature. Most of the proposed methodologies are multi-step processes. Their computation times are too high for making real-time predictions. This research aims to fill the mentioned gap in literature by proposing a stand-alone ANN model that will predict ship ETA in real time.

3   Automatic Identification System

Maritime ship tracking is done using the automatic identification system (AIS) system which provides regular updates on a ship's state to the port. It is generally encoded in specific formats which vary from authority to authority but the data attributes remain more or less the same. Some examples of encoded AIS messages are displayed in Fig. 2. For more details on AIS, see [9].

4   Methodology

The overall workflow is depicted in Fig. 3. These parts are discussed in detail in the following sections.

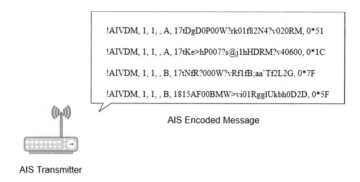

!AIVDM, 1, 1, , A, 17tDgD0P00W?rk01f82N4?v020RM, 0*51

!AIVDM, 1, 1, , A, 17tKe>hP007?s@j1hHDRM?v40600, 0*1C

!AIVDM, 1, 1, , B, 17tNfR?000W?vRf1fB;aa`Tf2L2G, 0*7F

!AIVDM, 1, 1, , B, 1815AF00BMW>vi01RgglUkbh0D2D, 0*5F

AIS Encoded Message

AIS Transmitter

Fig. 2 Examples of encoded AIS message

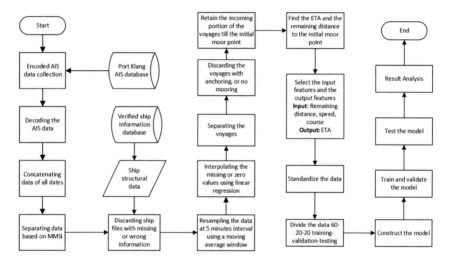

Fig. 3 Workflow for the prediction of ship arrival times

4.1 Step 1: Data Collection and Pre-processing

The data set is collected from the Port Klang authority. It consists of AIS encoded data from 1 January 2019 to 1 February 2019. The data is decoded and then separated based on the ships' Maritime Mobile Service Identities (MMSI) numbers. The ship distribution obtained from the data set is presented in Table 1.

For the pre-processing, first of all, the noisy and incomplete data points are removed. Next, the files are filtered to only retain the ship data sets with a considerable number of readings. Additional structural information related to the ships is collected from verified ship tracking websites, and based on it, improbable readings are removed. Then, the data is resampled at 5 min intervals using the moving average method. The intervals having no points are later imputed using linear interpolation

Table 1 Ship-type distribution

Type	Count
Cargo	87
Carrier	61
Container	355
Fishing	18
Passenger	11
Tanker	94
Tug	46
Other	20
Unspecified	3
Total	695

methods. The effect of different stages of pre-processing is presented in Figs. 4 and 5.

Some of the ships have made multiple voyages during the duration of the data set. These voyages are separated and considered independent. In this paper, only the

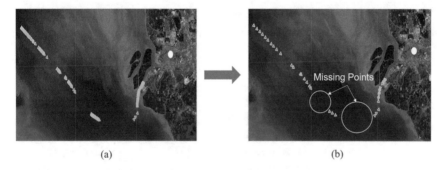

Fig. 4 a Before pre-processing, b after basic pre-processing and resampling

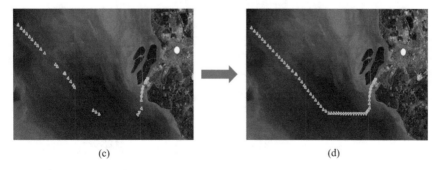

Fig. 5 c Before interpolation, d after interpolation

voyages without anchoring that reach mooring state at the port are considered as scheduling of anchoring time which is a different field of research. The final number of accepted voyages is 645 from 620 ships.

4.2 Step 2: Model Creation and Training

Three parameters are considered as input for the model—the distance remaining, the speed and the heading value at any given point—whereas the ETA is the output. All the values of training, validation and test sets are standardized. For the model, a sequential ANN is used. The layers and their varying hyperparameters are given in Table 2.

The general hyperparameters used are:

- Activation function: ReLU
- Epoch: 2187 using early stopping (patience: 100)
- Optimizer: Adam
- Loss Function: MAE
- Learning Rate: 0.001.

As the model is trained for a significant number of epochs, all intermediate layers have been regularized to avoid overfitting. The addition of an intermediate batch normalization layer is also for the same reason.

Table 2 ANN layers of the model

Layer type	No. of units	Regularizer
Fully connected	2048	–
Fully connected	1024	L2 regularizer (0.02)
Fully connected	512	L2 regularizer (0.02)
Fully connected	256	L2 regularizer (0.02)
Fully connected	128	L2 regularizer (0.02)
Batch normalization	–	–
Fully connected	128	L2 regularizer (0.02)
Fully connected	64	L2 regularizer (0.02)
Fully connected	64	L2 regularizer (0.02)
Fully connected	32	L2 regularizer (0.02)
Fully connected	16	L2 regularizer (0.02)
Fully connected	8	L2 regularizer (0.02)
Fully connected	4	L2 regularizer (0.02)
Fully connected	2	L2 regularizer (0.02)
Fully connected	1	–

4.3   Step 3: Model Evaluation

Four parameters have been used to determine the model's performance—mean absolute percentage error (MAPE), mean absolute error (MAE), root mean squared error (RMSE) and coefficient of determination (R^2) score.

5   Experimental Results and Discussions

Our model is trained on a data set containing 645 voyages from 620 ships from 1 January 2019 to 1 February 2019. On the test set, the model achieves an MAPE of 36.99%. The MAE value obtained is 4603.1367 s, and the RMSE value is 14029.6972 s. The model has an R^2 value of 78.67%, indicating that it is able to fit the data set to a satisfactory extent. The training and validation loss curves can be seen in Fig. 6. The training loss curve is closely followed by the validation loss curve, which indicates the model has not been overfitted. All the evaluation parameter values from the training, validation and test sets are given in Table 3. To further verify our model, using the same data set, the MAE and the RMSE of the general approach are also given in Table 3. The general approach calculates ETA by dividing the distance of the ship's path by its SOG.

Fig. 6 Learning loss curves

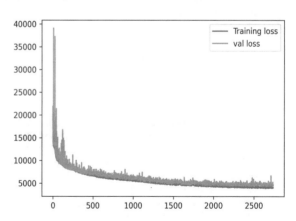

Table 3 Error metrics

	Training	Validation	Test	General
RMSE	13,014.96	13,391.1484	14,029.6972	423,443.7166
MAE	4082.1211	4606.2568	4603.1367	62,367.6629
MAPE (%)	31.56	34.613	36.99	–
R^2 (%)	–	–	78.67	–

The data given in Table 3 and the loss curves in Fig. 6 imply that our model is able to output adequate predictions without overfitting the data. This also means that the model will be able to produce reliable results with new data. Both the MAE and the RMSE of the general approach are significantly higher compared to that of our model, indicating that our model is a better system for predicting a ship's ETA. Once trained, the model's computation time is very low, making it suitable for real-time predictions.

6 Conclusion

In this paper, a data-driven ANN-based ship ETA prediction model for efficient port operation is proposed to face the challenges of 4th Industrial Revolution. The suggested model is an ANN-based model that is trained and evaluated on a real data set of AIS system communication gathered from the port authorities in Klang. The model performs admirably, with a MAPE of 36.99% and an MAE value of 4603.1367 s. The promising results and feasibility of the presented model can be utilized to create a real-time ship scheduling system for a smart port that increases port efficiency. Ports that house AIS receivers can easily implement our model to determine the ETA of incoming ships themselves, without relying on manual ETA notifications. Our future work will focus on other factors of an efficient port scheduling system, such as the unloading time of ships, the anchorage probability of a ship and the anchor time of ships.

References

1. United Nations Conference on Trade and Development. Review of maritime transport 2021. https://unctad.org/webflyer/review-maritime-transport-2021. Accessed 19 Feb 2022
2. Notteboom T (2006) The time factor in liner shipping services. Marit Econ Logist 8(1):19–39
3. Vernimmen B, Dullaert W, Engelen S (2007) Schedule unreliability in liner shipping: origins and consequences for the hinterland supply chain. Marit Econ Logist 9(3):193–213
4. Hasheminia H, Jiang C (2017) Strategic trade-off between vessel delay and schedule recovery: an empirical analysis of container liner shipping. Marit Policy Manag 44(4):458–473
5. Kwun H, Bae H (2021) Prediction of vessel arrival time using auto identification system data. Int J Innov Comput Inf Control 17(2):725–734
6. Park K, Sim S, Bae H (2021) Vessel estimated time of arrival prediction system based on a path-finding algorithm. Marit Transp Res 2:100012
7. Alessandrini A, Mazzarella F, Vespe M (2018) Estimated time of arrival using historical vessel tracking data. IEEE Trans Intell Transp Syst 20(1):7–15
8. Nguyen D, Le Van C, Ali MI (2018) Vessel destination and arrival time prediction with sequence-to-sequence models over spatial grid. In: Proceedings of the 12th ACM international conference on distributed and event-based systems. Association for Computing Machinery, Hamilton, New Zealand, pp 217–220
9. International Maritime Organization (2002) Guidelines for the onboard operational use of shipborne automatic identification systems (AIS). Resolution A.917(22)

A Critical Review on Search-Based Security Testing of Programs

Fatma Ahsan and Faisal Anwer

Abstract In the past decades, automated test case generation for detecting software vulnerabilities using the search-based algorithm has been a crucial research area for security researchers. Several proposed techniques partially achieve the automatic test case generation of detecting security issues that target specific programming languages or API types. This review paper performs a comparative analysis of research papers published during 2010–21 specific to the search-based techniques used for automatic test case generation. We have chosen five criteria for evaluating the proposed techniques. These are target vulnerability, search-based algorithms used, test case generation source, test case generation method, and target language. We focused primarily on the techniques detecting the four most dangerous vulnerabilities found in the modern distributed system, namely Denial of Service (e.g., program crash), SQL injection, Cross-Site Scripting, and XML injection. We also incorporated the fitness function comparison table for each method, including the merits and limitations. This work will assist security researchers in finding new directions in the field of search-based software security testing.

Keywords Security testing · Search-based software security testing (SBSST) · SBSE · Program crash · DoS · XMLI · SQLI · XSS

1 Introduction

Modern software systems are becoming complex and distributed in nature. These systems are usually developed in multiple programming languages and technologies to achieve scalability, security, high availability, and zero downtime. These software systems support various platforms like web browsers, mobile APPs, desktop APPs, rest API, and web services. With the increase in complexity, the system

F. Ahsan (✉) · F. Anwer
Aligarh Muslim University, Aligarh, UP 202002, India
e-mail: fahsan83@myamu.ac.in

F. Anwer
e-mail: faisalanwer.cs@amu.ac.in

© The Author(s), under exclusive license to Springer Nature Singapore Pte Ltd. 2023
A. Shukla et al. (eds.), *Computational Intelligence*, Lecture Notes in Electrical Engineering 968, https://doi.org/10.1007/978-981-19-7346-8_19

also exposes more attack surfaces for attackers or hackers with malicious intents. The attackers/hackers use the highly sophisticated mechanism to find system attack surfaces to enter or execute their malicious code.

Software testing [1] is essential for the software development life cycle. The security of software is critical to an application's success. It is necessary to think about software security from the beginning of the software development life cycle to develop secure software. Due to the difficulty of doing manual software security tests, it is preferable to employ automated solutions. The regular updating of vulnerability databases such as Common Vulnerabilities and Exposures (CVE) [2], Common Weaknesses and Enumeration (CWE) [3] necessitates the potential focus on security testing methodologies.

In his survey, McMinn [4] suggests that search-based techniques can be employed for security testing in the future. Researchers have created several tools and methods for comprehensive software security assessment based on search-based algorithms [5–29]. It is increasingly difficult for a practitioner or a new researcher to analyze and obtain an overview of the area as the number of articles in the SBSST field grows. Other techniques have also been proposed to address software testing, apart from search-based algorithms [30–32]. The purpose of this paper is to provide an overview of the current state of the art in SBSST so that practitioners can benefit from it. We have reviewed literature published during the last decade and included 25 such works in this work. We have concentrated on the four most common and dangerous vulnerabilities: Denial of Service (e.g., Program Crash) [5–13], SQL injection [14–16], Cross-Site Scripting [17–25], and XML injection [26–29]. This work targets comparative analysis based on five criteria: target vulnerability, evolutionary algorithm used, test case generation source, test case generation method, and target language. Moreover, we compared and contrasted the fitness functions and their merits and demerits.

The rest of the paper is organized as follows: Sect. 2 discusses some background information on program security, vulnerabilities, security testing, and SBSST. Section 3 compares various methods/tools published over the last decade. Section 4 discusses the work and gives future research direction, and Sect. 5 concludes the paper.

2 Background

Program Security. Program security refers to tools and techniques adopted during the design and development of software to prevent security flaws. It ensures that the software continues to function correctly under malicious attacks and guard against data and security breaches. CERT, a part of the Software Engineering Institute at Carnegie Mellon University [33], has created some security coding standards for programming languages such as C, C++, Java, Android, and Perl to guide the developers toward writing secure programs, which is the key to avoid bugs in the applications.

Vulnerability. Vulnerabilities are the faults in the system that allow attackers to exploit the system for malicious purposes. Denial of Service occurs when a software system is rendered inaccessible because of program crashes or overburdening the computing resources by attackers so that legitimate user requests cannot be processed [5–11]. SQL injection targets relational databases that record the system's data, e.g., SQL server, MySQL, Oracle, and any web application that uses these SQL databases. Once a hacker gains malicious access to the system, all database information can be inserted, modified, deleted, and updated, or the account's privileges elevated to gain control. The SQLI vulnerability has ranked third among the top ten OWASP vulnerabilities in 2021 [34]. The injection of malicious code into a web application front-end through the clever use of client-side scripting language JavaScript is known as XSS attack [17–25]. Once the attacker can execute its JavaScript code, it can gain access to the user session and cookies data stored in the browser, rewrite the web page's content, and redirect users to another malicious website for attackers gain without being noticed by the user. In a modern distributed system, XML is one of the possible choices for message communication between two systems. With the popularity of web services where two systems communicate using SOAP API, the XML is a de-facto standard to pass the message. An attacker can carefully craft a malicious message in an XML message format to achieve malicious intent. If a web API fails to validate user input strongly, XMLI vulnerability arises, altering the logic of the web services causing DoS attacks [35].

Security Testing. Security testing is performed to detect vulnerabilities before deploying to the production environment. It intends to find the security bugs so that developers can fix them before an attacker gets a chance to exploit them in the production environment. Ensuring confidentiality, integrity, availability, authentication, authorization, and non-repudiation are the essential components of security testing.

Search-Based Software Security Testing. SBSST is the subset of search-based software engineering (SBSE) [36] that uses evolutionary algorithms to solve many engineering problems. Evolutionary algorithms imitate the natural process of evolution to solve a particular problem. A fitness function guides search-based strategies, distinguishing between good and wrong solutions. Manual vulnerability detection is challenging due to software systems' size and complexity. SBSST reduces the software testing cost by helping automatic test case generation that detects vulnerability before production. It is specifically used to design new algorithmic techniques to tackle the subset of well-known security vulnerabilities [37].

3 SBSST Methods Comparison

Researchers have proposed several security testing methods using search-based techniques. The methods differ in vulnerability coverage, test case generation method, target language, fitness function, etc. This section compares different approaches that authors have proposed to detect different kinds of security vulnerabilities. We compare the SBSST methods on two different sets of criteria; one set is for general comparison based on target vulnerability, test case generation method, etc. The other set compares methods based on fitness function, limitations, and other criteria.

3.1 General Comparison Criteria of SBSST Work

We have proposed five criteria for comparing different methods of SBSST. These parameters are target vulnerability, evolutionary algorithm used, test case generation source, test case generation method, and target language. The classification is given in Table 1, with method/work in the first column and the remaining columns illustrating the comparison criteria outlined herein. We explained each of the parameters in detail and analyzes the methods.

Target Vulnerability. We have selected SBSST work based on the vulnerability it addresses. We have considered four vulnerabilities: DOS (program crash), XSS, SQLI, and XMLI. The second column of Table 1 gives vulnerability tackled by various SBSST works. In 25 SBSST work that we have considered, nine works address XSS [17–25], nine have addressed DOS (program crash) [5–13], four address XMLI [26–29], and the least addressed work is SQLI [14–16].

Evolutionary Algorithm Used. This column shows that the different types of search-best algorithms are used in SBSST work. The third column of Table 1 gives that most of the SBSST work uses genetic algorithms [5–11, 13, 14, 17–24, 26–28]. Other methods use genetic programming [15, 25], NSGA-II [12], differential evolution [16], and co-evolutionary algorithms [29].

Test Case Generation Source. The test case generation source defines from what the test cases are generated. The fourth column of Table 1 gives that test cases are derived from different sources like programming language source code [7, 13, 17–24], binary code [5, 6, 8–11], web API [14–16, 26–29], vulnerability instances [25].

Test Case Generation Method. The method defines how the test cases are generated. The fifth column of Table 1 illustrates numerous test case creation methods implemented in security testing. It is important to note that the standard software testing approaches have been used for security testing to perform test case generation. These include testability transformation [9], path constraint solver using the GA and Yices solver application using CFG [18, 19], and random testing using NSGA-II [12].

Target Language. This column shows the programming language for which the test case has to be generated to detect vulnerabilities. Many algorithms work on specific programming language source code [5–11, 13–25], and others are independent of programming language [12, 26–29]. The sixth column of Table 1 gives that few methods/tools are independent of any programming language. Out of 25 works, nine papers address DOS (program crash) [5–13], which targets three different languages: byte code [5, 6, 8–11], C [7], and language independent [12].

3.2   Fitness Function-Based Comparison of SBSST Work

In this paper, we see the majority of the technique used in automatic test case generation to detect vulnerabilities uses the genetic algorithm as a primary tool. GA relies on the fitness function to evaluate the fitness of the test cases. There are quite a few

Table 1 Summary of comparative analysis of search-based software security testing techniques

Method/work	Target vulnerability	Evolutionary algorithm used	Test case generation source	Test case generation method	Target language
Rawat and Mounier [7]	Buffer overflow	Genetic algorithm and evolutionary strategies	Source code	Generate vulnerability execution path (TDS) and generate inputs using EA to execute vulnerable paths	C
Avancini and Ceccato [17]	Reflected XSS	Taint analysis and genetic algorithm	Page source code	Test strings are generated for vulnerable paths using genetic algorithm	PHP
Cui et al. [6]	Integer overflow	IDA static analysis, genetic algorithm	Binary code	Test cases are generated using GA, taking input from static analysis	Binary language
Romano et al. [5]	Null pointer exception	Static analysis and genetic algorithm	Class files	Generating intraprocedural CFG identifies null paths and generates inputs (XSD) using GA	JAVA
Avancini and Ceccato [18]	XSS	Genetic algorithm	Page source code	Solve path constraint using GA and solver-based local search	PHP

(continued)

Table 1 (continued)

Method/work	Target vulnerability	Evolutionary algorithm used	Test case generation source	Test case generation method	Target language
Avancini [19]	XSS	Genetic algorithm	Source code	Potential candidate vulnerabilities are identified using static analysis tool; then test cases are generated using GA and solver-based local search strategy, in last, security Oracles are defined	PHP
Fraser and Arcuri [10]	Program crash	Genetic algorithm	Byte code (class file)	Test suites are generated using EVOSUITE	JAVA like byte code
Avancini and Ceccato [20]	XSS	Genetic algorithm and symbolic execution	Source code	HTTP response is used to calculate the fitness function by solving branch constraints, switching between GA and symbolic execution	PHP
BIOFUZZ [14]	SQLI	Genetic algorithm	Web pages	All the GET and POST parameters are collected using HTTP proxy that client and server exchange; SQLI attacks are generated for each of the target input parameters	SQL

(continued)

Table 1 (continued)

Method/work	Target vulnerability	Evolutionary algorithm used	Test case generation source	Test case generation method	Target language
Hydara et al. [21]	XSS	Genetic algorithm	Source code	Make control flow graphs (CFGs) using the white-box testing techniques from source code, then generate test cases for the vulnerabilities using GA	Java-based web applications
Galeotti et al. [11]	Program crash	Genetic algorithm, dynamic symbolic execution	Byte code, class file	It searches optimum solutions using genetic algorithm	JAVA
Hydara et al. [22]	XSS	Genetic algorithm	Source code	It creates CFG and solves constraints for the vulnerable path	JAVA
Fraser and Arcuri [9]	Program crash	Genetic algorithm	Byte code, class file	Testability transformation adds more lines in the source code to generate test cases	JAVA
Ahmed and Ali [23]	XSS	Taint analysis and genetic algorithm	Source code	Generating test string using GA	PHP, script
Aziz et al. [15]	SQLI	Genetic programming	Web API	Evolutionary computation system (ECJ) is used, search space is defined, and strings are searched in the search space using fitness functions	SQL

(continued)

Table 1 (continued)

Method/work	Target vulnerability	Evolutionary algorithm used	Test case generation source	Test case generation method	Target language
Jan et al. [26]	XMLI	Real-coded genetic algorithm (RGA)	Web API	First, it creates TOs (malicious XML messages) then aromatic generation of SUT input that generates TOs	XML
Wasef et al. [24]	XSS	Genetic algorithm	Source code	CFG is generated using Tant analysis; test strings are generated using genetic algorithm	PHP
Jan et al. [27]	XMLI	Genetic algorithm	Web API	Identification of test objectives and then generating inputs to attain those test objectives	Any language that manipulates XML
JCOMIX [28]	XMLI	Genetic algorithm	Web API	Generates malicious XML messages (TOs) using an attack grammar and then generates inputs that lead the system to generate malicious XML messages (TOs)	Any language that manipulates XML
Liu et al. [16]	SQLI	Differential evolution (DE)	Web page code	Inputs (SQL statements) are generated and compared with TO and then improved using DE	HTML, JavaScript, any other web page language

(continued)

Table 1 (continued)

Method/work	Target vulnerability	Evolutionary algorithm used	Test case generation source	Test case generation method	Target language
Jan et al. [29]	XMLI	Co-evolutionary algorithm	Web application	Uses linear complexity fitness function	Any language that manipulates XML
Anwer et al. [8]	Program crash in	Genetic algorithm	Byte code	The framework creates dummy environment that can generate pattern of exceptions to represent external resource abnormality	JAVA
Mao et al. [12]	Program crash	NSGA-II	Black-box testing	Test case is randomly generated and then improved using MOEA and fitness function	Language independent
Alyasiri [25]	XSS	Genetic programming	XSS in stances	Supervised learning approach based on the evolutionary computing	JavaScript
Iannone et al. [13]	Exploit of third-party vulnerabilities	Genetic algorithm	Source code and vulnerability description	Application code is instrumented, and vulnerability description is used to set goal and create fitness function then Evosuite is used to generate test cases for vulnerabilities	JAVA

cases where multiple fitness functions are used to assess and compared to find the best result [28]. Based on the results and efficiency of the generated test cases, the best fitness function has been opted. GA has a flaw in that it can get stuck in local optima, but it gives results fast [18]. Various methods have been used in conjunction with GA, such as dynamic symbolic execution to get out of the local optima and achieves better results [11, 18]. The results of the improved techniques when combined GA with other tools (hybrid algorithm) are outstanding.

We have proposed four criteria for comparing different methods in SBSST under fitness function evaluation. These parameters are evolutionary algorithm input parameter, fitness function, merits, and limitations. The classification is given in Table 2, with method/work in the first column and the remaining columns illustrating the comparison criteria outlined herein.

Evolutionary Algorithm Input Parameter. The second column of Table 2 gives what types of inputs are given to the fitness function to achieve the desired result of SBSST. In XSS, the parameter for EA is mostly CFG [17, 20–22, 24].

Fitness Function. The third column of Table 2 defines the different types of fitness functions used for the SBSST work. More number of works uses approach level fitness function [5, 17–20]. Other works have used linear distance fitness function [28, 29], branch distance [9–11].

Merits. In column fourth of Table 2, we compared the merits of the various techniques of SBSST taken in this review work.

Limitations. In the fifth column of Table 2, we have compared the limitations of the various SBSST.

4   Discussion and Research Direction

Testing an application for vulnerabilities is necessary to avoid exploiting and losing sensitive data after the application is live. Several studies have discussed application testing against four significant vulnerabilities, namely DoS [5–13], SQLI [14–16], XSS [17–25], and XMLI [26–29]. In his review, Afzal [38] included seven primary studies of search-based security testing and categorized them into one of the non-functional search-based software testing categories. We have compared the previous research with this study; our analysis reveals that some new evolutionary algorithms like NSGA-II, Differential evolution, and real-coded GA have been explored to generate test cases for vulnerabilities. Moreover, the evolutionary algorithm has also been combined with dynamic symbolic execution to develop more efficient test cases, and some advanced fitness functions have been explored in new primary studies. We have identified five criteria to conduct a comparative analysis of 25 such works in this review paper. The criteria are target vulnerability, evolutionary algorithm used, test case generation source, test case generation method, and target language.

Table 2 Summary of fitness function evaluation, merits, and limitations

Method/work	Evolutionary algorithm input parameter	Fitness function	Merits	Limitations
Rawat and Mounier [7]	Taint dependency sequences (TDS)	Fitness value for inputs is computed based on runtime dynamics $$F_i = \sum_{j=1}^{k} w_j * f_{ij}$$	It relies on dynamic string input generation to detect BoF attacks without knowing the constraint beforehand using GA. It performs better than random fuzzing	This approach takes more iteration as initially constraints are not specified on inputs
Avancini and Ceccato [17]	Control flow graph	Approach level (when it traverses 100% of the required branches)	It relies on static analysis using GA to find cross-site scripting vulnerability early in development cycles	Complex constraints input values branches are not reachable. They only consider the reflected XSS. They target only one path at a time
Cui et al. [6]	Control flow structure and basic blocks, danger function list	Multi-objective fitness function $$= w1 * bbcx + w2 * l\ covx + w3 * r\ Index * \log(Dx)$$	The approach identifies exceptions in the object program with more targeted test data	GA sometimes gives local optimum
Romano et al. [5]	NULL paths	Fitness function is based on approaching level and branch distance	It generates test input for programs with complex data structures because of tress-based input representation	Most of the bugs were artificially introduced in the system, and they might not represent the real vulnerability
Avancini and Ceccato [18]	Symbolic values and path constraints	Approach level	This approach generates test case for true positive and no test cases will be provided for false negative	This approach covers only path and does not provide actual attack, i.e., injecting html code in the web pages

(continued)

Table 2 (continued)

Method/work	Evolutionary algorithm input parameter	Fitness function	Merits	Limitations
Avancini [19]	Target branches	Approach level fitness function	The approach implements security Oracles which identifies actual attacks in the application under test	This work does not consider databases source domain, and it fails to generate test cases for those potential vulnerabilities that require databases inputs to be modified
Fraser and Arcuri [10]		Branch distance, $\text{fitness}(T) = |M| - |M_T| + \Sigma_{b_k \in B^d}(b_k, T)$	This method is also applicable for other languages that compiles to Java Bye-code (e.g., Groovy, Scala). Produce small test suits with high coverage. The effectiveness of the approach is not affected by varying number of targets	In this method, container classes are need to be identified manually, Evosuite covers only those codes that has no environmental dependencies
Avancini and Ceccato [20]	Control flow graph	Approach level fitness function	Concrete symbolic execution is combined for more vulnerability coverage	This is language-dependent approach
BIOFUZZ [14]		$s1 \times \text{NC}(I) + s2 \times \text{CC}(I) + s3 \times \text{IP}(I) + s4 \times \text{DD}(I)$	This is the first-time evolutionary testing has been used to detect SQL injections	Its effectiveness depends on the quality of the model produced by the crawler BIOFUZZ inherit all limitations of the underlying web crawler

(continued)

Table 2 (continued)

Method/work	Evolutionary algorithm input parameter	Fitness function	Merits	Limitations
Hydara et al. [21]		fitness function $= ((Cpaths\% + Diff) * XSSp\%)/100$	It uses better and improved GA operators to help in the detection and removal of XSS vulnerabilities and including all the three types of XSS	This does not remove infeasible paths from CFG
Galeotti et al. [11]	Control structure	Branch distance calculation	It is a fully automated tool. It integrates search algorithms to guide unit test generation to achieve high coverage	More work is required to achieve complete code coverage and expand its other programming language
Hydara et al. [22]	Context flow graph (CFG) of the java source files	fitness function $= ((Cpaths\% + Diff) * XSSp\%)/100$	It uses better and improved GA operators to help in the detection of XSS vulnerabilities as well as including all three types of XSS	This approach does not remove infeasible paths from the CFG
Fraser and Arcuri [9]	Program crash or undeclared exceptions	Evosuite fitness function [10]	In this work, it has been shown that SBSST can exercise automated Oracles and generate high coverage test suites. Testability transformation helps to trigger more failure	The transformations offer core guidance. The fitness function does not offer guidance toward making the reference null

(continued)

Table 2 (continued)

Method/work	Evolutionary algorithm input parameter	Fitness function	Merits	Limitations
Ahmed and Ali [23]	Attack patterns	Korel's distance function	This approach covers multiple paths simultaneously with as many patterns of XSS as possible; this approach has considered all three types of XSS vulnerabilities. Real XSS patterns are used	Certain paths in CFG do not perform at all; these paths are called infeasible paths. They have not been removed from the CFG
Aziz et al. [15]	Legitimate inputs for each of the injectable parameters	Koza fitness, fitness function $= \sum 1/(1+S)$	Proposed GP grammar corresponding to SQLIAs, successfully generates test cases for the known SQLI vulnerability for the subject application, which can be extended further to detect more future vulnerabilities	Generated test case inputs resulted in a high volume of syntax errors, which shows false positives and requires future work to be more effective
Jan et al. [26]	Test objectives (TOs)	Real-coded edit distance, $$d_R(A_n, B_m) = \begin{cases} d_R(A_{n-1}, B_m) + 1 \\ d_R(A_n, B_{m-1}) + 1 \\ d_R(A_{n-1}, B_{m-1}) + \frac{|a_n - b_m|}{1+|a_n - b_m|} \end{cases}$$	This approach detects more XMLI vulnerabilities in less time compared to other methods	This is a single target approach

(continued)

Table 2 (continued)

Method/work	Evolutionary algorithm input parameter	Fitness function	Merits	Limitations
Wasef et al. [24]	Control flow graph	FF by Moataz and Fakhreldin, $F(x) = ((\text{Miss}\% + D) *\text{Importance} * \text{DB}\%)/100$	This approach removes infeasible paths from the CFG to obtain better results	If the malicious inputs could not traverse the target paths, the results will consider these paths as safe
Jan et al. [27]	Test objectives	Edit distance between the TO and the XML message	It does not need to access the source code	This technique focuses on a single message at a time. It is a single target approach
JCOMIX [28]	Config file, proxy file, generated test objectives	Linear distance, $d(A, B) = \|n - m\| + \sum_{i=1}^{\min\{m,n\}} \frac{\|a_i - b_i\|}{\|a_i - b_i\| + 1}$	This approach targets all TOs at once and overcomes the limitations of budget allocation strategies	It does not support JSON data format
Liu et al. [16]	Test objectives, i.e., targeted SQL statements	Similarity matching distance (SMD), $d_{\text{SMD}}(S_l, \hat{s}_{l'}) = (1 - L \times p) \times d(S_l, \hat{s}_{l'})$	Differential evolutionary algorithm is used for test cases generation, which has not been used before	The semantic aspect of the SQL statements has not been considered
Jan et al. [29]	Test objectives	Linear distance, $d(A, B) = \|n - m\| + \sum_{i=1}^{\min\{m,n\}} \frac{\|a_i - b_i\|}{\|a_i - b_i\| + 1}$	This approach is multi-target, i.e., it handles more than one target at a time	More data exchange formats and vulnerabilities need to be covered
Anwer et al. [8]	PUT	Evosuite fitness function [9]	It detects vulnerabilities caused by exceptions due to the abnormality of external resources	The proposed framework has not been tested on real world applications

(continued)

Table 2 (continued)

Method/work	Evolutionary algorithm input parameter	Fitness function	Merits	Limitations
Mao et al. [12]	Random test case	$\text{Fitness}(x, E) = \min_{j=1}^{|E|} \text{dist}(x, tc_j)$	It uses NSGA-II for dispersed test case generation. This shows improved results than FSCS-ART, eART, RT	Diversification objectives could have been modified for improved results
Alyasiri [25]	Training dataset	fitness $= 1 - $ MCC	This method has used machine learning techniques to detect XSS vulnerability. The used GP effectively detects XSS and reduces the computational cost associated with the detection rules	Other types of search-based algorithm are not compared for the test result
Iannone et al. [13]	SUT	Fitness function [13]	This method is used to detect vulnerabilities in	It only deals with library vulnerabilities characterized by a single vulnerable source code line

Moreover, we have analyzed the methods for the fitness function. The criteria defined are evolutionary algorithm input parameter, fitness function, merits, and limitations.

- We have found that several recent studies have been done related to open-source libraries vulnerability.
- Out of all methods that we covered, most of the works are based on XSS and DoS vulnerability testing.
- Out of all papers, we found one method is programming language independent.
- Many techniques take input source code for test case generation, and few use byte code for test case generation.
- There are many different types of fitness functions that have been used by the author, among which branch distance, linear distance, and approach level are most common.
- In the last five years, more advanced fitness functions have been used like, similarity matching index (SMD) [16], real-coded edit distance [26], edit distance between the test objectives (TOs) and target message [27], Korel's distance function [23], and Koza fitness function [15].

5 Conclusion

This paper performed search-based software security testing (SBSST) research work during the last decades (2010–2021). We selected 25 papers based on their significance and effectiveness in targeting vulnerabilities and compared the algorithmic techniques used and achieved results. We found that none of the techniques discussed in these papers [5–29] proposes generic methods to address a particular vulnerability that applies to different programming languages. It mainly discussed improvements on specific parameters for a specific security vulnerability particular to a programming language or API. The majority of these papers propose improvements over previously proposed work for white-box and a few for black-box automated test case generation. One crucial research direction would be to find generic SBSST techniques independent of a programming language that detects many different security vulnerabilities in most of today's distributed systems.

We hope this work will provide one-stop holistic and comparative views of past decade works to help security researchers and practitioners choose the right approach for their future work.

References

1. Anwer F, Nazir M, Mustafa K (2017) Security testing. In: Trends in software testing, pp 35–66
2. CVE—common vulnerabilities and exposure. https://cve.mitre.org/. Accessed 30 Dec 2021
3. CWE—common weakness enumeration. https://cwe.mitre.org/. Accessed 30 Dec 2021
4. McMinn P (2004) Search-based software test data generation: a survey. Softw Test Verif Reliab 14(2):105–156

5. Romano D, Di Penta M, Antoniol G (2011) An approach for search based testing of null pointer exceptions. In: 2011 fourth IEEE international conference on software testing, verification and validation. IEEE, pp 160–169
6. Cui B, Liang X, Wang J (2011) The study on integer overflow vulnerability detection in binary executables based upon genetic algorithm. In: Foundations of intelligent systems. Springer, pp 259–266
7. Rawat S, Mounier L (2010) An evolutionary computing approach for hunting buffer overflow vulnerabilities: a case of aiming in dim light. In: 2010 European conference on computer network defense. IEEE, pp 37–45
8. Anwer F, Nazir M, Mustafa K (2019) Testing program crash based on search based testing and exception injection. In: International conference on security & privacy. Springer, pp 275–285
9. Fraser G, Arcuri A (2015) 1600 faults in 100 projects: automatically finding faults while achieving high coverage with Evosuite. Empir Softw Eng 20(3):611–639
10. Fraser G, Arcuri A (2012) Whole test suite generation. IEEE Trans Softw Eng 39(2):276–291
11. Galeotti JP, Fraser G, Arcuri A (2014) Extending a search-based test generator with adaptive dynamic symbolic execution. In: Proceedings of the 2014 international symposium on software testing and analysis, pp 421–424
12. Mao C, Wen L, Chen TY (2020) Adaptive random test case generation based on multi-objective evolutionary search. In: 2020 IEEE 19th international conference on trust, security and privacy in computing and communications (TrustCom). IEEE, pp 46–53
13. Iannone E, Di Nucci D, Sabetta A, De Lucia A (2021) Toward automated exploit generation for known vulnerabilities in open-source libraries. In: 2021 IEEE/ACM 29th international conference on program comprehension (ICPC). IEEE, pp 396–400
14. Thomé J, Gorla A, Zeller A (2014) Search-based security testing of web applications. In: Proceedings of the 7th international workshop on search-based software testing, pp 5–14
15. Aziz B, Bader M, Hippolyte C (2016) Search-based SQL injection attacks testing using genetic programming. In: European conference on genetic programming. Springer, pp 183–198
16. Liu M, Li K, Chen T (2019) Security testing of web applications: a search-based approach for detecting SQL injection vulnerabilities. In: Proceedings of the genetic and evolutionary computation conference companion, pp 417–418
17. Avancini A, Ceccato M (2010) Towards security testing with taint analysis and genetic algorithms. In: Proceedings of the 2010 ICSE workshop on software engineering for secure systems, pp 65–71
18. Avancini A, Ceccato M (2011) Security testing of web applications: a search-based approach for cross-site scripting vulnerabilities. In: 2011 IEEE 11th international working conference on source code analysis and manipulation. IEEE, pp 85–94
19. Avancini A (2012) Security testing of web applications: a research plan. In: 2012 34th international conference on software engineering (ICSE). IEEE, pp 1491–1494
20. Avancini A, Ceccato M (2013) Comparison and integration of genetic algorithms and dynamic symbolic execution for security testing of cross-site scripting vulnerabilities. Inf Softw Technol 55(12):2209–2222
21. Hydara I, Sultan ABM, Zulzalil H, Admodisastro N (2014) An approach for cross-site scripting detection and removal based on genetic algorithms. In: The ninth international conference on software engineering advances ICSEA
22. Hydara I, Sultan ABM, Zulzalil H, Admodisastro N (2015) Cross-site scripting detection based on an enhanced genetic algorithm. Indian J Sci Technol 8(30):1–7
23. Ahmed MA, Ali F (2016) Multiple-path testing for cross site scripting using genetic algorithms. J Syst Architect 64:50–62
24. Marashdih AW, Zaaba ZF, Omer HK (2017) Web security: detection of cross site scripting in PHP web application using genetic algorithm. Int J Adv Comput Sci Appl (IJACSA) 8(5)
25. Alyasiri H (2020) Evolving rules for detecting cross-site scripting attacks using genetic programming. In: International conference on advances in cyber security. Springer, pp 642–656
26. Jan S, Panichella A, Arcuri A, Briand L (2017) Automatic generation of tests to exploit xml injection vulnerabilities in web applications. IEEE Trans Softw Eng 45(4):335–362

27. Jan S, Nguyen CD, Arcuri A, Briand L (2017) A search-based testing approach for XML injection vulnerabilities in web applications. In: 2017 IEEE international conference on software testing, verification and validation (ICST). IEEE, pp 356–366

28. Stallenberg DM, Panichella A (2019) JCOMIX: a search-based tool to detect xml injection vulnerabilities in web applications. In: Proceedings of the 2019 27th ACM joint meeting on European software engineering conference and symposium on the foundations of software engineering, pp 1090–1094

29. Jan S, Panichella A, Arcuri A, Briand L (2019) Search-based multi-vulnerability testing of XML injections in web applications. Empir Softw Eng 24(6):3696–3729

30. Anwer F, Nazir M, Mustafa K (2016) Testing program for security using symbolic execution and exception injection. Indian J Sci Technol 9:19

31. Anwer F, Nazir M, Mustafa K (2014) Automatic testing of inconsistency caused by improper error handling: a safety and security perspective. In: Proceedings of the 2014 international conference on information and communication technology for competitive strategies, pp 1–5

32. Anwer F, Nazir M, Mustafa K (2013) Safety and security framework for exception handling in concurrent programming. In: 2013 third international conference on advances in computing and communications. IEEE, pp 308–311

33. SEI CERT coding standards—CERT secure coding—confluence. https://bit.ly/3FG7ota. Accessed 30 Dec 2021

34. OWASP top 10:2021. https://owasp.org/Top10/. Accessed 30 Dec 2021

35. Gupta C, Singh RK, Mohapatra AK (2020) A survey and classification of XML based attacks on web applications. Inf Secur J Glob Perspect 29(4):183–198

36. Jones BF, Yang HSX, Eyres D (1970) The automatic generation of software test data sets using adaptive search techniques. WIT Trans Inf Commun Technol 14

37. Harman M, Mansouri SA, Zhang Y (2012) Search-based software engineering: trends, techniques and applications. ACM Comput Surv (CSUR) 45(1):1–61

38. Afzal W, Torkar R, Feldt R (2009) A systematic review of search-based testing for non-functional system properties. Inf Softw Technol 51(6):957–976

Feature Selection Methods for IoT Intrusion Detection System: Comparative Study

Richa Singh and R. L. Ujjwal

Abstract With the advancement in the Internet of Things (IoT), more applications and services are deployed in information system and physical space. A huge volume of data is gathered from these devices through sensors embedded in the IoT environment. This massive volume of data can be exploited by malicious users and is vulnerable to various types of attacks. Therefore, various intrusion detection systems (IDS) for the IoT infrastructure have been developed to ensure the security of these devices. For improving the performance of IDS, feature selection (FS) methods play a crucial role. Feature selection methods reduce dimension of data and help machine learning (ML) classifiers in achieving better accuracy. In this paper, performance of four FS methods is compared for IoT-based IDS. The four bio-inspired FS methods used are whale optimization (WO), salp swarm algorithm (SSA), Harris hawk optimization (HHO), and gray wolf optimization (GWO). The classification of intrusive traffic is performed by two ML classifiers: K-nearest neighbor (KNN), and Naïve Bayes (NB). Experimental results show that GWO with KNN classifier outperforms other FS methods with the highest accuracy, precision, recall, *F*-score, and takes less execution time. The BoT-IoT dataset is used for system performance evaluation.

Keywords Internet of Things · BoT-IoT dataset · Feature selection · Intrusion detection system · Bio-inspired learning

1 Introduction

The IoT defines the things or objects embedded with the processing, sensing, and communicating ability that will allow them to communicate with each other and with other devices over the network [1]. The IoT systems are getting popularity in various

R. Singh (✉) · R. L. Ujjwal
University School of Information Communication and Technology, Guru Gobind Singh Indraprastha University, Delhi, India
e-mail: richa.singh081991@gmail.com

R. L. Ujjwal
e-mail: ujjwal@ipu.ac.in

fields including health care, agriculture, industries, smart homes, etc. According to Statista [2], the number of IoT smart devices will reach up to approximately 30.9 billion units by 2025. With the exponential growth of these smart devices, security risks involved with them are also constantly on the upsurge. Therefore, IoT system security is the concern area to protect the IoT infrastructure from malicious or unwanted activities. Due to insufficient built-in security features like other devices, IoT devices serve as an effortless target for intrusive activities. An IDS provides a security mechanism responsible for monitoring system activities and reports about any malicious or unauthorized activity. The IDS needs to handle a massive amount of data that contain both redundant as well as non-redundant features, which badly affects the performance of IDS. Therefore, there is a need for an efficient feature selection (FS) method which eliminates redundant features and improves the system's accuracy. FS [3] method selects features subset which will further be used for the classification task. FS methods are classified as filter FS, wrapper FS, and hybrid FS. In filter FS, features selected for classification are based on data characteristics such as correlation, distance, consistency, and information measure. In wrapper FS, learning algorithms are required for evaluating subsets of selected features, while hybrid FS combines the filter and wrapper techniques to provide better accuracy.

Contribution of this paper is summarized as follows:

- The bio-inspired algorithms such as whale optimization (WO), salp swarm algorithm (SSA), Harris hawk optimization (HHO), and gray wolf optimization (GWO) are used for FS, and their performance is compared.
- The BoT-IoT dataset is used for performance evaluation.
- Classification is performed using KNN and NB classifiers.

The remaining paper is organized as: literature review of several bio-inspired methods used for IDS is discussed in Sect. 2. In Sect. 3, proposed model is explained briefly, with the brief introduction of each FS method used. In Sect. 4, the results of each FS method are compared, and concluded in Sect. 5.

2 Literature Review

The IDS generally employs ML algorithms for classifying anomalies. The FS is an important part of IDS that employs an optimization process by reducing the dimension of the dataset. The optimum selected feature subset improves the performance of the classification algorithm, reduces the computational overhead, overfitting issues, and training time [4]. Metaheuristic bio-inspired methods are widely used for selecting optimal feature subset. Therefore, this section discusses the state-of-the-art wrapper-based FS methods based on metaheuristic algorithms for the IoT network.

The authors in [5] show the comparative study of various filter FS methods including information gain, Chi-square, and recursive feature elimination (RFE).

Davahli et al. [6] proposed a lightweight IDS for the IoT system. For selecting important features, the proposed system hybridizes genetic algorithm (GA) and GWO. The intrusive traffic is classified using the SVM classifier. The beneficial characteristics of both GA and GWO are combined, which simultaneously resolves the GWO premature convergence, and also improves the convergence of GA. Entire system is divided into three stages: data preparation, feature selection, and intrusion detection. The system performance is evaluated against the AWID dataset. The result shows that the system archives acceptable accuracy with low cost. However, they incurred high running time.

The authors in [7] used augmented WO algorithm (WOA) for selecting the optimal feature subset. Earlier, the WOA works for the continuous data. The authors modify the original WOA which deals with the binary problems. The premature convergence of WOA is avoided by the Elitist tournament method, and performance is evaluated using the N-BaIoT dataset. Furthermore, Alharbi et al. [8] use a modified bat algorithm (BA) for FS and hyperparameter tuning. The intrusive traffic is classified using deep artificial neural network (DNN). They used Gaussian distribution method for initializing bats population and bats velocity is also modified. The N-BaIoT dataset is used for system performance evaluation. Authors in [9] integrate HHO and harmony search (HS) for reducing the data dimensions. The intrusions for the IoT network are detected using a deep reinforcement learning classifier. The feature selection in [10] is performed by hybridizing SSA and ant lion optimization algorithm (ALO). To determine the intrusive traffic, KNN classifier is used. Local search is done using SSA, and global search is by using ALO. They used N-BaIoT dataset for performance evaluation. The algorithm reduced features by 90–96%. The authors in work [11] use particle swarm optimization (PSO) for FS of the NSL-KDD dataset. Random forest classifier is used for detecting intrusions. Another work in [12] uses GWO for selecting feature subset and hyperparameter tuning. They used one-class SVM as a classifier for identifying intrusive traffic. The performance is evaluated using the N-BaIoT dataset.

3   Proposed Model

The proposed model aims to determine an effective FS method which achieves better accuracy with minimal features for the IoT-based IDS. Proposed model architecture is shown in Fig. 1.

System performs the binary and multiclass classification. The entire system consists of the following stages including, data preprocessing, feature selection, classification, and followed by result evaluation. The following sections define the proposed model in detail.

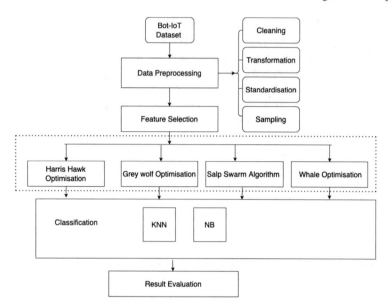

Fig. 1 Proposed model architecture

3.1 Dataset

The dataset is important for evaluating IDS performance. Here, the BoT-IoT dataset is used for system evaluation. This dataset was first introduced in 2019 by Koroniotis et al. [13]. It contains more than 72,000,000 records in 74 .csv files. We consider 5% of entire dataset for performance evaluation of FS methods, which contains 4.csv files. The dataset includes normal and intrusive traffic. The intrusive traffic contains four major categories: DoS, reconnaissance, DDoS, and data exfiltration. Dataset has 46 features with more than 99% of intrusive traffic instances and less than 1% normal traffic instances. Table 1 gives the instances of the BoT-IoT dataset used for system evaluation.

Table 1 Instances in BoT-IoT dataset

Category	Instances
DDoS	1,926,623
DoS	1,650,258
Reconnaissance	91,080
Information theft	78
Normal	475

3.2 Data Preprocessing

For providing effective data for the FS task, the BoT-IoT dataset has to be passed through several preprocessing tasks. Those task are identified as:

Data Cleaning: Data cleaning is an important task of preprocessing. This task enhances the data quality by eliminating redundant and unwanted attributes. The unwanted attributes are those values whose existence does not affect the detection process, while the redundant attributes are those values whose value is derived from other attributes. As the BoT-IoT dataset has 46 attributes but not all features are required for detecting intrusions. Therefore, features such as 'ltime,' 'daddr,' 'saddr,' 'sport,' 'dport,' and 'seq' are eliminated.

Transformation: During this phase, categorical features are transformed into numeric features. Many classifiers accept only numeric features so it is important to transform categorical to numeric one. Here, label encoder is used for transformation.

Standardization: The numerical data in the BoT-IoT dataset is in various ranges. These data values affect the working of the classifier adversely. Therefore, in order to compensate this, feature values are standardized so that the data values lie within a certain range. Features such as 'mean,' 'sum,' 'min,' 'max,' and 'rate' are standardized using the standard scalar technique.

Sampling: As the dataset is highly imbalanced, number of intrusive traffic is greater than the number of normal traffic. That is, intrusive traffic becomes the majority class, whereas normal traffic becomes the minority class such data is imbalanced in nature. This difference molds classifiers in favor toward the majority class which reduces the possibility of detecting normal traffic. Therefore, random sampling is performed to handle imbalanced data.

3.3 Feature Selection

After applying data preprocessing over the BoT-IoT dataset, feature selection is performed. The selection of an optimal feature subset is an important step of IDS implementation. The FS is performed using four metaheuristic bio-inspired optimization algorithms which include: SSA, GWO, WO, and HHO. The metaheuristic algorithms work on the concept of exploration and exploitation. The performance FS methods are compared in later sections. Therefore, this section provides an overview of each of them:

Harris Hawk Optimization: This algorithm is proposed by the authors [14] in 2019. It is motivated by the mutual hunting behavior of Harris hawk's birds, often known as 'surprise pounce.' Hunting activity takes time from a few seconds to hours, depending on the escaping pattern of prey. During exploration, Harris hawks can perform searching and tracking prey. They attentively keep an eye on prey. If they find difficulty in identifying prey, then they wait and monitor the site for observing prey. During exploitation, Harris hawk attacks on prey identified in the previous phase. However, if prey attempts to escape, then hawks follow the chasing strategy. In this algorithm, both phases are simultaneously performed. As in each iteration, some hawks perform exploration while some perform local exploitation and some others perform global exploitation. HHO is flexible and easy to implement. However, HHO is not an optimal algorithm for all optimization problems [15].

Gray Wolf Optimization: This optimization algorithm is proposed by the authors in [16]. This bio-inspired algorithm is motivated by the social and hunting behavior of gray wolves (search agents). They prefer living in a group. The average size of each group may range between 5 and 12. The gray wolves are classified into four categories based on the fitness value namely; alpha (α), beta (β), delta (δ), and omega (ω). The α wolves are the decision-makers and make all decisions regarding hunting, wake-up time, sleep time, and so on. They dominate all the other wolves in hierarchy. The β wolves assist α for making decisions. The δ wolves have the least ranking. They are the subordinate of β. Scouts, elders, hunters, caretakers, and sentinels belong to this category. However, they dominate the ω wolf. The ω is at the hierarchy's lowest level and acts as scapegoat. They are the last one allowed to eat. The algorithm includes the following steps: prey search, encircling prey, hunting, and attacking. GWO is feasible [17]. They need a few parameter adjustments. However, they have mathematical overhead and take more computation time. GWO easily falls into local optima and can face premature convergence.

Salp Swarm Algorithm: This algorithm was proposed by the authors [18] in 2017. This algorithm is motivated by flocking behaviors of salps while forging and navigating in oceans. Salps belong to the Salpidae family, they are alike jellyfishes in movement and texture. They form a swarm in the ocean which is known as the salp chain. Their cause of salp swarm behavior is not obvious yet, however, some researchers consider that this might be due to movement enhancement for food-seeking [19]. The salps are divided as a leader and followers. Leader is the salp in the front of the chain, while followers are remaining salps in the chain. SSA is flexible and efficient in providing global optima [20]. Furthermore, this algorithm rarely suffers from local optima issues. However, this algorithm might lead to premature convergence.

Whale Optimization: This optimization algorithm was proposed by the authors [21] in 2016. This algorithm is motivated by the foraging nature of humpback whales. They prefer to hunt small fish close to the water surface. This hunting is performed by creating distinctive bubbles along a circle, by swimming around a prey or '9'-shaped path. The WOA has exploitation and exploration as main phases. During the exploitation phase, prey encircling and attacking are performed. During the exploitation phase, prey searching is performed. The WOA does not fall into local optima. However, this algorithm does not have a good convergence speed.

3.4   Classification

Features selected from the FS phase are provided as an input to the classification phase. The classifiers classify the intrusive traffic from the incoming data.

K-Nearest Neighbor (KNN): This learning method is supervised ML algorithm. It is used for both regression and classification task. This classifier works based on the concept of similarity. The unlabeled data points are assigned based on the previously labeled data points. The performance of this classifier relies on the value of k. This classifier is easy to implement and work with multi-label classes. However, the performance of this classifier varies with data size and is sensitive to noise.

Naïve Bayes: This classifier relies on Bayes theorem. The NB classifier makes an assumption that a particular event occurrence is independent of other event occurrences. This classifier works efficiently with less training data. However, if data size grows, then classifier performance is affected adversely.

4   Experimentation

The experiment is done on a mac-OS Catalina with 8 GB RAM, and Intel Core i5 processor. Python 3.9 is used as implementing software. This section provides information about performance metrics used for system evaluation, and finally, the comparative study result is discussed.

Performance Metrics: The FS methods are compared based on the following metrics: number of features selected, execution time, accuracy, recall, F-score, and precision. Few formulas are given in Table 2.

Where, true positive (TP) indicates total attack instances labeled as malicious, false positive (FP) indicates total normal instances misclassified as malicious, true negative (TN) indicates total normal instances labeled as normal, and false negative (FN) indicates total attack samples misclassified as normal.

Performance of the four recent FS methods including SSA, GWO, WO, and HHO is compared using two classifiers: KNN and NB. The FS methods are provided with the same running environment. Tables 3 and 4 give the multiclass and binary classification using KNN. The performance of each FS method is compared. The result

Table 2 Metric formula

Metric	Formula
Accuracy	$\frac{TP+TN}{TP+TN+FP+FN}$
Precision	$\frac{TP}{TP+FP}$
Recall	$\frac{TP}{TP+FN}$
F-score	$\frac{Recall*Precision}{Recall+Precision} * 2$

shows that GWO achieves highest accuracy of 97.95% for multiclass and 99.98% for binary classification. Furthermore, the total features selected by GWO are lowest among all other FS methods, i.e., 8. Moreover, GWO takes less execution time of 73.29 s. for multiclass classification. However, WO has better execution time compared to GWO for binary classification. Both WO and GWO select an equal number of features, that is, 3 for binary classification as given in Table 4. GWO and SSA have the highest precision, recall, and F-score of 0.98 with KNN for multiclass classification. While for binary classification each FS method provides precision, recall, and F-score of 1.0 with KNN. Tables 5 and 6 give the comparative study of FS methods using NB classifier for multiclass, and binary classification, respectively. The result shows that WO achieves highest accuracy of 72.45% for multiclass classification. However, execution time of WO FS method is more compared to other methods. The GWO provides a best execution time of 67.83 s with accuracy of 61.25%. For the binary classification, NB classifier with HHO feature selection method achieves highest accuracy of 98.51%, but takes more execution time of 264.38 s. compared to other FS methods. NB classifier with GWO takes less execution time of 160.08 s. compared to other methods with an accuracy of 88.7%.

Table 3 Multiclass classification using KNN

FS method	Accuracy (%)	Precision	Recall	F-score	Feature selected	Time (s)
SSA	97.55	0.98	0.98	0.98	15	109.49
GWO	97.95	0.98	0.98	0.98	8	73.29
WO	96.1	0.96	0.96	0.96	15	159.45
HHO	97.25	0.97	0.97	0.97	10	117.91

Table 4 Binary classification using KNN

FS method	Accuracy (%)	Precision	Recall	F-score	Feature selected	Time (s)
SSA	99.93	1.0	1.0	1.0	6	227.14
GWO	99.98	1.0	1.0	1.0	3	186.51
WO	99.83	1.0	1.0	1.0	3	170.63
HHO	99.91	1.0	1.0	1.0	5	310.67

Table 5 Multiclass classification using NB

FS method	Accuracy (%)	Precision	Recall	F-score	Feature selected	Time (s)
SSA	61.45	0.68	0.61	0.60	15	136.61
GWO	61.25	0.68	0.61	0.59	9	67.83
WO	72.45	0.73	0.73	0.71	18	165.66
HHO	61.3	0.69	0.61	0.58	8	100.77

Table 6 Binary classification using NB

FS method	Accuracy (%)	Precision	Recall	F-score	Feature selected	Time (s)
SSA	85.23	0.87	0.87	0.87	6	198.56
GWO	88.7	0.89	0.89	0.89	3	160.08
WO	88.65	0.89	0.89	0.89	3	188.28
HHO	98.51	0.99	0.99	0.99	5	264.38

5   Conclusion

The effectiveness of IDS is dependent on FS methods and classifier used for classification. In this paper, four recent bio-inspired metaheuristic algorithms are applied for the FS phase of IDS. The comparative study of these FS methods is done using KNN and NB classifiers. The four algorithms used for FS are: SSA, GWO, WO, and HHO. Result shows that KNN with GWO provides better accuracy and takes less execution time for multiclass classification. However, when the NB classifier is used, then performance of GWO degrades. With NB classifier, WO feature selection method achieves highest accuracy for multiclass classification and HHO achieves highest accuracy for binary classification.

In the future work, we can modify any of these bio-inspired algorithms used for FS to achieve better accuracy and reduce the dimensionality of data. This will help in reducing the time taken for classification and will be efficient with real-time traffic. Further, we can experiment with other learning models to determine the efficacy of the system.

References

1. Whitmore A, Agarwal A, Xu LD (2015) The Internet of Things—a survey of topics and trends. Inf Syst Front 12(2):261–274
2. Vailshery LS. Statista, 8 Mar 2021. [Online]. Available: https://www.statista.com/statistics/110 1442/iot-number-of-connected-devices-worldwide/. Accessed 2021/12/09
3. Balasaraswathi LR, Sugumaran M, Hamid Y (2017) Feature selection techniques for intrusion detection using non-bio-inspired and bio-inspired optimization algorithms. J Commun Inf Netw 2(4):107–119
4. Maza S, Touahria M (2018) Feature selection algorithms in intrusion detection system: a survey. KSII Trans Internet Inf Syst (TIIS) 12(10):5079–5099
5. Thakkar A, Lohiya R (2021) Attack classification using feature selection techniques: a comparative study. J Ambient Intell Humaniz Comput 12(1):1249–1266
6. Davahli A, Shamsi M, Aba G (2020) Hybridizing genetic algorithm and grey wolf optimizer to advance an intelligent and lightweight intrusion detection system for IoT wireless networks. J Ambient Intell Humaniz Comput 11(11):5581–5609
7. Mafarja M, Heidari AA, Habib M, Faris H, Thaher T, Aljarah I (2020) Augmented whale feature selection for IoT attacks: structure, analysis and applications,. Futur Gener Comput Syst 112:18–40

8. Alharbi A, Alosaimi W, Alyami H, Rauf HT, Damaševičius R (2021) Botnet attack detection using local global best bat algorithm for industrial Internet of Things. Electronics 10(11):1341

9. Om Prakash P, Maram B, Nalinipriya G, Cristin R (2021) Harmony search hawks optimization-based deep reinforcement learning for intrusion detection in IoT using nonnegative matrix factorization. Int J Wavelets Multiresolut Inf Process 19(4):2050093

10. Khurma RA, Almomani I, Aljarah I (2021) IoT botnet detection using salp swarm and ant lion hybrid optimization model. Symmetry 13(8):1377

11. Tama BA, Rhee K-H (2018) An integration of PSO-based feature selection and random forest for anomaly detection in IoT network. MATEC Web Conf

12. Shorman AA, Faris H, Alja I (2020) Unsupervised intelligent system based on one class support vector machine and grey wolf optimization for IoT botnet detection. J Ambient Intell Humaniz Comput 11(7):2809–2825

13. Koroniotis N, Moustafa N, Sitnikova E, Turnbull B (2019) Towards the development of realistic botnet dataset in the Internet of Things for network forensic analytics: BoT-IoT dataset. Futur Gener Comput Syst 100:779–796

14. Heidari AA, Mirjalili S, Faris H, Aljarah I, Mafarja M, Chen H (2019) Harris hawks optimization: algorithm and applications. Futur Gener Comput Syst 97:849–872

15. Alabool HM, Alarabiat D, Abualigah L, Heidari AA (2021) Harris hawks optimization: a comprehensive review of recent variants and applications. Neural Comput Appl 1–42

16. Mirjalili S, Mirjalili SM, Lewis A (2014) Grey wolf optimizer. Adv Eng Softw 69:46–61

17. Davahli A, Shamsi M, Abaei G (2020) Hybridizing genetic algorithm and grey wolf optimizer to advance an intelligent and lightweight intrusion detection system for IoT wireless networks. J Ambient Intell Humaniz Comput 11(11):5581–5609

18. Mirjalili S, Gandomi AH, Mirjalili SZ, Saremi S, Faris H, Mirjalili SM (2017) Salp swarm algorithm: a bio-inspired optimizer for engineering design problems. Adv Eng Softw 115:163–191

19. Ibrahim HT, Mazher WJ, Ucan ON, Baya O (2017) Feature selection using salp swarm algorithm for real biomedical datasets. Int J Comput Sci Netw Secur 17(12):13–20

20. Abualigah L, Shehab M, Alshinwan M, Alabool H (2020) Salp swarm algorithm: a comprehensive survey. Neural Comput Appl 32(15):11195–11215

21. Mirjalili S, Lewis A (2016) The whale optimization algorithm. Adv Eng Softw 95:51–67

Reformed Binary Gray Wolf Optimizer (RbGWO) to Efficiently Detect Anomaly in IoT Network

Akhileshwar Prasad Agrawal⊙ and **Nanhay Singh**⊙

Abstract With extensive IoT ascendancy around globe and considering its wide range applications, it becomes important to detect any attacks aimed at the network. Machine learning algorithms are important giving efficient mechanism to detect such attacks. However, these techniques are resource intensive which against the resource scarce IoT nodes. To overcome this, the no. of attributes is required to be rationalized and optimized to save computation and resource efficiency. In this paper, reformed binary Gray wolf algorithm with fitness function is proposed and implemented to select the features subset instead of the whole set of features in the original dataset. Further, we aim to tune optimal parameters for SVM by comparing various kernels with their efficiency. In this paper, NSL-KDD dataset in highly imbalanced form is used to test the validity of the approach. The experiments done here prove validity of the approach in terms of both accuracy and no. of features. Moreover, the proposed work is compared to other relevant research work which additionally validates intended work.

Keywords IoT Gray wolf optimizer · Anomaly detection

1 Introduction

With increasing use of IoT, the threat of intrusion into IoT system is also growing which has the potential to compromise security privacy data. Various categories of attacks at the network layer can be distinguished into DoS attacks, U2R attacks, R2L attacks, etc. The IDS is placed inside network and it is tasked to analyze the various packets coming into the system, bifurcating the normal from the attack types. For securing the components of IoT, machine learning is very valuable. Various categories

A. P. Agrawal (✉)
NSUT East Campus, Guru Gobind Singh Indraprastha University (GGSIPU), Delhi, India
e-mail: kpw.ce08@gmail.com

N. Singh
NSUT East Campus, Guru Gobind Singh Indraprastha University, Delhi, India

© The Author(s), under exclusive license to Springer Nature Singapore Pte Ltd. 2023
A. Shukla et al. (eds.), *Computational Intelligence*, Lecture Notes in Electrical Engineering 968, https://doi.org/10.1007/978-981-19-7346-8_21

of expert techniques ML have developed for classify the data into 'normal' vis-a-vis 'attack' categories. One other major problem is the data dimensionality which involves analyzing huge sets of features and millions of the instances to train a machine learning model [1, 2]. In this background, the following focuses on the:

1. A reformed binary Gray wolf optimizer is proposed to used to find minimum no. of attributes required to efficiently classify the instances into normal and attack types.
2. Comparative analysis of SVM kernels is made to get best kernel value parameter for training and thereby testing of data in imbalanced dataset.
3. The RbGWO was tested using dataset (NSL-KDD) using various metrics like accuracy, $f1$ score, precision and recall.

2   Related Work

In one of the papers, the neural layer network classifier using bGWO did filtering and selected the significant features. However, the matrices like false alarm rate were not account into consideration [3].

The AdaBoost and ABC algorithms were intensively used to select and thereby evaluate features. The network-based IDS was tested on dataset-NSL and ISCXIDS 2012 datasets which gave high accuracy but could not lessen the number of features much [4].

In another study, the PSO was combined with entropy minimization using Naive Bayes classifier. The testing on NSL-KDD dataset did not revealed much improvement in accuracy; however, the number of features was reduced [5].

In one of the papers, in GWO, filter base principle was used to reduce redundancy caused by mutual information possessed by various nodes/entity; however, here also false positive matrices were not evaluated [6].

From the birth of SVM till date, various innovations have been researched in improving SVM technique to incorporate concepts like kernel functions to create SVM functions in a hypothesis space for high dimensionality problem. They draw heavily from optimization theory [7].

Compared to other techniques, SVM has performed better in classification and prediction. In this direction, Mukkamala et al. compared the SVM with the ANN, wherein he depicted how SVM discharged better to the ANN [8]. Kou et al. compared SVM with the decision tree, Naïve Bayes and found it much better than those techniques [9]. Shams et al. in 2018 paper applied SVM in vehicular ad hoc network for detecting attack types and found it to be efficient. To reduce the training time, subsampling techniques were experimented with to remove less useful training points. Vijayanand et al. applied IDS using SVM in mesh network and there also SVM was noted to be efficient machine learning algorithm [10].

However, these works in SVM suffer from non-comparison of various kernel parameters to find best kernel according to problem statement.

2.1 Binary Gray Wolf Optimizer

Inspired by the metaheuristic approach, it banks on the leadership positions and Gray wolves prey hunting way. The leader search wolves are three hunters best in pack, and based on their positions, the other wolves were to update their position till a near optimal solution is reached [11].

The update equation can be written as:

$$W_t(\text{iteration} + 1) = \frac{W_1 + W_2 + W_3}{3} \tag{1}$$

where W_1, W_2, W_3 are the vector positions relative to the alpha, and other leader wolves.

2.2 Binary Gray Wolf Optimizer

As the attributes of dataset cannot be described by the continuous variable, hence, in this algorithm, the mapping is done from continuous variable to binary $\in \{0, 1\}$ [12].

For this,

$$W_t(\text{iteration} + 1) = \text{Crossover}(W_1, W_2, W_3) \tag{2}$$

3 Reformed Binary Gray Wolf Optimizer

The reformed binary Gray wolf optimizer uses the two best hunters alpha and beta to guide the new positions of other wolves. Figure 1 reveals algorithm of the RbGWO. This is done to reduce the impact of less efficient wolves in deciding the next iteration results for updating the other wolves positions. This has helped increase the quality of decision making by excluding the less accurate results in decision of updating the positions in space. The crossover approach of bGWO is used to map the positions into binary form and change the positions of wolves in the position space. The mathematical model for initializing the position of wolves and their subsequent updation is as:

$$W_t(\text{iteration} + 1) = \text{Crossover}(W_1, W_2) \tag{3}$$

Initialize the grey wolf population as W$_i$ (1......n)

Initialize x, Y, Z

Calculate the fitness of all the search agents (W$_1$......W$_n$)

Sort the fitness of all search agents

Assign W$_\alpha$: best search agent

Assign W$_\beta$: second best search agent

Initialize iter as 1

While (iter< maximum iteration)

      For each search agent

      Calculate W$_1$, W$_2$

      Calculate W$_{up}$ as in equation ( )

      End for loop

      Update the parameters x, Y, Z

      Calculate the fitness of all search agents

      Update newly found W$_\alpha$ and W$_\beta$ by sorting again the values in previous step

Iter=iter+1

End the while loop

Return W$_\alpha$ in the end

Fig. 1 Reformed binary gray wolf optimizer algorithm

$$W_1 = \left\{ \begin{array}{ll} 1 & \text{if}\left(W_{\text{alpha}} + S1_{\text{alpha}}\right) \geq 1 \\ 0 & \text{otherwise} \end{array} \right\} \tag{4}$$

$$W_2 = \left\{ \begin{array}{ll} 1 & \text{if}(W_{\text{beta}} + S1_{\text{beta}}) \geq 1 \\ 0 & \text{otherwise} \end{array} \right\} \tag{5}$$

$$S1_{\text{alpha}} = \left\{ \begin{array}{ll} 1 & \text{if} S2_{\text{alpha}} \geq \text{rand} \\ 0 & \text{otherwise} \end{array} \right\} \tag{6}$$

$$S1_{\text{beta}} = \left\{ \begin{array}{ll} 1 & \text{if} S2_{\text{beta}} \geq \text{rand} \\ 0 & \text{otherwise} \end{array} \right\} \tag{7}$$

W_{alpha}, W_{beta} are alpha, beta location vector, $S2_{\text{alpha}}$, $S2_{\text{beta}}$ are step functions.

3.1 Reformed Fitness Function

Fitness function plays much important role in providing correct measure of a model/algorithm applicability on a problem dataset. In the literature, the emphasis is more on the accuracy and no. of features selected by the algorithm. However, the accuracy cannot be compensated with reduction of no. of features. Thus, the fitment function which is used in the reformed binary Gray wolf optimizer algorithm emphasis on the accuracy level only during the execution of algorithm; however, the no. of features will be considered while comparing with others' work in same field. In essence, the trade-off between no. of features and accuracy is not always justified especially in sensitive security matters where accuracy is most important. The reformed fitness function is as:

$$\text{Fitness} = 1 - \frac{c}{T} \qquad (8)$$

where c is correctly identified instances and T is the total no. of test cases. In this case, the goal of the algorithm is to minimize the fitness function. The no. of features will separately be compared in the experimental sections to get the goal to keep the no. of features as low as possible without compromising the accuracy to a large extent.

4 Methodology and Experimental Results

Intel® Core™ i5-8265U CPU @1.80 GHz system and python language were employed for implementation. Figure 2 depicts the methodology. Most of the datasets based on network traffic collection are imbalanced where the quantity of 'normal' is usually higher than the attack types. This creates the problem of efficiently detecting the various attack types because the training data for a particular attack is insufficient. In literature, the NSL-KDD dataset is widely used for investigating the efficiency of various algorithms and techniques. However, this dataset is inflicted from imbalanced ratios. SVM has shown immense potential using the technique of creating hyperplane to separate the classes. In SVM, multi-classification is possible through 'one versus rest' and 'one versus one' strategy. Similarly, another problem is the features number which are 41 in number. Figure 3 shows the flowchart of the reformed GWO.

4.1 Data Specification and Pre-processing

In this research article, NSL-KDD is experimented with and used for experimental results. Each instance has 41 independent features and one dependent class label [1]. Firstly, SVM can only process numeric attributes; hence, various symbolic data like 'protocol_type,' 'flag' are converted into numeral form. In this work, ordinal encoding is utilized for the conversion purpose. After this, minimum–maximum technique of scaling is used to scale the data within uniform range. After scaling is done, the next forth step was splitting combined dataset into training and test dataset. The performance was evaluated using metrics as in Eqs. (9)–(11).

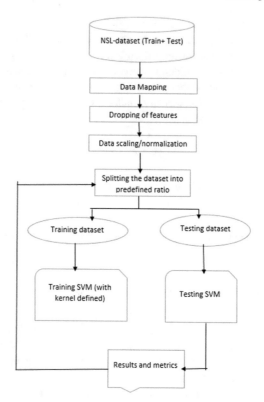

Fig. 2 Methodology adopted

Here, $F1_{\text{score}}$ is $F1$ score, P_{re} stands for Precision, T_{po} stands as True-Positive, T_{ne} stands as True-Negative, F_{po} stands as False-Positive and F_{ne} stands as False-negative, R_{call} stands for Recall.

$$P_{\text{re}} = \frac{T_{\text{po}}}{T_{\text{po}} + F_{\text{po}}} \tag{9}$$

$$F1_{\text{score}} = \frac{2(P_{\text{re}})(R_{\text{call}})}{P_{\text{re}} + R_{\text{call}}} \tag{10}$$

$$\text{Accuracy} = \frac{T_{\text{po}} + T_{\text{ne}}}{T_{\text{po}} + T_{\text{ne}} + F_{\text{po}} + F_{\text{ne}}} \tag{11}$$

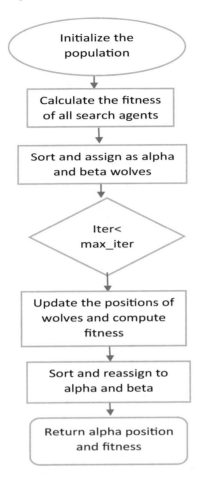

Fig. 3 Flowchart of RbGWO

4.2 Comparison with Related Work

The presented study was also compared to other work in the related domain using the same dataset. The results have been listed in the form of Tables 1 and 2.

Figure 4 gives the accuracy and other measures for comparing our approach and algorithm with other established algorithms or techniques in the literature. This algorithm shows better accuracy and with it number of features are reduced to an extent. Thus, in addition to improving the efficiency of classification, the computation required for processing of the dataset is reduced. Figure 5 shows the comparison of kernel functions of SVM for detection.

Table 1 Comparison of proposed work with other algorithms

Methods	Precision	F1-score	Accuracy
SVM	96.08	96.69	96.71
NB	96.29	96.62	96.63
DT	97.02	96.73	96.73
k-NN	97.24	97.11	97.18
LR	95.07	95.82	95.85
Proposed approach	99.41	99.28	99.32

Table 2 Comparison of accuracy and features

IDS	Number of features	Accuracy
Naïve Bayes [13]	41	76.56
Random forest [13]	41	80.67
AdaBoost [14]	41	90.31
FDR + kernel PCA [15]	23	90
SIGMOID_PIO [16]	18	86.90 ± 0.6
PSO [17]	37	78.2 ± 0.8
MBGWO [18]	14	99.22
Proposed approach	12	99.32

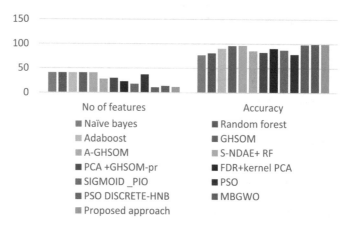

Fig. 4 Comparison of different algorithms' accuracy with features

5   Conclusion and Future Study

In present task, the focus was on two fold things—first improvement of accuracy of classification and second reduction of computation load by eliminating not so useful

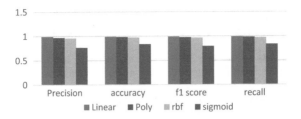

Fig. 5 Comparison of performance of various SVM kernels in detection

independent features from the entire dataset. The proposed reformed binary Gray wolf optimization algorithm along with various SVM kernels was implemented for evaluating the results. Through the experiments, it is clear that this work performs better in measures of metric like accuracy and reducing computation effort. The further research can be done for class balancing using techniques like SMOTE or others. This will help in reducing the class imbalance in the dataset thereby improving the efficiency of classification of minority classes.

References

1. Rathore S, Park JH (2018) Semi-supervised learning based distributed attack detection framework for IoT. Appl Soft Comput J 72:79–89
2. Gu J, Lu S (2021) An effective intrusion detection approach using SVM with naïve Bayes feature embedding. Comput Secur 103:102158
3. Mazini M, Shirazi B, Mahdavi I (2019) Anomaly network-based intrusion detection system using a reliable hybrid artificial bee colony and AdaBoost algorithms. J King Saud Univ Comput Inf Sci 31:541–553
4. Sailaja M, Kumar RK, Murty PSR, Prasad P (2012) A novel approach for intrusion detection using swarm intelligence. In: Proceedings of the international conference on information systems design and intelligent application. Springer, India, pp 469–479
5. Elngar AA, EI Mohamed A, Ghaleb FFM (2013) A real-time anomaly network intrusion detection system with high accuracy. Inf Sci Lett 2(2):49–56
6. Devi R, Suganthe RC (2017) Feature selection in intrusion detection grey wolf optimizer. Asian J Res Soc Sci Human 7(3):671–682
7. Gauthama Raman MR, Somu N, Kirthivasan K, Liscano R, Shankar Sriram VS (2017) An efficient intrusion detection system based on hypergraph—genetic algorithm for parameter optimization and feature selection in support vector machine. Knowl Based Syst 134:1–12
8. Mukkamala S, Janoski G, Sung A (2002) Intrusion detection using neural networks and support vector machines. In: Proceedings of the international joint conference on neural networks, vol 2, pp 1702–1707
9. Kou G, Peng Y, Chen Z, Shi Y (2009) Multiple criteria mathematical programming for multi-class classification and application in network intrusion detection. Inf Sci 179(4):371–381
10. Vijayanand R, Devaraj D, Kannapiran B (2018) Intrusion detection system for wireless mesh network using multiple support vector machine classifiers with genetic-algorithm-based feature selection. Comput Secur 77:304–314
11. Mirjalili S, Mirjalili SM, Lewis A (2014) Grey wolf optimizer. Adv Eng Softw 69:46–61
12. Emary E, Zawbaa HM, Hassanien AE (2016) Binary grey wolf optimization approaches for feature selection. Neurocomputing 172:371–381

13. Tavallaee M, Bagheri E, Lu W, Ghorbani AA (2009) A detailed analysis of the KDD CUP 99 data set. In: IEEE symposium on computational intelligence for security and defense applications. IEEE, pp 1–6
14. Panda M, Abraham A, Patra MR (2010) Discriminative multinomial naive Bayes for network intrusion detection. In: Sixth international conference on information assurance and security. IEEE, pp 5–10
15. Hoz ED, Ortiz A, Ortega J, Hoz ED (2013) Network anomaly classification by support vector classifiers ensemble and non-linear projection techniques. In: International conference on hybrid artificial intelligence systems. Springer, pp 103–111
16. Alazzam H, Sharieh A, Sabri KE (2020) A feature selection algorithm for intrusion detection system based on pigeon inspired optimizer. Expert Syst Appl 148:113249
17. Tama BA, Comuzzi M, Rhee K (2019) TSE-IDS: a two-stage classifier ensemble for intelligent anomaly-based intrusion detection system. IEEE Access 7:94497–94507
18. Alzubi QM, Anbar M, Alqattan ZN, Al-Betar MA, Abdullah R (2019) Intrusion detection system based on a modified binary gray wolf optimisation. In: Neural computing and applications, pp 1–13

A Feature-Based Recommendation System for Mobile Number Portability

Yugma Patel, Vrukshal Patel, Mohammad S. Obaidat, Nilesh Kumar Jadav, Rajesh Gupta, and Sudeep Tanwar

Abstract Mobile phones have become an inherent part of people's lives in contemporary days. Mobile numbers have become the digital identity of a person these days so, changing phone number requires one to change their phone number at all its registered places. Mobile number portability (MNP) enables users to switch their network operator without changing the phone number. After the emergence of MNP services in India, the number of MNP requests has risen significantly since the last decade. This provides with the exemption from changing their number at every registered place. Despite such benefits of MNP service, people still face problems while switching to a new mobile network operator (MNO). Network strength varies from MNO to MNO and from area to area, making it difficult for users to choose MNO. With the diversity of plans in terms of daily data, price, and additional benefits, users are confounded and it becomes difficult to go through all the plans and decide the one that fulfills all their requirements. Thus, authors came up with a k-medoids clustering-based recommendation system that takes users' requirements as input and recommends plans to choose from.

Y. Patel · V. Patel · N. K. Jadav · R. Gupta · S. Tanwar (✉)
Department of Computer Science and Engineering, Institute of Technology, Nirma University, Ahmedabad, Gujarat 382481, India
e-mail: sudeep.tanwar@nirmauni.ac.in

Y. Patel
e-mail: 19BCE204@nirmauni.ac.in

V. Patel
e-mail: 19BCE203@nirmauni.ac.in

N. K. Jadav
e-mail: 21ftphde53@nirmauni.ac.in

R. Gupta
e-mail: 18ftvphde31@nirmauni.ac.in

M. S. Obaidat
College of Computing and Information, University of Sharjah, Sharjah, UAE
e-mail: m.s.obaidat@ieee.org

Department of Computer Science and Engineering, Indian Institute of Technology, Dhanbad 826004, India

Keywords Mobile number portability · Clustering · Mobile network operator · Recommendation system

1 Introduction

With an annual sales of 162 million mobile phones in 2021, India has marked a 12% percent increase compared to 2020 [1]. Such an increase in numbers indicates the rise in mobile phone usage, resulting in an increase in MNP requests. Mobile number portability is a service using which mobile number users can change their telecom company without changing their phone number. Before November 2010, when mobile number portability (MNP) was introduced in India, users were forced to stick to the same telecom company despite the dwindling service standards. Shifting to another telecom company is cumbersome as it requires one to change their phone number. After the advent of MNP services in India, many MNP requests increased. Figure 1 shows the number of MNP requests in India; based on the growth observed in recent past years. We infer that number of requests is likely to be doubled by 2027 [2] due to the advent of the 5G and beyond the network in 2022. India is predicted to have 500 million subscribers of 5G by 2027 [3]. This significant change from 4G to 5G might produce many MNP requests as all network operators try to provide different quality services so that a user stays with the same network operator.

People primarily use mobile networks to communicate, align their business operations online, and access the Internet quickly. Individuals rely extensively on network connectivity in this era of the Internet for various reasons, including education, businesses, work-from-home, etc. Hence, people want robust, reliable, and affordable network access for many of these reasons. For an individual, affordable pricing is more important than robust network connectivity. As a result of the COVID-19 pan-

Fig. 1 Number of MNP requests in India

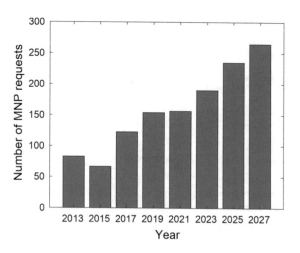

demic, businesses have shifted to work-from-home, requiring their employees to work in remote areas, and the education industry has shifted to e-learning, requiring students to study online. Therefore, the demand for a reliable and robust Internet connection has skyrocketed [4]. People want to choose the best mobile network operator (MNO) that can provide a network connection that is fast, cost-effective, and reliable.

There are various MNOs in India, and within these network operators, there is a wide range of plans available to the users. Even though all the information regarding the plans is available online, people still face a huge dilemma in changing their current mobile network operator (MNO) [1, 5, 6]. The user needs to browse through each MNO's website, compare all the available plans, and choose the best MNO and plan according to their requirement. Sometimes, for better clarity regarding network plans, the customer needs to visit the nearby telecom office or connect with the customer care of a particular MNO. Moreover, users who require international calls in their plan need to compare the rates of all the MNOs and then decide which plan and provider are best suitable for them. Additionally, the strength of a particular MNO varies with the area and state, in which the customer is residing, making it difficult for them to decide which MNO to opt for. For example, [7] shows the speed of Indian MNO's on different states, and one can clearly see that there is a significant difference in speed between other states. This makes mobile number portability tedious and the selection of a plan confounding. Even after selecting a specific plan, the user is not satisfied with the quality of service and experience.

Furthermore, the users are also utilizing the benefits of the Over The Top (OTT) platform subscription other than strong and affordable connections. There is a significant rise of OTT platform's global subscriber count from 104 million to 116 million at the end of the third quarter of 2021 [8] and thus stating the rise in the number of people interested in watching shows on such OTT platforms. Thus, many MNOs have started providing free subscriptions of such OTT platforms, benefiting users. However, it incurs a lot of challenges when one has to obtain the best plans; this is because all MNO's plans are very close, and it is difficult for one to choose the best plan efficiently. Many MNO's also provide post-data consumption speed; if a user consumes their entire data for the day, they can continue using the Internet at the provided speed (generally 64 kbps). On the other hand, specific MNO's offers unlimited data for the time 12:00 AM to 6:00 AM in some of their plans. Some users also wish to have these extra services included in their plans.

Several researchers have delivered solutions to pave the way to provide the best recommendation of plans to the users. For example, the research community has focused on predicting churn behavior by analyzing the customer's data. [9] focused on the predictive factors and predicting churn behavior for MNP. In [10], authors have proposed a game theory-based machine learning model to predict potential subscribers who are likely to move to other MNO [11]. In [12], authors have focused on analyzing factors that lead to customer satisfaction when using telecom services. They concluded that reliability, price, and coverage were the three most important customer preferences for an MNO. Kumaravel and Kandasamy [13] conducted a study to analyze user switching behavior and concluded that there is more number of unhappy customers in India in terms of services received by MNO. In [5], authors have proposed a system that analyzes user call rate, data usage, and a number of messages sent. Further, they mentioned that, due to unlimited call services and the emergence of IP messaging, the rage of call rate and simple messages, i.e., short messaging service (SMS), are getting down. However, they have not studied the most critical parameter, i.e., the signal strength of an MNO. Since MNP is available across many countries in the world, there is a strong requirement for a system that provides the potential parameter that helps in recommending suitable plans according to the user's needs. Therefore, in this paper, we considered all the essential and desirable parameters for users to select the best MNO and plan. The essential characteristics of machine learning can assist a user in conveying the best MNO and their optimal plans according to the user's requirement. Firstly, a dataset is formed using the data provided on the official websites of all the Indian MNOs with all the parameters such as price, data per day, etc. Then, the dataset is divided in 4 clusters using k-medoids clustering. Then, user's requirements are taken as input in order to assign a cluster to them, followed by extraction of the plans satisfying user's requirements. Finally, the user is recommended plans from this extracted plans which are ranked on the basis of signal strength and data per day.

1.1 Motivations

The motivation of this article can be defined as follows

- The number of MNP requests has increased since the last decade. In 2021, 156 million MNP requests were registered in India; moreover, the diversity of plans available, along with the variety of MNOs, makes the task of MNP difficult for the user. Hence, reflecting the demand for a recommendation system that can assist this huge demand of MNP.
- Even after choosing a plan, the user is not satisfied with the service provided against the money paid for the service. Requirements such as signal strength worthiness of the plan remain unsatisfied even after paying the price. Therefore, the recommendation system has considered the signal strength, price, etc., to recommend the user the most suitable plan.
- Features such as data rollovers OTT platform subscriptions are some of the complementary features provided by the service provider; many users are unaware of the network operator's features and blindly choose an obscure plan. Therefore, there is a requirement for a system that can provide a flexible and reliable recommendation of MNO's plan.

Table 1 List of symbols

Symbol	Meaning
N	Set of users desiring the MNP
A	Set of areas
K	Set of MNOs
P	Set of plans offered by all MNOs
λ	set of clusters
μ	Set of price per day
$\rho_r^{(w)}$	Signal strength of network provider K_r in area A_w
χ_i	Rank of a plan on the basis of ρ
ψ_i	Rank of a plan on the basis of μ
Υ	All plans better than the existing plan the user has.
θ_{N_h}	Daily data used by user N_h
θ_{P_j}	Daily data provided by plan P_j

1.2 Contributions

Choosing a suitable plan is a troublesome task due to the variety and diversity of plans and varying signal strength of MNO's across different places. Previously researchers have not focused on a comprehensive view of the parameters that shape a user's choice in choosing a plan. Given this, the following are the objectives of the paper.

- We prepared the dataset consisting of the plans along with all features, such as price and data per day, provided by mobile network operators.
- Clustering-based recommendation system is presented that takes into account user's requirement and suggests plans with optimal price and reliable Internet connection.
- Evaluation of the recommendation system is performed using a testing dataset made by varying requirements from the users and analysis of the result is performed.

1.3 Organization

The organization of the paper is as follows, Sect. 2 consists of a system model which describes the various attributes and variables used in the paper followed by the problem formulation that generalizes the problem solved by the proposed model. Section 3 describes the architecture of the proposed clustering-based recommendation system. Section 4 consists of the analysis of the result of the proposed system. Finally, Sect. 5 concludes the paper. Table 1 gives the symbols and their corresponding representations used in the paper.

2 System Model and Problem Formulation

2.1 System Model

Figure 2 shows the proposed architecture to provide suitable recommendation of plans to the user. In this architecture, there are (N) users such as $\{N_1, N_2, \ldots, N_i\} \in$ N, residing in a particular area (A) such as $\{A_1, A_2, \ldots, A_m\} \in A$, and poses network operators (K) such as $\{K_1, K_2, \ldots, K_z\} \in K$ that provides telecom services. We have gathered a dataset (D) from different network operators, consisting of rows (P) such as $\{P_1, P_2, \ldots, P_r\}$ that represents the plans of various network operators and columns (C) such as $\{C_1, C_2, \ldots, C_q\}$ represents various attributes of a particular plan. We have divided plans into clusters (λ) such as $\{\lambda_1, \lambda_2, \ldots, \lambda_t\} \in \lambda$ using k-medoids clustering algorithm. A new user $N_h \notin N$ residing in area $A_i \in A$ who is using network operator $K_x \in K$ is facing problems with the current network operator and wants to port his/her number to some other network operator K_y where $y \neq x$. Based on N_h's requirements, we choose a cluster $\lambda_g \in \lambda$ and recommend plans from that cluster only. After choosing a cluster for the N_h, in order to recommend the best possible plan, we extract plans that satisfy N_h's requirements like daily cellular data and OTT subscription, etc. We get a set of plans E where $E = \{E_1, E_2, E_3, \ldots, E_v\} \subseteq$ D ($v \leqslant p$) is the set of plans that satisfy N_h's requirements. The recommendation of the extracted plans depends on two features price per day (μ) and probability of strong signal strength ($\rho_r^{(w)}$). $\rho_r^{(w)}$ represents signal strength of network operator K_r in area A_w, θ_{N_h} represents the daily data used by user N_h and θ_{P_j} represents the daily data provided by plan P_j. Then the plans are sorted on the basis of $\rho_r^{(w)}$ (in decreasing order) and μ (in increasing order). Consequently, the plans are ranked. The rank of plan i on the basis of $\rho_r^{(w)}$ is χ_i and on the basis of μ is ψ_i. Therefore, in the proposed architecture, we have described the system that optimizes our recommendation.

2.2 Problem Formulation

Consider a user N_h residing in area A_x encountering problems with his current service provider and wishes to port their number toward a new plan. Additionally, he is wondering which MNO and plan to choose to obtain the best service alongside fulfilling their requirements.

$$N_h \xrightarrow[\text{in}]{\text{resides}} A_x \tag{1}$$

where $A_x \in A$. Further, the user N_h has a current plan $P_i \in P$ which has price per day μ_i, signal strength indicator $\rho_r^{(x)}$, daily data usage of the user θ_{N_h} and daily data provided by plan θ_{P_j} are the differentiator features compared with other plans of an area A_x.

$$N_h \xrightarrow{\text{has}} P_i(\mu_i, \rho_r^{(x)}, \theta_{N_h}, \Theta_{P_j}) \tag{2}$$

Due to the current obscure plan P_i, the N_h chooses to switch to some another plan p_j

$$N_h \xrightarrow{\text{currently has}} P_i \qquad (3)$$

$$N_h \xrightarrow{\text{switches to}} P_j \qquad (4)$$

Every plan P has a set of feature Υ such that, $\{\Upsilon_1, \Upsilon_2, \ldots, \Upsilon_j\} \in \Upsilon$, where a Υ_j represents a specific benefit in terms of higher $\rho_r^{(x)}$, lower (μ_i), etc., to make the plan P_j better than P_i.

$$\forall\, P \ni \Upsilon, \text{ where } \Upsilon = \{\Upsilon_1, \Upsilon_2, \ldots \Upsilon_j\} \subseteq P \qquad (5)$$

$$P_i(\Upsilon_j) \leq P_j \qquad (6)$$

$$P_i(\Upsilon_j) \geq P_j \qquad (7)$$

We want to recommend P_j such that N_h's requirements are satisfied with minimal price and best possible Internet connection using higher signal strength $\rho_r^{(x)}$.

$$P_j(\mu_j, \rho_r^{(x)}, \theta_{N_h}, \Theta_{P_j}) = \begin{cases} \max(\rho_r^{(x)}) \\ \min(\mu_j) \\ \Theta_{P_j} \geq \theta_{N_h} \end{cases} \qquad (8)$$

3 Proposed Architecture

3.1 MNP Recommendation System

Figure 2 shows the proposed architecture for the recommendation system, where we have acquired prepaid mobile plans from the official website of different MNO to form a dataset. All the plans included in the dataset comprise unlimited local talk time. Further, there are trivial plans on the website that are less important in the dataset. Therefore, to reduce the complexity of the dataset, we have not used such plans in our dataset. Additionally, most users do not want recommendations for such plans as they only have one feature. The inclusion of these plans would have resulted in the non-uniform dataset. So, formally we have 96 plans in our final dataset, and each plan has a competent feature by which we have created a recommendation system. Consider that the daily data used by the user N_h is θ and daily data provided by a particular plan is Ω; then, the recommender system should recommend the plans such that $\Omega \geq \theta$ a. Because the user can afford to have more data after his daily requirement, however, we cannot compromise his daily requirement. The system removes all the plans that do not satisfy the user's requirement of daily data.

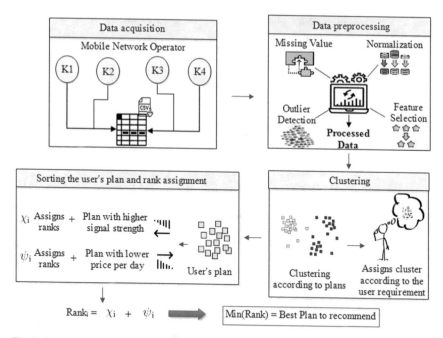

Fig. 2 Proposed architecture to recommend the best plan

Algorithm 1 MNP recommendation system.

Input: $D_1, D_2, ..., D_p \in D$

Output: Best plan (E_b)

1: **procedure** RECOMMENDATION($D_1, D_2, ..., D_p$)
2:    Form a dataset consisting of all the plans .
3:    Remove irrelevant features
4:    Form clusters of the dataset
5:    Extract plans on the basis of user requirement for example OTT subscription, data per day, etc.
6:    Sort plans in descending order on the basis of signal strength($\rho_r^{(w)}$)
7:    Sort plans in ascending order on the basis of price per day(μ)
8:    (χ_i) ← ranks for signal strength of plan i
9:    (ψ_i) ← ranks for price per day of plan i
10:   $Rank_i$ ← $\chi_i + \psi_i$
11:   Sort plans in increasing order on the basis of $Rank$ assigned to each plan
12:   Best plan (E_b) = Plan with min(Rank)

Many plans provide subscriptions to the OTT platform for a year. If N_h wants the subscription of OTT with his plan, then the system should only recommend the plans that comprise the subscription of OTT. On the other hand, if the user does not want a subscription to OTT, then the system should only recommend the plans that do not provide an additional subscription to the OTT because the subscription of OTT costs extra than the base price of the plan. The recommendation should strictly emphasize the user's choice; thus, we remove the plans that do not match N_h's choice of OTT subscription. After picking out the plans fulfilling the user's basic requirements, we have focused on the extra benefits, the relative strength of a network provider in the user's area, and the price per day for a particular plan E_b. If the validity of a plan is τ and price is Φ, then the price per day μ is calculated using the formula.

$$\mu = \frac{\Phi}{\tau} \tag{9}$$

dividing the price of the plan by its validity. In order to calculate signal strength $(\rho_r^{(w)})$ in an area w, system uses the below formula

$$\rho_r^{(w)} = \frac{\Phi_j^{(w)}}{\sum_{i=1}^h \Phi_j^{(w)}} \tag{10}$$

where $\Phi_j^{(w)}$ represents the number of users who chose operator j to be best network provider in area w. Then, the final rank for recommending the best plan can be calculated through,

$$\delta_i = \chi_i + \psi_i \tag{11}$$

For finding the plan that satisfies the user's requirement, δ needs to be sorted in increasing order and the plan with minimum value of δ is the best plan for the user.

3.2 Clustering Analysis

As the validity of a plan increases, the price increases proportionally. A user cannot spend more than his budget on buying a plan. Therefore, the plans in our dataset were for one month, two months, three months, and up to one year. Using the price and validity of the plans as input parameters, we have divided the plans into 4 clusters using a k-medoids clustering algorithm as shown in Fig. 3. The reason behind choosing the k-medoids clustering algorithm is that we know the number of clusters we want to form in the dataset. Moreover, k-medoids also handle outliers, unlike k-means, and thus improve the clusters formation [6]. When a user asks for a recommendation using his previous subscriptions, we assign one of the four clusters to a user and only recommend plans from the given cluster. If a new user asks for a recommendation, we can assign the user one of the clusters based on validity and price (Fig. 3).

Fig. 3 Clustering of plans
on the basis of their duration
and price

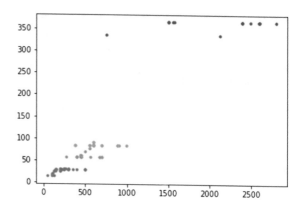

4 Result Discussion

This section discusses the performance evaluation of the proposed scheme.

4.1 Dataset Description

Dataset used in the research was taken from the official website of the Indian mobile
network operators. There were many plans whose price changed state wise. In the
dataset, only plans that are common across all states were included. The data collected
has seven features, which are as follows.

- *Price*: Price of the plan
- *Validity*: Duration for which the mentioned service is provided
- *Operator*: Network operator which provides the plan
- *Data per day*: Amount of cellular data user can utilize in one day
- *OTT subscription*: Represents the complimentary subscriptions provided in addition to the plan
- *Post-data speed*: Data rate(in Kbps) at which user can continue using Internet after completion of the daily data usage
- *Total data*: Total cellular data for the validity of the plan.

4.2 Implementation Interface

The MNP recommendation system has a user-friendly interface, in which the user
can match their requirements and corresponding plans are displayed as shown in
Fig. 4. The user simply needs to select the duration of plan and the plans that suit

him/her best would be displayed rank-wise along with its price, data per day, and signal strength.

4.3 Analytic-Based Results

In order to test the results of our recommendation system, the authors made a testing dataset of varying requirements and tested the output of the system. The recommendation differed significantly according to a user's requirement.

Figure 5a shows the number of times an MNO was recommended for a particular value of data per day. As we can see in Fig. 5a, K1 is recommended for the highest amount of times when the daily data requirement is 1, 1.5, and 2 GB. Next, when the daily data requirement is 2.5 or 3 GB, the K2 operator is recommended multiple times. Figure 5b shows the number of times an MNO is recommended for a particular validity of the plan. Moreover, we can see in Fig. 5b nearly half of the people chose a

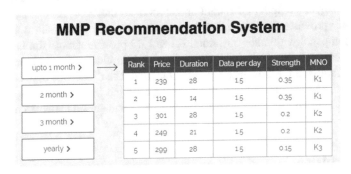

MNP Recommendation System

	Rank	Price	Duration	Data per day	Strength	MNO
upto 1 month >	1	239	28	1.5	0.35	K1
2 month >	2	119	14	1.5	0.35	K1
3 month >	3	301	28	1.5	0.2	K2
	4	249	21	1.5	0.2	K2
yearly >	5	299	28	1.5	0.15	K3

Fig. 4 User Interface for plan recommendation

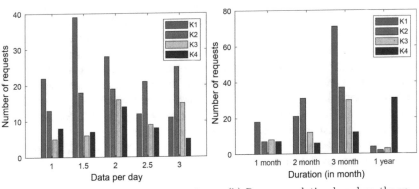

(a) Recommendations based on the value of data per day.

(b) Recommendation based on the validity of plan.

Fig. 5 Number of user data request

Fig. 6 Area wise signal strength of MNOs

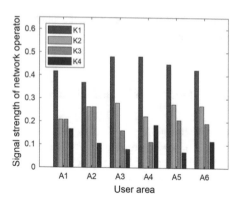

validity of 3 months, and hence, K1 is recommended multiple times when the validity is one and three months. K4 is recommended almost always in the case of yearly plans because K4 provides very low prices for yearly plans. This signifies that the recommendation of a plan depends on the duration of the plan required by the user. Figure 6 shows the signal strength of different operators in all the areas. As one can observe clearly, K1 provides the best network strength across all areas with a significant margin compared with other operators. K2 and K3 provide almost similar network strength across all the areas. On the other hand, K4 provided inferior signal strength that is not suitable for a qualified plan for the user. The difference between recommendations of different MNO is not very large, which indicates that no company dominates in providing plans with affordable prices and strong network connection.

5 Conclusion

Choosing a suitable plan with the best possible network operator is a difficult task due to the variety of plans and differing network strength of MNOs. One plan is not the optimal choice for everyone because requirement differs from user to user, making it challenging to decide on the best-personalized plan for users. Thus, this paper proposed a clustering-based recommendation system that chooses plans that satisfy users' requirements and provide users with the best possible network connectivity. We have observed from the graphs that no single MNO satisfies all of the criteria for a perfect MNO. The proposed recommendation system remedies this problem of identifying the most suitable plan and MNO for the users.

In the future, we will make the proposed system universal worldwide and also incorporate machine learning algorithms for higher accuracy.

References

1. Number of mobile phone purchases in 2021 (2021). https://bit.ly/3uA4XWg. Accessed 24 Nov 2021
2. Number of mnp requests in india from 2011 to 2021 (2021). https://www.trai.gov.in/release-publication/reports/telecom-subscriptions-reports. Accessed 26 Dec 2021
3. Increase in 5g services by 2027 (2021). https://www.financialexpress.com/industry/technology/5g-in-india-26-of-mobile-subscribers-in-india-to-use-5g-network-by-2026-end-says-report/2274389/. Accessed 26 Dec 2021
4. Need of strong internet during covid-19 pandemic (2020). https://blogs.worldbank.org/voices/covid-19-reinforces-need-connectivity. Accessed 25 Nov 2021
5. Achyuth K, Kutty SN, Bharathi B (2015) Recommender system for prepaid mobile recharging using apis. In: 2015 international conference on innovations in information, embedded and communication systems (ICIIECS). IEEE
6. Kanika, Rani K, Sangeeta, Preeti (2019) Visual analytics for comparing the impact of outliers in k-means and k-medoids algorithm. In: 2019 Amity international conference on artificial intelligence (AICAI), pp 93–97. https://doi.org/10.1109/AICAI.2019.8701355
7. Difference of mobile data speed across different states (2021). https://myspeed.trai.gov.in/. Accessed 20 Nov 2021
8. Increase in ott subscriptions (2021). https://www.exchange4media.com/digital-news/disney-hotstar-adds-almost-1173-million-paid-subscribers-in-q3-114968.html. Accessed 24 Nov 2021
9. Kaur G, Sambyal R (2016) Xploring predictive switching factors for mobile number portability. Vikalpa 41(1):74–95. https://doi.org/10.1177/0256090916631638
10. Ouyang Y, Yang A, Zeng S, Meng F (2019) Mnp inside out: a game theory assisted machine learning model to detect subscriber churn behaviors under china's mobile number portability policy. In: 2019 IEEE international conference on big data (big data), pp 1878–1886. https://doi.org/10.1109/BigData47090.2019.9006459
11. Patel K, Mistry C, Mehta D, Thakker U, Tanwar S, Gupta R, Kumar N (2021) A survey on artificial intelligence techniques for chronic diseases: open issues and challenges. Artif Intel Rev 1–44. https://doi.org/10.1007/s10462-021-10084-2
12. Paulrajan R, Rajkumar H (2011) Service quality and customers preference of cellular mobile service providers. J Technol Manage Innov 6:38–45. http://www.scielo.cl/scielo.php?script=sci_arttext&pid=S0718-27242011000100004&nrm=iso
13. Kumaravel V, Kandasamy C (2011) Impact of mobile number portability on mobile users switchover behavior-indian mobile market. Res World 2(4):200

An Efficient Deep Learning Model FVNet for Fingerprint Verification

G. Jayakala and L. R. Sudha

Abstract Fingerprint which is an impression made by the ridges on a finger is the most generally utilized biometric trait for human verification in many applications such as forensic investigation, law enforcement, custom access etc., for more than 100 years. Though different machine learning and deep learning approaches have been proposed for unique mark confirmation, there is still room for performance improvement. So, in this paper we have proposed an efficient Convolutional Neural Network (CNN) model namely FVNet by fine-tuning the hyper-parameters such as activation function, batch size and dropout of CNN. Performance of FVNet is evaluated by conducting experiments on FVC2000_DB4_B dataset furthermore the effectiveness of the proposed model is demonstrated by contrasting its exhibition and the presentation of pre-trained deep CNN models namely ResNet50 and VGG16.

Keywords Fingerprint verification · VGG16 · Resnet50 · Hyper-parameters · Fine-tuning

1 Introduction

Finger impression distinguishing proof is the method involved with looking at the grinding edge impression of a known print to an obscure one to check whether they match. It is a broadly utilized biometric when contrasted and the different biometric strategies due to many reasons like simplicity of catch, exceptionally uniqueness, tirelessness over the long run, and furthermore the finger impression sensors are more modest and less expensive [1]. The unique finger impression is a blend of large numbers of edges and a considerable lot of valleys on the fingertip's surface [2] as displayed in Fig. 1.

G. Jayakala (✉)
Department of Computer and Information Science, Annamalai University, Chidambaram, India
e-mail: gunajaya2015@gmail.com

L. R. Sudha
Department of Computer Science and Engineering, Annamalai University, Chidambaram, India

© The Author(s), under exclusive license to Springer Nature Singapore Pte Ltd. 2023   261
A. Shukla et al. (eds.), *Computational Intelligence*, Lecture Notes in Electrical Engineering 968, https://doi.org/10.1007/978-981-19-7346-8_23

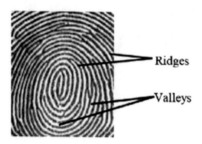

Fig. 1 Fingerprint image with ridges and valleys

Everybody's finger impression is novel, in any event, for indistinguishable twins. In this way, the finger impression is a broadly acknowledged biometric technique that is utilized in different businesses including law requirement offices, schools, medical clinics, banks, private associations and so on. It has a long history since 1893 [3]. There are different methodologies of programmed unique mark coordinating with that have been proposed which incorporate particulars based methodologies, furthermore picture based methodologies [4]. Be that as it may, minutia-based methodologies require broad pre-preparing activities to lessen the quantity of bogus details, incorrectly distinguished in loud finger impression pictures. Picture put together methodologies with respect to the next hand normally applied straightforwardly onto the dark scale finger impression picture without pre-preparing, and henceforth they might accomplish higher computational productivity than details based strategy [5].

Since the finger impression is novel, it resembles a mark. There are four fundamental kinds of fingerprints: curve, circle, risen curve, and whorl. The edges on curve fingerprints look like delicate slopes. Risen curve prints look like extremely steep slopes. They are basically the same as curve prints, yet the curve is higher in a risen curve print. The lines on a circle unique finger impression bend around and structure designs that look like circles. The fourth kind of unique finger impression is known as a whorl. Whorl fingerprints look like circles inside circles [6] (Fig. 2).

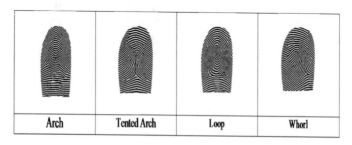

Fig. 2 Types of fingerprints

2 Literature Survey

Unique mark check is as of now the most famous strategy of biometric individual ID. Ali et al. [7], had proposed an arrangement based details—matching calculation for the check of human finger impression which has been considered as an exceptional mark guaranteeing one's personality. As a biometric evidence of ID, bas been generally examined, notwithstanding, very few have wandered into the universe of unique finger impression distinguishing proof utilizing picture based check. Picture based methodology doesn't utilize the details highlights for finger impression coordinating [8].

Patil and Zaveri, Fingerprints are graphical stream like edges present on human fingers. They have been extensively used in near and dear distinctive verification for quite a while. The legitimacy of their utilization has been grounded. Intrinsically, utilizing current innovation finger impression recognizable proof is significantly more solid than different sorts of well-known individual ID techniques dependent on mark, face, and discourse [10].

In Image Based Fingerprint Verification Hong et al. [11], researched the reasonableness of the finger impression examined picture to be confirmed by means of converse Fast Fourier Transform after a diminishing cycle. The procedure might be applied straightforwardly onto a dark scaled finger impression picture without pre-handling [12]. This connection coefficient approach is equipped for discovering the correspondences between the information unique finger impression picture and the put away enlisted format at higher computational proficiency than details based technique.

Seow et al. [13] carried out an Automatic Fingerprint Identification System AFIS with the utilization of unique mark grouping and particulars design coordinating. They have broken down the different benefits of the generally utilized particulars based coordinating in unique mark acknowledgment frameworks. Prior techniques center just around territorial recognition yet this cycle includes edge identification and particulars extraction [14].

Almedia and Dutta [15], proposed a delicate biometric framework for programmed age recognition from facial pictures. The initial step is pre-processing, to improve include extraction. The subsequent stage is definition, where strategies like wavelet is utilized. Liu et al. [16], proposed a clever methodology in their work, by changing the yield layers, combining arrangement and relapse age assessment techniques, and featuring significant components by pre-processing pictures and information expansion strategies. Aguilar et al. [17], proposed a unique finger impression check framework by joining Fast Fourier Transform (FFT) and Gabor channels to improve the caught finger impression picture.

In this paper we have proposed a proficient Convolutional Neural Network model FVNet by fine-tuning the hyper boundaries of traditional CNN for individual check from unique finger impression pictures. The engineering and exploratory consequences of these models are examined in the accompanying segments.

3 Methodology

The general block diagram for fingerprint verification system is discussed in this paper and is given in Fig. 3. Every deep learning architecture performs two tasks one is feature extraction and another one is classification. Firstly, fingerprint input images are fed into two pre-prepared CNN models VGG16, Resnet50 and the proposed FVNet model for feature for highlight extraction. The removed highlights are extraction. The extracted features are then used for training the model for verification process. After training the models the test image is given to the models and the highlights are removed. Then, at that point, the separated test highlights are coordinated with the prepared elements to decide if the unique mark is coordinated or not. The architecture of the models used in this paper are discussed briefly in the following sessions.

A. ResNet50

Resnet50 is otherwise called as Residual network which is a commonly used model in CNN to identify mapping by shortcuts [11]. Leftover Networks includes different resulting lingering modules, which are the fundamental establishment squares of ResNet. As the organization goes progressively deep, the preparation is more troublesome and it is difficult to unite. For the most part, the info highlight guide will be trailed by the convolutional channel, non-straight actuation work and a pooling activity lastly the result layer. Here, back proliferation calculation is executed for preparing the model.

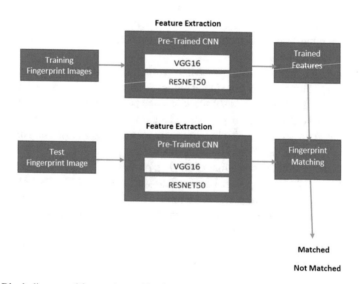

Fig. 3 Block diagram of fingerprint verification

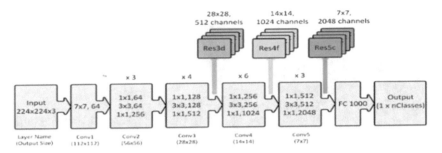

Fig. 4 Architecture of ResNet50

The design of ResNet50 is portrayed in Fig. 4. The information size of the picture is $224 \times 224 \times 3$. Each ResNet50 structure makes the principal convolution and max pooling utilizing 7×7 and 3×3 part estimates unmistakably. Then, first phase of the organization begins and it includes 3 Residual squares containing 3 layers each. The size of the pieces used to play out the convolution activity with each of the 3 layers of the square of the main stage is 64, 64 and 128 individually. The bended bolts allude to the character association. The associated bolt addresses that the convolution activity in the Residual square is executed with step 2. Subsequently, the size of info will be diminished to half corresponding to stature and width however the channel width will be multiplied.

For every leftover capacity F, 3 layers are stacked one over the other. The three layers are 1×1, 3×3, 1×1 convolutions. The 1×1 convolution layers are liable for diminishing and afterward supplanting the aspects. The 3×3 layer stays as a bottleneck with less info/yield aspects. At last, the organization has a normal pooling layer by an associated layer with 1000 neurons since this model is pre-prepared for various ImageNet data set with 1000 classes.

B. VGG16

VGG16 is one of the famous deep Convolutional Neural Network model submitted to ILSVRC-2014 [12]. It makes the improvement over AlexNet by supplanting huge piece measured channels (11 and 5 in the first and second convolutional layer, individually) with various 3×3 bit estimated channels consistently. Figure 5 addresses the engineering of VGG16 model.

The contribution to cov1 layer is a decent size 224×224 RGB picture. The picture is gone through a pile of convolutional (conv.) layers, where the channels were utilized with a tiny open field: 3×3 (which is the littlest size to catch the idea of left/right, up/down, focus). In one of the setups, it additionally uses 1×1 convolution channels, which can be viewed as a direct change of the info channels (trailed by non-linearity). The convolution step is fixed to 1 pixel; the spatial cushioning of convolutional layer input is to such an extent that the spatial goal is safeguarded after convolution, for example the cushioning is 1-pixel for 3×3 convolutional layers. Spatial pooling is completed by five max-pooling layers, which follow a portion of the conv. layers

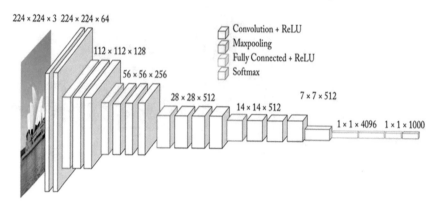

Fig. 5 Architecture of VGG16 model

(not all the conv. layers are trailed by max-pooling). Max-pooling is performed over a 2 × 2-pixel window, with step 2.

Three Fully-Connected (FC) layers follow a heap of convolutional layers: the initial two has 4096 channels each, the third performs 1000-way ILSVRC order and in this manner contains 1000 channels (one for each class). The last layer is the delicate max layer. The arrangement of the completely associated layers is something very similar in all organizations. In this pre-prepared model, we have freeze initial 10 layers in the model and enact just the last square of convolutional and completely associated layers.

C. FVNet Model

Deep Learning comprises a wide variety of algorithms which depend on numerous hyper-parameters that influence performance. Careful optimization of hyper-parameter values is a critical step to avoid overfitting either by maximizing the predictive accuracy or minimizing the error. FVNet Model is an optimized CNN model, with a stack of three convolution layer and max-pooling layers.

The proposed FVNet model was optimized by varying the values of the following three parameters:

- Activation Function: Tanh, Sigmoid and ReLU
- Batch Size: 4, 8, 12 and 16
- Dropout Probabilities: 0.2, 0.3, 0.4 and 0.5.

We have obtained 48 different combinations from the above three optimized parameters (4 × 4 × 3).

4 Experimental Result

A. Dataset

The FVC2000_DB4_B dataset are collected from the Kaggle dataset repository. Sample thumb fingerprint images from the dataset are shown in Fig. 6. It contains a total of 6400 images of size 103 × 103, 6000 images are used for training the model and 400 for validation. Fivefold validation is used to determine the optimal hyper parameters.

B. Five-fold Cross Validation

This approach randomly divides the data into five folds of equal size as shown in Fig. 7. The first fold is treated as a validation set, and the remaining four folds are used to train the model.

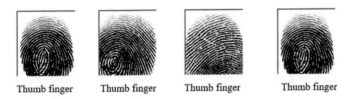

Thumb finger   Thumb finger   Thumb finger   Thumb finger

Fig. 6 Sample input images

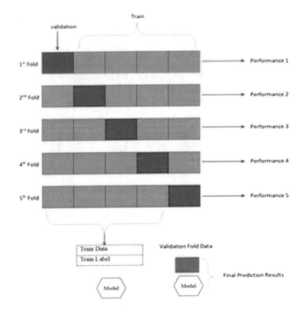

Fig. 7 Flow diagram fivefold validation

C. Performance Metrics

Execution measurements like Precision, Recall, Accuracy and $F1$-score are calculated using Eqs. (1)–(4).

$$\text{Accuracy} = \frac{\text{Total no. of correct prediction}}{\text{No. of input samples}} \tag{1}$$

$$\text{PRE}_i = \frac{\text{TP}_i}{\text{TP}_i + \text{FP}_i} \tag{2}$$

$$\text{REC}_i = \frac{\text{TP}_i}{\text{TP}_i + \text{FN}_i} \tag{3}$$

$$F_1^i = \frac{\text{PRE}_i \times \text{REC}_i}{\text{PRE}_i + \text{REC}_i} \tag{4}$$

Genuine positive (TP) alludes to the quantity of expectations where the classifier accurately predicts the positive class as certain. Genuine Negative (TN) alludes to the quantity of expectations where the classifier accurately predicts the negative class as negative. Bogus positive (FP) alludes to the quantity of expectations where the classifier mistakenly predicts the negative class as certain. Bogus Negative (FN) alludes to the quantity of forecasts where the classifier erroneously predicts the positive class as negative. The performance comparison of various hyper-parameters is shown in Table 1. The optimizable parameters and the corresponding search ranges and the selected values are displayed in Table 2.

Table 1 Performance comparison of various hyper-parameters

Activation function	Batch size	Dropout	Accuracy (%)
Relu	32	0.5	94
	64	0.5	97
Tanh	32	0.5	52
	64	0.5	65
Sigmoid	32	0.5	50
	64	0.5	53

Table 2 Parameter optimization for FVNet model

S. No.	Optimizable	Values of parameter	Selected value
1	Batch size	[4, 8, 12, 16]	[4]
2	Dropout probabilities	[0.2, 0.3, 0.4, 0.5]	[0.3]
3	Activation function	[tanh, Sigmoid, ReLU]	[ReLU]

Summary of the optimized FVNet Model

Layer Name	Details	Output shape	Parameters
Input	RGB images	222, 222, 3	–
Conv_2D	Conv (32)	222, 222, 32	896
Activation	ReLU	222, 222, 32	–
MaxPooling_2D	Pool size (2, 2)	111, 111, 32	0
Conv_2D	Conv (32)	111, 111, 32	18,496
Activation	ReLU	111, 111, 32	–
MaxPooling_2D	Pool size (2, 2)	55, 55, 32	0
Conv_2D	Conv(64)	55, 55, 64	36,928
Activation	ReLU	55, 55, 64	–
MaxPooling_2D	Pool size (2, 2)	27, 27, 64	0
Flatten	Convert 2D to 1D	46, 656	–
Dense	Input class = 2	2	93,312
Total parameters:		149, 632	
Trainable parameters:		149, 632	
Non trainable parameters:		0	

D. Error Rates:

To assess the exhibition of confirmation framework two mistake rates: FAR (False Accept Rate) and FRR (False Reject Rate) are determined utilizing the accompanying equation.

FAR = Number of fingerprints accepted falsely divided by

the number of impostors.

$$FAR = 4/200$$

$$= 0.02$$

FRR = Number of fingerprints rejected falsely divided by

the number of genuine fingerprints.

$$FRR = 3/200$$

$$= 0.015$$

E. Sample Output

In this work we have tweaked the hyper-parameters of initiation works specifically ReLu. Figure 8 shows one anticipated result of the proposed model. Exactness and misfortune bend of the proposed model is given in Fig. 9.

Disarray lattice in Fig. 10 is utilized for summing up the presentation of our arrangement calculation where the quantity of right and wrong expectations are summed up.

Use one Table 3 shows the upsides of execution measurements of the proposed approach and Fig. 11 is the comparing outline.

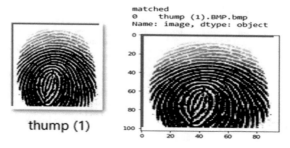

Fig. 8 Sample output

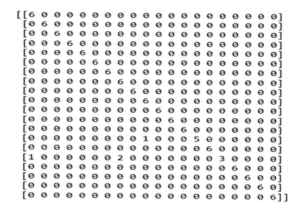

Fig. 9 Accuracy and loss

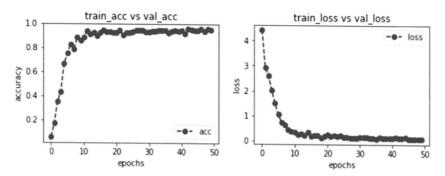

Fig. 10 Confusion metrics

Table 3 Performance metrics of proposed approach

Deep learning models	Accuracy	Precision	Recall	$F1$-score
VGG16	86	85	83.6	86
Resnet50	94	91.5	93	92
FVNet	97	95	95	96

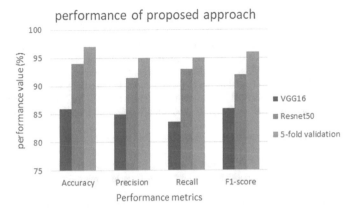

Fig. 11 Chart of performance metrics

5 Conclusion

In this paper, an improved model in particular FVNet is proposed to verify a person using fingerprint. We have trained the model using images from FVC2000-DB4 database. Performance of FVNet is compared with two deep learning models Resnet50, VGG16 and observed that FVNet outperforms the other two. To assess the presentation of check framework two error rates FAR and FRR are calculated. Low error rates and high exactness show the adequacy of the proposed framework.

References

1. Ali MMH, Mahale VH, Yannawar P, Gaikwad AT (2016) Fingerprint recognition for person identification and verification based on minutiae matching. In: IEEE 6th international conference on advanced computing, pp 332–339
2. Patil AR, Zaveri MA (2010) Novel approach for fingerprint matching using minutiae. In: Fourth Asia international conference on mathematical/analytical modelling and computer simulation 2010. IEEE 978-0-7695-4062-7/10
3. Hong L, Wan Y, Jain AK (1998) Fingerprint image enhancement: algorithms and performance evaluation. IEEE Trans PAMI 20(8):777–789
4. Seow BC, Yeoh SK, Lai SL, Abu NA (2002) Image based fingerprint verification. In: Student conference on research and development proceedings. Shah Alam, Malaysia, pp 58–61
5. He K, Zhang X, Ren S, Sun J (2016) Deep residual learning for image recognition. In: Proceedings in IEEE conference on computer vision and pattern recognition, pp 770–778
6. Jain A, Hong L, Bolle R (1997) On-line fingerprint verification. IEEE Trans Pattern Anal Mach Intell 19(4):302–314
7. Mohan P, Ananda S, Varghesea RB, Aravinth P, Raveena Judie Dolly D (2019) Analysis on fingerprint extraction using edge detection and minutiae extraction. In: International conference on signal processing and communication, pp161–164
8. Agarwal V, Sahai A, Gupta A, Jain N (2017) Human identification and verification based on signature, fingerprint and iris integration. IEEE, pp 456–461

9. Huang P, Chang C-Y, Chen C-C (2007) Implementation of an automatic fingerprint identification system. In: IEEE EIT proceedings, pp 412–415
10. Almedia V, Dutta MK (2016) Automatic age detection based on facial images. In: 2nd international conference on communication control and intelligent system (CCIS), pp 110–114
11. Liu X, Zuo Y, Hailan (2020) Face image age estimation based on data augmentation and lightweight convolutional neural network. In: 2nd international conference, vol 12, pp 1–17
12. Rattani A, Reddy N, Zena (2017) Convolutional neural network for age classification smart phone based on ocular images. In: IEEE international joint conference on biometrics, pp 756–761
13. Saxina AK, Sharma S, Vijay (2014) Fingerprint based human age estimation. In: IEEE India conference, pp 1–6
14. Sharma S, Shaurasiya K (2015) Neural network based on human age estimation domain curvelet. In: Eleventh international multi-conference on information processing, vol 54, pp 781–789
15. Michelsandi D, Antreea, Guichi Y (2017) Fast fingerprint classification with deep neural networks. In: International science and technology publication, pp 202–209
16. Merkel R, Nasrollahi K (2016) Resource-efficient latent fingerprint age estimation for Ad-hoc crime scene forensics: quality assessment of flatbed scans and statistical features. IEEE Trans 16:1–11
17. Modi SK, Stephen J (2007) Impact of age groups on fingerprint recognition performance. IEEE Trans 7:19–23
18. Lin CY, Wu M, Bloom JA, Cox IJ, Miller M (2001) Rotation, scale, and translation resilient public watermarking for images. IEEE Trans Image Process 10(5):767–782

Microcalcification Detection Using Ensemble Classifier

S. Vidivelli and S. Sathiya Devi

Abstract Breast cancer diagnosis and classification utilizing a CAD system with mammography pictures as input remains a difficult task in healthcare management systems. Microcalcification (MCs) is a microscopic deposit of calcium particles that acts as an early indicator of breast cancer. Without the assistance of an expert radiologist, automatic recognition of the MC region is a difficult task for a CAD system. MCs appear as bright small spots embedded in normal tissues on mammography images, and classification into normal, benign, and malignant is difficult. Clinical investigations demonstrate that benign zones are much denser than malignant regions, and that the malignant are more clustered than the benign. In this paper, we offer an entropy technique-based framework for autonomously identifying cancer regions. Fractal, topological, and statistical properties are retrieved to classify cancer as benign or malignant. An ensemble classifier comprising a combination of K-Nearest Neighbours (KNN), Support Vector Machine (SVM), Random Forest (RF), and Naive Bayes (NB) classifiers is introduced for successful output prediction. Our suggested model's performance was evaluated using several metrics such as accuracy, specificity, sensitivity, precision, and recall, and it was found to be superior to existing techniques.

Keywords Entropy · Fractal · Topology · Graph · Vertex degree · Ensemble learner

S. Vidivelli (✉)
School of Computing, SASTRA Deemed to be University, Thanjavur, Tamil Nadu, India
e-mail: vidivelli@cse.sastra.ac.in

S. Sathiya Devi
Department of Computer Science and Engineering, University College of Engineering-BIT Campus, Trichy, India

1 Introduction

Early detection of breast cancer is a challenging task because one in eight women is affected by breast cancer in USA. In India also, the mortal rate increased day by day due to breast cancer among women. Nearly 30% of mortality can be avoided with the help of X-ray mammography. Screening mammography is one of the widely used techniques to find early symptoms of breast cancer. MCs are difficult to detect because it is embedded in the mammographic background and overlaps with each other, and the size is range from 0.05 to 1 mm. Even an expert radiologist can face difficulties in the detection and classification of MCs. Hence, in this paper, we propose a framework to detect the MCs and classify it with high accuracy when compared to existing methodologies. In our work, we have extracted morphological features through Graph Theory approach and fractal features by box counting method [1]. Describes the property of any object in terms of self-similarity based on the fractal properties. The fractal dimension is an important characteristic of fractals because it contains information about their geometric structure. Fractal dimension is a statistical quantity which measures the degree of irregularity when the surface is zoomed down into a finer scale. Box counting is the simplest method to find fractal dimension with $O(N)$ computational cost. Many researchers concentrate on the morphological approaches to extract the MC region and a topological method to examine the MCs. We are motivated by the work of [2], they have analysed connectedness of the MCs and extract topological features and produce a good results in the classification of clusters. Kumari et al. [3] suggested a soft voting classifier model based on a combination of three machine learning algorithms: Random Forest, logistic regression, and Naive Bayes. The proposed model was first tested on the Pima Indians diabetes dataset, following which it was used with the breast cancer dataset. The ensemble soft voting classifier has produced positive results, and an effort has been made in this study to improve the prediction accuracy by using the ensemble learning method. In our proposed framework, the above two methods like topological features, statistical features, and fractal features are extracted, and MCs are classified using ensemble of classifier such as RF, SVM, KNN, and NB as benign and malignant. The remaining of the paper is organized as follows. Section 2 presents related work, Sect. 3 describes the proposed approach. Section 4 discusses the classification approach, Sect. 5 describes the dataset, result and discussions are made in Sect. 6, and finally, Sect. 7 concludes the paper with future work.

2 Related Work

We are inspired by numerous works in the literature related to pre-processing of mammogram images, feature extraction, and classification, and some of them are reviewed in this section. For effective enhancement and segmentation of MCs,

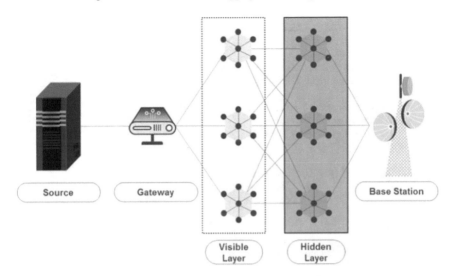

Fig. 5 GCA and HVM hybrid architecture

D_a starts for data aggregation, Node i represent data transmission $Node_N$ is a total number of nodes packet i denotes transferring data, and P_n denotes the ending packet of the transmission.

The data is aggregated through the intermediate nodes. GCA mechanism is connected with two-hop neighboring nodes, and HVM mechanism is a two layered techniques. The interconnection of two-hop neighboring node is changed as hidden and visible nodes. The visible layer gets the input from source and send to the hidden layer, the hidden layer send the data to the basestation after check the redundancy of the data. The hidden and visible nodes are also monitoring the traffic condition of the transmission path. The traffic is inside the path, then the hidden node informs to the gateway node, after gateway changes the transmission path. The GHVM is explained in Algorithm 1.

Algorithm 1

Step 1: Start
Step 2: // *Gateway Selection*
     if $Node_i = T_E \leq TH_E$ // T_E: *Transmission Energy*
                                 // TH_E: *Maximum amount of Energy*
     then
       Node i = Gateway Node
     else if $Node\ i + 1 = T_E \leq TH_E$
       Node i + 1 = Gateway Node
     end if
Step 3:
     if $C_N = D_T \leq Q_L$ // C_N: *Congested Node*
                        // D_L: *Data Transmission*
                        // Q_L: *Queue Length*
     then
       Congested route
     else
       The path is available to transmit the data
     else if
Step 4: // *Data Aggregation*
Step 5: Visible Layer send the data to the hidden layer
Step 6: Hidden layer Analyze the data and transfer the data to basestation
Step 7:
     if $D_i \neq D_N$ // D_i = *Selected data*
                  // D_N = *No. of data*
     then
       Send to basestation
     else
       Data redundancy available
       Skip the particular data
     end if
Step 8: End

4 Performance Metrics

The proposed mechanism hybrid GHVM is examined using NS2 in this section. The network life is distinct as the time it takes for the gateway node.

In WSN, the packet transmission consumes energy. Figure 6 shows the energy consumption based on the node's transmission.

The transmission takes time to send the data. Figure 7 shows the delay based on the node's transmission.

Packet delivery ratio is calculated by packet transmission successfully. Total packets and received packets are computed and compared with proposed mechanisms. In Fig. 8, the packet delivery ratio is compared with the proposed mechanism.

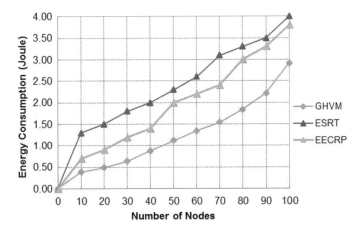

Fig. 6 Node versus energy consumption

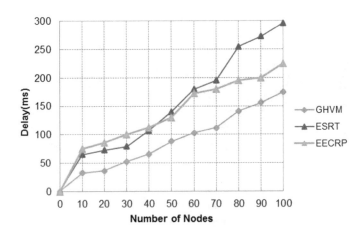

Fig. 7 Node versus delay

Figure 9 shows the number of data that can be sent from the source to destination throughput of around packets per second increased the nodes of the network using the GHVM mechanism.

Figure 10 shows the various results based on the speed. The proposed GHVM reduces energy consumption.

Figure 11 shows an speed and delay for previous and proposed mechanism. The proposed GHVM reduces the end-to-end delay.

In Fig. 12, the proposed mechanisms give a different result, it gives the speed of the packet delivery ratio.

Figure 13 shows an extreme throughput on pkts/s with the extended lifetime of the network.

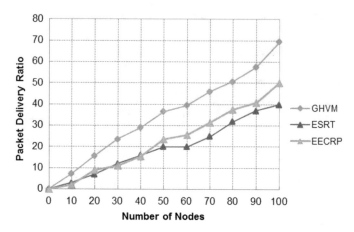

Fig. 8 Node versus packet delivery ratio

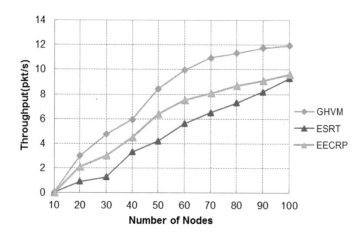

Fig. 9 Node versus throughput

5 Conclusion

The proposed mechanism is used to reduce the congestion and data aggregation. Congestion is reduced by gateway node and data aggregation is formulated and identified by the hidden and visible layer. The gateway node is added into the hidden and visible layer process. This hybrid mechanism is reduced the congestion and data redundancy. In the future, the study of CH selection from the selected CH and take steps to consume less energy with less delay.

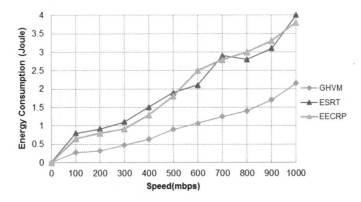

Fig. 10 Speed versus energy consumption

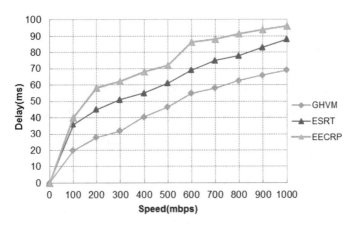

Fig. 11 Speed versus delay

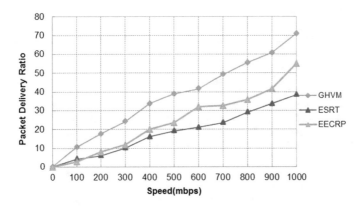

Fig. 12 Speed versus packet delivery ratio

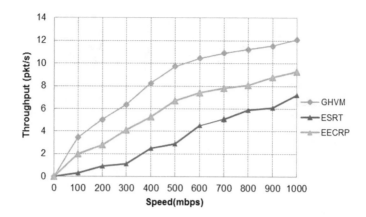

Fig. 13 Speed versus throughput

References

1. Matin MA, Islam MM (2012) Overview of wireless sensor network. Wirel Sens Netw Technol Protocols. https://doi.org/10.5772/49376
2. Obaidat MS, Misra S (2014) Introduction to wireless sensor networks. In: Principles of wireless sensor networks
3. Jones J, Atiquzzaman M (2007) Transport protocols for wireless sensor networks: state-of-the-art and future directions. Int J Distrib Sens Netw 119–133
4. Nawaz B, Mahmood K, Khan J, ul Hassan M, Shah AM, Saeed MK (2019) Congestion control techniques in WSNs: a review. (IJACSA) Int J Adv Comput Sci Appl 10(4)
5. Dagar M, Mahajan S (2013) Data aggregation in wireless sensor network: a survey. Int J Inf Comput Technol 3(3)
6. Patil NS, Patil PR (2010) Data aggregation in wireless sensor network. In: 2010 IEEE international conference on computational intelligence and computing research
7. Floyd S, Jacobson V (1993) Random early detection gateways for congestion avoidance. IEEE/ACM Trans Netw
8. Wan C, Eisenman S, Campbell A (2003) CODA: congestion detection and avoidance in sensor networks. In: The 1st international conference on embedded networked sensor systems. ACM Press, Los Angeles, pp 266–279
9. Sankara Subramaniam Y, Akan OB, Akyildiz I (2003) ESRT: event-to-sink reliable transport in wireless sensor networks. In: IEEE Mobi Hoc 2004. ACM Press, Annapolis, pp 177–188
10. Yang X, Chen X, Xia R, Qian Z (2018) Wireless sensor network congestion control based on standard particle swarm optimization and single neuron PID. Sensors 18(4)
11. Yadav SL, Ujjwal RL, Kumar S, Kaiwartya O, Kumar M, Kashyap PK (2021) Traffic and energy aware optimization for congestion control in next generation wireless sensor networks
12. Paek J, Govindan R (2010) RCRT: rate-controlled reliable transport protocol for wireless sensor networks. ACM Trans Sens Netw 7(3)
13. Shaikh FK, Khelil A, Ali A, Suri N (2010) TRCCIT: tunable reliability with congestion control for information transport in wireless sensor networks
14. Wang C, Sohraby K, Lawrence V, Li B, Hu Y (2006) Priority-based congestion control in wireless sensor networks. In: Proceedings of the IEEE international conference on sensor networks, ubiquitous, and trustworthy computing (SUTC'06)
15. Haoxiang W, Smys S (2020) Soft computing strategies for optimized route selection in wireless sensor network. J Soft Comput Paradigm (JSCP) 2(01):1–12

16. Pradhan S, Sharma K (2016) Cluster head rotation in wireless sensor network: a simplified approach. Int J Sens Appl Control Syst 4(1):1–10
17. Larochelle H, Mandel MI, Pascanu R, Bengio Y (2012) Learning algorithms for the classification restricted Boltzmann machine. J Mach Learn Res 13(1):643–669
18. Chaubey NK, Patel DH (2016) Energy efficient clustering algorithm for decreasing energy consumption and delay in wireless sensor networks (WSN). Int J Innov Res Comput Commun Eng (An ISO 3297: 2007 Certified Organization) 4(5)
19. Revathi A, Santhi SG (2022) Gateway based congestion avoidance using two-hop node in wireless sensor networks. In: 3rd international conference on mobile computing and sustainable informatics (ICMCSI 2022)

Relevance Vector Machine Tools for Evaluating the Strength Parameters of HPC Incorporating Quarry Dust and Mineral Admixtures with Fibers

D. Maruthachalam, M. Kaarthik, and S. C. Boobalan

Abstract This study work validates the validity of the entire association on the relevance vector machine (RVM) to regulate the high-performance concrete's mechanical properties focused on quarry dust. This study work presents the results arrived from experimental investigations on concrete integrating quarry dust with mineral admixtures and fibers. The RVM is based on a stochastic model that implements a priori prototype parameters governed in the collection of weights-related hyperparameters in which the most likely values are determined mostly from data sequentially. 3 RVM methodologies have been familiarized for training and validation of material characteristics utilizing MATLAB software. With above 70% of the total data sources, RVM models are being educated and with about 30% of the total databases evaluated. The estimated results of the RVM models are verified and identified has acceptance accordance as that like related measured results.

Keywords Relevance vector machine · Quarry dust · Silica fume · Steel fibers · Polypropylene fibers

1 Introduction

Because of reasonable rates and consistency, Concrete is the most broadly used building material. The manufacture of cement and concrete uses vast quantities of natural materials and aggregates, causing major climate and energy losses [1, 2]. This outcome further contributes a lot to the release of CO_2, a greenhouse gas that happens normally. To tackle these ecological and financial problems, changes and improvements to the current concrete production methods are necessary. These have motivated concrete technology and innovative methods, researchers to examine and

D. Maruthachalam (✉) · S. C. Boobalan
Civil Engineering, Sri Krishna College of Engineering and Technology, Coimbatore, Tamil Nadu, India
e-mail: maruthachalamd@skcet.ac.in

M. Kaarthik
Civil Engineering, Coimbatore Institute of Technology, Coimbatore, Tamil Nadu, India

© The Author(s), under exclusive license to Springer Nature Singapore Pte Ltd. 2023
A. Shukla et al. (eds.), *Computational Intelligence*, Lecture Notes in Electrical
Engineering 968, https://doi.org/10.1007/978-981-19-7346-8_29

classify substitute waste-product resources that can be cast-off in concrete production as replacements for material properties [3–5].

In traditional concrete methods, because of its high fineness, the incorporation of quarry dust spread is minimal [6]. The introduction of fresh concrete would surge the water demand and hence the water-cement ratio for the basic requirements including mechanical and durability properties. Another potential benefit is the rate savings include the use of quarry dust [7, 8]. Depending on the sources, the material expenses will vary. This paper adopts RVM to evaluate the concrete stability and durability property in concrete incorporated with quarry dust, silica fume, steel fibers/polypropylene fibers and examines its applicability [9].

2 Literature Review

Nataraja et al. produced concrete with coarse aggregates utilizing large-size quarry waste [10], while Ho et al. utilized quarry dust in self-compacting concrete applications as a substitute material for structural concrete. Besides that, the adequacy of quarry waste as aggregate to manufacture the flowable concrete was examined [11]. Safiuddin et al. and Kumar et al. assessed the flexural strength of beams made with HPC with integrating of coarse and fine aggregates using powdered sandstone and concluded by using powdered materials will surge the strength factors. In recent times for the development of granite crusher dust at varying ratios with steel fibers of miscellaneous thickness and density percentages [12, 13]. Eren and Marar examined the strength development with concrete, while the grinding dust is being used as a substitute for the fine aggregates (< 5 mm). To address Support Vector Machine (SVM) constraints [14]. Tipping implemented a new approach called Relevance Vector Machine (RVM). The key aspect that the primary role including its SVM aim is to reduce a variety of test set errors while simultaneously increasing its 'margin' among the two classes (implicitly specified by the kernel in the feature space) [15–17].

3 Experimental Investigation

The Mechanical strength of the cubes, cylinder and prism have been tested with concrete incorporated quarry dust, silica fume, steel fibers or polypropylene fibers [18]. The various combinations of concrete proposed to evaluate mechanical properties are given in Table 1 with their strengths. The characteristic properties of the various raw materials taken for preparing concrete of different blends are shown in Table 2.

From the sieve analysis, the properties of QD was calculated and they are in coincidence with Zone II of IS 383:1970. Compressive strength studies have been tested on a cube of 150 mm, split tensile strength studies on the cylinder of size

Table 1 Mechanical properties of concrete for various combinations

Combination		Age	Water binder ratio (w/b)	% Silica fume	% Fibers		Mechanical properties		
MIX					Steel	Polypropylene	Comp. strength, MPa	Split, MPa	Flexural strength, MPa
QD	FA								
0	100	3	0.3	–	–	–	32.43	3.12	4.23
40	60	3	0.3	–	–	–	31.24	3.03	4.18
50	50	3	0.3	–	–	–	34.54	3.21	4.28
60	40	3	0.3				34.65	3.32	4.30
70	30	3	0.3				31.26	3.01	4.20
80	20	3	0.3	–	–	–	29.42	2.86	4.16
100	0	3	0.3	–	–	–	26.42	2.78	3.83
0	100	7	0.3	–	–	–	45.33	3.74	4.93
40	60	7	0.3	–	–	–	45.60	3.73	4.99
50	50	7	0.3	–	–	–	47.54	3.81	5.12
60	40	7	0.3	–	–	–	51.18	3.86	5.23
70	30	7	0.3	–	–	–	44.30	3.74	4.82
80	20	7	0.3	–	–	–	40.21	3.63	4.66
100	0	7	0.3	–	–	–	35.23	3.43	4.41
0	100	28	0.3	–	–	–	67.56	4.38	6.20
40	60	28	0.3	–	–	–	68.22	4.45	6.30
50	50	28	0.3	–	–	–	69.32	4.53	6.38
60	40	28	0.3	–	–	–	70.80	4.74	6.49
70	30	28	0.3	–	–	–	64.78	4.10	6.42

(continued)

Table 1 (continued)

| Combination | | Age | Water binder ratio (w/b) | % Silica fume | % Fibers | | Mechanical properties | | |
QD	FA				Steel	Polypropylene	Comp. strength, MPa	Split, MPa	Flexural strength, MPa
80	20	28	0.3	–	–	–	60.25	3.82	6.30
100	0	28	0.3	–	–	–	55.38	2.83	6.10
60	40	3	0.3	0.10	–	–	35.43	3.41	4.12
60	40	3	0.3	0.15	–	–	36.21	3.46	4.14
60	40	3	0.3	0.20	–	–	33.21	3.26	4.01
60	40	3	0.3	0.30	–	–	27.82	3.12	3.83
60	40	7	0.3	0.10	–	–	51.50	3.89	4.96
60	40	7	0.3	0.15	–	–	56.56	4.03	5.08
60	40	7	0.3	0.20	–	–	53.10	3.92	4.99
60	40	7	0.3	0.30	–	–	43.65	3.68	4.76
60	40	28	0.3	0.10	–	–	73.00	4.86	6.70
60	40	28	0.3	0.15	–	–	75.86	5.14	6.96
60	40	28	0.3	0.20	–	–	73.20	4.76	6.76
60	40	28	0.3	0.30	–	–	63.21	4.63	6.10
60	40	3	0.3	0.15	0.5	–	36.26	3.63	4.24
60	40	3	0.3	0.15	1.0	–	37.01	3.66	4.31
60	40	3	0.3	0.15	1.5	–	36.72	3.81	4.46

(continued)

Table 1 (continued)

Combination					% Fibers		Mechanical properties		
MIX		Age	Water binder ratio (w/b)	% Silica fume	Steel	Polypropylene	Comp. strength, MPa	Split, MPa	Flexural strength, MPa
QD	FA								
60	40	3	0.3	0.15	2.0	–	35.01	3.99	4.89
60	40	7	0.3	0.15	0.5	–	56.83	4.76	5.73
60	40	7	0.3	0.15	1.0	–	57.11	4.81	5.88
60	40	7	0.3	0.15	1.5	–	56.72	4.93	6.12
60	40	7	0.3	0.15	2.0	–	55.12	5.12	6.35
60	40	28	0.3	0.15	0.5	–	76.33	6.12	7.86
60	40	28	0.3	0.15	1.0	–	77.90	6.66	8.44
60	40	28	0.3	0.15	1.5	–	76.06	6.86	8.62
60	40	28	0.3	0.15	2.0	–	75.01	7.23	9.12
60	40	3	0.3	0.15	–	0.10	36.74	3.53	4.46
60	40	3	0.3	0.15	–	0.15	37.87	3.58	4.51
60	40	3	0.3	0.15	–	0.20	36.54	3.46	4.42
60	40	3	0.3	0.15	–	0.30	34.34	3.33	4.30
60	40	7	0.3	0.15	–	0.10	56.72	4.23	4.93
60	40	7	0.3	0.15	–	0.15	57.04	4.26	4.98
60	40	7	0.3	0.15	–	0.20	56.81	4.21	4.91
60	40	7	0.3	0.15	–	0.30	57.32	4.06	4.76

(continued)

Table 1 (continued)

Combination							Mechanical properties		
MIX		Age	Water binder ratio (w/b)	% Silica fume	% Fibers		Comp. strength, MPa	Split, MPa	Flexural strength, MPa
QD	FA				Steel	Polypropylene			
60	40	28	0.3	0.15	–	0.10	76.95	6.00	7.76
60	40	28	0.3	0.15	–	0.15	77.50	6.40	8.11
60	40	28	0.3	0.15	–	0.20	76.83	6.32	7.82
60	40	28	0.3	0.15	–	0.30	74.53	6.10	7.10

Table 2 Property of the materials

Cement grade—53 (OPC) Particle size range — 31–7.5 µm Compressive strength at 28 days—57 MPa	Steel fiber Length = 25 mm Diameter = 1 mm Aspect ratio = 25 Specific gravity = 7.8 Yield strength = 850 MPa	Quarry dust (QD) Particle size range— 0.5–0.09 mm (Grade 3 of IS:650) Water absorption—2.50 Fineness modulus—2.71 Specific gravity—2.62
Polypropylene fibers SG = 0.89 g/cc Tensile strength = 500–690 MPa E = 3.5 GPa, length = 12 mm Elongation at failure (%) = 21%	Silica fume The range of particle size lies between 0.15 and 20 µm	Fine aggregates (FA) The range of particles size lies between 0.5 and 0.09 mm Coarse aggregate ranges between 4.75 and 20 mm

150 mm × 300 mm size and flexural strength on prisms of 500 mm × 100 mm × 100 mm size. With the integration of Mix design procedure, the normal concrete mix ratio was calculated in the following ratio 1:1.212:2.40 (C:FA:CA). The output for various combinations are highlighted in Table 1.

4 Relevance Vector Machine

RVM begins in idea concerning model parameters that are widely adopted at several correlation methods, i.e., $y(x)$ function must be estimated at a certain arbitrary point x following a sequence of $t = (t_1, y, t_N)$ variable observations, also at other development terms $x = (x_1, y, x_N)$:

$$t_i = y(x_i) + \varepsilon_i \tag{1}$$

in this condition, ε_i represents linear prediction variable for normal distribution 0 and divergence σ^2. The orbital $y(x)$ represents the weighted sum of several identified variable functions, often under the polynomial regression hypothesis, i.e.,

$$y(x) = \sum_{i=1}^{M} w_i \varphi_i(x) \tag{2}$$

At which $w_i = (w_1, ..., w_M)$ is a test specimen of the parameters of the weighted sum and $y(x)$ is a linear sum of M, normally variational and static, reference parameters, $\varphi_i(x) = (\varphi_1(x), \varphi_2(x), \ldots, \varphi_M(x))^T$.

The main advantage of this system is that it provides good output for simplification; the implied determinants are usually unique since they conclude comparatively few parameters of non-zero weight (w_i).

During the learning process, the maximum of variables is immediately set to zero, offering a method that is highly successful in discerning certain specific activities that are 'important' to make good decisions [15, 16].

4.1 RVM-Based Analysis

The three different RVM systems have been presented to assess the mechanical characteristics of the concrete. To build RVM models, MATLAB software is used.

It was identified from the laboratory results in the mechanical properties of the tested specimens were influenced by quarry dust, fine aggregate (river sand), the time required for curing, the addition of silica fume, the addition of steel or polypropylene fibers. The parameters, namely, quarry dust, fine aggregate (river sand), curing period, silica fume, steel fibers, and polypropylene fibers. The input vector is generated by these variables. As shown in Table 1, this input data has multiple statistical limits. Thus further, until the data is provided with the sequence to any numerical machine learning, regularization of the data is necessary. For the static regularization of results to data, values lie between 0 and 1 arrived from Eq. 3.

$$x_i^n = \frac{x_i^a - x_i^{min}}{x_i^{max} - x_i^{min}} \tag{3}$$

Thus, with and without standardization where, x_i^a and x_i^n are the ith parameters of the activation function and x_i^{min} are the higher and lower positions of all the parameters of both the activation function until preprocessing.

5 Implementation Results of RVM Models

RVM development was focused at choosing the kernel width (σ) in Eq. 3. Therefore, for computing thermal-based performance is minimal to match the various properties for that a post modeling analysis is needed [1]. Training and testing R values were evaluated after modeling depending on the number of related functions participating in the system, also that particular function is worked and identify the differences in the main functional condition.

The σ factor was originally thought to be 0.13 and the equation is determined for the estimated σ value. The modeling approach provides the use NRVs with their particular related functions (w_i). The consistency about its method produced is evaluated as the basis on relation coefficient (R) function with its computed with the help of Formula 4.

$$R = \frac{\sum_{i=1}^{n} \left(E_{ai} - \overline{E}_a\right)\left(E_{pi} - \overline{E}_p\right)}{\sqrt{\sum_{i=1}^{n} \left(E_{ai} - \overline{E}_a\right)}\sqrt{\sum_{i=1}^{n} \left(E_{pi} - \overline{E}_p\right)}} \quad (4)$$

E_{ai} and E_{pi} denotes measured and expected variables, respectively, $\overline{E}_a$ and $\overline{E}_p$ are the mean of observed and expected E samples belonging to n sequences.

Unless its r coefficients were near 1, therefore this methodology was concluded, otherwise, the value of σ is modified and the designing testing is carried out again. Figure 1 was developed by Yuvaraj et al. [8] shows the operation process responsible for the creation of RVM prototypes was represented.

This was found that perhaps the test R-value reached its limit for the corresponding models at the kernel width tabulated in Table 4, despite a minimum number of appropriate vectors. Table 3 presents the preparation with its test sets R results arrived from the developed prototypes.

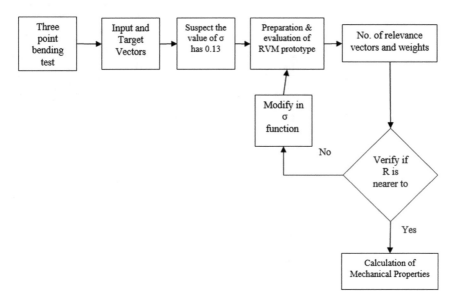

Fig. 1 Simplified sketch of RVM model process

Parameter	Correlation ship coefficient (r)		width (σ)
	Training	Evaluation	
Compressive strength	0.95	0.975	0.6
Split tensile strength	0.98	0.95	0.6
Flexural strength	0.98	0.97	0.6

Table 3 Output of RVM models created

Table 4 Weightage factors

Train set No.	Comp. strength	Split. strength	Flexural strength
	Weights (w_i)		
1	0	0	0
2	0	0	0
3	0	0	0
4	0	0	0
5	0	0	0
6	0.068932	0.093541	0.02689
7	0.119869	0	0.102451
8	0	0	0
9	0	0.060093	0
10	0	0	0
11	0	0	0
12	0.715377	0.235079	0.414792
13	0	0.173698	0
14	0	0	0
15	0.627313	0	0.138321
16	0.007928	0	0.262099
17	0	0	0
18	0	0	0
19	0	0	0
20	0	0	0
21	0	0	0
22	0	0	0
23	0	0	0
24	0.575528	0.246063	0.247049
25	0	−0.10537	0
26	0	0	0
27	0	0	0
28	0	0	0
29	0.353335	0.397616	0
30	0	0	0.200873
31	0	0	0
32	0	0.239276	0.213445
33	0.327824	0	0
34	0.566122	0.792891	0.803778
35	−0.04572	0	0

(continued)

Table 4 (continued)

Train set No.	Comp. strength	Split. strength	Flexural strength
	Weights (w_i)		
36	0	0	0
37	0	0	0
38	0	0	0
39	0	0	0
40	0.426148	0.175362	0.072424
41	0.282186	0.366191	0.497726
42	0.611104	0.417182	0.193352

Therefore, by trying to replace Eq. 4 in Eq. 3 by modifying the related σ values at Table 3 and w_0 will be 0, thus obtain their required formula with the help of the RVM method proposed.

$$y = \text{Comp. Strength} = \sum_{i=1}^{42} w_i \exp\left\{ -\frac{(x_i - x)^T (x_i - x)}{2 * (0.6)^2} \right\} + 0$$

$$y = \text{Comp. Strength} = \sum_{i=1}^{42} w_i \exp\left\{ -\frac{(x_i - x)^T (x_i - x)}{0.72} \right\} \tag{5}$$

$$y = \text{Split Strength} = \sum_{i=1}^{42} w_i \exp\left\{ -\frac{(x_i - x)^T (x_i - x)}{0.72} \right\}$$

$$y = \text{Flexural Strength} = \sum_{i=1}^{42} w_i \exp\left\{ -\frac{(x_i - x)^T (x_i - x)}{0.72} \right\} \tag{6}$$

The appropriate weightage factors "w_i" are discussed in Table 4 for the creation of models.

Compared to any other statistical app, one of RVM's major benefits is that it includes either the testing set or the research dataset. Below is the variance of the training and validation database for the built RVM models (Figs. 2 and 3). To evaluate the complexity of the system, the received variation might be used.

The prototype was checked with the existing research datasets after satisfactory implementation of RVM training. A normalized data is the position to generate derived from the RVM template and then, with the support of Eq. 7, the structured information is converted to its actual purpose.

$$x_i^a = x_i^n \left(x_i^{\max} - x_i^{\min} \right) + x_i^{\min} \tag{7}$$

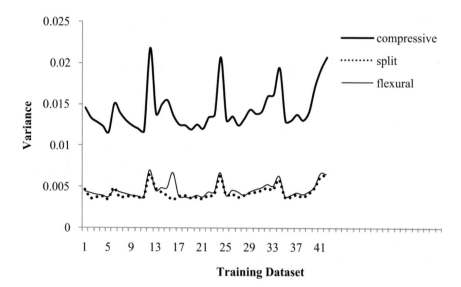

Fig. 2 Variance of training datasets

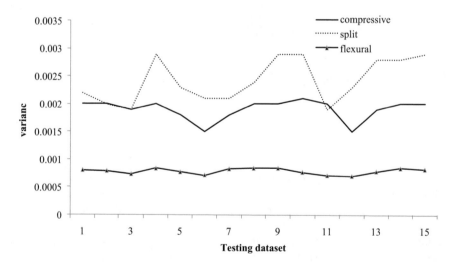

Fig. 3 Variance of testing datasets

Figures 4, 5 and 6 show the comparison among predicted and experimental mechanical strength properties. From Figs. 4, 5 and 6, it could be noted that the functional prototypes are effective and accurate.

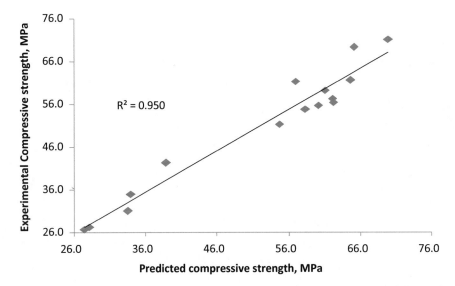

Fig. 4 Experimental versus predicted compressive strength

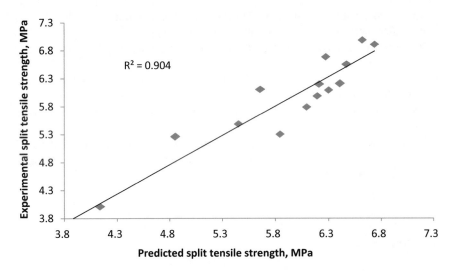

Fig. 5 Experimental versus predicted split tensile strength

6   Conclusions

- Concrete containing quarry dust, silica fume, steel fibers or polypropylene fibers mix proportions were tested for mechanical properties.
- Using MATLAB software, 3 RVM methods have been created for the classification and validation of mechanical properties.

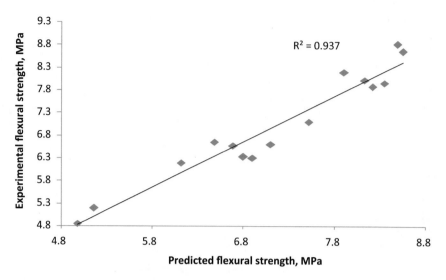

Fig. 6 Experimental versus predicted flexural strength

- The approved standard of the multilayer perception function's kernel width (σ) was calculated with the help of interpretation concepts.
- It was observed that the various strength of concrete cast is in strong relationship to the laboratory test results.
- The simulated results could be used by properly keeping track of the "w_i" weight values for the structural applications.
- The R values are near to 1 including all produced models, suggesting good model consistency.
- With the results obtained, the optimum form has been recognized and that can be carried out in the construction field to lessen the cost and instantaneously increase the durability factors.

References

1. Caesarendra W, Widodo A, Yang B-S (2010) Application of relevance vector machine and logistic regression for machine degradation assessment. Mech Syst Signal Process 24:1161–1171
2. Liu K, Xu Z (2011) Traffic flow prediction of highway based on wavelet relevance vector machine. J Inform Comput Sci 8(9):1641–1647
3. Han D, Cluckie I, Kang W (2002) Flow modelling using relevance vector machine (RVM). In: Hydroinformatics 2002: proceedings of the fifth international conference on hydroinformatics, Cardiff, UK © IWA Publishing and the authors
4. Das SK, Samui P (2008) Prediction of liquefaction potential based on CPT data: a relevance vector machine approach. In: 12th international conference of international association for computer methods and advances in geomechanics (IACMAG)

5. Caesarendra W, Widodo A, Yang B-S (2009) Application of relevance vector machine and logistic regression for machine degradation assessment. J Mech Syst Signal Process 24:1161–1171
6. Wang X, Ye M, Duanmu CJ (2009) Classification of data from electronic nose using relevance vector machines. Sens Actuators B 140:143–148
7. Dehwah HAF (2012) Corrosion resistance of self-compacting concrete incorporating quarry dust powder, silica fume and fly ash. Constr Build Mater 37:277–282
8. Yuvaraj P, Ramachandra Murthy A, Iyer NR, Samui P, Sekar SK (2014) Prediction of fracture characteristics of high strength and ultra-high strength concrete beams based on relevance vector machine. Int J Damage Mech 23(7):979–1004. https://doi.org/10.1177/1056789514520796
9. Ghosh S, Mujumdar PP (2008) Statistical downscaling of GCM simulations to streamflow using relevance vector machine. Adv Water Resour 31(1):132–146
10. Nataraja MC, Nagaraj TS, Reddy A (2008) Proportioning concrete mixes with quarry wastes. Cem Concr Aggr 23(2):81–87
11. Ho DWS, Sheinn AMM, Ng CC, Tam CT (2002) The use of quarry dust for SCC applications. Cem Concr Res 32(4):505–511
12. Safiuddin M, Raman SN, Zain MFM (2007) Flowing concretes with quarry waste fine aggregate. J Civil Eng Res Pract 4(1):17–25
13. Kumar PS, Mannan MA, Kurian VJ, Achuytha H (2007) Investigation on the flexural behaviour of high-performance concrete beams using sandstone aggregates. Build Environ 42(7):2622–2629
14. Eren Ö, Marar K (2009) Effects of limestone crusher dust and steel fibres on concrete. Constr Build Mater 23(2):981–988
15. Tipping ME (2000) The relevance vector machine. In: Solla SA, Leen TK, Muller KR (eds) Advances in neural information processing systems, vol 12, pp 652–658
16. Tipping ME (2001) Sparse Bayesian learning and the relevance vector machine. J Mach Learn 1:211–244
17. Wei L, Yang Y, Nishikawa RM, Wernick MN, Edwards A (2005) Relevance vector machine for automatic detection of clustered microcalcifications. IEEE Trans Med Imag 24(10)
18. Raman SN, Ngo T, Mendis P, Mahmudc HB (2001) High-strength rice husk ash concrete incorporating quarry dust as a partial substitute for sand. Constr Build Mater 25:3123–3130

AI-Enabled Circuit Design to Detect Walking Pattern for Humanoid Robot Using Force Sensor

Sandip Bhattacharya, Subhajit Das, Shubham Tayal, J. Ajayan, and L. M. I. Leo Joseph

Abstract Based on current work, we designed an AI-enabled recognition circuit for walking-pattern detection of a humanoid-robot walking on several indoor surfaces. The nearest-neighbor searching (NNS) algorithm is used for walking-pattern recognition. The NNS algorithm is finally converted into a circuit to test the walking pattern in real time. The force sensor is applied to generate a walking pattern, which is attached underneath the humanoid-robot feet. Our designed circuit performs 93.98% average recognition accuracy with fast recognition time of 2.91 ms. This circuit may be useful for real-time human identification by detecting the walking pattern instead of gesture recognition.

Keywords Artificial intelligence · Force sensor · Pattern recognition · Humanoid robot

1 Introduction

With the advancement of AI technologies, machine learning is acting an important role in humanoid robotics [1]. Nowadays, humanoid robotics is the latest research area in different research domains like space exploration, industrial operation, rescue management, medical application, military application, personal assistant, etc. In the early days, most robotics researchers are working in some specific fields like mechanical modeling of humanoid robot, kinematics, and inverse-kinematics simple pendulum model for stable walking on smooth and non-smooth surfaces with less number of application purposes. In recent days, robotics research becomes more challenging as per industry requirements. Most of the robotics industry wants to convert their existing system into an advanced system with autonomous or AI-enabled

S. Bhattacharya (✉) · S. Tayal · J. Ajayan · L. M. I. Leo Joseph
SR University, Warangal, Telangana, India
e-mail: 1983.sandip@gmail.com

S. Das
IIEST, Shibpur, India

features (i.e., a transformation from Industry-3.0 to Industry-4.0). The recognition-system design is a challenging research area nowadays. A number of research works have been conducted based on walking-patterns recognition by people identification that was reported in several research works [2–4]. In our research work, we designed an NNS algorithm-enabled circuit and implemented it on an FPGA platform to detect the robot walking pattern dynamically [3]. The tactile sensor (or force sensor) data is used to train the NNS algorithm [4–7]. The NNS algorithm is also implemented in the form of an AI chip using cadence virtuoso tools [7–9]. The main purpose to design an AI chip is because it is lower in dimension and easy to attach to the humanoid-robot body. In this research, we subdivided the work into three different modules, (a) robot learning and recognition framework design, (b) hardware architecture of NNS algorithm and its layout design, and (c) circuit implementation in FPGA platform to test walking patterns for recognition purposes.

2 Robot System Design

2.1 Robot Learning and Recognition Framework Design

Here, we discussed learning and walking-pattern recognition for a humanoid-walking robot. The whole system is subdivided into different units in Fig. 1a. In this research work, we considered KHR-3HV humanoid robot designed by KONDO-Kagaku Company ltd., two force sensors attached underneath of the robot feet, different walking surfaces (i.e., rough and smooth surface), A/D converter, learning algo-rithm circuit, pattern matching and decision-making circuit (NNS), and memory for reference data storage. When the humanoid-walking robot is walking on both rough and smooth surfaces, the force sensor which is attached underneath the robot's feet will generate an electrical signal. The force sensor is an electro-mechanical device, which converts mechanical force to an electrical signal. Force sensor generated elec-trical signal basically a raw-sensor data applied for a walking-pattern generation. A/D converter is present inside the Arduino-Uno board is used to convert analog-signal to digital-signal. This step is known as the data preprocessing step. The main purpose of these preprocessing steps is to generate filtered walking patterns so that the recognition system is capable of identifying each walking step correctly with higher accuracy. More than 1000 pre-pressing data (i.e., reference walking pattern) needs to be transferred into a learning circuits which is implemented on another Arduino Uno embedded board. Learning circuits will store these 1000 numbers of walking samples inside the memory (64 KB internal ROM). Information stored within the memory is known as the learning or training steps of a humanoid robot. For data processing on a large scale this internal memory that is present inside the learning circuit is not sufficient, for that reason, we interfaced one external memory separately with the learning circuit. Once the learning data or reference is available, this can be used to match with real-time walking patterns or online data patterns for testing purposes. In this experiment, two different types of the dataset are considered, i.e., reference dataset (R), stored inside flash memory, and testing data (T) known as real-time pattern. The reference datasets are represented as $R = [R\text{-}(1), R\text{-}(2), R\text{-}(3), R\text{-}(4) \ldots R\text{-}(n)]$, and testing datasets are represented as $T = [T\text{-}(1), T\text{-}(2), T\text{-}(3), T\text{-}(4) \ldots T\text{-}(n)]$. The NNS algorithm stores different walking patterns, which are trained from a reference dataset, inside a memory during the learning or training process. For walking steps or pattern recognition, a Euclidean distance calculation is implemented to compare each testing sample (T) with 1000 numbers of reference samples (R) to finds out the closest reference pattern with the associated robot walking action. Euclidean distance defined in Eq. (1)

$$E_D = \sqrt{\sum_{j=1}^{n} (R_j - T_j)^2} \tag{1}$$

The proposed work is implemented using a parallelized and pipelined hardware circuit (see Fig. 2a) for executing the distance metric, i.e., Euclidean distance $(E_D)^2$, to avoid the excess resources for square root calculation, to get fast pattern matching.

2.2 Hardware Architecture of Nearest-Neighbor Search (NNS) Algorithm and Its Layout

The NNS algorithm is implemented as hardware circuitry using Euclidean distance ED. Here, left and right foot force-sensors data are used as inputs of the humanoid robot and two surfaces, (i.e., smooth surface and rough surface). All four conditions can be recognized by this architecture (e.g., left leg smooth surface (LLSS), left leg rough surface (LLRS), right leg smooth surface (RLSS), and right leg rough surface (RLRS) in a parallel way. Here, T refers to the testing or online data (or real-time data), while $R\text{-}1$ to $R\text{-}N$ refers to the reference data. A number of different pipeline levels can be used to execute the data processing steps. We group 20 interest points originating from pressure-sensor data into basic detection units (or detection vectors), such as for the right foot on a rough surface, to prepare n vectors (20) for calculating the Euclidean distance for the following process in the first level. Data length n is a variable parameter. In the circuit, first and third pipelines calculate distance components between the N reference numbers, and the online force sensor data can be determined independently. In the fourth level of the pipeline, N multiplexors are used as data selectors, and they are used to estimate Euclidean distance between test data set and N reference data sets using all n components squared in parallel. The 2:1 multiplexer circuit is controlled by a 7-bit counter, and the short-distance search is calculated by a 32-bit comparator circuit. With the help of the cadence virtuoso simulation environment shown in Fig. 2b, the NNS hardware circuitry is converted into a layout.

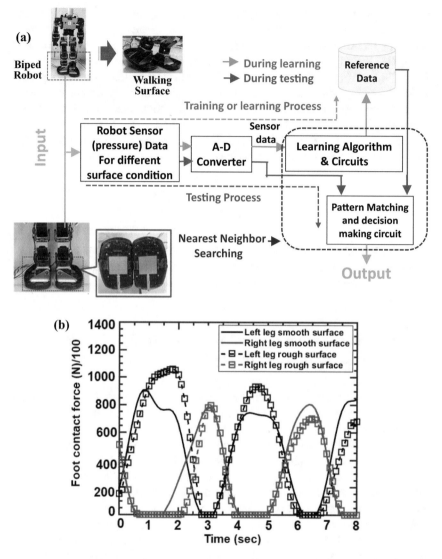

Fig. 1 **a** Robot learning and recognition framework. **b** Force sensor extracted walking-pattern generation for different footsteps and walking surfaces

2.3 Circuit Implementation in FPGA Platform

The general hardware architecture for pattern recognition is illustrated in Fig. 3. The entire NNS circuit is designed using Intel Quarts-II software and FPGA.

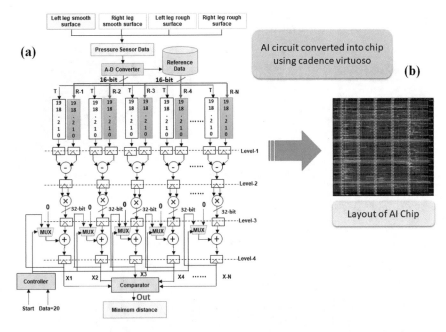

Fig. 2 a Walking-pattern matching circuit implemented using nearest-neighbor algorithm.
b Hardware circuit corresponding layout design using cadence virtuoso simulation environment

3 Result and Discussion

For real-time pattern recognition, we used a test bench shown in Fig. 3b, where four ON–OFF switches are used as online data or test data whereas internal memory of FPGA is used to store reference datasets. If both data are matched, then the LCD screen will display a red-shaded figure (see Fig. 3a). Using this hardware setup, we measured the walking-pattern recognition accuracy as well as recognition time. A 93.98% recognition accuracy is observed on average, whereas the average recognition time is 2.91 ms. Our experimental results are compared with some existing work, and it is shown that our analysis result shows better performance in terms of recognition accuracy and recognition time [10].

4 Conclusion

In this paper, we study how humanoid robots can accurately recognize walking patterns and their recognition times on different indoor surfaces. The force sensor generated offline walking pattern or data is used for real-time surface recognition purposes. The FPGA board is used for walking-pattern testing with a

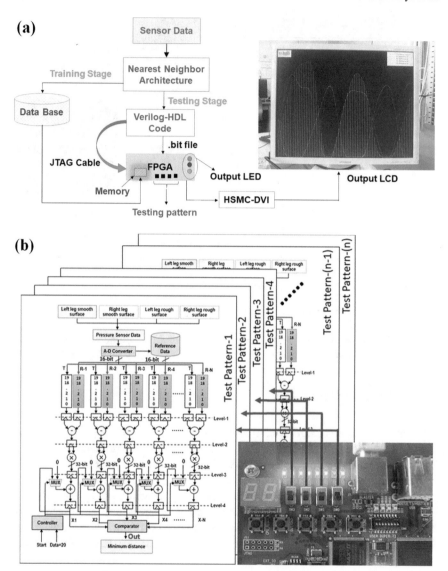

Fig. 3 **a** General hardware architecture for pattern recognition. **b** Hardware circuit implementation on FPGA

real robot walking environment. An approach such as this will be beneficial for future human identification instead of the face or Irish or any type of biometric recognition (Table 1).

Table 1 Measurement result for different walking pattern

Walking pattern on a different surface	Recognition accuracy (%)	Recognition time (ms)
LLSS	95.93	2.45
LLRS	92.45	3.33
RLSS	93.44	2.38
RLRS	94.12	3.51
	Avg-93.98	Avg-2.91

References

1. Wang S et al (2012) Machine learning algorithms in bipedal robot control. IEEE Trans Syst Man Cybern Part C (Appl Rev) 42(5):728–743
2. Farrell TR et al (2008) A comparison of the effects of electrode implantation and targeting on pattern classification accuracy for prosthesis control. IEEE Trans Biomed Eng 55(9):2198–2211
3. Intel—FPGA. https://www.altera.com
4. Dahiya RS, Metta G, Valle M, Sandini G (2010) Tactile sensing-from humans to humanoids. IEEE Trans Robot 26(1):1–20
5. Worth AJ, Spencer RR (1992) A neural network for tactile sensing: the Hertzian contact problem. IEEE Trans Syst Man Cybern 22(1):177–182
6. Jamali N, Sammut C (2010) Material classification by tactile sensing using surface textures. In: Proceedings IEEE international conference on robotics and automation, pp 2336–2341
7. Giguere P, Dudek G (2011) A simple tactile probe for surface identification by mobile robots. IEEE Trans Robot 27(3):534–544
8. Cover T et al (1967) Nearest neighbor pattern classification. IEEE Trans Inf Theory 13(1):21–27
9. Chen TW (2011) Flexible hardware architecture of hierarchical K-means clustering for large cluster number. IEEE Trans VLSI Syst 19(8):1336–1345
10. Wu XA et al (2016) Integrated ground reaction force sensing and terrain classification for small legged robots. IEEE Robot Autom Lett 1(2):1125–1132

A Review on Geo-location-Based Authentication with Various Lossless Compression Techniques

Vivek Kumar, Gursharan Singh, and Iqbal Singh

Abstract Data transactions on the Internet are rising every day, posing a significant security concern. Consequently, the development of techniques such as steganography and cryptography were required. These approaches change the raw data into a different format or into an unreadable format so that it can be safeguarded from attack. The attackers, on the other hand, are keeping up with the times. An old technique of geo-location is also a potent defender against attack of real-time information. It helps in providing a safe delivery of data at remote location. As a result, for increased security, numerous techniques combining these notions have been created. A review of several compression techniques and combinations of these and geo-location approaches is presented here. The complexity may degrade the quality of result for which Huffman is used which reduce the redundancy. Public-key cryptography algorithm can be used through which we can share key with the receiver. Algorithms like RSA and elliptical curve cryptosystem are such potent types. The review founds that a combined algorithm with authentication can make much composite and strong way to protect the data.

Keywords Huffman coding · Steganography · Cryptography · Elliptical curve cryptosystem · RSA · Geo-location

V. Kumar · G. Singh (✉)
School of Computer Science and Engineering, Lovely Professional University Jalandhar, Phagwara, Punjab, India
e-mail: gursharan.16967@lpu.co.in

V. Kumar
e-mail: kr.vivekumar8@gmail.com

I. Singh (✉)
School of Computer Science and Engineering, Modern Group of Colleges, Mukerian, Punjab, India
e-mail: Iqbal_306@rediff.com

© The Author(s), under exclusive license to Springer Nature Singapore Pte Ltd. 2023
A. Shukla et al. (eds.), *Computational Intelligence*, Lecture Notes in Electrical Engineering 968, https://doi.org/10.1007/978-981-19-7346-8_31

1 Introduction

In today's digital age, data is more significant than it has ever been for Internet communication. Encryption, steganography, and cryptography are some of the important principles that keep our information secure. Encryption transformed the raw data into cipher format using a variety of complicated and powerful algorithms. Because many of the algorithms are breakable, this strategy is ineffective on its own [1]. Using public–private key techniques like as elliptical curve cryptosystem (ECC) may, however, assist to strengthen it. Then, there's steganography, which encrypts data of one sort (text, audio, video, and picture) and hides in another. Image steganography, for example, is a technique for disguising plain text within an image. The least significant bit (LSB) is a popular method for embedding plain text (conversion into bits) at the least most bit of the pixels value [2].

Geo-location is becoming increasingly popular for real-time security in complicated scenarios. Previously primarily employed for military purposes, we now need to provide more complicated security in situations like COVID-19, where we are working from home. Another layer of protection is the use of a geo-location-based method when sending private data to a remote server or owner. It authenticates and sends the data after first encoding the receiver's location (latitude and longitude). The information to be deciphered is decoded using the location as the key. As a result, [3] incorporating geo-location will be a significant benefit for communications conveyed to a remote permanent site.

As steganography advances, new methods for ensuring more security are devised. For example, choose where to incorporate data. It will be easy for a hacker to obtain all the data if it is embedded in a continuous manner. As a result, selecting a random point (with the same unpredictability as at the time of extraction) is required. Some approaches, such as using chaotic map for pseudo random generator, offer a quick way to generate a random number [4]. Other steganography techniques, such as pixel mapping, DWT-DCT coefficient [5], and format-based techniques, are being researched for improved security. Combining, we can add extra layer of protection with encryption and steganography (Fig. 1).

Another method for safeguarding data transfer is cryptography. Despite of hiding the data, it converts it into an unreadable format, making it difficult for a hacker to access the information. The unreadable format, although, attracts more attackers than the steganography approach, which hides the possibility of cover media manipulation. AES is a great cryptography option. With increasing study in this domain, several combination approaches such as steganography and cryptography, encryption, steganography, and cryptography, and so on have been published [6].

When using these techniques and principles, several metrics like as PSNR, MSE, resilience, distortion, degree of quality, and accuracy is added into account [1]. Like, for geo-location, we can use GPS for assigning the location of receiver, however, it may happen that it deciphers the data the border of boundary or outside location. These all are concerned for making any algorithm successful and usable. To be more specific, preserving these values necessitated the use of additional methods such as

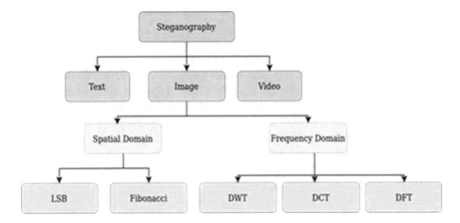

Fig. 1 Different types of steganography

compression, lossy, and lossless calculation. Humming coding is one such way; it improves in data compression by decreasing or ignoring unnecessary bits. Before getting into the full examination and algorithms of similar approaches, it is important to grasp the definitions of related methods.

Huffman Coding: A lossless data compression technique that generates a Huffman tree that may be used to restore data to its original state after compression. Prefixes are variable-length codes (bit sequences) that are assigned as input with proper care that they are not the prefix of any other character, thereby eliminating ambiguity [7].

Elliptical Curve Cryptosystem: ECC is a public-key cryptography based on the elliptical curve structure over a finite field. This method may be used in a variety of key management and security techniques, such as digital signatures, pseudo random generators, and more. This can also be used for encryption. ECC reduces the complexity of pairing supplied by other algorithms by providing a scalar product on an elliptical curve, resulting in lower memory consumption and resource requirements for users. Automated Encoding: The process of translating addresses (such as a street address) into geographic coordinates (latitude and longitude) that may be used to set markers on a map or position the map is known as automated geo-coding.

In this article, the reviews of various research outcomes have been mentioned in Sect. 2. Decoding the review, the problem statement is mentioned in the next section, i.e., Sect. 3. A methodology is proposed related to problem statement in Sect. 3 which is mentioned in Sect. 4 of this article. The conclusion is discussed in Sect. 4. References are then added in the last as per template.

2 Literature Review

For data security, Uttam et al. used a mix of visual cryptography and picture steganography technique. Visual cryptography works by splitting an image file into a few shares (noise image) at the sender's end. An eavesdropper may simply brute force the data. Less overhead in terms of space and computing complexity, as well as improved security. Bhardwaj et al. describe a novel privacy-aware framework for detecting and mitigating phishing attacks using cutting-edge strategies to maximize impact on unwary victims and the Internet of Things. The authors developed a new taxonomy for phishing. The suggested framework is capable of detecting numerous phishing assaults, according to the results.

Pranati et al. used a mix of visual cryptography with audio steganography to ensure safe data delivery. To obtain the picture, visual cryptography first divides the image file into a number of shares and requires the same or k (predefined) values of each share. However, because this approach works on the (n, n) number of shares on both the receiver as well as sender sides, it might be a weakness, making it easy for a hacker to extract the secret picture via eavesdropping. In many cases, the architecture connects to the unsafe Internet. This has exposed a number in the vast majority of cases, the architecture connects to the unsafe Internet. This has exposed a number of key flaws that have resulted in cyber-security assaults on IoT devices. IoT connections, protocols, and architecture were never designed to withstand modern-day cyber-security threats. The computing, storage, network, and memory capabilities of IoT devices are restricted. The authors offer IAF, a novel IoT attack architecture that focuses on the impact of IoT assaults on IoT applications and service levels, in this study. Bhardwaj et al. also developed an attack taxonomy that would categories various assaults against IoT networks.

For the hidden text embedding approach, Osama et al. employed Huffman coding for text compression. To eliminate data redundancy while retaining quality, Huffman is utilized. RSA also offers a fresh way to developing a more complicated and robust data security algorithm. It is a dependable solution since it requires less space and transmits data quickly. By Agrawal et al., a realistic comparison of AES, DES, and RSA is carried out in order to evaluate their performance criteria in terms of space complexity, encryption time, and throughput. Experiments are conducted on various types of data files in order to assess the performance of three. As a result, we can estimate the extent of security given by these three and, as a result, we may construct a layered structure in our system for higher and more complicated algorithms. Also, we may utilize them in different ways for distinct functions, such as using RSA for key sharing and AES for encrypting or converting information into an unreadable format.

Mritha et al. proposed employing the affine transformation approach to hide data. The video is first divided into 8×8 blocks, further split into 4×4 blocks to create sub bands. The IWT coefficient is derived using the HH band, multiplied by the identity matrix, and the MSB replaced by data. During decryption, the procedure is reversed. This approach aids in the reduction of calculation costs and noise in the final result.

Identifying the characteristics and discovering the stegno-images are two of the most common steganalysis issues. Rani et al. have used a Gaussian distribution to create a steganalysis metric for picture modeling. The distribution of DCT coefficients is quantified using a Gaussian distribution model and a ratio of two Fourier coefficients of the distribution of DCT coefficients. This resulting steganalysis metric is compared to three steganographic methods: least significant bit (LSB), spread spectrum image steganography (SSIS), and steg-hide (graph theoretic approach). Different classification approaches, such as SVM, are used to classify the picture features data set.

Sabyasachi et al. have studied data transmission security by union of steganography and encryption. A digital signature is also used to verify the sender's identity at the receiving end. It is more dependable and safer when the RSA algorithm or public keys and private keys are used. It also reduces distortion and increases invisibility.

For understanding biometric identification and user geo-location, Oleh et al. recommend a change to the customer identification approach. The various steps for identification are observed, with an array of features taken into account (tuple of geo-location data, i.e., object motion in space as well as time, and parameters of their buying). The table of payment statistical likelihood is constructed based on the previously specified facts, and the user is recognized using the table. The distance between the user's location coordinates and the time of their movement to the outskirts of existing clusters is compared to the distance between the user's location coordinates and the time of their transaction, allocating the user to either "identified," "unidentified," or "needs the identification of the PIN." High levels of security, ease of use, dependability, flexibility, scalability, and efficiency are all requirements for adopting a contact less payment system.

Akoramurthy et al. has used the user's cellphone number and geo-location are utilized to authenticate the user in addition to the existing two-factor authentication approach employing user ID, password, and OTP. The geo-location informs the bank of the location from which the transaction will be carried out, assisting banks in ensuring secure transactions. The network provider obtains the user's geo-location, eliminating the requirement for GSM. Geo-locations multi-factor authentication improves security when conducting mobile transactions and protects users from a variety of threats.

Hikiya et al. developed a method that uses a mix of DWT and DCT techniques, with DWT modification of the host picture followed by DCT on the selected DWT sub bands. In this case, vectors are also employed to enhance security of DWT and DCT. It improves degree of quality and safety while providing resilience and imperceptibility. Using the ECC algorithm, Joseph et al. improved the security of a steganography and cryptography combo. This technique generates the public for sender and private keys for receiver to perform plain text encryption. A secret image is also used as a cover for encrypted text, with 40% of the common combined with the base image.

Baziyad et al. have presented an alternative for obtaining the image's superfluous space to integrate additional data into it. The video is turned into a three-dimensional vector, with the first two dimensions being the rows and columns of pixels, and the last being the time variable and redundant for embedding data. The redundant coefficient is obtained by finding the connection between the moving frames of video and performing DCT on it using the motion vector. The suggested approach takes advantage of redundant video segments to build a high-quality stego picture with a high hiding capacity ratio.

3 Problem Formulation

The backdrop of some of the paper is the use of discrete concepts to provide security and for others achieving more than one layer of security is an advantage. Like, if a text is encrypted using Caesar cipher, it could be easily cracked. But if it is followed with addition of these, it can provide a complex and hard to crack scenario. Some of the existing researches have used single approach with strong ways like visual cryptography but these are directly prone to strong brute force attack. Audio steganography could be eavesdropped if it is implemented alone.

Although, various existing research have used more than one algorithm to provide higher level of security, but with combination of different algorithms makes it complex and destroy the quality of information and sometimes it is lost to complexity. That means the combination advantage is also harmful. Although, if we take care about noise distortion while processing of data, it can be reduced. Several compression techniques like DWT, DCT are available to provide data compression such that the retrieval is smooth, and quality is maintained [8]. After reviewing various topics related to steganography and cryptography, we concluded that a combination of these is good to go and geo-location encoding is enough to provide stable security.

4 Proposed Methodology

The plain text is first enciphered using elliptical curve cryptosystem of which the key had to be share in advance. The major approach would be as follow:
Sender Side:

- Encryption of raw data using public key of receiver.
- At this stage, we can embed the receiver's coordinate into the cipher text as header or combined it.
- Embedding of plain text into a base image using Huffman coding technique.
- The encoded image would be then combined with cover image using DWT-DCT compression technique.
- This combined image could be inserted into a video as a frame using pseudo random place technique.

Receiver Side:

- Extraction of embedded image from the video.
- Extraction of base image from the combined image.
- Extraction of cipher text from the base image.
- The receiver coordinates will be already checked up and can proceed for next step.
- Applying the private key decrypt, the plain text.

The steps involved in putting the suggested strategy into practice depicts in Fig. 2, which might result in better results and data security. The first module or step is to convert plain text (data) into encrypted text using the receiver's public key (RSA or elliptical curve cryptosystem). At this stage, the cipher text is embedded with the coordinates data of receiver like a header in any data packet transfer. Here, we can provide the range, also boundary in which this data should be opened. The encrypted text would then be embedded into the basic medium (steganography) using Huffman coding, which is a lossless data compression method that allows us to return the image to its original state without losing any data. Furthermore, the base picture is embedded (or hidden) into a cover media utilizing a mix of DWT-DCT compression techniques providing an additional tier or layer of protection. This combination of algorithms and steganography tiers may give increased security. The final step is to turn the cover picture into a video frame (which will be delivered to the recipient), and the frame is picked using the powerful pseudo random algorithm.

The receiver could reverse the procedure after receiving the secret data from the transmitter in order to recover the genuine data. Figure 3 depicts the many phases of retrieving our data. To begin, the video or data will be separated into frames (the division will be same as it was divided on sender side, this value should be known while key sharing). The shared key would be used to recover the secret picture that was embedded in frames (generated by pseudo random generator). The receiver must now choose an image from the frames that correspond to the specified number or

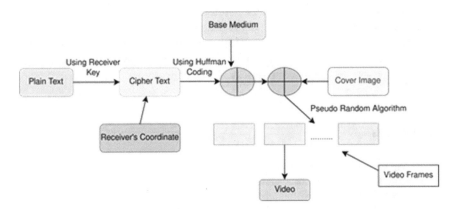

Fig. 2 Encryption at sender's side

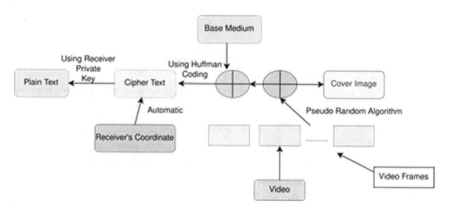

Fig. 3 Decryption at receiver side

key. The picture retrieved is a composite of the base and cover images, which were split using steganography (image). The encrypted text (lossless) would be extracted from the base picture using Huffman coding after separation. The encrypted text will be provided, which can be readily decrypted to plain text (data) using the receiver's private key. Here, the coordinate will be automatically verified and if the receiver is in provided range or boundary, it would be open, but it will fail outside the boundary.

5 Conclusion

The final picture will be less distorted if several steganographic compression techniques, such as Huffman lossless compression and DWT-DCT compression, are used together. By reducing superfluous data, the former gives ample room for embedding data, while the latter gives higher security with complicated embedding. Elliptical curve cryptosystems, often known as public-key cryptosystems, aid in providing security at the very last minute, i.e., at the plain text level. Geo-location adds authorization level for more user-friendly structure. Consequently, the algorithm will become more secure along with quality of data is maintained. Hence, a combined approach is suggested with taking quality as in picture.

References

1. Kumar, V, Kaur H, Paul S (2021) CryptMe: a cryptographic framework with steganography for securing data. In: 2021 2nd international conference on intelligent engineering and management (ICIEM). IEEE, pp 294–299
2. Rakshit P, Ganguly S, Pal S, Aly AA, Le D (2021) Securing technique using pattern-based lsb audio steganography and intensity-based visual cryptography. Comput Mater Continua 67(1):1207–1224

3. Akoramurthy B, Arthi J (2017) GeoMoB—a geo location based browser for secured mobile banking. In: 2016 8th international conference on advanced computing (ICoAC). IEEE, pp 83–88
4. Gahan AV, Devanagavi GD (2020) A secure steganography model using random-bit select algorithm. In: 2020 3rd international conference on advances in electronics, computers and communications (ICAECC). IEEE, pp 1–5
5. Korgaonkar VV, Gaonkar MN (2017) A DWT-DCT combined approach for video steganography. In: 2017 2nd IEEE international conference on recent trends in electronics, information & communication technology (RTEICT). IEEE, pp 421–424
6. Mondal UK, Pal S, Dutta A, Mandal JK (2021) A new approach to enhance security of visual cryptography using steganography (VisUS). arXiv Preprint arXiv:2103.09477
7. Gopinath A, Ravisankar M (2020) Comparison of lossless data compression techniques. In: 2020 international conference on inventive computation technologies (ICICT). IEEE, pp 628–633
8. Patel R, Lad K, Patel M (2021) Study and investigation of video steganography over uncompressed and compressed domain: a comprehensive review. Multimedia Syst 27(5):985–1024
9. Bhardwaj A, Al-Turjman F, Sapra V, Kumar M, Stephan T (2021) Privacy-aware detection framework to mitigate new-age phishing attacks. Comput Electr Eng 96:107546
10. Bhardwaj A, Kumar M, Stephan T, Shankar A, Ghalib MR, Abujar S (2021) IAF: IoT attack framework and unique taxonomy. J Circ, Syst Comput, p 225002
11. Wahab OFA, Khalaf AA, Hussein AI, Hamed HF (2021) Hiding data using efficient combination of RSA cryptography, and compression steganography techniques. IEEE Access 9:31805–31815
12. Agarwal J, Kumar M, Srivastava AK (2021) Estimation of various parameters for AES, DES, and RSA. In: Emerging technologies in data mining and information security. Springer, Singapore, pp 275–283
13. Ramalingam M, Isa NAM, Puviarasi R (2020) A secured data hiding using affine transformation in video steganography. Procedia Comput Sci 171:1147–1156
14. Rani A, Kumar M, Goel P (2016) Image modelling: a feature detection approach for steganalysis. In: International conference on advances in computing and data sciences. Springer, Singapore, pp 140–148
15. Pramanik S, Bandyopadhyay SK, Ghosh R (2020) Signature image hiding in color image using steganography and cryptography based on digital signature concepts. In: 2020 2nd international conference on innovative mechanisms for industry applications (ICIMIA). IEEE, pp 665–669
16. Zolotukhin O, Kudryavtseva M (2018) Authentication method in contactless payment systems. In: 2018 international scientific-practical conference problems of infocommunications. Science and technology (PIC S&T). IEEE, pp 397–400
17. Joseph H, Rajan BK (2020) Image security enhancement using DCT & DWT watermarking technique. In: 2020 international conference on communication and signal processing (ICCSP). IEEE, pp 0940–0945
18. Gladwin SJ, Gowthami PL (2020) Combined cryptography and steganography for enhanced security in suboptimal images. In: 2020 international conference on artificial intelligence and signal processing (AISP). IEEE, pp 1–5
19. Baziyad M, Rabie T, Kamel I (2020) Directional pixogram: a new approach for video steganography. In: 2020 advances in science and engineering technology international conferences (ASET). IEEEE, pp 1–5
20. Mukherjee S, Roy S, Sanyal G (2018) Image steganography using mid position value technique. Procedia Comput Sci 132:461–468
21. Rehman A, Saba T, Mahmood T, Mehmood Z, Shah M, Anjum A (2019) Data hiding technique in steganography for information security using number theory. J Inf Sci 45(6):767–778
22. Gutub A, Al-Ghamdi M (2019) Image based steganography to facilitate improving counting-based secret sharing. 3D Res 10(1):1–36
23. Muhammad W, Sulaksono DH, Agustini S (2020) Message security using Rivest-Shamir-Adleman cryptography and least significant bit steganography with video platform. Int J Artif Intell Robot (IJAIR) 2(2):52–61

24. Mulya M, Arsalan O, Alhaura L, Wijaya R, Ramadhan AS, Yeremia C (2020) Text steganography on digital video using discrete wavelet transform and cryptographic advanced encryption standard algorithm. In: Sriwijaya international conference on information technology and its applications (SICONIAN 2019). Atlantis Press, pp 141–145
25. Saravanan M, Priya A (2019) An algorithm for security enhancement in image transmission using steganography. J Inst Electron Comput 1(1):1–8
26. Fotohi R, Firoozi Bari S, Yusefi M (2020) Securing wireless sensor networks against denial-of-sleep attacks using RSA cryptography algorithm and interlock protocol. Int J Commun Syst 33(4):e4234
27. Jenifer JM, Ratna SR, Loret JS, Gethsy DM (2018) A survey on different video steganography techniques. In: 2018 2nd international conference on trends in electronics and informatics (ICOEI). IEEE, pp 627–632
28. Thakur RK, Saravanan C (2016) March. Analysis of steganography with various bits of LSB for color images. In: 2016 international conference on electrical, electronics, and optimization techniques (ICEEOT). IEEE, pp 2154–2158
29. Taqa A, Zaidan AA, Zaidan BB (2009) New framework for high secure data hidden in the MPEG using AES encryption algorithm. Int J Comput Electr Eng 1(5):1793–8163

Fine-Tuning BART for Abstractive Reviews Summarization

Hemant Yadav, Nehal Patel, and Dishank Jani

Abstract Abstractive text summarization is a widely studied problem in sequence-to-sequence (seq2seq) architecture. BART is the state-of-the-art (SOTA) model for sequence-to-sequence architecture. In this paper, we have implemented abstractive text summarization by fine-tuning the BART architecture which improves the model significantly resulting in the optimized overall summarization quality. Here, use of Sortish sampling allows the model to become smoother and faster. Also usage of weight decay increments the performance by adding regularization to the model. For the improvement of the input data, BartTokenizerFast is used for tokenization. The paper concludes with the comparison of the proposed optimization strategy with previously published models. The effectiveness is evaluated using a ROUGE score which is giving similarity between two documents. Experiments indicate that the fine-tuned BART model proposed achieves significant improvement in ROUGE scores on Amazon reviews data.

Keywords Abstractive summarization · BART · Reviews summarization · Transformers · Bidirectional encoder · Seq2Seq

1 Introduction

People are becoming overwhelmed by the vast amount of information and documents available on the Internet as the Internet and big data continue to grow in popularity. As a result, various investigators are motivated to develop a technological approach that can automatically summarize texts. Text summarizing began in the late 1950s [1], and it has progressed significantly since then. In this field of study, a wide variety of approaches have been created [2]. The technique of creating a brief text that covers the

H. Yadav (✉) · N. Patel · D. Jani
K D Patel Department of Information Technology, Faculty of Technology & Engineering (FTE), Chandubhai S. Patel Institute of Technology (CSPIT), Charotar University of Science and Technology (CHARUSAT), Changa, Gujarat, India
e-mail: hemantyadav.it@charusat.ac.in

N. Patel
e-mail: nehalpatel.it@charusat.ac.in

© The Author(s), under exclusive license to Springer Nature Singapore Pte Ltd. 2023
A. Shukla et al. (eds.), *Computational Intelligence*, Lecture Notes in Electrical Engineering 968, https://doi.org/10.1007/978-981-19-7346-8_32

major points of a longer document is known as automatic text summarization. Readability, coherency, syntax, non-redundancy, sentence ordering, conciseness, information diversity, and analysis are all significant components of a decent summary [3]. Summaries in the extractive approach concatenate the most essential sentences from the source text with no alterations, but summaries in the abstractive method are wholly new formulations that convey the same sense as the concepts in the original document. Abstractive summarization is more difficult because it involves higher-level considerations including semantic representation, surface manifestation, and content structure [4].

The most generally utilized way for implementing sequence-to-sequence models is deep encoder–decoder architecture. However, this models has a number of difficulties. For instance, they are unable to effectively manage long-term addictions. Furthermore, due to their sequential character, they cannot be parallelized. Low novelty, exposure bias, mismatch, and lack of generalization are among the other concerns.

A self-attention transformer [5] architecture is the first self-attention transduction model to describe its source data and results without using sequence-based recurrent neural networks or convolutional neural networks. The earlier challenges with sequence-to-sequence models have been solved by this architecture. It has a parallel-based design that supports parallelization, which substantially enhances computing speed and gives more accurate summaries with higher ROUGE scores [6], which is the most generally used abstractive automatic text summarization measure. Many large-scale pre-trained language models, such as Facebook's BART, Google's BERT, PEGASUS, T5, and Switch, Open AI's generative pre-training (GPT), GPT-2, and GPT-3, are trained on massive corpora of texts created by big businesses using this architecture [7–13].

For pretraining sequence-to-sequence models, BART is a denoising auto-encoder. BART is learned by using an arbitrary noising function to distort text and then building a model to recover the original text. It employs a typical transformer-based neural machine translation architecture that, despite its simplicity, generalizes BERT—due to the bidirectional encoder, GPT—due to the left-to-right decoder [7].

This paper focuses on fine-tuning of the BART pre-trained model for abstractive summarization of Amazon reviews. BART is basically a combination of bidirectional and auto-regressive transformers trained for creating denoising auto encoders [7]. It is pre-trained on corrupted text generated by an arbitrary noising function to learn a model to reconstruct the original text. As BART has an autoregressive decoder, it can be fine-tuned for sequence generation tasks such as abstractive text summarization [7]. It is observed that setting the weight decay very small for regularization has significantly improved the performance. Also the Sortish sampling of tokens made the model smoother and faster.

2 Background

Deep learning has applications in variety of NLP tasks because it allows for the learning of multilevel hierarchal data representations employing many nonlinear data processing layers [14–17]. RNNs, CNNs, and sequence-to-sequence deep learning models are used for text summarization. In 2015, deep learning was employed for the first time in abstractive text summarization, and the model proposed, was based on the encoder-decoder architecture [18]. From the perspectives of network architectures, training methodologies, and summary generation algorithms, Shi et al. [19] give a comprehensive literature analysis on multiple sequence-to-sequence models for abstractive text summarization.

The sequence-to-sequence model is used in the RNN encoder–decoder architecture. This model converts the neural network's input sequence into a similar sequence of characters, words, or phrases. Several NLP applications, such as machine translation and text summarization, use mathematical problems in engineering. The input sequence in the document that needs to be summarized, and the output is the summary [20, 21]. Lopyrev [22] developed a simpler attention method that was used to produce headlines for news items using an encoder-decoder RNN.

Unidirectional RNNs can only capture context based on the token's history, and therefore do not take into account all of the context needed in the future. This could result in erroneous predictions and, as a result, improper data creation. In human perception, the context of both sides of a sentence is taken into consideration when predicting an unfamiliar word in a sentence. As a result, utilizing bidirectional RNN improves results by allowing each concealed state to be aware of contextual information from both directions, i.e., past and future contexts. Al-Sabahi et al. [23] suggested a bidirectional encoder–decoder model that significantly enhances the results of generated summaries and eliminates the issue of creating incorrect data. For abstractive summarization of tiny datasets, Khandelwal [24] used a sequence-to-sequence model consisting of an LSTM encoder–decoder. After reading the encoder's hidden representations and transferring them to the softmax layer, the decoder develops the summary of output. As it is known fact that the sequence-to-sequence model does not memorize information, it cannot be generalized. To take the decision to select the golden token or the previously developed output at each step, the model given here uses imitation learning.

Bidirectional encoder representations from transformers (BERT) [8] has potential, and it can be fine-tuned with additional output layer for tasks like question answering which means less task-specific architecture modifications. Liu et al. [25] have fine-tuned BERT for abstractive and extractive summarization. CNN/Daily Mail and NYT datasets are used for experiments without exploiting any language generation capabilities. Liu et al. used the encoder as a BERT-pre-trained document-level encoder, while the decoder was a transformer that was randomly initialized and taught from the ground up.

In 2015, 2017, and 2018, Lopyrev [22], See et al. [26], Paulus et al. [27], Xia et al. [28], Ling et al. [29], Zhang et al. [30], Celikyilmaz et al. [31], Lin et al. [32], and Pasunuru et al. [33] used the XENT, RL, and hybrid-objective training model to implement LSTM to LSTM framework. With the Gigaword, CNN/DM, Newsela, NYT, LCSTS, WikiSmall, and WikiLarge datasets, they used several optimizers such as RMSProp, Adadelta, Adam, and SGD. They have also used BLEU, ROUGE, and human measures.

Chen et al. [34], Gulcehre et al. [35], Gu et al. [36], Zhou et al. [37] and Li et al. [38] deployed the XENT and RL training model with Adadelta and SGD optimizer to construct the GRU to GRU framework using CNN, LCSTS, DUC, Gigaword, and MSR-ATC datasets with ROUGE metrics in 2016, 2017, and 2018. Gehring et al. [39], Fan et al. [40], and Wang et al. [41] built a CNN to CNN framework using DUC, Gigaword, CNN/DM, and LCSTS datasets with ROUGE and human metrics in 2017, using the XENT and RL training model and Adam optimizer.

In 2019, Zhang et al. [42] suggested a scenario in which they want to find the best RL objective that balances the factual accuracy, linguistic likelihood, and overlap with the target summary in the model summary. In the year 2020, three well-known datasets, CNN, Daily Mail, and their merged form CNN/Daily Mail, were used to evaluate the proposed model by Mohsen et al. [43]. Experiments demonstrated that the model outperformed SOTA models in terms of ROUGE scores on all three datasets.

3 Proposed Model

3.1 Dataset and Data Preprocessing

Most natural language processing architectures require an adequate collection of input text corpus and label summaries for which Amazon Fine Food reviews dataset [44] is used in this paper. Amazon Reviews Dataset consists of ten attributes as ID, product ID, user ID, profile name, helpfulness numerator, helpfulness denominator, score, time, text, and summary. In text summarization only two columns text and summary are considered which are text corpus and label summary. The dataset contains 568,427 data entries with text and its summary from over 256,059 users. There are approximately 568,427 data entries with text and its summary from over 256,059 users. Here, the significant amount of reviews is duplicated, thus it is important to limit the source review tokens and dropping duplicates. Ax length limit of sequences is set to 1024 tokens as BART can take a 1024 token sequence [7].

The pre-processing of Amazon reviews data is performed by elimination of unnecessary columns and dropping null values. Bart fast tokenizer of Huggingface library is used for tokenizing the training data in a BART specific input format before passing it to the model [45]. The advantage of using Huggingface fast tokenizer over base BART tokenizer is that it has capacity to handle all the shared methods for

Fig. 1 BART architecture
[7]

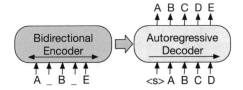

tokenization and special tokens automatically. Here, it is not necessary to specify the decoder attention mask. Input attention mask, input IDs, and labels are encoded as input tokens. Train data and validation data maps these tokens and batch size for fine-tuning is set to 256.

3.2 Model Architecture

BART is the SOTA architecture for seq2seq models which can be termed as generalizing BERT and GPT, where BERT is used as bidirectional encoder. BART employs sequence-to-sequence transformer architecture from [5], with activation function GeLUs [46] instead of ReLU. Core model of BART is used which has six encoder and decoder layer. Each layer of the decoder has cross-attention to the last hidden layer of the encoder. It results in BART has in total has around 10% more parameters, then in BERT [7]. Figure 1 shows the BART architecture.

Experiments suggests that weight decay becomes more effective with layer dropout for efficient pruning. Weight decay is set as 0.01 which helps in regularization to avoid overfitting. Decoder layer dropout is set as 0.2. One more observation is that during fine-tuning decreasing the batch size, the ROUGE score decreases, thus batch size for fine-tuning is set to 256. Once model is established, batch size is set to 4 during training of fine-tuned model and maximum length of 1024 maximizes the ROUGE scores on Amazon reviews. Model is constructed with the additional decoder layer drop and attention dropout.

Sortish sampling is employed for selecting tokens which makes the model smoother and faster by generating accurate predictions on unsampled data. Using Sortish sampling the padding tokens can be minimized as it sorts the text based on the input text length. Thus by doing this, the batch will contain sequences of words of similar length which will lead to minimization of pad tokens and makes the BART model much smoother.

3.3 Training of Model

The preprocessed input training data which consists of over 550,000 reviews and label summaries is passed to the auto-regressive encoder and training is instantiated on the sequence-to-sequence trainer with sequence-to-sequence training arguments. Sequence-to-sequence training arguments class is developed to fine-tune the training arguments on pre-trained BART model to improve performance of BART on reviews summarization, where batch size is set to 4 and logging steps is set to 1000. Also, encoder and decoder layer dropout is set to 0.2. Validation set consists of 5000 data entries, Training set consists of 550,000 reviews to train the BART model efficiently, and testing set is considered with almost 2000 reviews, and their summaries to evaluate on trained model. The hyper parameters were updated manually. Warm up steps are set to 1000 to use a low learning rate for initial 1000 steps and after that the model is trained on a regular learning rate scheduler. This helps to fine-tune the attention mechanisms. Number of beams set to 4 which is used to select the best options for text generation. No repeat n-gram size is set to 3 so that all the n-grams of size 3 can only occur once to avoid the phrase repetitions. The learning rate of BART is set to 0.15 with linear learning rate scheduler which increases the learning rate to tune the model with faster.

Here, AdamWoptimizer [47] is used for training the model to increase the rate of gradient descent algorithm to accelerate the performance of the model. It is necessary to assign learning rate dynamically for each step thus linear learning rate scheduler is used here due to sequence-to-sequence training. The model length penalty is assigned to 2 which gives advantage to long input reviews and increases the efficiency.

4   Results

The results in Table 1 demonstrate sample generated summaries with the original label summary. The results are performed on the Amazon reviews. The summary generated by the model looks efficient and much closer to the label summary.

We study that the predicted summary and label summary are close to each other. While in some of the cases such as in the fourth example, we can observe that the semantics of sentences changes as the model is trained on just 16 epochs which indicates more training. The evaluation of our model is performed on ROUGE [48] metrics. It is better to use ROUGE score for abstractive text summarization because it indicates the overlapping of machine generated summary with original summary without taking semantics into consideration. The ROUGE technique compares the predicted summary with the original labeled summary and generates a score based on similarity and context. It basically overlaps N-grams thus acting like a proxy metric for text summarization. ROUGE 1 evaluation is calculation of measuring overlapping of each word (unigram), ROUGE N refers to measurement of higher order n-grams. ROUGE L measures in context with longest common subsequences. The model is evaluated on ROUGE 1 and ROUGE 1 computer metrics. The result is also computed on ROUGE 1 precision, recall, and F1-score.

Table 1 Comparison of original and predicted summary

Text	Original summary	Predicted summary
This is a product used in my house for the past 60 years….need to say more…	Lavazza or espresso	Great product
Before I purchased this product, I read one of the reviews which said the Lavazza black can was better. After buying and tasting the Lavazza qualityor, I have to agree. The black can is better. The qualityor leaves a bitter aftertaste, and has an acidic quality that doesn't appeal to me. In my opinion, the Lavazza black can of coffee has a better aroma and a smother taste	Smooth	Good cup of coffee
The coffee is fine. But, the first box I received was badly damaged, with four of the eight cans having beed crushed open. The reason: poorly packed with no bubble wrap or protection. I asked for replacements, which were sent, but again, poorly packed and some cans were damaged. I talked to customer service, who just did not seem to get the fact that they need to pay attention to how they pack and ship. Otherwise fine	Coffee was fine; packaging was not	Damaged goods
Excellent flavor. Arabica beans is the only way to go for an espresso or cappuccino having the same flavor as the ones made in Italy	Authentic Italian flavor	Excellent flavor

5 Analysis

Generally, it has been found that ROUGE score for reviews summarization on Amazon reviews tends to be lower compared to news articles summarization on CNN/DM data as it contains the reviews of the customer which contains various stop words and grammatical errors. The experimental findings of several models using ROUGE 1 assessment metrics for $F1$-score, recall, and precision are provided in Table 2.

Evaluation of the proposed model is tested on Amazon reviews. The main aim is to show the results of the proposed architecture for $F1$-score, recall, and precision. It is clear that T5 architecture performs well on recall and $F1$-score, while the proposed model works well for precision. The model took 9 h to get trained for 16 epochs for beam size as 4 and vocabulary size as 55,000, most of the model were trained on

Table 2 Comparison of summarization models on ROUGE metrics from scale 0 to 1

Abstractive summarization model	R 1—precision	R 1—recall	R 1—F1-score	Dataset
Proposed model	0.698	0.511	0.482	Amazon reviews
Seq2Seq model [49]	0.388	0.324	0.313	BBC news data
LSTM + Attention [50]	0.333	–	–	Amazon reviews
Transformers architecture [51]	0.58	–	–	COVID-19 dataset
T5 transformers model [49]	0.467	0.48	0.473	BBC news data
BART Model (abs) [52]	N/A	0.4697	0.4350	CL–LaySumm 2020
BART + multi-label [52]	N/A	0.5013	0.4600	CL–LaySumm 2020

Table 3 Comparison of models on ROUGE score of scale 0–100

Model	ROUGE 1	ROUGE L
Transformers [53]	17.87(F1-score)	–
Transformers + Pointer generator [54]	22.10	14.66
Transformers + Pointer generator + n–grams [54]	25.31	15.99
Proposed model	28.85	28.11

nearly 24–30 h for a many of epochs. From the graph, we can see that on increasing the number of epochs the ROUGE metrics score increases, thus training the model for a large number of epochs can result in even better score and accuracy. Table 3 indications the evaluation on ROUGE 1 and ROUGE L metrics for proposed model of Amazon reviews summarization.

Here, deaton [55] summarization model is performed on CNN/Daily Mail dataset, while transformer model [6] is performed on Amazon reviews dataset. Figure 2 shows that on increasing the epochs the ROUGE score increases, thus we can conclude that if the number of epochs increases overall ROUGE 1 and ROUGE L score increases.

6 Conclusion and Future Work

On fine-tuning BART architecture, we have enhanced the performance of the model and showed better results on ROUGE compute metrics evaluation. The comparison with the previous SOTA models on Amazon Fine Food reviews dataset is also shown in an efficient manner. Here, it is also found that if the model is trained on an even larger value of epochs and an increasing number of steps, then the performance can further be enhanced also. It also concluded that implementation and improving

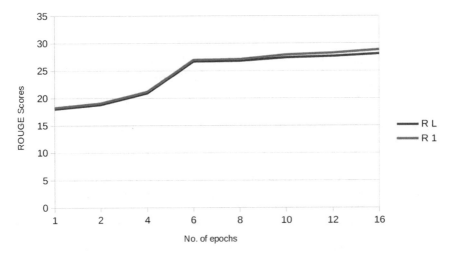

Fig. 2 Rouge score vs number of epochs

Sortish sampling has made the model smoother and modifying weight decay has helped to increase the performance. The Bart fast tokenizer has also helped to eventually improve overall efficiency. The experimental results evaluated on ROUGE 1 and ROUGE L metrics improves for $F1$-score, precision, and recall. In the future, the BART model for Amazon reviews with multi-label can be implemented to help improve the performance. Another work which can improve the efficiency is addition of pointer generator networks with BART architecture can help to accurately reproduce the information and keep track of what has been summarized.

References

1. Luhn HP (1958) The automatic creation of literature abstracts. IBM J Res Dev 2(2):159–165
2. Jones KS (2007) Automatic summarising: the state of the art. Inf Process Manage 43(6):1449–1481
3. Gambhir M, Gupta V (2017) Recent automatic text summarization techniques: a survey. Artif Intell Rev 47(1):1–66
4. Yao J-g, Wan X, Xiao J (2017) Recent advances in document summarization. Knowl Inf Syst 53(2):297–336
5. Vaswani A, Shazeer N, Parmar N, Uszkoreit J, Jones L, Gomez AN, Kaiser L, Polosukhin I (2017) Attention is all you need. Adv Neural Inf Process Syst 5998–6008
6. Lin, Chin-Yew (2004) Rouge: a package for automatic evaluation of summaries. In Text summarization branches out, pp 74–81
7. Lewis M, Liu Y, Goyal N, Ghazvininejad M, Mohamed A, Levy O, Stoyanov V, Zettlemoyer L (2019) Bart: denoising sequence-to-sequence pre-training for natural language generation, translation, and comprehension. arXiv Preprint arXiv:1910.13461
8. Devlin J, Chang MW, Lee K, Toutanova K (2018) Bert: pre-training of deep bidirectional transformers for language understanding. arXiv Preprint arXiv:1810.04805

9. Zhang J, Zhao Y, Saleh M, Liu P (2020) Pegasus: pre-training with extracted gap-sentences for abstractive summarization. In: International conference on machine learning. PMLR, pp 11328–11339

10. Raffel C, Shazeer N, Roberts A, Lee K, Narang S, Matena M, Zhou Y, Li W, Liu PJ (2019) Exploring the limits of transfer learning with a unified text-to-text transformer. arXiv Preprint arXiv:1910.10683

11. Fedus W, Zoph B, Shazeer N (2021) Switch transformers: scaling to trillion parameter models with simple and efficient sparsity. arXiv Preprint arXiv:2101.03961

12. Radford A, Narasimhan K, Salimans T, Sutskever I (2018) Improving language understanding by generative pre-training

13. Radford A, Wu J, Child R, Luan D, Amodei D, Sutskever I (2019) Language models are unsupervised multitask learners. OpenAI blog 1(8):9

14. LeCun Y, Bengio Y, Hinton G (2015) Deep learning. Nature 521(7553):436–444

15. Wang H, Zeng D (2020) Fusing logical relationship information of text in neural network for text classification. Math Probl Eng 2020:1–16

16. Yi J, Zhang Y, Zhao X, Wan J (2017) A novel text clustering approach using deep-learning vocabulary network. Math Probl Eng 2017:1–13

17. Young T, Hazarika D, Poria S, Cambria E (2018) Recent trends in deep learning based natural language processing [review article]. IEEE Comput Intell Mag 13(3):55–75

18. Rush AM, Chopra S, Weston J (2015) A neural attention model for abstractive sentence summarization. arXiv Preprint arXiv:1509.00685

19. Shi T, Keneshloo Y, Ramakrishnan N, Reddy CK (2021) Neural abstractive text summarization with sequence-to-sequence models. ACM Trans Data Sci 2(1):1–37

20. Lopyrev K (2015) Generating news headlines with recurrent neural networks. https://arxiv.org/abs/1512.01712 9

21. Song S, Huang H, Ruan T (2018) Abstractive text summarization using LSTM-CNN based deep learning. Multimedia Tools Appl

22. Lopyrev K (2015) Generating news headlines with recurrent neural networks. arXiv Preprint arXiv:1512.01712

23. Al-Sabahi K, Zuping Z, Kang Y (2018) Bidirectional attentional encoder-decoder model and bidirectional beam search for abstractive summarization. arXiv Preprint arXiv:1809.06662

24. Khandelwal U, Qi P, Jurafsky D. Neural text summarization

25. Liu Y, Lapata M (2019) Text summarization with pretrained encoders. arXiv Preprint arXiv:1908.08345

26. See A, Liu PJ, Manning CD (2017) Get to the point: summarization with pointer-generator networks. In: Proceedings of the 55th annual meeting of the association for computational linguistics (Volume 1: Long Papers). Association for Computational Linguistics, pp 1073–1083

27. Paulus R, Xiong C, Socher R (2017) A deep reinforced model for abstractive summarization. arXiv Preprint arXiv:1705.04304

28. Xia Y, Tian F, Wu L, Lin J, Qin T, Yu N, Liu TY (2017) Deliberation networks: sequence generation beyond one-pass decoding. Adv Neural Inf Process Syst 1782–1792

29. Ling J (2017) Coarse-to-fine attention models for document summarization. PhD diss

30. Zhang X, Lapata M (2017) Sentence simplification with deep reinforcement learning. arXiv Preprint arXiv:1703.10931

31. Celikyilmaz A, Bosselut A, He X, Choi Y (2018) Deep communicating agents for abstractive summarization. In: Proceedings of the 2018 conference of the North American chapter of the Association for Computational Linguistics: human language technologies, (Volume 1 Long Papers), pp 1662–1675

32. Lin J, Sun X, Ma S, Su Q (2018) Global encoding for abstractive summarization. In: Proceedings of the 56th annual meeting of the Association for Computational Linguistics (Volume 2: Short Papers). Association for computational linguistics, pp 163–169

33. Pasunuru R, Bansal M (2018) Multi-reward reinforced summarization with saliency and entailment. arXiv preprint arXiv:1804.06451

34. Chen Q, Zhu XD, Ling ZH, Wei S, Jiang H (2016) Distraction-based neural networks for modeling documents. In: Proceedings of the 25th international joint conference on artificial intelligence. AAAI Press, pp 2754–2760
35. Gulcehre C, Ahn S, Nallapati R, Zhou B, Bengio Y (2016) Pointing the unknown words. In: Proceedings of the 54th annual meeting of the association for computational linguistics (volume 1: Long Papers), pp 140–149
36. Gu J, Lu Z, Li H, Li VO (2016) Incorporating copying mechanism in sequence-tosequence learning. In Proceedings of the 54th annual meeting of the association for computational linguistics (volume 1: Long Papers), Vol 1, pp 1631–1640
37. Zhou Q, Yang N, Wei F, Zhou M (2017) Selective encoding for abstractive sentence summarization. arXiv Preprint arXiv:1704.07073
38. Li P, Bing L, Lam W (2018) Actor-critic based training framework for abstractive summarization. arXiv Preprint arXiv:1803.11070
39. Gehring J, Auli M, Grangier, Yarats D, Yann N. Dauphin. 2017. Convolutional sequence to sequence learning. In: Proceedings of the International Conference on Machine Learning. 1243–1252.
40. Fabbri AR, Li I, She T, Li S, Radev DR (2019) Multi-news: A large-scale multi-document summarization dataset and abstractive hierarchical model. arXiv Preprint arXiv:1906.01749
41. Wang L, Yao J, Tao Y, Zhong L, Liu W, Du Q (2018) A reinforced topic-aware convolutional sequence-to-sequence model for abstractive text summarization. In: Proceedings of the 27th international joint conference on artificial intelligence. AAAI Press, pp 4453–4460
42. Zhang Y, Merck D, Tsai EB, Manning CD, Langlotz CP (2019) Optimizing the factual correctness of a summary: a study of summarizing radiology reports. arXiv Preprint arXiv:1911.02541
43. Mohsen F, Wang JKamal Al-Sabahi (2020) A hierarchical self-attentive neural extractive summarizer via reinforcement learning (HSASRL). Appl Intell 1–14
44. McAuley JJ, Leskovec J (2013) From amateurs to connoisseurs: modeling the evolution of user expertise through online reviews. In: Proceedings of the 22nd international conference on World Wide Web
45. https://huggingface.co/transformers/model_doc/bart.html
46. Hendrycks D, Gimpel K (2016) Gaussian error linear units (gelus). arXiv Preprint arXiv:1606.08415
47. Loshchilov I, Hutter F (2017) Decoupled weight decay regularization. arXiv Preprint arXiv:1711.05101
48. Lin CY, Och FJ (2004) Looking for a few good metrics: ROUGE and its evaluation. In: Ntcir Workshop
49. Zolotareva E, Tashu TM, Horváth T (2020) Abstractive text summarization using transfer learning. In: ITAT, pp 75–80
50. Deorukhkar K, Jaison D, Hippurgikar S, Ambilkar S (2020) Text summarization on Amazon food reviews. Int J Emerg Trends Technol Comput Sci (IJETTCS) 9(2):010–012. ISSN: 2278-6856
51. Hayatin N, Ghufron KM, Wicaksono GW (2021) Summarization of COVID-19 news documents deep learning-based using transformer architecture. Telkomnika 19(3)
52. Yu T, Su D, Dai W, Fung P (2020) Dimsum@ laysumm 20: bart-based approach for scientific document summarization. arXiv preprint arXiv:2010.09252
53. Sanjabi N (2018) Abstractive text summarization with attention-based mechanism. Master's thesis, UniversitatPolitècnica de Catalunya
54. Deaton J, Jacobs A, Kathleen K, See A (2019) Transformers and pointer-generator networks for abstractive summarization
55. Brown T, Mann B, Ryder N, Subbiah M, Kaplan J, Dhariwal P, Neelakantan A Shyam P, Sastry G, Askell A, Agarwal S (2020) Language models are few-shot learners. arXiv preprint arXiv:2005.14165

Cell Segmentation of Histopathological Images of Glioma Using Voronoi Tessellation and Quadtree Representation

V. Brindha and **P. Jayashree**

Abstract Automatic cell segmentation is a challenging task in histopathological image analysis which is responsible for examining tissues and cells for diagnosing the severity of cancer in patients. Cell segmentation is the technique of breaking down a microscopic image area into sections that reflect individual cell occurrences. Finding the high density of cells and finer edge detection of cells is complicated due to the overlapping regions. To overcome this difficulty, a novel method of segmenting Whole Slide Images (WSI) of Low Grade Glioma (LGG) and High Grade Glioma (HGG) using voronoi tessellation of polygon approximation is proposed. The suggested approach consistently segments images of distinct cell types growing in dense cultures that were captured using various morphological techniques. The edge detection of the cells obtained from voronoi is finer compared to other existing edge detection methods. Quadtree image representation is implemented using proposed approach which is utilized to calculate the cell count and density estimate of tumor from voronoi tessellation in $O(\log_4 n)$. The factors gleaned from the histopathological analysis helps the pathologist in finding the densely populated brain tumors cells in WSI and treating the patient accordingly.

Keywords Histopathology · WSI · Quadtree · Voronoi · Glioma · LGG · HGG

1 Introduction

Medical diagnosis depends intensely on the histopathological assessment of tissue tests. Histopathology is a discipline of medicine concerned with the examination and diagnosis of disease in tissues that are examined using a microscopic biopsy approach. Compare to other diagnostic methods like MRI, CT, and PET to detect cancer, biopsy determines the origination and aggressive of cancer, and it increases the significant amount of cancer detection. Various regions of the tissue can be visualized through microscope where the areas are dyed with (H&E) stains which

V. Brindha (✉) · P. Jayashree
Department of Computer Technology, Anna University—MIT Campus, Chennai, India
e-mail: brindhamaha@gmail.com

© The Author(s), under exclusive license to Springer Nature Singapore Pte Ltd. 2023   387
A. Shukla et al. (eds.), *Computational Intelligence*, Lecture Notes in Electrical Engineering 968, https://doi.org/10.1007/978-981-19-7346-8_33

are blend of hematoxylin and Eosin. Eosin stains are represented by pink color, while hematoxylin stains are indicated by blue color. Gliomas are the main type of tumor currently attracting the interest of brain tumor researchers. Glioblastoma Multiforme (GBM) which is the High Grade Glioma (HGG) is the most forceful kind of malignant growth of cancer which emerges in the brain. The fatality rate due to brain tumor is highest in Asia. The number of deaths due to brain tumors has increased by 300% over the past years.

Image segmentation includes edge detection. It is a technique for segmenting an image into disconnected sections. Detecting significant edges with high precision is important to the success of several image processing applications. Evaluating the images of GBM provides more detail as compared to analysis of various imaging modalities. The detailed information is necessary to aid pathologists in deciding the kind of treatment for the cancer. A pathologist's manual tissue analysis obtained from a biopsy is time-consuming and prone to subjective errors. Thus, methods of computer-assisted diagnosis involving graphical representation of cell contours and strategies for image processing is being explored.

In this work, novel method for finding cell contours of histopathological images is done through voronoi tessellation which separates the patched image into polygons with each polygon contains one point which is close to given set of objects. Distinguishing cells in an image (cell segmentation) is critical. A wide variety of morphological applications are applied for analysis the medical image in this research. As a result, accuracy in histopathological analysis is paramount, and since humans are prone to making subjective errors, polygon approximation methods of simplifying this method have been proposed. Due to the complexity of cell shape and overlapping of cells in the patches makes automatic cell segmentation more complex. Compare to manual and semi-automatic cell segmentation, automatic segmentation using voronoi tessellation and Quadtree representation to detect the densely populated region of cells results better accuracy.

2   Related Works

The literature employs a variety of methodologies which possess benefits and drawbacks. Authors of [1] proposed extracting features by using activation maps from a neural network guided by nuclei centroid detection. The centroids were detected using color deconvolution followed by Gaussian blurring. A proposed optimization was to use only rectangular patches surrounding nuclei instead of using all patches in the image, for convolution. In paper [2], different feature extraction techniques such as Histogram of Oriented Gradients (HOG) and Local binary patterns (LBP) have been compared. In HOG, weighted histograms of gradient directions in localized areas of the image are obtained, with the values corresponding to the gradient magnitudes. This approach is quite good at detecting objects, as seen by its performance in applications like pedestrian detection [3]. In this paper [4], to improve the form of the image, Karhunen-Loève (KL) transform histogram equalization method is used. By emphasizing the luminance variation among foregroundand background, the contrast enhancement method stiffens the image boundary and ensures complete. In this paper [5], median filter is suitable for preprocessing images. This method removes the outlier while preserving the image's crispness. In paper [6], watershed algorithm is implemented to obtain an over-segmented image. Gravitational search is performed to find out the segments must merged to obtain the final segmented image. The paper's drawback is that the search space is relatively wide (huge sets of coordinates) and hence takes a long time to converge.

In [7], Persistent Homology Profile (basically a 1D statistical distribution describing the features of the image) is computed from each patch in the image. Normal and tumor patches are identified using a CNN, and their PHPs are computed. The similarities between the PHPs of input and exemplar patches are computed to identify the label for the input patch. In this paper [8] to create a binarized image, adaptive thresholding is used. Watershed algorithm is then applied to coarsely segment the image, with no differentiation between clustered nuclei. The gray level results are assigned and merged with normalized distance map values to enable segmenting clustered nuclei. In paper [9], adaptive thresholding is used to get a rough partition between the foreground (nuclei) and the background. The threshold image is subjected to a space conversion to obtain regional maxima to act as seeds for watershed algorithm. Watershed algorithm is applied and an over-segmented image is applied. Finally, algorithm of hole-filling is implemented to obtain a properly segmented image.

In paper [10], author proposed that voting is done on the gradient image by using tensors parallel and perpendicular to the gradient at each point. Using the images obtained by voting, centers of nuclei are found. Each nucleus center is used to obtain a nucleus boundary by using a Markov random field inside a window centered at the seed point. In paper [11], author proposed that Glioma or Glioblastoma is more aggressive kind of tumor. Image fusion based on discrete wavelet transformation (DWT) is used to decompose the image. The sub-bands of low frequency are combined using weighted average rule and sub-bands of high frequency are merged with the maximum selection rule. In paper [12], proposed that the utility of bright spot analysis in discrimination of GBM at boundary zone is investigated. GBM can be distinguished from normal tissue using bright spot analysis. The bright spot area in tumor regions is much lower than in normal regions is used in classification.

3 Proposed Work

Cell contour detection of Glioma from histopathological images is proposed. The proposed method involves stages like preprocessing and cell contour detection using polygon approximation of voronoi tessellation and Quadtree image representation. Overall architecture of the proposed work is shown in Fig. 1.

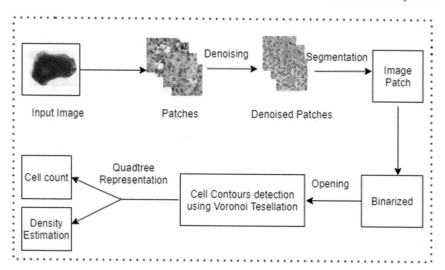

Fig. 1 Overall architecture

3.1 Preprocessing

Initially, each Whole Slide Image (WSI) of histopathological image of GBM is splitted into patches of 256 × 256 pixels. Hematoxylin and eosin (H&E) images are transformed into grayscale patched images and denoised using median filters. Every image pixel is taken by the median filter, and it considers the neighboring pixel values to determine whether or not it is reflective of its surroundings. The window size of filter is taken as 3. This particular filtering approach was chosen because it can blur noise while maintaining edges. Images in grayscale are transformed to binary. Thus, a binary image containing segmented cell nuclei and tiny spots (noise) is obtained. Using morphological opening, the disruptive pixels are cleaned.

3.2 Cell Contour Detection Using Polygon Approximation of Voronoi Tessellation

Automatic cell segmentation is still a difficult task to find out the edges of the cells with overlapping areas. Edge detection is the approach of detecting the contours (outlines) of any form or item in an image in order to distinguish it from the backdrop or other objects. So, the suggested approach is initiated by applying voronoi tessellation on the denoised patched images to find the cell contours. Image segregation and establishing the contour of each cell are required for cell identification from the context. It includes segmenting the image and determining the contour of each cell to distinguish the cells from the backdrop. Polygons are matched on the contours by using the directional shift of the gradient between adjacent contour points. A threshold is set and if a shift in perspective of the gradient in the grayscale image at a point swap by increasing this threshold, it is marked as polygon's corner. Using this method, the quantity of cells having a set number of polygon edges is counted.

From the contours, the centroid of each cell is detected, and the voronoi tessellation is applied by using these centroids as the seed points. Given a series of points (seeds), the surface is split into as many regions as there are points in the voronoi, with quarter holding all coordinates in the plane that seem to be nearer to the seed in that region than any other area's seed. Mathematically, it is defined as

$$R_k = x \in X/d(x, P_k) \leq d(x, P_j) \forall_j \neq k \tag{1}$$

where R_k denotes the region belonging to point P_k and all vertices are represented by X. The interval between two points $d(x, P_k)$, $d(x, P_y)$, using Euclidean function chosen as a factor in this work in Eq. 1. The range between P_k and voronoi is equal to the range between the voronoi cell and the other sites P_j.

3.3 Quadtree Image Representation

A Quadtree is a geographic data structure with hierarchical levels. It's a tree with each level denoting a refinement of the area under examination. Followed by voronoi, partitioning the territory into quadrants proportionately, Quadtree image representation is employed and recursively redividing into quartile which is shown in Fig. 2 Finally, using the ultimate points approach on the edges, the cell counts in every quadrant were computed. Maximum number of cells in the quadrant is found out and again it is subdivide into subquadrants and so on. In Fig. 2, level 0 is the input image. It is divided into quadrants in level 1 with high density of cells is marked as red color which is again subdivided into four. High density of cells in level 2 is marked as green color. Quadtree structure is deploy to quickly identify the most clustered cell region in $O(\log_4 n)$ over an image have n cells.

4 Performance and Result Analysis

4.1 Dataset and Implementation

Dataset is collected from Genomic Data Commons Data Portal (GDC) hosted by the National Cancer Institute. The Cancer Genome Archive (TCGA) dataset is utilized to gather ten quality images. The images obtaining from two different datasets collection, namely Low Grade Glioma (LGG) images from TCGA-LGG and the

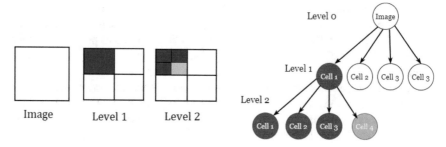

Fig. 2 Quadtree image representation

High Grade Glioma (HGG) that is glioblastoma images from TCGA-GBM, with five images in each classes. The solution was coded in MATLAB 2018a for image processing functionality. Each image was tiled into around 1000 patches, with each of dimensions 256 × 256 using Qupath script. The input image is in the format of svs file which is patched into tiles and converted into jpg format. Patches with too much of air space and blank patches were removed manually before being processed.

4.2 Results of Cell Contour Detection

Results of the voronoi tessellation leads to finer cell edges compared to other traditional contour detecting methods like Sobel, Robert, Log, ZeroCross, Prewitt, and Canny which is shown in Fig. 3. Comparison table of existing and proposed method is illustrated in Fig. 4. Performance metrics of contour detection is calculated using PSNR, MSE, and maxerr. Peak Signal to Noise Ratio (PSNR) value is high, the image quality is good. The mean square error (MSE) is low and less error introduced to the image. Quality metrics of the image is measured using maxerr approximation. Figure 5 demonstrates the proposed cell contour method using voronoi.

4.3 Results of Quadtree Representation of HGG and LGG

Each patch of HGG sample image is divided into quadrant, total cell count as 135, The orange colored values indicate maximum cell count with 37 is subquadrant into 4, get maximum cell count as 10 and again subquadrant into 4 to get 4 as the highest value.

Quadtree structure is adapted to efficiently recognize the sectors having greater cell density. Each cell shows up as a dark spot in the image, pixels belonging to clusters with the maximum intensity is used as the marker for marking foreground objects (cells) which are dark in color and the remaining cells are considered as background pixels. In patches, cell count is higher in the densely packed region, where

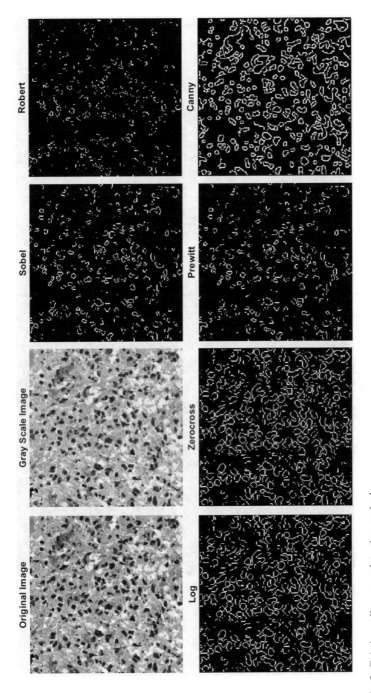

Fig. 3 Existing cell contour detection methods

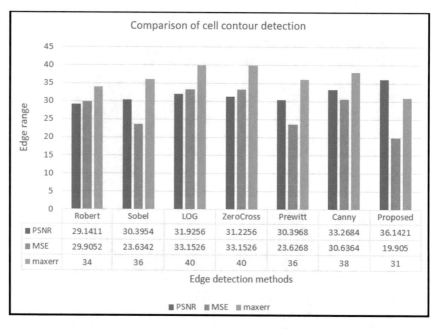

Fig. 4 Comparison of existing and proposed method for cell contour detection

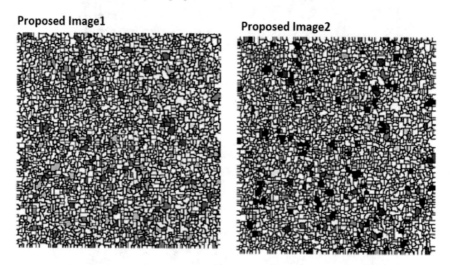

Fig. 5 Proposed cell contour method of sample HGG image with voronoi tessellation

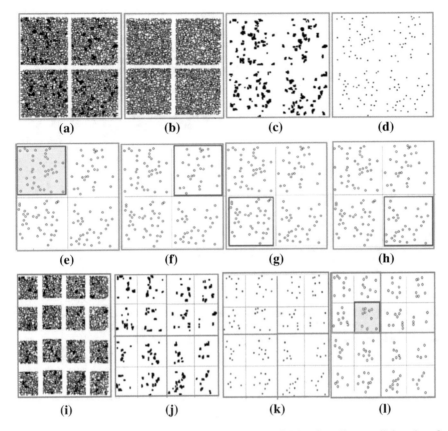

Fig. 6 Quadtree image representation of HGG: **a** quadrant division, **b** outline, **c** cell detection, **d** ultimate points, **e** high density of cells found in the top left corner in level 1, **f** cell count in top right, **g** bottom left, **h** bottom right, **i** subquadrant, **j** cell detection, **k** ultimate points and, **l** high density cells in bottom right of level 2

as lower in the sparsely packed region, which is represented in Figs. 6 and 7. Thus, the density of cells in an image patch is quantified using Quadtree representation, as illustrated in Tables 1 and 2 gives how the Quadtree structure is utilized to distinguish LGG from HGG based on cell density.

4.4 Effective Discriminant of LGG and HGG

It is inference that in Low Grade Glioma (LGG), minimum cell count is 10, maximum cell count is 20 in level 1 and for level 2, minimum cell count is 1, and maximum cell count is 10 in level 1. For High Grade Glioma (HGG), minimum cell count is

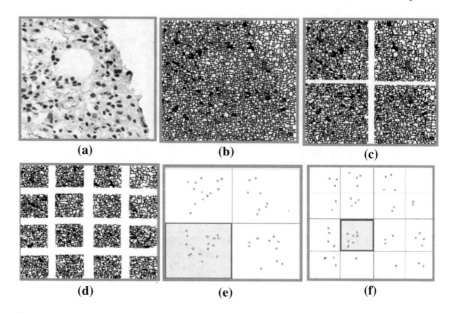

Fig. 7 Quadtree image representation of LGG: **a** Grayscale of LGG, **b** contour detection using LGG, **c** quadrant division, **d** subquadrant division, **e** high density of cells in the third quadrant of level 1, and **f** high density cells in the third quadrant top left of level 2

Table 1 Quadtree representation of sample HGG image

Quadrant	Total cell count = 135	Subdividing 1st quadrant	Cell count with maximum values = 37	Subdividing 4th quadrant	Cell count with maximum values = 10
1	37	1	9	1	4
2	29	2	9	2	3
3	36	3	9	3	2
4	33	4	10	4	1

Table 2 Effective discriminant of HGG and LGG

Glioma	Average cell count at level 1				Average cell count at level 2			
	Cell 1	Cell 2	Cell 3	Cell 4	Cell 1	Cell 2	Cell 3	Cell 4
LGG	20	10	10	10	10	5	4	1
HGG	39	30	36	45	13	15	10	7

30 and maximum cell count is 45 in level 1 and for level 2, minimum cell count is 7 and maximum cell count is 15 in level 1 given in Table 2.

5 Conclusion

Cell segmentation is essential since it aids medical research and diagnostics. Proper diagnosis of tumors is crucial since human lives depend on it. Analysis of brain tumors using histopathological images provides better results when compared to the analysis of other imaging modalities. Compared to other existing methods for cell contour detection, a mathematical approach of voronoi tessellation is proposed and is applied by using the centroid's seed points. Histopathological image segmentation approach was presented to finding the better contour and cell density using voronoi and Quadtree structures. The proposed model will be capable of finding the density of tumor cells with considerable speed and accuracy.

References

1. Mukundan R (2018) Image features based on characteristic curves and local binary patterns for automated HER2 scoring. J Imaging 4(2):35
2. Alhindi TJ, Kalra S, Ng KH, Afrin A, Tizhoosh HR (2018) Com-paring LBP, HOG and deep features for classification of histopathology images. In: 2018 international joint conference on neural networks (IJCNN). IEEE, pp 1–7
3. Chee KW, Teoh SS (2019) Pedestrian detection in visual images using com-bination of HOG and HOM features. In: 10th international conference on robotics, vision, signal processing and power applications. Springer, pp 591–597
4. Hoshyar AN, Al-Jumaily A, Hoshyar AN (2014) The beneficial techniques in preprocessing step of skin cancer detection system comparing. Procedia Comput Sci 42:25–31
5. Khuriwal N, Mishra N (2018) Breast cancer detection from histopathological images using deep learning. In: 2018 3rd International conference and workshops on recent advances and innovations in engineering (ICRAIE). IEEE, pp 1–4
6. Mittal H, Saraswat M (2019) An automatic nuclei segmentation method using intelligent gravitational search algorithm based superpixel clustering. Swarm Evol Comput 45:15–32
7. Qaiser T, Tsang YW, Taniyama D, Sakamoto N, Nakane K, Epstein D, Rajpoot N (2019) Fast and accurate tumor segmentation of histology images using persistent homology and deep convolutional features. Med Image Anal 55:1–14
8. Abdolhoseini M, Kluge MG, Walker FR, Johnson J (2019) Segmenta-tion of heavily clustered nuclei from histopathological images. Sci Rep 9(1):1–13
9. Paulik R, Micsik T, Kiszler G, Kaszál, P., Székely J, Paulik N, Várhalmi E, Prémusz V, Krenács T, Molnár B (2017) An optimized image analysis algorithm for detecting nuclear signals in digital whole slides for histopathology. Cytometry Part A 91(6):595–608
10. Paramanandam M, O'Byrne M, Ghosh B, Mammen JJ, Manipadam MT, Thamburaj R, Pakrashi V (2016) Automated segmentation of nuclei in breast cancer histopathology images. PloS one 11(9)
11. Brindha V, Jayashree P (2019) Fusion of radiological images of glioblastoma multiforme using weighted average and maximum selection method. In IEEE 11th international conference on advanced computing (ICoAC) 328–332
12. Yoneyama T, Watanabe T, Tamai S, Miyashita K, Nakada M (2019) Bright spot analysis for photodynamic diagnosis of brain tumors using confocal microscopy. Photodiagn Photodyn Ther 463–471

Comparative Analysis of LSTM, Encoder-Decoder and GRU Models for Stock Price Prediction

Parul Arora and Abhigyan Balyan

Abstract The stock market is a venue where shares of publicly traded enterprises can be bought and sold on a regular basis. It is a vital component of any country's economy. Some individuals consider it to be a barometer of the country's economy. Many traders and investors use various strategies to forecast the stock's future price, for example, traders use technical analysis to forecast the stock's future price. However, improvements in deep learning and machine learning algorithms have allowed researchers to test new algorithms for predicting price of the stocks. Long short-term memory, encoder-decoder, and gated recurrent unit prediction results are compared in this study.

Keywords Deep learning · LSTM · Encoder-decoder GRU · Nifty 50 index · NSE

1 Introduction

Stock exchanges have a fairly unpredictable structure, and there are various reasons responsible for variation in stock values. Economists have attempted many times to anticipate the future of a stock by offering ideas such as efficient market hypothesis, random walk theory challenge, etc. [1, 2]. The market volatility sometimes depends upon crisis. For example, in COVID-19, the market was quite volatile [3]. As a result, in such a volatile market, retail investors and traders will need to rely more on artificial intelligence techniques to achieve greater returns. The availability of huge data and increased computational capacity allows for more ease and efficiency in addition to competitive advantage. Authors have attempted to forecast market direction and make improved decisions and give better returns for retail investors and traders by presenting advanced deep learning algorithms for instance long short-term memory (LSTM), encoder-decoder, as well as gated recurrent unit (GRU). Many techniques such as autoregressive integrated moving average (ARIMA) or seasonal autoregressive integrated moving average (SARIMA), and other machine learning

P. Arora (✉) · A. Balyan
Department of Electronics and Communication, Jaypee Institute of Information Technology, Noida, India
e-mail: parul.arora@jiit.ac.in

© The Author(s), under exclusive license to Springer Nature Singapore Pte Ltd. 2023
A. Shukla et al. (eds.), *Computational Intelligence*, Lecture Notes in Electrical Engineering 968, https://doi.org/10.1007/978-981-19-7346-8_34

algorithms have demonstrated very well in the past to predict stock prices, but none of them can predict the future when the market is highly erratic. So, whether it's simple neural network (NN) models, generative adversarial networks (GAN) [4] or stacked LSTM [5], academics currently prefer deep learning over machine learning techniques. The LSTM, encoder-decoder model, and GRU were employed in this research project to forecast the price of Nifty 50 indexes. All three methods are cutting-edge deep learning algorithms that have been applied on the Nifty 50 index. The Nifty 50 is an Indian stock market index that is a weighted average of the top 50 businesses listed on the National Stock Exchange (NSE). So we are not only estimating the closing price of the index by employing deep learning models but also analyzing the outputs of all three models. LSTM, encoder-decoder, and GRU are more advanced versions of recurrent neural networks (RNN). RNNs might have been employed as well, however, they face the issue of vanishing gradients, which makes learning large data sequences difficult. This data necessitates the use of models that do not suffer from vanishing gradients. R2 score (also known as coefficient of determination) and root mean squared error (RMSE) scores were used to examine the final outcome. This study will make the retail investors and traders to employ the best models and implement them in the real market, allowing them to make improved decisions and increase their overall accuracy. These models can be used with several timeframes, which improves accuracy even further. Methods are not only limited to the Nifty 50 index but can also be implemented on the derivative market data which is also highly volatile and has a huge trading volume in the Indian market. And the livelihood of thousands of traders depends on trading in derivatives.

1.1 Problem Statement

As many times retail investors and traders suffer big losses in the market because of high volatility, whereas big investors and financial institutions do not suffer such big losses. Big sharks like these institutions are more dependent on artificial intelligence to generate better returns in the financial markets. So now it is time for small investors and traders to increase their dependency on artificial intelligence in the market. Keeping this problem in mind, new techniques involving deep learning algorithms have been proposed in this paper intending to help retail investors and traders.

1.2 Contributions

The main contribution in this work is the comparison of the outcome of three deep learning techniques namely GRU, LSTM, and encoder-decoder on Nifty 50 dataset. These models together have never been compared in the past on Nifty dataset. The remainder paper is filtered through as: Sect. 2 consists of the literature survey.

Section 3 includes methodology and brief explanation of these algorithms. In Sect. 4, the results are discussed and in the end the work has been concluded.

2 Literature Survey

The pursuit of coherent stock price prediction approaches is intense among time series technique researchers. The authors [6] compared the modeling accuracy of artificial neural networks, linear optimization, and genetic algorithms (GAs) with time series data. In their work, linear optimization models provided the best esti- mations when the parameters were kept constant, whereas simple neural networks performed the worst of the other models. In the paper [7], performance of SARIMA and basic neural network models was examined for forecasting the Korean stock index. SARIMA outperformed basic neural networks in this study. The researchers developed a model dubbed feature fusion in their publication [8], which blends convolutional neural networks (CNN) and long short-term memory (LSTM). These algorithms were used to extract temporal features and image features, and they have compared their feature fusion model with single-layer CNN and LSTM and the fusion model has outperformed the CNN and LSTM model. Authors have also analyzed the change in price not on a daily time frame but on a weekly time frame and have used a deep NN as a classifier to classify true and fake golden crossover [9]. In a similar paper on financial forecasting [10], it was demonstrated that a basic NN model's performance was at par as compared to the ARIMA model in forecasting. This paper [11] provides a technique for forecasting financial time series by using high-order structures, motifs. Convolutional neural networks are used to learn the patterns of the reconstructed sequence, which give essential information for ups and downs prediction. In terms of computing complexity, this strategy is quite efficient. Additionally, researchers compared machine learning to deep learning techniques and discovered that deep learning models outperformed regular machine learning algorithms. Numerous machine learning models (SVC, Adaboost, XGBoost, random forest, decision tree, logistic regression, ANN, Naive Bayes, and KNN) were used as predictors in the paper [12], as well as two deep learning approaches (RNN and LSTM). Being motivated by a huge amount of valuable knowledge available in social media, researchers have used different approaches to forecast market like sentiment analysis [13] for stock market movement prediction.

3 Methodology

3.1 LSTM

Recurrent neural networks (RNN) encounter the issue of vanishing gradients and are incapable of dealing with the long-term dependencies of the input data. In 1997, a distinct sort of RNN called LSTM was introduced, and it is now being redefined by many researchers. The LSTM's entire construction is depicted in Fig. 1, including all three gates, forget gate, input gate, and output gate. LSTM is purpose-built to avoid long-term dependence. Simple RNN is composed of recurrent neural network modules. LSTM [14] also features a chain-like structure, but the repeating module of LSTM is structured differently than the recurring module of RNN.

LSTM comprises three gates: forget gate, input gate, and output gate, which are briefly discussed in the following section.

Forget gate

The internal construction of the forget gate is shown in Fig. 2, where input is routed through the sigmoid layer. The LSTM model's initial step is to determine which information should be discarded. This conclusion is drawn by the forget gate's sigmoid layer. It takes two values, $y(t-1)$ and $x(t)$ and after applying sigmoid function, returning a 0 or 1 for each value from cell state $z(t-1)$. The numbers 0 and 1 represent "get rid of the information" and "keep the information," respectively. The function passing from the sigmoid layer is shown in Eq. 1.

$$U_t = \sigma\left(w_f.\left[y_{(t-1)}, x_t\right]\right) + b_f \tag{1}$$

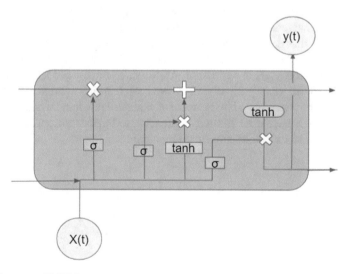

Fig. 1 Structure of LSTM

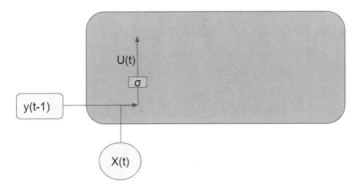

Fig. 2 Forget gate of LSTM

Input gate

After selecting which information should be saved and which should be discarded. The next stage is for the input gate to figure out new information that must be saved in the cell's state. This process has two stages, as illustrated in Fig. 3. First, the sigmoid layer of the input gate layer determines which values need to be updated, and then the tanh layer provides a vector of new values, which is added to the state, and then the two values together will provide an updated state.

The equations of q_t and Z'_t are as follows (Fig. 4):

$$q_t = \sigma\left(w_i * \left[y_{(t-1)}, x_t\right]\right) + b_i \tag{2}$$

$$Z'_t = \tanh\left(w_c * \left[y_{(t-1)}, x_i\right]\right) + b_c \tag{3}$$

Now update our old cell state $Z_{(t-1)}$ to Z_t using the values of q_t, Z'_t and U_t. Equation 4 is the equation of the new cell state:

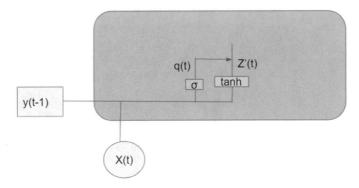

Fig. 3 Input gate layer of LSTM

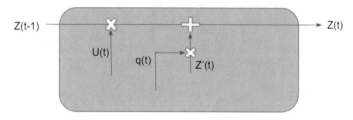

Fig. 4 New cell state path

$$Z_t = U_t * Z_{(t-1)} + q_t * Z'_t \tag{4}$$

Output Gate

The output must now be determined. The LSTM's output gate is made up of two layers. The internal construction of the output gate, which has two layers, is shown in Fig. 5. Layer 1 a sigmoid layer selects which portion of the cell state is going to be the output. The second layer is the tan$h$ layer, which generates values between − 1 and 1 and multiplies them with the sigmoid layer's output. The multiplied values will be our output.

Where Op$_t$ is

$$Op_t = \sigma (w_o[y_{(t-1)}, x_t]) + b_o \tag{5}$$

And output y_t will be

$$y_t = Op_t * \tanh(Z_t) \tag{6}$$

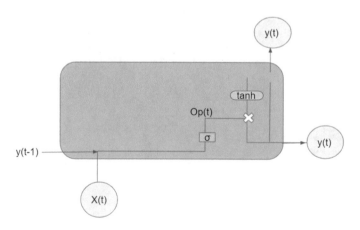

Fig. 5 Output gate of LSTM

3.2 Encoder-Decoder

The encoder-decoder [15] model in RNN solves the problem of seq-to-seq mapping models and is commonly used in applications like time series forecasting, language translations, and chatbots. The model, in Fig. 6, is made up of three parts: encoders, context vectors, and decoders.

Encoder

It is a pile of several recurrent units; units can be of LSTM, GRU, or simple RNN. The task of each unit is to accept input sequence, and then it accumulates the input sequence's information and propagates the information forward.

$$m(t) = f((w * m(t - 1) + Z(t)))) \tag{7}$$

The $m(t)$ is computed with use of weights, previous unit information and with the function (f). The final vector, i.e., $m(t)$ contains the encoded information.

Context Vector

The $m(t)$ is passed to the context vector where it sums up the information of the input elements and propagates the information to the decoder in order to help it in making better forecasts. And it is also the final hidden state produced from the encoder part.

Decoder

The decoder also consists of the pile of the recurrent units. The units accepts the hidden state $n(t - 1)$

from the previous state and produces the output $y'(t)$

the $n(t)$ is computed by the equation

$$n(t) = f(w * m(t - 1)) \tag{8}$$

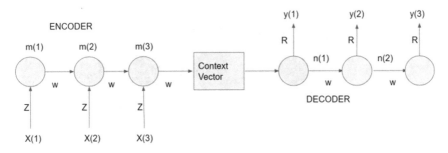

Fig. 6 Encoder-decoder architecture

and the final output is

$$y'(t) = \text{softmax}(R(t)). \tag{9}$$

3.3 GRU

Gated recurrent unit (GRU) [16] is the next model employed. It differs slightly from LSTM. The main distinction between LSTM and GRU is that GRU integrates forget and input gates in a single gate known as the update gate. It combines the cell's visible and concealed states. And the GRU model that results is significantly less sophisticated than a normal LSTM model. Because GRU contains fewer equations than LSTM, forward and backward propagation will take less time due to the lower number of weights. GRUs will be substantially faster as a result of this enhancement. The GRU model's equations are as follows (Fig. 7):

$$Z_t = \sigma\left(W_z * \left[y_{(t-1)}, x_t\right]\right) \tag{10}$$

$$R_t = \sigma\left(W_r * \left[y_{(t-1)}, x_t\right]\right) \tag{11}$$

$$Z'_t = \tanh\left(W * \left[R_t * y_{(t-1)}, x_t\right]\right) \tag{12}$$

$$y_t = (1 - Z_t) * y_{(t-1)} + Z_t * Z'_t \tag{13}$$

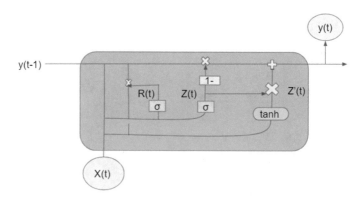

Fig. 7 Structure of GRU

4   Results and Discussion

4.1   Dataset Description

Nifty 50 is a benchmark Indian stock market exchange listed on National Stock Exchange (NSE). It consists of the top 50 companies of India. It was incorporated in April 1996. It is considered to be one of the two main benchmark indexes, the other being SENSEX. Most of the Indian traders earn their livelihood by trading in Nifty 50 index. The dataset that is being used is the closing price of Nifty from the year 2000–2020. The dataset consists of total 5208 values and has been divided, 70% data into training data whereas 30% as test data.

4.2   Discussion

We have to preprocess our data in order to arrive at the final results. Because the values in our data span a wide range, we used the sklearn toolkit to normalize it before building deep learning models. We calculated the RMSE and R2 score of our models after running them on our dataset and plotted the graphical results of all three models, where the blue, orange, and the green line represents the training, testing, and the prediction made on the testing dataset, respectively. Figure 8 shows the LSTM model's prediction result, whereas Fig. 9 shows the encoder-decoder model's prediction result, and Fig. 10 depicts the GRU model's prediction result. By assessing a vast dataset and obtaining little figures, it is tough to determine which model is the most accurate. As a result, we must rely on the models' RMSE and R2 scores. After examining the scores of all three models in Table 1, it can be determined that GRU is the finest deep learning model out of all three models because it has the lowest value of RMSE and the largest value of R2 score. As a result, retail traders and investors can utilize the GRU model for market forecasting, make smarter decisions, and increase their accuracy in high-volatility markets.

5   Conclusion

Many time series prediction models exist today, although they are less accurate than LSTM, encoder-decoder, and GRU, as these models are specifically built to take in to account the data's long-term dependencies. As a result, it can also generate reliable predictions on enormous datasets. So, based on our findings, we can confidently say that GRU outperformed encoder-decoder and LSTM on our Nifty 50 data. These models can assist many stock market traders and investors in making better judgments and increasing their profits. Despite the fact that we have provided deep learning approaches, retail traders and investors can also employ other techniques, such as

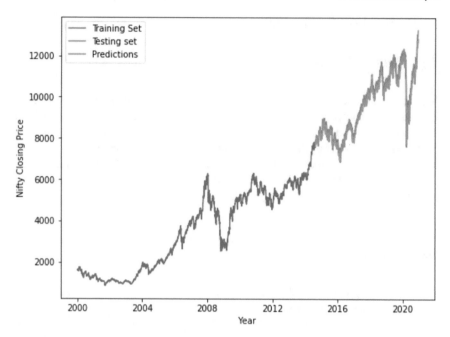

Fig. 8 LSTM output

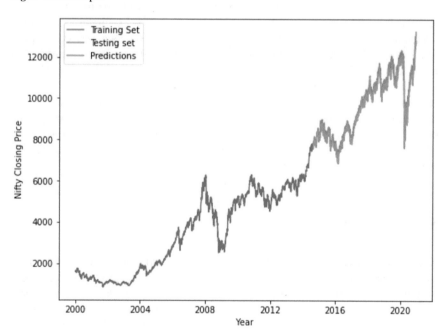

Fig. 9 Encoder-decoder output

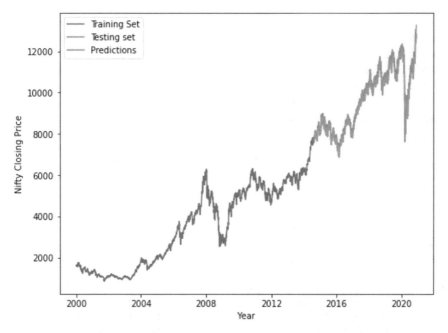

Fig. 10 GRU output

Table 1 RMSE, R2 score of all models

Models	LSTM	Encoder-decoder	GRU
RMSE score	639.6622	398.4574	209.3483
R2 score	0.8935	0.9342	0.9742

technical analysis procedures, to make money. For example, the golden crossover technique can be used in conjunction with our recommended methodologies, which will increase the accuracy of trading even further. While for investing strategies, models such as the CANSLIM models and the price to earnings ratio approach can be merged with the methodologies, allowing investors to create profits over the long run. The key advantage of these strategies is that they can be backtested on historical data, allowing retail traders and investors to select the optimal strategy that will perform best for their particular situation. And which technique earned the most amount of profit. Furthermore, similar to how the Nifty 50 data has been used, the Banknifty data, which is also an important indicator, can be employed in the future research. Collaboration with various stock market brokers can result in the creation of a platform, which can then be accessed by shops. High frequency trading systems, which make decisions in milliseconds and are now only available to large financial institutions, can also be employed with the use of superior computer power, as well. However, as they become more widely available to the general public, they will be better equipped to make timely decisions and minimize financial losses. The fact that traders will no longer make decisions based on their emotions is another advantage of employing these deep learning-based trading systems. Traders who make decisions based on their emotions are the ones that incur the greatest losses. Because of this, when computer algorithms are used, the emotional component is decreased, and the choice is more data driven. Moreover these models are not only confined to stock price forecasting, but they may be used in various disciplines such as sales projection, weather forecasting, and so on.

References

1. Atanasov V, Pirinsky C, Wang Q (2018) The efficient market hypothesis and investor behavior
2. Rehman S, Chhapra IU, Kashif M, Rehan R (2018) Are stock prices a random walk? an empirical evidence of asian stock markets. Etikonomi 17(2):237–252
3. Jindal R, Bansal N, Chawla N, Singhal S (2021) Improving traditional stock market prediction algorithms using covid-19 analysis. In: 3rd international proceedings on emerging smart computing and informatics (ESCI), pp 374–379, IEEE, Pune, India
4. Zhang K, Zhong G, Dong J, Wang S, Wang Y (2019) Stock market prediction based on generative adversarial networks. Procedia Comput Sci 147:400–406
5. Ojo SO, Owolawi PA, Mphahlele M, Adisa JA (2019) Stock market behavior prediction using stacked LSTM networks. In: International proceedings on multidisciplinary information technology and engineering conference (IMITEC), pp 1–5, IEEE, Vanderbijlpark, South Africa
6. Tansel IN, Yang SY, Venkataraman G, Sasirathsiri A, Bao WY, Mahendrakar N (1999) Modeling time series data by using neural networks and genetic algorithms. In: Dagli CH, Buczak AL, Ghosh J, Embrechts MJ, Erosy O (eds) Proceedings of the intelligent engineering systems through artificial neural networks, vol 9, pp 1055–1060, ASME Press, New York
7. Lee CK, Sehwan Y, Jongdae J (2007) Neural network model versus SARIMA model in forecasting Korean stock price index (KOSPI). Iss Inform Syst 8(2):372–378
8. Kim T, Kim HY (2019) Forecasting stock prices with a feature fusion lstm-cnn model using different representations of the same data. PLoS ONE 14(2):1–23
9. Shi M, Zhao Q (2020) Stock market trend prediction and investment strategy by deep neural networks. In: 11th international proceedings on awareness science and technology (iCAST), pp 1–6, IEEE, Qingdao, China
10. Lahane AG (2008) Financial forecasting: comparison of ARIMA, FFNN and SVR. http://www.it.iitb.ac.in/~ashishl/files/MtechProjectPresentation.pdf
11. Wen M, Zhang L, Chen Y (2019) Stock market trend prediction using high-order information of time series. IEEE Access 7:28299–28308
12. Nabipour M, Nayyeri P, Jabani H (2020) Predicting stock market trends using machine learning and deep learning algorithms via continuous and binary data; a comparative analysis. IEEE Access 8:150199–150212
13. Bouktif S, Fiaz A (2020) Augmented textual features-based stock market prediction. IEEE Access 8:40269–40282
14. Siami-Namini S, Tavakoli N, Namin AS (2018) A comparison of ARIMA and LSTM in forecasting time series. In: 17th international proceedings on machine learning and applications (ICMLA), pp 1394–1401, IEEE, Orlando
15. Phandoidaen N, Richter S (2020) Forecasting time series with encoder-decoder neural networks. arXiv preprint arXiv:2009.08848
16. Zhang X, Shen F, Zhao J, Yang G (2017) Time series forecasting using GRU neural network with multi-lag after decomposition. In: International conference on neural information processing, pp 523–532

Prediction of Software Vulnerabilities Using Random Forest Regressor

Navirah Kamal and Supriya Raheja

Abstract The launch of new software technologies with new features helps the developer in software development but it can be prone to vulnerabilities. Software vulnerabilities are still a critical issue for software security as they can negatively impact the organization and the end user. To mitigate this problem, various techniques have been adopted, machine learning is one of them. The main objective of this paper is to predict the severity of software vulnerabilities using a random forest regressor algorithm. To evaluate the performance, two other machine learning algorithms are also implemented for the same task. A dataset of the National Vulnerabilities database is used for the present work. The efficacy of the models has been evaluated and compared using four different performance metrics namely mean absolute error, mean square error, root mean square error, and R2 score. Random forest regressor performed the best out of the applied machine learning algorithms with a root mean square error of 0.01945.

Keywords Vulnerabilities · Software development · Software security · Machine learning techniques · Random forest · Decision tree · K-nearest

1 Introduction

The analysis of software systems' security has gained importance in recent years as the software systems even from the top companies are seen to have some sort of vulnerabilities that requires detection and fixing. Errors in specification, development, or configuration of a software can cause these vulnerabilities [1]. If these are executed, it may violate the security policy. In other words, these vulnerabilities are the various bugs that can provide an attacker with a window to alter or access the software system. Overlooking such vulnerabilities can be a serious security threat to the software system and the database connected to it. While some bugs may grant unauthorized access to users, some may affect the functionality of the software system.

N. Kamal · S. Raheja (✉)
Amity University, Sector 125, Noida 201313, India
e-mail: supriya.raheja@gmail.com

© The Author(s), under exclusive license to Springer Nature Singapore Pte Ltd. 2023   411
A. Shukla et al. (eds.), *Computational Intelligence*, Lecture Notes in Electrical
Engineering 968, https://doi.org/10.1007/978-981-19-7346-8_35

Despite the work done towards improving the security of software, vulnerability analysis and fixing are the focus points for security analysts.

Often these vulnerabilities are observed after the launch of the software, when the multiple users are already using it. It becomes critical to tackle them as soon as they are found. To deal with such situations, there are various 'Vulnerability Databases' available like the National Vulnerability Database (NVD). These allow developers to report vulnerabilities found in their software systems and in turn benefit other developers who may check their software for the reported vulnerability. In the NVD, the software vulnerabilities are stored in common vulnerability enumeration (CVE) [2] lists. Approximately 7000 new vulnerabilities get registered in the NVD every year [2]. Using the attributes of a reported vulnerability, a score is calculated to describe the severity of the software vulnerability. The severity depends on many factors and determines the extent of damage that exploitation of that vulnerability will cause to the software system.

This paper aims to analyse the software vulnerabilities by using machine learning algorithms. The vulnerabilities are studied to determine performance metrics namely score and severity. There are mainly two scoring standards, CVSS v2.0 and CVSS v3.X. These are standards of common vulnerability scoring system and each of them use different metrics to score the vulnerabilities. This paper uses CVSS v3.X scores to calculate scores for un-scored software vulnerabilities. In addition to this, a performance comparison is drawn among different machine learning algorithms namely random forest, K-nearest neighbours, and decision tree in terms of mean absolute error, mean square error, root mean square error, and R2 score.

There are six sections in this paper. After the introduction, Sect. 2 presents the literature review. Data and methods used are presented in Sect. 3, followed by the analysis of software vulnerabilities in Sect. 4. Finally, the results are presented, and observations are discussed in Sect. 5, before concluding the paper.

2 Literature Review

This section highlights and discusses the work available in open literature related to software vulnerabilities. Ghosh et al. [3] presented an approach for identifying the vulnerabilities in software. Authors considered that a large percentage of security breaches are due to errors in software source code.

Baker et al. in 1999 [4] proposed the idea to create CVE lists to increase access to software vulnerabilities to the public. In the same year, Bishop presented a model for classification of vulnerabilities in computer system [5]. In 2012, Shahriar and Zulkernine [6] presented a wide review on different methodologies to alleviate security vulnerabilities which were published between 1994 and 2010.

Number of authors introduced static software analysis methods to track the likely indicators of vulnerabilities [7–10]. Few authors also presented a manual software security testing approach in the literature [5, 11]. Liu et al. [12] discussed methods such as fuzzing and penetration along with static analysis. It was observed that static analysis worked well while developing the software and others worked well post software development. Kuang et al. [13] described a dynamic analysis approach that attempted to test the behaviour of the application when an abnormal value was returned by the called function.

In addition to these approaches, in recent years, authors have been using the machine learning methodologies for different applications. The machine learning approach is also appropriate for analysing software vulnerabilities.

Ghaffarian and Shahriari [14] discussed briefly about the software vulnerabilities and machine learning approaches to address them. Their work is split into three main categories: vulnerability prediction based on software metrics, anomaly detection, and vulnerable code pattern recognition. In last many years, authors have been proposing different machine learning-based approaches to analyse and detect software vulnerabilities. Jie et al. [2] provided a comprehensive analysis of many such approaches for software vulnerability analysis as well as detection. Some approaches train the classifiers using vulnerable dataset. In 2020, Singh and Chaturvedi [15] used the deep learning approach to diminish the process of manual selection of feature. Deep learning techniques were also used to recognize patterns in code that can be vulnerable [16]. Furthermore, the use of pattern recognition was also seen to detect behaviour-based attacks in industrial control systems. The enhancements in the system caused these systems to have Internet access and consequently made them vulnerable to such attacks [17].

Kronjee et al. [18] presented a machine learning approach to detect SQL injection and other vulnerabilities in the PHP application. Authors combined the machine learning approach with the static analysis. Ucci et al. [19] presented a survey for the analysis of malware using different machine learning methods. Authors found machine learning very useful. Sumanth et al. [20] presented a machine learning-based analysis on false positives for software vulnerabilities. Authors predicted that the vulnerability was real, or if it was a false positive.

Many classification techniques have been used to analyse software vulnerabilities. Raheja and Munjal [21] opted for supervised learning techniques such as Bayes classification and k-nearest neighbours (KNN) to classify Microsoft-Office related vulnerabilities. Furthermore, the vulnerabilities were analysed based on availability, confidentiality, and integrity using such classification techniques [22]. When confidentiality is considered, data security is of importance. Kumar et al. [23] discussed the potential data attack on video data and the implemented histogram bit shifting-based reversible data hiding to perform encryption and safeguard vulnerable data.

In 2018, Chernis and Verma [24] used multiple classifiers to analyse and detect vulnerable functions in a C source code. They first extracted the features from 100 °C source codes and upon dividing it into training and testing data, trained the classifiers using it.

Vulnerabilities are observed in all kinds of systems. Each of them posing some sort of threat to the working of the system. Machine learning algorithms like the decision tree and the KNN are popular. Their application has been seen from analysing power systems to analysing level of infection in a body [25, 26].

The objectives of this paper are as follows:

- To predict the two factors: score and severity for software vulnerabilities and analyse these vulnerabilities.
- To present the comparative analysis of three important machine learning algorithms namely random forest, k-nearest, and decision tree.
- Results are computed in terms of mean absolute error, mean square error, root mean square error, and R2 score.
- Based on score and severity, the software developers can avoid and focus on a particular vulnerability while developing the software.

3 Data and Methods

This section presents the dataset and methodology used for the implementation the current work.

3.1 Dataset Used

The dataset used for current work has been picked from the NVD (https://nvd.nist. gov/) CVE data base [27] for the years 2019 and 2020. Snippet for few of the entries of dataset is shown in Table 1.

Table 1 Snippet of NVD-CVE dataset

CVE_data_type	CVE_data_format	CVE_data_version	CVE_data_numberOfCVEs	CVE_data_timestamp	CVE_Items
CVE	MITRE	4	15,393	2020-04-28T07:06Z	{'cve': {'data_type': 'CVE', 'data_format': 'MITRE', ...
CVE	MITRE	4	15,393	2020-04-28T07:06Z	{'cve': {'data_type': 'CVE', 'data_format': 'MITRE' ...
CVE	MITRE	4	15,393	2020-04-28T07:06Z	{'cve': {'data_type': 'CVE', 'data_format': 'MITRE'...

3.2 Methodology

The dataset used to train all three regressor algorithms was curated out of two NVD-CVE feeds for year 2019 and 2020. After preprocessing the data feeds are combined and then passed to random forest regressor, K-nearest, and decision tree for training (Fig. 1).

Data Preprocessing. The dataset taken from NVD-CVE Data Feeds contains various information about each software vulnerability [28]. It contains timestamp data with information about both CVSS v2 and CVSS v3 metrics. As this paper is considering only CVSS v3 metrics, so CVSS v3 data have filtered from the original datasets. To convert the 'string' values assigned to each metric for each vulnerability, these two files for year 2019 and 2020 are processed again to obtain datasets which can be combined and used for model training. Values by which the 'string' values are replaced are in respect to each metric's effect on 'base score'. A snippet for dataset after preprocessing for year 2020 is as given in Table 2.

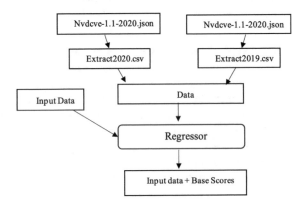

Fig. 1 Methodology adopted for prediction of vulnerabilities

Table 2 Snippet of 'Extract2020.csv'

	ExploitabilityScore	ImpactScore	AttackVectorVal	AvailabilityImpactVal	IntegrityImpactVal	ScopeVal
0	1.8	5.9	2	3	3	1
1	2.8	5.9	4	3	3	1
2	0.8	5.9	2	3	3	1
3	1.8	3.6	2	3	1	1
4	0.8	5.9	2	3	3	1
5	2.8	3.6	4	1	1	1
6	1.8	3.6	2	1	1	1
7	1.0	3.6	2	1	1	1
8	1.8	3.6	2	1	3	1

	UserInteractionVal	AttackComplexityVal	ConfidentialityImpactVal	PriviledgesRequiredVal	BaseScore
0	1	3	2	3	7.8
1	2	3	2	1	8.8
2	2	2	2	3	6.7
3	1	3	1	3	5.5
4	1	3	2	2	6.7
5	2	3	2	1	6.5
6	1	3	2	3	5.5
7	1	2	2	3	4.7
8	1	3	1	3	5.5

Random Forest Regressor. The current work is using the supervised learning algorithm: Random forest to analyse the vulnerabilities. 'Random forests' or 'random decision forests' are a complete self-sufficed learning method for 'classification', 'regression' and similar other functions and tasks that perform by a unique way of 'constructing a multitude of decision trees at training time and outputting the class that is the mode of the classes or mean prediction of the individual trees' [29].

Random forest regressor randomly distributes the dataset among different decision trees. It considers the summation of all predictions done by each decision tree for an output. This behaviour of random forest regressor makes it a 'additive model'. This can formally be represented as given in Eq. 1.

$$g(x) = f_0(x) + f_1(x) + f_2(x) + \cdots \tag{1}$$

where 'g' is base model summation and 'fi' is base classifier (decision tree). In this paper, random forest regressor is implemented using the Scipy library in Python. Authors analysed the regressor for different values of 'n Estimator'. To finely tune random forest regressor and train the model effectively to have maximum 'model score' or least errors, an appropriate value of 'n Estimators' is to be decided by authors. 'n Estimators' defines the number of decision trees that are going to be used by the model to come to a final decision. The resultant maximum value is

N Estimator: 120

Model score: 0.999934

The pseudo code for the same is mentioned as Pseudo Code 1.

```
Pseudo Code 1: Random Forest
#TRAIN THE RANDOM FOREST REGRESSOR AND PREDICT THE
OUTCOME.
   def train(self,X,y):
      ran = range(10,140,10)
      plotd = {'n_estimators':[],'model_score':[]}
   #CALCULATING  THE  MOST  EFFICIENT  VALUE  FOR
N-ESTIMATOR
         for i in ran:
            X_train, X_test, y_train, y_test = train_test_split(X, y, test_size =
0.3, random_state = 0)
            regressor = RandomForestRegressor(n_estimators = i, random_
state = 0)
            regressor.fit(X_train, y_train)
            modelsc = regressor.score(X_test,y_test)
            plotd['n_estimators'].append(i)
            plotd['model_score'].append(modelsc)
```

To find the most effective value of 'n Estimators' authors plotted a 'model score' vs 'n Estimators' graph to visually see which value gives the highest 'model score' as illustrated in Fig. 2.

K-nearest neighbours (KNN). KNN is an algorithm used for classification as well as regression. It works on similarity-based learning. Values for new data points are predicted by using the similarity between the data point that is new and other already existing points. The first step of this algorithm is the calculation of distance between the new point and each of the existing points [30]. The KNN algorithm implemented to analyse the software vulnerability uses the standard Euclidean metric. It is also the default distance metric used in nearest neighbour algorithms. The formula to calculate the Euclidean distance between points (x, y) and (a, b) is as in Eq. 2.

$$ED((x, y), (a, b)) = \sqrt{\left([\![(x - a)]\!]^2 + [\![(y - b)]\!]^2\right)} \tag{2}$$

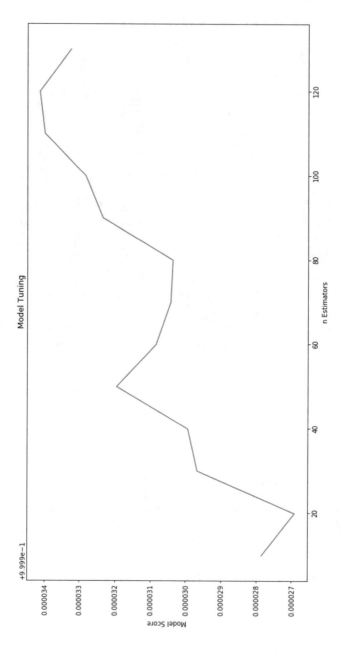

Fig. 2 'Model score' versus 'n Estimators'

The second step comprises of choosing the optimum value of 'k'. The value of 'k' represents the number of neighbours to be considered while assigning the data point that is new its value. Its optimum value can be chosen based on error calculation.

```
Pseudo Code 2: KNN
#TRAIN THE K NEAREST NEIGHBOURS REGRESSOR AND PREDICT
THE OUTCOME
    def train(self,X,y):
        X_train, self.X_test, y_train, self.y_test = train_test_split(X, y, test_
size = 0.3, random_state = 0)
        scaler = StandardScaler()
        scaler.fit(X_train)
        X_train = scaler.transform(X_train)
        self.X_test = scaler.transform(self.X_test)
        self.knnr = KNeighborsRegressor(n_neighbors = 10)
        self.knnr.fit(X_train, y_train)
        self.y_pred = self.knnr.predict(self.X_test)
        return self.knnr
```

Decision Tree The decision tree algorithm performs a series of tests on the values of the descriptive feature. A descriptive feature is a factor influencing the outcome. The structure of a decision tree comprises of four types of components, a root node, interior nodes, leaf nodes, and branches. The root node is the starting node that is connected to the interior or leaf nodes via the branches. Leaf nodes are nodes without any further nodes attached to it. Number of branches coming downwards from a non-leaf node depends upon the number of levels that can be possibly taken by the descriptive feature. The significance of each leaf node is that it specifies all possible outcomes that can be predicted for the target feature [30].

```
Pseudo Code 3: Decision Tree
#TRAIN THE DECISION TREE REGRESSOR AND PREDICT THE
OUTCOME
    def train(self,X,y):
        X_train, self.X_test, y_train, self.y_test = train_test_split(X, y, test_
size = 0.4, random_state = 0)
        self.regressor = DecisionTreeRegressor()
        self.regressor.fit(X_train,y_train)
```

```
self.y_pred = self.regressor.predict(self.X_test)
return self.regressor
```

4 Analysis of Software Vulnerabilities

In this paper, a model is trained to predict values for two important metrics of vulnerabilities namely 'base score' and 'base severity'. There are eight vulnerability metrics involved in CVSS v3 standard as follows:

1. Attack complexity
2. Attack vector
3. Privileges required
4. User interaction
5. Confidentiality impact
6. Integrity impact
7. Availability impact
8. Scope.

Each of these eight metrics have a set of values that can be marked for each software vulnerability. The 'base score' is affected by the values marked for these metrics and changes accordingly. Combination of these values is used to give a score to each vulnerability from 0 to 10. These scores determine the vulnerability's severity, higher the score, more severe the software vulnerability is. The snippet of prediction results for both 'base score' and 'base severity' is presented in Table 4.

Two other machine learning algorithms KNN and decision tree were also implemented to check the efficacy of prediction results of random forest technique. To compare all these three techniques, performance metrics are computed namely mean absolute error (MAE), mean square error (MSE), root mean square error (RMSE), and R2 score.

5 Results and Observations

After training the random forest regressor, the decision tree regressor, and the k-nearest neighbours regressor on the same CVE datasets were applied. Table 3 contains the calculated values of mean absolute error, mean square error, root mean square error, and R2 score for each of the three machine learning algorithms used.

The random forest regressor model was found to have performed better with a root mean square error of 0.01945, while the other two had a value greater than 0.02. While the 'R2 score' for random forest regressor and decision tree regressor models had

Table 3 Prediction results

Model	Mean absolute error	Mean square error	Root mean square error	R2 score
Random forest	0.00147	0.000378	0.01945	0.99986
Decision tree	0.00148	0.000723	0.02689	0.99974
K-nearest	0.02673	0.018618	0.13644	0.99347

very little difference, K-nearest regressor model was evidently lesser. Consequently, random forest regressor was applied to calculate the score and severity of un-scored vulnerabilities. The result is shown in Table 4. From Table 4, it can be observed that

- The set of software vulnerabilities reported recently fall towards mainly 'medium', 'high', or 'critical' severity category.
- Absence of vulnerabilities falling in 'none' or 'low' categories can be observed.
- It can be evaluated that in recent times, more harmful vulnerabilities have been detected.

6 Conclusion

Software vulnerabilities can affect systems by enabling attackers to breach its confidentiality, integrity, availability, etc. This can cause violation of security policies and cause serious security threats to the data and software. Analysing these vulnerabilities provide an insight into the severity of consequences if said vulnerabilities are exploited. Base score and base severity were calculated against various software vulnerabilities by considering CVSS v3 vulnerability metrics. Machine learning algorithms, namely random forest, decision tree, and K-nearest neighbour, were employed for this task, and random forest regressor was observed to have provided with the best R2 score amongst the three. Further, work can include encompassing more factors affecting the severity of a software vulnerability and performing analysis for specific platforms.

Table 4 Prediction values of base score and base severity by random forest regressor algorithm

	Unnamed: 0	ExploitabilityScore	ImpactScore	AttackVectorVal	AvailabilityImpactVal	IntegrityImpactVal	ScopeVal
0	0	1.8	5.9	2	3	3	1
1	1	3.9	5.9	4	3	3	1
2	2	1.8	5.9	2	3	3	1
3	3	2.2	5.2	4	3	3	1
4	4	3.9	5.9	4	3	3	1
5	5	2.8	5.9	4	3	3	1

	UserInteractionVal	AttackComplexityVal	ConfidentialityImpactVal	PrivilegesRequiredVal	BaseScore	BaseSeverity
0	2	3	2	1	7.799999999999960	HIGH
1	1	3	2	1	9.799999999999986	CRITICAL
2	1	3	2	3	7.799999999999960	HIGH
3	1	2	1	1	7.394999999999990	HIGH
4	1	3	2	1	9.799999999999986	CRITICAL
5	1	3	2	3	8.799999999999995	HIGH

References

1. Krsul IV (1998) Software vulnerability analysis. Purdue University, West Lafayette
2. Jie G, Xiao-Hui K, Qiang L (2016) Survey on software vulnerability analysis method based on machine learning. In: 2016 IEEE first international conference on data science in cyberspace (DSC), pp 642–647. IEEE
3. Ghosh AK, O'Connor T, McGraw G (1998) An automated approach for identifying potential vulnerabilities in software. In: Proceedings. 1998 IEEE symposium on security and privacy (Cat. No. 98CB36186), pp 104–114. IEEE
4. Baker DW, Christey SM, Hill WH, Mann DE (1999) The development of a common enumeration of vulnerabilities and exposures. In: Recent advances in intrusion detection, vol 7, p 9
5. Bishop M (1999) Vulnerabilities analysis. In: Proceedings of the recent advances in intrusion detection, pp 125–136
6. Shahriar H, Zulkernine M (2012) Information-theoretic detection of sql injection attacks. In: 2012 IEEE 14th international symposium on high-assurance systems engineering, pp 40–47. IEEE
7. Evans D, Larochelle D (2012) Improving security using extensible lightweight static analysis. IEEE Softw 19(1):42–51
8. Larus JR, Ball T, Das M, DeLine R, Fahndrich M, Pincus J, Rajamani SK, Venkatapathy R (2004) Righting Software. IEEE Softw 21(3):92–100
9. Ayewah N, Pugh W, Hovemeyer D, Morgenthaler JD, Penix J (2008) Using static analysis to find bugs. IEEE Softw 25(5):22–29
10. Bessey A, Block K, Chelf B, Chou A, Fulton B, Hallem S, Henri- C, Kamsky A, McPeak S, Engler D (2010) A few billion lines of code later: using static analysis to find bugs in the real world. Commun ACM 53(2):66–75
11. Arkin B, Stender S, McGraw G (2005) Software penetration testing. IEEE Secur Priv 3(1):84–87
12. Liu B, Shi L, Cai Z, Li M (2012) Software vulnerability discovery techniques: a survey. In: 2012 fourth international conference on multimedia information networking and security, pp 152–156. IEEE
13. Ghaffarian SM, Shahriari HR (2017) Software vulnerability analysis and discovery using machine-learning and data-mining techniques: a survey. ACM Comput Surv (CSUR) 50(4):1–36
14. Kuang C, Miao Q, Chen H (2006) Analysis of software vulnerability. WSEAS Trans Comput Res 1(1):45–50
15. Singh SK, Chaturvedi A (2020) Applying deep learning for discovery and analysis of software vulnerabilities: a brief survey. Soft Comput Theor Appl 649–658
16. Zeng P, Lin G, Pan L, Tai Y, Zhang J (2020) Software vulnerability analysis and discovery using deep learning techniques: a survey. IEEE Access
17. Bhardwaj A, Al- F, Kumar M, Stephan T, Mostarda L (2020) Capturing-the-invisible (CTI): behavior-based attacks recognition in IoT-oriented industrial control systems. IEEE Access 8:104956–104966
18. Kronjee J, Hommersom A, Vranken H (2018) Discovering software vulnerabilities using dataflow analysis and machine learning. In: Proceedings of the 13th international conference on availability, reliability and security, pp 1–10
19. Ucci D, Aniello L, Baldoni R (2019) Survey of machine learning techniques for malware analysis. Comput Secur 81:123–147
20. Sumanth R, Bhanu KN (2020) Raspberry Pi based intrusion detection system using k-means clustering algorithm. In: 2020 second international conference on inventive research in computing applications (ICIRCA), Coimbatore, India, pp 221–229
21. Raheja S, Munjal G (2021) Classification of microsoft office vulnerabilities: a step ahead for secure software development. In: Bio-inspired neurocomputing. Springer, Singapore, pp 381–402

22. Raheja S, Munjal G (2016) Shagun: analysis of linux kernel vulnerabilities. Ind J Sci Technol 9:12–29
23. Kumar M, Aggarwal J, Rani A et al (2021) Secure video communication using firefly optimization and visual cryptography. Artif Intell Rev
24. Chernis B, Verma R (2018) Machine learning methods for software vulnerability detection. In: Proceedings of the fourth ACM international workshop on security and privacy analytics, pp 31–39
25. Aliyan E, Aghamohammadi M, Kia M, Heidari A, Shafie-khah M, Catalão JP (2020) Decision tree analysis to identify harmful contingencies and estimate blackout indices for predicting system vulnerability. Electr Power Syst Res 178:106036
26. Salam A, Prasetiyowati SS, Sibaroni Y (2020) Prediction vulnerability level of dengue fever using KNN and random forest. Jurnal RESTI (Rekayasa Sistem dan Teknologi Informasi) 4(3):531–536
27. CVE details: the ultimate security vulnerability data source—vulnerabilities by type. https://www.cvedetails.com/index.php
28. Sanguino LAB, Uetz R (2017) Software vulnerability analysis using CPE and CVE. Comput Secur 1–29
29. Li X, Chang X, Board JA, Trivedi KS (2017) A novel approach for software vulnerability classification. In: 2017 annual reliability and maintainability symposium (RAMS), pp 1–7. IEEE
30. Kelleher JD, Namee BM, D'Arcy A (2015) Fundamentals of machine learning for predictive data analysis. The MIT Press, London

Forward Solver for Electrical Impedance Tomography System for Detection of Breast Cancer Using Simulation

Priya Tushar Hankare⊙, **Alice N. Cheeran**⊙, **Prashant Bhopale**⊙, **and Ashitosh Joshi**⊙

Abstract Electrical impedance tomography is a harmless and safe imaging modality that estimates the electrical properties at the interior of an object from measurements made on its surface. To obtain the electrical properties of an object, electrodes are placed on the surface of the object, current is passed between a pair of electrodes, and voltage is measured across other electrodes, and the electrical conductivity distribution is calculated and imaged. Since this paper deals with detection of breast cancer, the object (breast) is simulated as a phantom considering the conductivity distribution/relative permittivity as that of breast tissues. Also benign and malignant growth mass of different sizes is simulated with different conductivities, and relative permittivity is simulated to be present in different locations inside the phantom. This is carried out using COMSOL Multiphysics software platform. Further to obtain the current/voltage data, algorithm of forward solver is applied to the two-dimensional simulated EIT system. The data obtained can be used for image reconstruction applying the inverse problem.

P. T. Hankare (✉)
Veermata Jijabai Technological Institute, Matunga, Mumbai 400019, India
e-mail: priya.h@somaiya.edu; pthankare_p18@el.vjti.ac.in

K.J. Somaiya Institute of Engineering and Information Technology, Sion East, Mumbai 400022, India

P. T. Hankare · A. N. Cheeran · P. Bhopale · A. Joshi
University of Mumbai, Mumbai, India
e-mail: ancheeran@ee.vjti.ac.in

P. Bhopale
e-mail: psbhopale@el.vjti.ac.in

A. Joshi
e-mail: aajoshi@et.vjti.ac.in

A. N. Cheeran · A. N. Cheeran · P. Bhopale · P. Bhopale · A. Joshi · A. Joshi
Electrical Engineering Department, Veermata Jijabai Technological Institute, Matunga, Mumbai 400019, India

© The Author(s), under exclusive license to Springer Nature Singapore Pte Ltd. 2023
A. Shukla et al. (eds.), *Computational Intelligence*, Lecture Notes in Electrical Engineering 968, https://doi.org/10.1007/978-981-19-7346-8_36

425

Keywords Electrical impedance tomography (EIT) · Breast tissue phantom ·
Forward problem · COMSOL multiphysics · Breast electrical conductivity · Breast
relative permittivity

1 Introduction

Breast cancer is most common in older women since cells undergo multiple genetic
alterations leading to malignancy. There is also an increased risk of breast cancer in
case of family hereditary and relapse may also happen [1]. If it is detected at an early
stage, then chances of cure increase, leading to extended life. Various techniques are
used for breast cancer detection like mammography, MRI imaging, etc. [2].

EIT imaging is one of the medical tomographic imaging techniques which is
harmless, low cost, radiation free, and non-invasive and can be developed into a
portable device. Since EIT has numerous advantages, it has been widely explored in
several fields of science and engineering, like medical imaging, industrial process
tomography, chemical engineering, civil engineering, geosciences, oceanography,
manufacturing technology, microbiology, biotechnology, etc. [3]. EIT in medical
imaging works on the principle that human body tissues offer impedance to flow
of electric current. It images the interior electrical properties of the human body by
obtaining measurements made on the body's exterior surface. This is usually accomplished by infusing current into the object through electrodes placed on its surface
of the object to be imaged and measuring the resulting voltages at the electrodes.
Human body consists of different types of tissues, and each tissue will have its own
conductivity value due to the variation in tissue composition, tissue physiology, and
tissue health (e.g., swelling, disease, and infection) [4].

For ethical reasons, experiments cannot be performed on human body. In this
paper on early detection of breast cancer, EIT experiments are simulated. A phantom
representing the breast is simulated using COMSOL Multiphysics software platform
taking the breast tissue conductivity and permittivity into consideration. Further
objects with various conductivity and permittivity with different sizes representing
malignant and benign tumor are inserted into the phantom. These are imaged. Further
synthesis is carried out on this model as forward solver, to obtain the data which
corresponds to the reading obtained by practically performing EIT experiments.

2 Methodology

2.1 EIT Technique

Humans are not allowed to participate in biomedical experiments for ethical concerns.
As a result, a hypothetical model (phantom) resembling a breast is needed to be built
for conducting tests [5]. Electrodes/sensors will be positioned around the phantom's

Fig. 1 Adjacent pattern

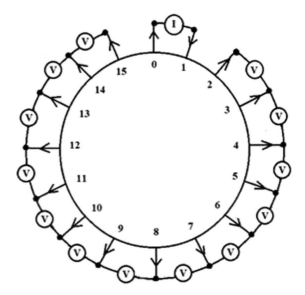

circle. Pairs of electrode will receive current, and voltage will be monitored across the remaining electrode pairs as shown in Fig. 1. Pairs of electrode will receive current, and voltage will be monitored across the remaining electrode pairs Experiments are repeated changing the electrode pairs for sending the current and monitoring voltages. Since the current is to be supplied to the human body for tissue impedance measurement, a low amplitude ac current is necessary.

In adjacent pattern method, as shown in Fig. 1, current is injected through electrodes 0 and 1, and the voltages across the other 13 electrode pairs 2–3, 3–4, and 14–15 are monitored. In the second projection, current is injected through electrodes 1 and 2, and voltages are measured across the remaining 13 electrode pairs (3–4, 4–5, etc.). This method is repeated until all 16 neighboring pairs of electrodes have received current. The process produces a total of $16 \times 13 = 208$ measurements [2].

2.2 Forward Solver

Maxwell's equations are used to characterize the forward problem in EIT applications. Here in the mathematical analysis of EIT, magnetic field is not considered since bioelectrical impedance is characterized by a pure resistor. The calculation of voltage measurements based on the known conductivity distribution of a specific object and the applied current value is known as a forward problem. It plays an important role since precision of results depends on forward model design and geometry.

Mathematically, it can be represented as Eq. (1).

$$V = [T][\sigma] \tag{1}$$

where, $V =$ electric potential produced by applying low frequency and low magnitude AC current I,

$\sigma =$ tissue conductivity vector, and

$T =$ transformation matrix which depends on the applied algorithm.

2.3 Block Diagram of Forward Problem of an EIT System Using COMSOL Multiphysics

COMSOL Multiphysics is a FEM solver and Multiphysics simulation software [6]. To solve and simulate the forward problem in COMSOL Multiphysics software, we need to first define the geometry of model as 2D or 3D and define the type of electrodes, materials associated with the geometry, etc.

Figure 2 shows the block diagram of simulation of forward problem system. It is a virtual system which is a replica of an actual system. The procedure starts with COMSOL Multiphysics software initiation. After defining the geometry, electrodes and material associated current will be given to terminal electrode and ground will be assigned to another electrode. Voltage measurements will be carried out.

Here while simulating the forward problem, the electrical conductivity and relative permittivity of malignant/benign tissues are considered to represent the breast tumor [7].

Table 1 shows the various parameters defined for building 2D geometry of a breast. Also the details of material of electrodes, current parameters, and the various conductivity and permittivity of the various tissues considered are presented in Table 1.

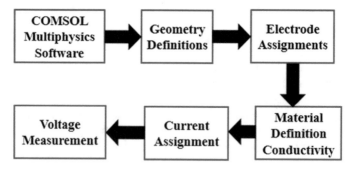

Fig. 2 Block diagram of simulation of solution toward forward problem system

Table 1 Parameters defined for design of a 2D breast model

S. No.	Geometry parameters	Values
1	Phantom size (radius)	6 cm
2	Number of electrodes	16
3	Type of electrodes	Circular (arc)
4	Material of electrode	Silver
5	Current injection pattern	Adjacent
6	AC current frequency	10, 50, and 200 kHz
7	Applied AC current	10 mA
8	Electrical conductivity of breast tissue	5 S/m
9	Relative permittivity of breast tissue	10
10	Electrical conductivity of benign tissue	30 S/m
11	Relative permittivity of benign tissue	50
12	Electrical conductivity of malignant tissue	0.77
13	Relative permittivity of malignant tissue	80
14	Size of benign and malignant tissue	0.25, 0.4, and 0.5 cm

3 Results

Electrical conductivity and relative permittivity of real breast benign and malignant tissue are considered [8, 9], and forward model is simulated. Figure 3a shows the forward model with benign and malignant tumor. Figure 3b shows 10 mA current with 10 kHz frequency that is applied between Electrode 1 and Electrode 2, and Fig. 3c shows forward mesh consisting of 3110 domain elements and 312 boundary elements.

The types of mesh available are extremely fine, extra fine, finer, fine, normal, coarse, coarser, extra coarse, and extremely coarse. In this case, finer mesh is used. Figure 4 shows forward model images at different frequencies, tumor sizes, and location.

Once the model is executed properly, voltage values are generated at the output. These voltage readings can be saved in CSV or XLS format. Figure 5 shows the voltage values for 10 kHz frequency. Since number of electrodes used are 16, it is giving an output matrix of 16 × 16 voltage values.

Figure 4a shows homogeneous model at 10 kHz frequency, Fig. 4b shows the forward model with circular-shaped benign and malignant tissue mass of 0.25 cm at 50 kHz frequency, Fig. 4c shows the forward model with circular-shaped benign and

Fig. 3 **a** Forward model with benign and malignant tumor. **b** 10 mA current with 10 kHz frequency between Electrode 1 and Electrode 2. **c** Forward mesh consisting of 3110 domain elements and 312 boundary elements

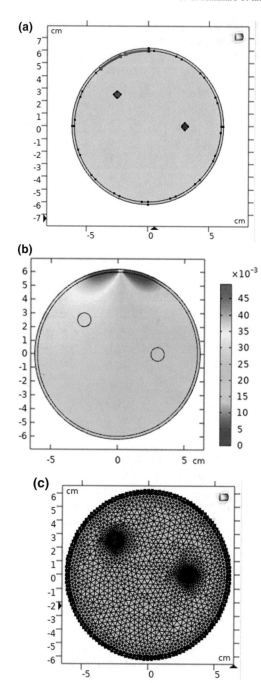

Fig. 4 Forward model images at different frequencies, tumor sizes, and location of tumor. **a** Homogeneous model at 10 kHz frequency. **b** With benign and malignant tissue of 0.25 cm at 50 kHz frequency. **c** With benign and malignant tissue of 0.5 cm at 10 kHz frequency. **d** With benign and malignant tissue of 0.4 cm at 200 kHz frequency

Forward Model Images

(a) Homogeneous Model at 10 KHz frequency

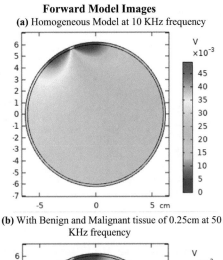

(b) With Benign and Malignant tissue of 0.25cm at 50 KHz frequency

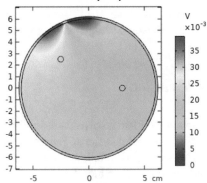

(c) With Benign and Malignant tissue of 0.5cm at 10 KHz frequency

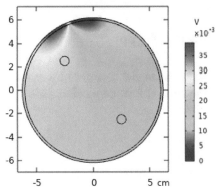

Fig. 4 (continued)   **(d)** With Benign and Malignant tissue of 0.4cm at 200
KHz frequency

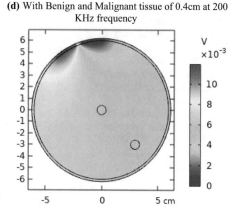

malignant tissue of 0.5 cm at 10 kHz frequency, and Fig. 4d shows the forward model
with circular-shaped benign and malignant tissue of 0.4 cm at 200 kHz frequency.

4   Conclusion

The forward problem is solved and simulated using COMSOL Multiphysics software
platform for two-dimensional geometry with 16 silver sensor electrodes. Since the
platform is computationally very fast, results are obtained quickly with respect in
obtaining the voltage values. Further, we can change the material associated with the
electrodes to steel or gold. Also number of electrodes can be changed from 16 to 32
or 64 and simulate the model.

5   Future Scope

Inverse problem can be solved by using various image reconstruction algorithms
from solution of forward problem [10].

	E0	E1	E2	E3	E4	E5	E6	E7	E8	E9	E10	E11	E12	E13	E14	E15
E0	0.012867	0.002799	0.001418	0.000926	0.000653	0.000529	0.000531	0.000529	0.000529	0.000661	0.000856	0.001275	0.002366	0.007487	0.006146	-0.03961
E1	-0.03966	0.012886	0.002809	0.001429	0.000879	0.000627	0.000579	0.000534	0.000534	0.000566	0.000661	0.000858	0.001277	0.00237	0.007497	0.006159
E2	0.006164	-0.03964	0.012893	0.002827	0.001373	0.000825	0.000686	0.000584	0.000584	0.000528	0.000566	0.000662	0.000858	0.001278	0.002371	0.007496
E3	0.007497	0.006161	-0.03958	0.012923	0.002764	0.001257	0.000899	0.000694	0.000583	0.000533	0.000528	0.000567	0.000663	0.00086	0.00128	0.002373
E4	0.002387	0.00753	0.00617	-0.03953	0.012976	0.00249	0.00134	0.000908	0.000692	0.000582	0.000534	0.00053	0.000569	0.000667	0.000865	0.001288
E5	0.001287	0.002404	0.007627	0.006256	-0.0393	0.012688	0.002526	0.001311	0.000869	0.000663	0.000561	0.000517	0.000516	0.000558	0.000658	0.000839
E6	0.000784	0.001175	0.002218	0.007354	0.006257	-0.03838	0.012974	0.002695	0.00133	0.000851	0.000635	0.00053	0.000483	0.000479	0.000514	0.000602
E7	0.000646	0.000827	0.001203	0.002143	0.007104	0.006307	-0.03934	0.013039	0.002843	0.001432	0.000925	0.000693	0.000578	0.000527	0.00052	0.000556
E8	0.000569	0.000665	0.000854	0.001222	0.002102	0.007524	0.006202	-0.03956	0.012916	0.002815	0.001423	0.000924	0.000696	0.000583	0.000534	0.00053
E9	0.00053	0.00057	0.000667	0.000845	0.001139	0.002334	0.00755	0.006175	-0.03966	0.0129	0.002806	0.001418	0.00092	0.000693	0.000581	0.000533
E10	0.000532	0.000529	0.00057	0.000658	0.000773	0.001236	0.002394	0.007509	0.006171	-0.03966	0.012883	0.002801	0.001414	0.000918	0.000691	0.00058
E11	0.000579	0.000532	0.000529	0.000565	0.000601	0.000819	0.00129	0.002376	0.007492	0.00616	-0.03962	0.012861	0.002797	0.001412	0.000916	0.00069
E12	0.000689	0.000579	0.000533	0.000527	0.000517	0.000634	0.000865	0.001281	0.002369	0.007486	0.006147	-0.0396	0.012864	0.002794	0.001411	0.000915
E13	0.000917	0.000692	0.000582	0.000533	0.000485	0.000529	0.000667	0.000862	0.001279	0.00237	0.007498	0.00616	-0.03967	0.012885	0.002799	0.001413
E14	0.001414	0.000919	0.000694	0.000583	0.000493	0.000488	0.000569	0.000665	0.000859	0.001277	0.002369	0.007495	0.006166	-0.03967	0.012875	0.002799
E15	0.002796	0.001413	0.00092	0.000696	0.000542	0.000488	0.000528	0.000567	0.000662	0.000856	0.001274	0.002365	0.007483	0.006153	-0.0396	0.012855

Fig. 5 Output voltage values for 50 kHz frequency

References

1. Breast cancer section overview [Online]. Available: http://www.cancerresearchuk.org/cancer-help/type/breast-cancer/. accessed on August 2020
2. Hankare PT, Cheeran AN (2020) An automated portable and noninvasive approach for breast tumor findings using 2D electrical impedance tomography in early stage. In: 2020 international conference on computer science, engineering and applications (ICCSEA), Gunupur, India, pp 1–5. https://doi.org/10.1109/ICCSEA49143.2020.9132864
3. Bera TK, Nagaraju J (2014) Electrical impedance tomography (EIT): a harmless medical imaging modality. In: Research developments in computer vision and image processing: methodologies and applications, pp 235–273. https://doi.org/10.4018/978-1-4666-4558-5.ch013
4. Bera TK (2014) Bioelectrical impedance methods for noninvasive health monitoring: a review. J Med Eng. https://doi.org/10.1155/2014/381251
5. Sarode V, Patkar S, Cheeran AN (2013) Comparison of factors affecting the detection of small impurities in breast cancer using EIT. Int J Eng Sci Technol (IJEST). 5(6):1267–1271. ISSN: 0975-5462
6. Fouchard A, Bonnet S, Herve L, David O (2015) Flexible numerical platform for electrical impedance tomography. In: Excerpt from the proceedings of the 2015 COMSOL conference in Grenoble
7. Hankare PT, Dongaonkar TN, Naik N, Cheeran AN (2021) Simulation of forward problem for non-invasive two dimensional electrical impedance tomography system using EIDORS and COMSOL multiphysics. In: 4th international conference on advances in science and technology (ICAST-2021). https://ssrn.com/abstract=3867446
8. Joines WT, Zhang Y, Li C, Jirtle RL (1994) The measured electrical properties of normal and malignant human tissues from 50 to 900 MHz. Med Phys 21:547–550
9. O'Halloran M, Byrne D, Conceicao RC, Jones E, Glavin M (2016) Anatomy and dielectric properties of the breast and breast cancer. In: An introduction to microwave imaging for breast cancer detection, pp 5–16. https://doi.org/10.1007/978-3-319-27866-7_2
10. Hankare PT, Cheeran AN, Bhopale PS (2021) Comparison of image reconstruction algorithms for finding impurities utilizing EIT for clinical application in breast cancer. Int J Med Eng Inform. https://doi.org/10.1504/IJMEI.2021.10040190

Smart Farming System Based on IoT for Precision Controlled Greenhouse Management

Ashay Rokade and Manwinder Singh

Abstract Smart farming with accurate greenhouses needs to be implemented for better farming growth management, and therefore, precision agriculture monitoring in diverse conditions is required. The Internet of Things (IoT) is a new era in computer communication that is gaining traction due to its vast range of applications in project development. The IoT provides individuals with smart and remote approaches, such as smart agriculture, smart environment, smart security, and smart cities. This is the most recent technology that is making things easier these days. The Internet of Things (IoT) has fundamentally expanded remote distance control and the diversity of networked things or devices, which is an intriguing element. The IoT comprises the hardware as well as the Internet connectivity to the real-time application. Sensors, actuators, embedded systems, and an Internet connection are the key components of the Internet of Things. As a result, we are interested in creating a smart farm IoT application. In greenhouse agriculture, this study presented a remote sensing of parameters and control system. The objective is to manage CO_2, temperature, soil moisture, humidity, and light, with regulating actions for greenhouse windows/doors dependent on crops being carried out once a quarter throughout the year. The major goal is to properly regulate greenhouse conditions in accordance with plant requirements, in order to enhance output and provide organic farming. The results show that the greenhouse may be controlled remotely for CO_2, temperature, soil moisture, humidity, and light, resulting in improved management.

Keywords Smart farming · Greenhouse management · IoT · Precision farming

1 Introduction

Due to the fact that plant growth and strength are increasingly crucial factors, whether for food or cash crops, plant development has proven to be a test of innovation. One of the most significant difficulties in modern agriculture is the absence of knowledge

A. Rokade · M. Singh (✉)
School of Electronics and Electrical Engineering, Lovely Professional University, Phagwara, Punjab, India
e-mail: manwinder.25231@lpu.co.in

© The Author(s), under exclusive license to Springer Nature Singapore Pte Ltd. 2023   435
A. Shukla et al. (eds.), *Computational Intelligence*, Lecture Notes in Electrical Engineering 968, https://doi.org/10.1007/978-981-19-7346-8_37

about agricultural factors and information about emerging advances [1]. In the past, our ancestors avoided using a specialized development for specific plant growth, preferring instead to utilize a general wonder for all plants. Plants may be developed under unusual typical natural conditions as a result of technology advancements in agriculture, and specialized plants can be developed under specific conditions, resulting in more yield and less compost [2].

Precision agriculture in greenhouses for plant growth is becoming increasingly popular as a result of lower-cost technologies enabling agriculturists to re-arrive output. The greenhouse is a transparent house-like structure that can maintain a regulated temperature, needed moisture level, light infiltration, as well as other aspects optimal plant development. Precision agriculture is a framework for detecting, monitoring, and responding to environmental modifications. It is a technology for detecting greenhouse climate, after which the recognized data is transferred to the cloud, and the agriculturist takes the necessary action based on the obtained data. This may be seen by the current breakthrough known as the IoT, which is a technology that connects everything or any device to the Internet via web-based approaches. In view of the novel advancement in wireless sensor networks (WSN), which is nothing more than an IoT, the precise agricultural framework is moving toward improvement. Precision agriculture has lately emerged as the most developed cultivating technology with a large framework. It includes recognizing, estimating, and communicating nursery data to ranchers in the event of a rapid shift in the greenhouse [3, 4].

A greenhouse is an arrangement with simple materials to maintain a microclimate for solid plant development, such as water stream management, directed temperature range, and so on. As a result, it avoids excessive light infiltration, extreme temperatures, diseases, and creepy crawlies, among other things. By maintaining environmental conditions, a farmer may produce any plant in any season. Going back to the areas of interest, there is the relevance of greenhouse cultivation, which demonstrates the truth of why they have become so popular. Greenhouses require a considerably smaller water supply than traditional cultivation since they trap the moisture. It shortens the editing time and broadens the types of harvests. Humidity and temperature are effectively managed in accordance with the needs of the plants. It is also possible to create slow-growing crops using nurseries. Pests may be easily managed. Harvests may be filled in many different ways of environmental circumstances, making it extremely adaptable. The plants' development will be influenced by the fluctuating climatic conditions in the greenhouse, resulting in lower yield near the conclusion of the cultivation. As a result, greenhouse factors including CO_2, soil moisture, temperature, and light must be controlled and monitored. This problem may be handled by using an IoT innovation in precision agriculture, which includes a precise application for certain greenhouse factors, such as temperature management, water flow control, light radiation, and so on, for optimal plant development [5–7].

2 Related Work

An adaptable stage ready to adapt to soilless culture needs in full distribution nurseries utilizing tolerably saline water is proposed [1]. An exceptionally versatile perceptive framework controlling, and observing nursery temperature [3] utilizing IoT innovations is presented. Author contributes [2] toward the ongoing IoT innovations in the horticulture area, alongside the improvement of equipment and programming frameworks. An independent fuzzy logic controller (FLC) with IoT abilities is created [5] for investigation and etymological dynamic about fertigation (manures + water) in a nursery. New proposition for agrarian farmland cautiousness is presented [4]. The framework utilizes Raspberry Pi board to recognize any pernicious exercises or movement in the ranch land and triggers the PiCam to take image of the scene picture. In this all-encompassing dynamic, a minimal effort, secluded, and energy-effective IoT stage for SA, indicated as VegIoT Garden, in view of commercial off-the-shelf (COTS) gadgets [8], receiving short- and long-range correspondence conventions (IEEE 802.11 and LoRa), and targeting improving the administration of vegetable nurseries through the assortment, observing, and examination of sensor information, identified with significant boundaries of developing plants (i.e., air and soil mugginess and temperature), is introduced. Contrasted with the current IoT-based horticulture and cultivating arrangements, the proposed arrangement decreases [6] network inactivity up partially. A detecting organization to assemble the field information of certain harvests (potatoes, tomatoes, and so on), at that point took care of these information to an AI calculation to receive an alert and finally, a graphical user interface is being used to display both the information and the warning message [9]. The framework [10] means to build up a PC vision-based mechanical weed control framework (WCS) for constant control of weeds in onion fields. An autonomous farming system [7] dependent on Fog processing worldview and LoRa innovation. The exploration rotates [11] around the structure and plan of a web empowered measured cultivating framework that tends to the requirement for individuals to tend vigorously on their developing harvests. RiceTalk venture [12] uses non-picture IoT gadgets to recognize rice impact. AgriTalk [13] is an economical IoT stage for accuracy cultivating of soil development. They direct trials on turmeric development, which demonstrates that the turmeric quality is altogether upgraded through AgriTalk. A connection quality-arranged course (LQOR) convention [14] for versatile IoT networks is proposed. Karim et al. [15] created and tested cloud-IoT-based late scourge choice emotionally supportive network. They introduced a choice emotionally supportive network to forestall potato sickness.

A structure that gives a framework to screen a greenhouse, poultry, and a fish tank is proposed [16]. Utilizing Raspberry Pi, they controlled and displayed the climate. IoT actuators and sensors are the critical segments to screen and respond for the climate. The improvement of a portable LoRaWAN passage gadget that can be used for expand nurseries' profitability and precision is introduced [17]. A low-power and adaptable IoT-based engineering for home ranchers and logical purposes that empowers to confirm the ecological effect on plants improvements by observing the dirt dampness and temperature is proposed [18]. Study proposed [19] a gathering framework dependent on the Internet of Things innovation and keen picture acknowledgment. A MQTT convention-based savvy cultivating arrangement [20] has capacity to assemble data from the field atmosphere for consistent examination and furthermore to create and convey a brilliant innovation for horticulture area for improving natural and rural supportability to improve crop discernibility and to expand generally yield. The exploration work proposed has [21] the plan of a conventional IoT system for improving agribusiness yield by successfully planning water system and preparation dependent on the harvests flow necessities, natural conditions, and climate figures. A sun oriented fueled brilliant farming observing framework with IoT gadgets is introduced [22]. S. Sarangi et al. have created [23] a convenient IoT stage on the edge including a versatile and a sensor hub that can be customized, which permits us to locally address logically important detecting needs of the rancher. P. Sureephong et al. examine [24] a prototyping of incorporated arrangement of Internet of Things-based wetting front locator (IoT-WFD) which centers on how to improve the IoT-based wetting front identifier plan for brilliant water system framework.

3 Proposed Work

Figure 1 depicts the exact smart farming system for greenhouses. Sensor layer, edge layer, and cloud layer are the three primary architectural layers. The noteworthy element of the proposed system is that it helps farmers by offering an IoT-based precision agricultural framework for greenhouse management. The objective is to provide agriculturists with field data that is remotely controlled greenhouse agricultural factors, for example, soil moisture, CO_2, light, and temperature from afar, and based on the soil moisture values, a controlling move for the greenhouse windows/doors to roll on/off may be made. This prevents agriculturists from physically visiting the fields.

3.1 Sensor Layer

We chose gerbera and broccoli for testing in the greenhouse, which has a mostly climate-sensitive habitat. Gas sensor, DHT11 sensor for temperature and humidity, light sensor, gas sensor, and moisture sensor are used to monitor parameters in the greenhouse environment. Actuators will be chosen and deployed to regulate parameters through relay to control devices such as fans and pumps. Controlled conditions of RH, temperature, and light, protection from rains, storms, and searing sunshine, as well as pest control, and diseases are all advantages of employing a greenhouse management system for such a crop.

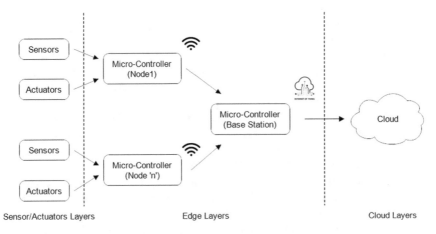

Fig. 1 Smart farming framework for greenhouse management

3.2 Edge Layer

Sensors installed in the field at various locations, referred to as nodes or edges, are coupled to a low-power microcontroller suited for IoT. In our experiment, we employed a NodeMCU ESP 32, which has the potential to gather and analyze data from sensors before sending it to the edge layer's base station. Sensors gather data in analog or digital form according to specifications and must be calibrated and verified against a standard value. Data is collected for diverse climate circumstances, both healthy and unhealthy, in order to understand all conceivable environmental conditions and to make it accurate for crop management to ensure their survival.

3.3 Cloud Layer

Data from each node in the edge layer, which is then processed and managed at the base station, will be visualized at the cloud layer using a user interface (UI)-based application that will assist farmers in monitoring crop cultivation status.

4 Results

The suggested experimental plan is performed on a prototype that has been thoroughly tested on various crops and farming methods. Creation of preliminary models for plant growth and nutrition needs two most important stages in experimentation i.e. development of a sensor net for smart greenhouse monitoring and automation for actuators. The suggested system uses an embedded system to analyze green-house execution factors such as CO_2, soil moisture, temperature, and light for plants, resulting in realistic results. The greenhouse doors/windows may also be rolled on/off dependent on the soil moisture levels. The plant photosynthesis process needs a high level of CO_2 concentration and water in the evenings, as opposed to the daytime; with the assistance of these two energies, the photosynthesis method maintains the plant cool and aids in rapid growth. After observing a CO_2 concentration level experiment in a greenhouse, it maintains a CO_2 level maximum at night as indicated in Fig. 2, since the greenhouse consumes CO_2 from daytime to nighttime. As a consequence, the CO_2 level throughout the day is lower; Fig. 2 shows an example for this.

Water quantity in the soil is essential because a fungal infection in the plant might be caused by too much water, while too little water causes the plant to become dry or even die. As a result, the plant's necessary quantity of water is critical. Plants require more water with CO_2 throughout the night for photosynthesis. When the soil moisture sensor returns a negative value, it implies that the plants have completely covered the water, as depicted in Fig. 3, and the greenhouse windows/doors will close automatically with the aid of a DC motor. The soil is dry if the value is positive. As illustrated in Fig. 3, it needs to be re-wet.

Among the most popular essential parameters in a greenhouse is temperature; the temperature should be kept as high as possible. Because temperature aids blooming, fruiting, photosynthesis, seed germination, and other processes Fig. 4 shows the inside and outside temperature of a greenhouse which can be useful to maintain the temperature inside the greenhouse.

As a result, the humidity and temperature range in the greenhouse was maintained to the fullest extent possible, as shown in Fig. 5, in comparison with the humidity and temperature range in the outside greenhouse environment, as shown in Fig. 5. The various hues of sunlight are beneficial to the photosynthetic process, which is

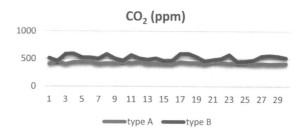

Fig. 2 CO_2 representation, daytime (type A) and nighttime (type B)

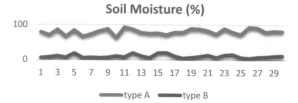

Fig. 3 Soil moisture representation, dry soil (type A) and wet soil (type B)

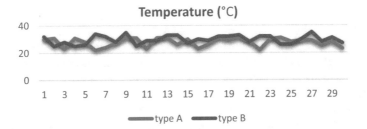

Fig. 4 Temperature representation, outside greenhouse (type A) and inside greenhouse (type B)

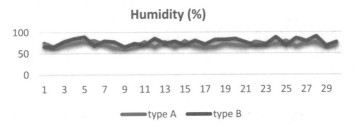

Fig. 5 Humidity representation, outside greenhouse (type A) and inside greenhouse (type B)

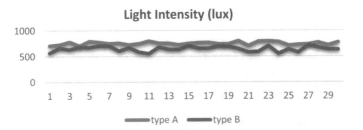

Fig. 6 Light intensity representation, outside greenhouse (type A) and inside greenhouse (type B)

found in the green portion of plants and is responsible for plant growth, blooming, and form.

As a result, as depicted in Fig. 6, inside the greenhouse, a constant degree of light penetration was maintained, as opposed to typical light penetration outside the greenhouse.

5 Conclusion

A smart greenhouse system is a combination of IoT and agriculture that is driven by market demands and based on an optimization schedule. Monitoring and regulating the greenhouse environment require automation and great efficiency. The detailed information was discovered, allowing them to be fully included in the gap analysis based on their limitations and the likelihood of a job extension. Growers can regulate environmental factors in typical greenhouses using a proportional control system that requires manual intervention, which often results in output loss, energy waste, and higher labor costs. To address these issues, an automated greenhouse monitoring and control system based on the IoT comes to the rescue. IoT is a commonly utilized technology for connecting objects and gathering data. The technology is meant to remotely monitor greenhouse factors including CO_2, soil moisture, temperature, and light. This data can be gathered by farmers using a cloud account and an Internet connection. Greenhouse windows/doors roll on/off dependent on soil moisture levels, which is another automated regulating activity. Thus, using IoT, the system will assist farmers in avoiding physical visits to the field while also increasing production by maintaining precise parameters like CO_2, soil moisture, temperature, and light in the greenhouse. The IoT kit and an Internet connection are used to complete the project. With the help of graphical representation based on the practical data obtained by the IoT kit, the findings for greenhouse factors such as CO_2, soil moisture, temperature, and light for broccoli and gerbera plants are examined. Work on more specific agriculture crops can be done in the future.

References

1. Zamora- M, Santa J, Martinez J, Martínez V, Skarmeta A (2019) Smart farming IoT platform based on edge and cloud computing. Biosys Eng 2019:4–17. https://doi.org/10.1016/j.biosys temseng.2018.10.014
2. Kour V, Arora S (2020) Recent developments of the internet of things in agriculture: a survey. IEEE Access 8:129924–129957. https://doi.org/10.1109/ACCESS.2020.3009298
3. Subahi A, Bouazza KE (2020) An intelligent IoT-based system design for controlling and monitoring greenhouse temperature. IEEE Access 1–1. https://doi.org/10.1109/ACCESS.2020.3007955
4. Angadi S, Katagall R (2019) Agrivigilance: a security system for intrusion detection in agriculture using raspberry Pi and Opencv. Int J Sci Technol Res 8(11)
5. Carrasquilla-Batista A, Chacón-Rodríguez A (2019) Standalone fuzzy logic controller applied to greenhouse horticulture using internet of things. In: 2019 7th international engineering, sciences and technology conference (IESTEC), Panama, Panama, pp 574–579. https://doi.org/10.1109/IESTEC46403.2019.00108
6. Ahmed N, De D, Hussain I (2018) IoT for smart precision agriculture and farming in rural areas. IEEE Internet Things J 5(6):4890–4899. https://doi.org/10.1109/JIOT.2018.2879579
7. Baghrous M, Ezzouhairi A, Benamar N (2019) Towards autonomous farms based on fog computing. In: 2019 2nd IEEE Middle East and North Africa COMMunications Conference (MENACOMM), Manama, Bahrain, pp 1–4. https://doi.org/10.1109/MENACOMM46666.2019.8988547
8. Codeluppi G, Cilfone A, Davoli L, Ferrari G (2019) VegIoT Garden: a modular IoT management platform for urban vegetable gardens. https://doi.org/10.1109/MetroAgriFor.2019.8909228
9. Araby AA et al (2019) Smart IoT monitoring system for agriculture with predictive analysis. In: 2019 8th international conference on modern circuits and systems technologies (MOCAST), Thessaloniki, Greece, pp 1–4. https://doi.org/10.1109/MOCAST.2019.8741794

10. Arakeri MP, Vijaya Kumar BP, Barsaiya S, Sairam HV (2017) Computer vision based robotic weed control system for precision agriculture. In: 2017 international conference on advances in computing, communications and informatics (ICACCI), Udupi, pp 1201–1205. https://doi. org/10.1109/ICACCI.2017.8126005

11. Belista FCL, Go MPC, Luceñara LL, Policarpio CJG, Tan XJM, Baldovino RG (2018) A smart aeroponic tailored for IoT vertical agriculture using network connected modular environmental chambers. In: 2018 IEEE 10th international conference on humanoid, nanotechnology, information technology, communication and control, environment and management (HNICEM), Baguio City, Philippines, pp 1–4. https://doi.org/10.1109/HNICEM.2018.8666382

12. Chen W, Lin Y, Ng F, Liu C, Lin Y (2020) RiceTalk: rice blast detection using internet of things and artificial intelligence technologies. IEEE Internet Things J 7(2):1001–1010. https:// doi.org/10.1109/JIOT.2019.2947624

13. Chen W et al (2019) AgriTalk: IoT for precision soil farming of turmeric cultivation. IEEE Internet Things J 6(3):5209–5223. https://doi.org/10.1109/JIOT.2019.2899128

14. Farhan L, Kharel R, Kaiwartya O, Quiroz-Castellanos M, Raza U, Teay SH (2018) LQOR: link quality-oriented route selection on internet of things networks for green computing. In: 2018 11th international symposium on communication systems, networks & digital signal processing (CSNDSP), Budapest, pp 1–6. https://doi.org/10.1109/CSNDSP.2018.8471884

15. Karim F, Fathallah K, Frihida A (2019) A cloud-IOT based decision support system for potato pest prevention. Procedia Comput Sci 160:616–623. https://doi.org/10.1016/j.procs. 2019.11.038

16. Gnanaraj AA, Jayanthi JG (2017) An application framework for IoTs enabled smart agriculture waste recycle management system. In: 2017 world congress on computing and communication technologies (WCCCT), Tiruchirappalli, pp 1–5. https://doi.org/10.1109/WCCCT.2016.11

17. Gutiérrez S, Martínez I, Varona J, Cardona M, Espinosa R (2019)Smart Mobile LoRa agriculture system based on internet of things. In: IEEE 39th Central America and Panama Convention (CONCAPAN XXXIX), Guatemala City, Guatemala, pp 1–6. https://doi.org/10.1109/CON CAPANXXXIX47272.2019.8977109

18. Hirsch C, Bartocci E, Grosu R (2019) Capacitive soil moisture sensor node for IoT in agriculture and home. In: 2019 IEEE 23rd international symposium on consumer technologies (ISCT), Ancona, Italy, pp 97–102. https://doi.org/10.1109/ISCE.2019.8901012

19. Horng G, Liu M, Chen C (2020) The smart image recognition mechanism for crop harvesting system in intelligent agriculture. IEEE Sens J 20(5):2766–2781. https://doi.org/10.1109/JSEN. 2019.2954287

20. Pooja S, Uday DV, Nagesh UB, Talekar SG (2017) Application of MQTT protocol for real time weather monitoring and precision farming. In: 2017 international conference on electrical, electronics, communication, computer, and optimization techniques (ICEECCOT), Mysuru, pp 1–6. https://doi.org/10.1109/ICEECCOT.2017.8284616

21. Prabha R, Sinitambirivoutin E, Passelaigue F, Ramesh MV (2018) Design and development of an IoT based smart irrigation and fertilization system for Chilli farming. In: 2018 international conference on wireless communications, signal processing and networking (WiSPNET), Chennai, pp 1–7. https://doi.org/10.1109/WiSPNET.2018.8538568

22. Sadowski S, Spachos P (2018) Solar-powered smart agricultural monitoring system using internet of things devices. In: 2018 IEEE 9th annual information technology, electronics and mobile communication conference (IEMCON), Vancouver, BC, pp 18–23. https://doi.org/10. 1109/IEMCON.2018.8614981

23. Sarangi S et al (2019) An affordable IoT edge platform for digital farming in developing regions. In: 2019 11th international conference on communication systems & networks (COMSNETS), Bengaluru, India, pp 556–558. https://doi.org/10.1109/COMSNETS.2019.8711388

24. Sureephong P, Wiangnak P, Wicha S (2017) The comparison of soil sensors for integrated creation of IOT-based Wetting front detector (WFD) with an efficient irrigation system to support precision farming. In: 2017 international conference on digital arts, media and technology (ICDAMT), Chiang Mai, pp 132–135. https://doi.org/10.1109/ICDAMT.2017.790 4949

Performance Analysis of QuickMerge and QuickInsertionMerge Algorithms

Naresh Poloju

Abstract An algorithm is a procedure for solving a problem using a finite sequence of actions. A sorting algorithm is applied to arrange the given array or list of elements into ascending or descending order. In computer science, sorting is a critical issue. Many researchers are working on sorting algorithms, because the best sorting technique enables working of other algorithms to be optimized. This paper proposed a novel algorithm which integrates quicksort, insertion sort, and merge sort and breaks the input dataset into different sub-lists. The suggested hybrid algorithm outperforms the original quicksort algorithm.

Keywords Quicksort · Merge sort · QuickMerge sort · QuickInsertionMerge · Algorithm

1 Introduction

The sorting algorithm is enticing research domains in computer science. Sorting algorithms takes an array or list of elements as inputs and produces the elements in ascending or descending order to be easier to insert, delete, and search. At present, many sorting algorithms are available such as selection sort, insertion sort, bubble sort, quicksort, merge sort.

N. Poloju (✉)
Department of Information and Communication Engineering, Anna University, Chennai, Tamilnadu, India
e-mail: naresh.poloju56@gmail.com

© The Author(s), under exclusive license to Springer Nature Singapore Pte Ltd. 2023
A. Shukla et al. (eds.), *Computational Intelligence*, Lecture Notes in Electrical Engineering 968, https://doi.org/10.1007/978-981-19-7346-8_38

As per the experiments conducted on insertion, merge, quick, and bubble sort, quicksort beats all the other sorting algorithms and the worst algorithm is bubble sort [1]. If excess amount of data is present in the list, then quicksort or merge sort performs well. The insertion or selection sort performs better if less elements are present [2]. For sorting bulky data elements, quicksort can deliver the best result on merge sort, but the time complexity of the quicksort in worst case is $O(n^2)$, which is worse than the merge sort worst-case time complexity. Above-mentioned algorithms have their own pros and cons; hence, hybrid algorithms are introduced to improve performance.

Sorting algorithm is divided into two categories such as comparison based and non-comparison based (radix sort). Out of all the comparison-based algorithms, quicksort is better if the size of the dataset is large enough [3].

In [4], the author introduced QuickMerge Sort. The QuickMerge sorting algorithm sorts the list by first dividing the extensive list into 'n' sub-lists, applying quicksort on each sub-lists and merge the individually sorted sub-lists using the merge method of merge sort. This approach reduced the worst-case comparisons of quicksort.

2 Literature Work

In [4], the author focused only on comparison minimization in quicksort's worst case. However, the fact is, the worst case of quicksort will only come into picture if input list is already sorted, and this method does not provide any information about the number of sub-lists we have to create. The time taken by the QuickMerge algorithm varies according to how many sub-lists are generated. This method can be further optimized in two ways. By analyzing the algorithm's time complexity with various sub-list sizes, the first optimization is to fix exact and proper size for the sub-list to reduce the time complexity. The second optimization can be the insertion sort performs best if list contains small size of input dataset. Therefore, the QuickMerge algorithm can be further optimized by introducing insertion sort.

3 Proposed Method

The proposed algorithms aim to fix the sub-list size in the QuickMerge algorithm to provide better performance. The QuickInsertionMerge sorting algorithm further improvises the QuickMerge sorting algorithm.

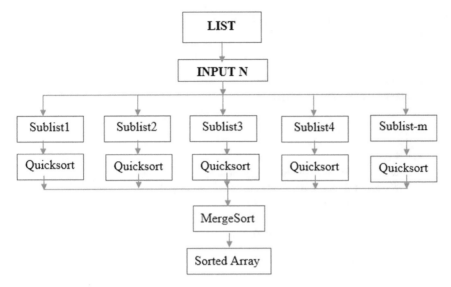

Fig. 1 QuickMerge algorithm

3.1 QuickMerge Sort

QuickMerge algorithm is performed by:

- Selecting the value for n (size of sub-list).
- Split the input data into size n sub-lists.
- Apply quicksort on each sub-list.
- Merge the individually sorted sub-lists using the merge method of merge sort.
    Figure 1 depicts the architecture of the QuickMerge sorting algorithm.

3.2 QuickInsertionMerge Sort

The QuickInsertionMerge algorithm is performed by:

- Selecting the value for n (size of sub-list).
- Divide the input dataset into the size n sub-list.
- Apply quicksort on each sub-list.
- If the input data of quicksort is less than 10, apply insertion sort.
- Merge the individually sorted sub-lists using the merge method of merge sort.

QuickInsertionMerge Sort Algorithm. The algorithm for QuickInsertionMerge sort is given below.

```
Algorithm InsertionSort(a,low,high):
    // Traverse through 1 to len(a)
    {
    fori=low+1 to high+1 do
        {
    key = arr[i]
            # Move elements of a[0..i-1], that are
            # greater than the given key, to one place ahead
            # of their current position
            j = i-1
    while ((j >=low) and (key <arr[j])) do
            {
    arr[j+1] = arr[j]
                j -= 1
            }
    arr[j+1] = key
        }
    }
    Algorithm Partition (a, m, p)
    //within a[m],a[m+1]....,a[p-1] the elements are rearranged in such a manner that
    //if at start t=a[m] then , after execution a[q]=t for some q between m and p-1,
    //a[k]<=t for m<=k<q, and a[k]>=t for q<k<p. q is returned.
    {
    i = m
      j=p
      v=arr[m]
    while(i<j) do
        {
    i=i+1
    while(arr[i]<v) do
            {
    i=i+1
            }
            j=j-1
```

```
while(arr[j]>v)  do
        {
            j=j-1
        }
if(i<j) then
        {
            t=arr[i]
arr[i]=arr[j]
arr[j]=t
        }
    }
arr[m]=arr[j]
arr[j]=v
return j
}
Algorithm QuickSort(a, low, high)
// sorts the elements from a[low]..a[high] of array a into ascending order
//a[high+1] is bigger than any element of a[low:high]
{
if (low < high) then// if there exists at least two element
    {
if (high-low>10)  then // if input size is more than 10
        {
pi = partition(arr, low, high+1)// position of partitioning element
QuickSort(a, low, pi-1)
QuickSort(a, pi+1, high)
        }
else
  //if the number of elements is less than 11
    {
InsertionSort(a,low,high)
    }
}}
Algorithm Merge(x)
// x is an array of sublists of size n
{
while(len(x)>1) do  // this process will be continued until there is only one sub-list
in x
    {
i = j = k = 0
        arr1=[]
        L=x[i].copy()
        R=x[i+1].copy()
        //Copy data into temporary arrays L[] and R[]
x.remove(x[i+1])
```

```
x.remove(x[i])          // remove the sub-lists from sub-lists array x
while (i<len(L)) and (j <len(R)) do
        {
if (L[i]<R[j]) then
            {
arr1.append(L[i])
i += 1
            }
else
            {
                arr1.append(R[j])
                j += 1
                k += 1
            }
        }
        # checkingfor the left over elements
while (i<len(L)) do
        {
arr1.append(L[i])
i += 1
        }
while (j<len(R)) do
        {
                arr1.append(R[j])
                j += 1
        }
x.append(arr1) // add the sorted sub-lists to x
}}
Algorithm Quick_Insertion_Mergesort(arr,n)
// arr is the input array and it is the size of n sub-list
//let max is a number
{
part=n
    x = [arr[i * part:(i + 1) * part] for i in range((len(arr) + part - 1) // part )]
    // x is the list of sublists of size n
for l in x do   // l is the sublist taken from list of sublists x
    {
l.append(max)
        n=len(l)
QuickSort(l,  0, n-2)
l.remove(max)
    }  Merge(x)            // combining the sorted sublists
}
```

4   Results

The language and the implementation machine affect sorting algorithm's runtime [5]. Python programming language has been used for the implementation of the above-discussed algorithms, and the platform used is Windows 10 operating system with Intel Core i3 7th generation processor.

4.1   QuickMergeSort Implementation with Various Sub-list Sizes

The algorithms were examined based on different sized sorted datasets, and comparative analysis has been done for the pure quicksort, QuickMerge used in [4] and QuickMergeSort with various sub-list sizes are given below:

Figure 2 shows the observed number of comparisons; from this, it is evident that the proposed method performs far better when the sub-list is of different size [4]. When sub-list size decreases, the number of comparisons also decreases. The algorithm's time complexity with various sorted sub-list sizes is shown in Fig. 3.

The same algorithm was applied to unsorted data, and the outcomes are observed in Figs. 4 and 5.

From this analysis, it is known that the QuickMerge algorithm with sub-list size 300 has less average computation time, and it is much close to the average running time of pure quicksort.

4.2   QuickInsertionMerge Sort

The QuickInsertionMerge Sort algorithm was given with a sorted and unsorted set of sorting elements. This algorithm has been executed 100 times, and the analysis of the results is shown in Figs. 6 and 7.

The average comparison count and running time of QuickMerge sort and QuickInsertionMerge sort algorithms were compared, and analysis is shown in Figs. 8, 9, 10, and 11.

From the above observations, it is found that the QuickMerge sorting algorithm is giving the best performance when sub-list size is 300. The QuickInsertionMerge-sorting algorithm is found better than QuickMerge sorting algorithms for sorted and unsorted elements.

Data set size	Pure Quick Sort	Quick Merge Used IN[4]	QuickMerge Sublist Size									
			400	350	300	250	200	150	100	70	50	20
500	127245	88370	88144	75594	68044	65494	48743	38842	29241	21778	17486	10941
700	248145	172320	129244	126694	100343	87593	70342	56041	41439	31486	24982	15871
1000	504495	349490	187493	174743	147692	132492	108491	81839	59986	45691	36476	23386
1500	1131745	783240	296342	263241	237741	200490	158638	127486	91531	70014	56266	36811
2000	2008995	1389480	416991	355040	308689	267988	219986	169232	122976	94347	75956	49776
3000	4513495	3119750	607288	539487	479986	405484	333081	259476	187566	145263	117036	78126
4000	8017995	5539960	839986	726184	633482	541980	445976	346919	251956	195628	157916	105556
5000	12522495	8649330	1033483	918531	795079	682476	559671	435112	317146	245584	199596	136226

Fig. 2 Comparison count for pure quicksort and QuickMerge used in [4] and QuickMerge with a sorted sub-list of different sizes

Data Set Size	Pure Quick Sort	QuickMerge Sublist Size									
		400	350	300	250	200	150	100	70	50	20
500	0.0827	0.0554	0.0483	0.0533	0.0404	0.0292	0.0208	0.0184	0.0211	0.0102	0.0066
700	0.2522	0.0764	0.0895	0.0658	0.0462	0.0499	0.0287	0.0283	0.0153	0.0121	0.0089
1000	0.3134	0.1120	0.1105	0.1006	0.0769	0.0698	0.0579	0.0351	0.0281	0.0274	0.0129
1500	0.7184	0.1980	0.1599	0.1233	0.1075	0.1174	0.0898	0.0677	0.0523	0.0369	0.0301
2000	1.3032	0.2759	0.2177	0.2019	0.1684	0.1383	0.0913	0.0939	0.0626	0.0627	0.0343
3000	2.7184	0.3581	0.3002	0.2965	0.2814	0.2053	0.1491	0.1850	0.1410	0.0961	0.0715
4000	5.1721	0.5592	0.4610	0.4035	0.3367	0.3233	0.2408	0.1507	0.1164	0.1283	0.0939
5000	8.2288	0.6805	0.5819	0.4899	0.4257	0.3508	0.3282	0.2094	0.1735	0.1572	0.0960

Fig. 3 Time complexity of pure quicksort and QuickMerge with a sorted sub-list of different sizes

Data Set Size	Pure Quick Sort	QuickMerge Sublist Size									
		400	350	300	250	200	150	100	70	50	20
500	8753	8995	8777	8674	8691	8958	8764	8741	8773	8770	8893
700	12916	12932	12815	13462	13014	12926	13008	12943	12959	12956	13113
1000	19445	20025	19734	19639	19469	19577	19618	19553	19584	19588	19777
1500	31156	30803	31480	31067	31035	30893	31011	30848	31177	30929	31413
2000	43330	43184	43403	43101	42888	43089	43175	43135	43005	43094	43568
2500	55505	56179	55665	55626	55258	55494	55131	55495	55335	55468	55785
3000	68108	67756	68058	68110	67967	67941	68007	67774	68124	67899	68858
3500	81297	80966	81224	81182	80941	80970	81164	80717	81025	80909	82021
4000	94251	93832	94676	94392	93392	94118	94032	94138	94028	94327	95166
4500	107440	108379	107447	106864	107230	107624	106848	107415	106984	107770	108348
5000	122816	121672	121469	120652	120829	120880	120574	120777	120720	121009	121547
5500	135390	134497	133732	134887	134632	134491	134437	134188	134543	134527	135548

Fig. 4 Comparison count of pure quicksort and QuickMerge with unsorted sub-lists

Data Set Size	Pure Quick Sort	QuickMerge Sublist Size									
		400	350	300	250	200	150	100	70	50	20
500	0.0051	0.0054	0.0051	0.0053	0.0056	0.0053	0.0052	0.0055	0.0058	0.0061	0.0064
700	0.0074	0.0077	0.0083	0.0079	0.0075	0.0075	0.0088	0.0078	0.0090	0.0090	0.0100
1000	0.0100	0.0129	0.0118	0.0116	0.0121	0.0134	0.0132	0.0128	0.0138	0.0143	0.0143
1500	0.0167	0.0186	0.0215	0.0201	0.0227	0.0197	0.0192	0.0204	0.0202	0.0218	0.0233
2000	0.0244	0.0273	0.0277	0.0282	0.0282	0.0279	0.0282	0.0289	0.0304	0.0289	0.0330
2500	0.0309	0.0393	0.0358	0.0369	0.0373	0.0370	0.0364	0.0387	0.0405	0.0392	0.0451
3000	0.0396	0.0447	0.0454	0.0469	0.0443	0.0444	0.0444	0.0473	0.0487	0.0479	0.0532
3500	0.0469	0.0535	0.0542	0.0548	0.0545	0.0562	0.0579	0.0554	0.0612	0.0606	0.0642
4000	0.0523	0.0623	0.0665	0.0624	0.0660	0.0643	0.0647	0.0639	0.0663	0.0671	0.0709
4500	0.0594	0.0734	0.0740	0.0726	0.0800	0.0754	0.0784	0.0821	0.0831	0.0826	0.0945
5000	0.0693	0.0819	0.0873	0.0830	0.0828	0.0827	0.0843	0.0868	0.0888	0.0942	0.0983
5500	0.0855	0.0998	0.0931	0.0939	0.0959	0.0976	0.1009	0.0985	0.1044	0.1024	0.1107
6000	0.0863	0.1039	0.1068	0.0992	0.1044	0.1072	0.1069	0.1091	0.1062	0.1139	0.1217

Fig. 5 Running time of pure quicksort and QuickMerge with unsorted sub-lists of different sizes

Data Set Size	QuickInsertionMerge Sublist Size									
	400	350	300	250	200	150	100	70	50	20
500	7639	7425	7329	7333	7609	7406	7371	7390	7390	7472
700	11039	10923	11573	11119	11031	11114	11036	11048	11045	11130
1000	17334	17039	16917	16769	16850	16896	16841	16833	16843	16956
1500	26738	27435	27000	26947	26811	26944	26792	27065	26834	27161
2000	37790	38021	37692	37482	37668	37737	37675	37548	37645	37909
2500	49424	48892	48882	48463	48724	48359	48681	48449	48616	48705
3000	59625	59938	59999	59835	59837	59844	59608	59950	59662	60365
3500	71585	71765	71718	71505	71508	71648	71243	71502	71279	72133
4000	83015	83854	83587	82528	83274	83179	83242	83087	83314	83855
4500	96200	95257	94729	95078	95461	94672	95179	94670	95411	95653
5000	108148	108014	107154	107271	107375	106986	107197	107028	107276	107400
5500	119657	118888	120051	119728	119634	119483	119264	119533	119408	120047
6000	131389	132235	131964	131980	131435	131956	131153	132077	131394	132667

Fig.6 Comparison count of QuickInsertionMerge algorithm

Data Set Size	QuickInsertionMerge Sublist Size									
	400	350	300	250	200	150	100	70	50	20
500	0.00504	0.0046	0.004901	0.005105	0.005237	0.004911	0.004913	0.005432	0.00563	0.006158
700	0.006298	0.007736	0.007888	0.006815	0.007264	0.007904	0.007631	0.008024	0.008721	0.00914
1000	0.011378	0.010772	0.01141	0.011015	0.012645	0.011614	0.01287	0.012632	0.012415	0.014611
1500	0.019353	0.019538	0.019229	0.020834	0.019675	0.01858	0.019503	0.022168	0.021163	0.022671
2000	0.027666	0.027105	0.026693	0.026521	0.027798	0.027328	0.02867	0.029445	0.028539	0.032081
2500	0.034626	0.034826	0.03917	0.036569	0.03549	0.034871	0.034358	0.037729	0.039712	0.038878
3000	0.044294	0.042338	0.040184	0.041451	0.0443	0.045613	0.04494	0.046819	0.050153	0.050991
3500	0.051666	0.048532	0.051921	0.050525	0.052889	0.057759	0.054422	0.057049	0.055023	0.060206
4000	0.063497	0.064886	0.060793	0.064495	0.061249	0.059201	0.061294	0.061727	0.064551	0.070011
4500	0.071121	0.070046	0.06894	0.070012	0.071892	0.075584	0.07582	0.08158	0.082045	0.089512
5000	0.081384	0.080816	0.077482	0.081765	0.080056	0.082201	0.079813	0.085343	0.085939	0.091502
5500	0.089869	0.086947	0.091101	0.090538	0.091569	0.093355	0.100313	0.097008	0.100641	0.099569
6000	0.096638	0.100352	0.095901	0.097002	0.102716	0.106251	0.104522	0.105856	0.104698	0.116546

Fig.7 Running time of QuickInsertionMerge algorithm

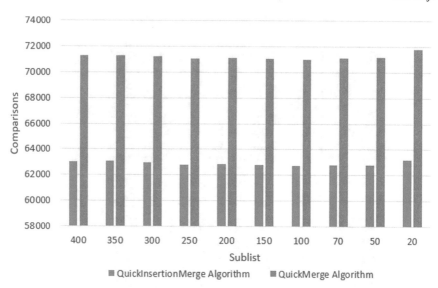

Fig.8 Average comparisons of QuickInsertionMerge versus QuickMerge using unsorted elements

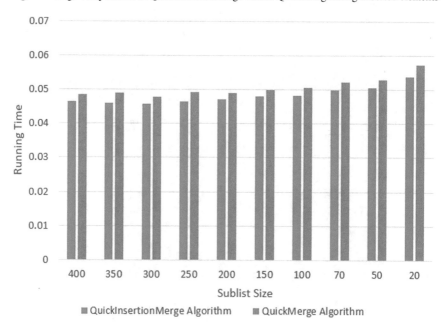

Fig. 9 Average running time of QuickInsertionMerge versus QuickMerge using unsorted elements

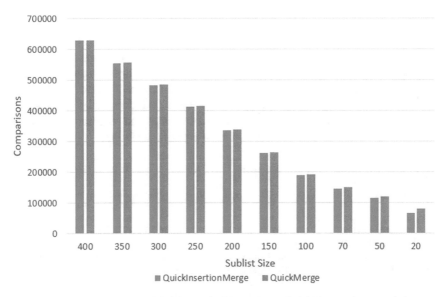

Fig. 10 Average comparisons of QuickInsertionMerge versus QuickMerge using sorted elements

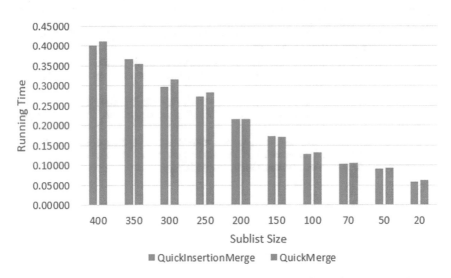

Fig. 11 Average running time of QuickInsertionMerge versus QuickMerge using sorted elements

5 Conclusion and Future Scope

In this research paper, performance analysis of the QuickMerge sorting and Quick-InsertionMerge sorting algorithms was done with different sub-list sizes. The performance analysis is performed based on several comparisons and running time.

The proposed algorithm shows better performances than the existing algorithms. According to the experimental analysis, efficiency of QuickMerge sorting was improved if the given input dataset is divided into sub-lists of different sizes, and the sub-list size 300 is giving the average performance even if the list is either sorted or unsorted. The QuickInsertionMerge sort performs better than QuickMerge sort on sorted and unsorted inputs. The analysis of this algorithm is done in the Windows10+ Intel Core i3 7th Gen platform. It could be implemented and analyzed for different platforms to find a higher performance environment in the future.

References

1. Usman M (2015) Performance analysis of sorting algorithms with C#. Int J Res Appl Sci Eng Technol (IJRASET) 201–204
2. Yang Y et al (2011) Experimental study on the five sort algorithms. In: Second international conference on mechanic automation and control engineering, pp 1314–1317
3. Hema AM, Performance analysis of sorting algorithms. TEJAS Thiagarajar College J
4. Suresh, George AK (2018) Performance analysis of various combination sorting algorithms for large dataset to fit a multi-core architecture. In: 2018 second international conference on inventive communication and computational technologies (ICICCT), pp 51–56. https://doi.org/10.1109/ICICCT.2018.8472956
5. Crişan DA, Simion GF, Moraru PE (2015) Run-time analysis for sorting algorithms. Rom Econ Bus Rev Rom Am Univ 9(1):147–156
6. Sareen P (2013) Comparison of sorting algorithms (based on average-case). IJARCSSE 3(3)
7. Horowitz E, Sahni S, Rajasekharan S, Fundamentals of computer algorithms, 2nd edn. Universities Press
8. Dave PH, Dave HB, Design and analysis of algorithms, 2nd edn. Pearson Education
9. Goodrich MT, Tomassi R, Algorithm design: foundations, analysis and internet examples. Wiley
10. Cormen TH, Leiserson CE, Rivest RL, Stein C, Introduction to algorithms, 3rd edn. PHI
11. Lambert KA (2011) The fundamentals of python: first programs. Cengage Learning
12. Downey AB, Think python, 1st edn. Orielly Publishing

Classification of Slow and Fast Learners Using Deep Learning Model

V. A. Bharadi, K. K. Prasad, and Y. G. Mulye

Abstract Cognitive learning strategies are focused on the improvement of the learner's ability to analyze information more deeply, efficiently handle new situations by transferring and applying the knowledge. These techniques result in enhanced and better-retained learning. To cater to the needs of different students having different levels of cognitive learning, it is very important to assess their learning ability. In this paper, a method based on deep learning is presented to classify the earners based on their past performance. This technique is taking the student's past semester marks, their total failures in subjects/passing heads, and their current semester attendance. The proposed method classifies the learners into three categories, namely slow, fast, and average learners. A deep learning classifier with multilayer perceptron-based nodes is built for the classification. The proposed method is fully automatic and robust. A final accuracy of 90% is achieved in the classification of the learners in their cognitive learning level.

Keywords Deep learning · Multilayered perceptron · Artificial neural networks · Neural networks · Slow learners · Fast learners · Classification · Remedial work

1 Introduction

The learning ability of a student plays a crucial role in academics. Some students have got outstanding cognitive skills, and they learn the concepts at a faster rate as compared to others. Some students face difficulties to cope up with the teaching–learning process due to reasons such as low cognitive ability, low attendance, less time for the preparation due to preparation for the examination due to failure in other passing heads. Teachers have to give special attention and remedial work to such slow learners. If they are identified in the early stages of the academic year, their performance can be improved significantly [1].

V. A. Bharadi (✉) · K. K. Prasad · Y. G. Mulye
Finolex Academy of Management and Technology, P60,60-1, MIDC, Mirjole Block, Ratnagiri, Maharashtra 415639, India
e-mail: vinayak.bharadi@famt.ac.in

© The Author(s), under exclusive license to Springer Nature Singapore Pte Ltd. 2023   461
A. Shukla et al. (eds.), *Computational Intelligence*, Lecture Notes in Electrical Engineering 968, https://doi.org/10.1007/978-981-19-7346-8_39

In this research work, the case study of an engineering college affiliated with Mumbai University is considered; the institute follows a semester pattern. To prepare students for a better understanding of the concepts as well as to prepare them for the examinations and further hone their expertise, the teacher has to know the students learning level. This process is quite tedious, and after knowing students more deeply, the teachers can accordingly plan the activities for the students. In the case of the semester pattern, there is less time for this acquittance period, and it becomes a challenge to identify the cognitive learning level of the students [2].

Besides this, various statutory and regulatory bodies such as National Assessment and Accreditation Council (NAAC) and the National Accreditation Board (NBA) which are accrediting the higher education institutions also stress the identification of the learning levels of the students and accordingly steer the teaching–learning process for them. If the weak learners are identified at the start of the semester, the respective subject teachers can plan their academic activity for a better understanding of the subject matter by such students, and their results improve as they go to higher semesters. To improve the students' performance by tailored teaching–learning activities for the slow and fast learners is the key outcome of this research.

Machine learning enables computing devices to learn without being explicitly programmed. It is a subdomain of artificial intelligence (AI) that makes software programs or applications capable of precisely predicting the outcome. Deep learning methods analyze the patterns in the data, and these patterns are modeled as complex multilayered networks; it is the most general way to model a problem. It makes deep learning capable of solving complex problems; which are otherwise not addressable by modular programming logic [3, 4]. Deep learning refers to machine learning using deep (artificial) neural networks. A number of algorithms are proposed in the literature that implements deep learning using hidden layers other than conventional neural networks. The concept of "artificial" neural networks dates back to the 1940s. It consists of a network of artificial neurons programmed out of interconnected threshold switches. This network is referred to as an artificial neural network, and it can learn to recognize patterns like the human brain does [5].

For the classification of the learner as per their learning ability, conventionally, a method based on a mathematical formulation was used in the institute under consideration. This method simply weights the Semester Grade Point Average Score and the Internal Assessment Marks. This method is quite simple and linear. A technique based on the deep learning technique is presented here. The proposed method is based on deep learning technique, which uses a deep neural network to classify the students in slow and fast learners. The method generally gives 15–20% students in the slow and advanced learners category; the remaining students are in a general class or the average learners' class. The performance of the proposed deep learning classifier is compared with other types of classifiers such as linear regression, Naive Bayes, decision trees, and support vector machines.

2 Slow and Fast Learners Classification

The teaching–learning process is highly dependent on the cognitive levels of the students. The methods followed, the assignments, as well as the remedial work given to the students, should be different for different cognitive levels of the students. In the report published by Hacettepe Üniversitesi Eğitim Fakültesi Dergisi (H. U. Journal of Education) [6], the influence of the constructivist learning approach on the cognitive learning levels of the students while learning trigonometry and on their attitudes toward mathematics was analyzed. The constructivist teaching–learning process based on the students learning level was found to be better as compared to the conventional method.

Sehar [6] performed a study that measured the cognitive levels of examinations questions to evaluate students' learning. Question papers from various teachers related to different subjects were collected for analyzing the targeted cognitive levels. The analysis concluded that there are several teaching methodologies for teaching–learning, but the examinations were confined to the lower level of learning.

Koparan et al. [8] studied the effect of the project-based learning approach on the secondary-school students' statistical literacy levels, and it was found that the project-based learning approach increases students' level of cognitive learning.

In the current research, the cognitive-level assessment of students is performed to give a specific type of remedial work for the slow learners so that their understanding of the subject contents will improve; the fast learners will be given creative assignments to prepare them for the higher cognitive-level assignments. The overall process is to boost student's understanding of the subject and prepare them for the examinations as well as placements.

3 Deep Learning for Pattern Recognition

Deep learning is a popular research domain with a variety of applications. Several studies support the prominence of deep neural networks (DNNs) that exceed the performance of the previous leading standards in diverse machine learning applications [9–13].

Deep learning is a set of machine learning methods designed to model data with a high level of abstraction without being explicitly programmed. Deep learning is derived from the articulation of architectures of various transformations in the nonlinear space [9, 14]. The rising interest in deep learning research is mainly because of its conceptual as well as its technological advances. Factually, the available deep learning solutions based on model learning need an immense reservoir of computing power. This huge compute capacity is made available through actual modern computers, as well as requesting the main processor (CPU) and the graphic dedicated processors (GPU) as well as cloud-based deployments [15–18].

Deep learning, is considered to be the most significant innovation in the past decade in the field of pattern recognition and machine learning, has influenced the methodology of related fields like computer vision and attained enormous progress in both academy and industry. Deep learning solutions accomplished an end-to-end pattern recognition, merging previous steps of preprocessing, feature extraction, classifier design and post-processing [19, 20].

4 Proposed System and Implementation

The problem under consideration is to classify a student in either of slow, average, or fast learner, given the cumulative performance, attendance in the current semester, and total failures in the past semesters. For the classification problem of this kind, a training dataset is needed. The dataset used for this research work is explained below.

4.1 Description of the Dataset

For the current research, a dataset of 1082 students is prepared; the dataset consists of the Cumulative Grade Point Average (CGPA) of the student, the attendance (ATTN) in the current semester till point of the evaluation, and the number of failures (KT) till the current semester. This data are collected from the semester progression records of second, third, and final year students of an engineering college (FAMT, Ratnagiri) for the academic year 2019–20 and 2020–21. This data are collected from examination committee and the attendance management system of the institute.

The fourth column in this dataset is the class; for the training purpose, the cognitive level of the students assessed by the teachers based on his/her performance is added. This data will be used for training as well as testing; using this data, the class of the student has to be predicted as "Class 1—Slow Learner," "Class 2—Fast Learner;" the slow and fast learners are extreme ends of the student's population. Special care has to be taken to improve the performance of slow learners so that they come up to average learners which is the bigger section we call it as "Class 0—Average Learner." Hence, the main task is to identify the slow and fast learners, and that is the research objective of this work (Table 1).

The dataset elements and their correlation are shown in Fig. 1; it can be seen. Three classes and their candidate distribution can be seen there.

4.2 Deep Learning Model

Deep learning overcomes the challenges of classification problem's workflow management by simplifying workflows while also improving accuracy, at least in

Table 1 Description of dataset

Parameter	CGPI	ATTN	KT	CLASS
Count	1081	1081	1081	1081
Mean	6.5211	6.1173	0.67068	0.59112
std	1.23849	20.2052	1.37395	0.73237
Min	1.47	0.05	0	0
25%	5.785	0.73	0	0
50%	6.446	0.8	0	0
75%	7.22	0.9	1	1
Max	9.66667	100	8	2

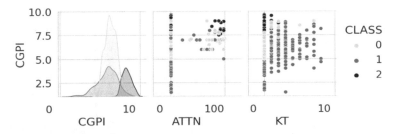

Fig. 1 Correlation of slow and fast learners dataset

many contexts. For the classification of slow and fast learners using deep learning, a deep neural network using multilayered perceptron nodes is used [21]. MLP is universal approximators, and they are very good at modeling nonlinear functions. The multilayer perceptron finds common application in the classification and regression applications in many fields, such as pattern recognition, image, speech, and biometric recognition and classification problems. The choice of the architecture has a great impact on the convergence of these networks [22].

For the implementation of the deep learning model, the Keras deep learning library is used [23]. Keras deep learning application programming interface is a high-level ANN library programmed in Python to build neural network models. The researchers do not have to work about the numerical techniques, tensor algebra, and the mathematical aspects of optimization methods. The Keras library is built on TensorFlow 2.0 and can scale to entire tensor processing units (TPUs) pods or a large cluster of graphics processing units (GPUs). The focus of Keras library is to facilitate experimentations by allowing researchers to go for quick prototyping. They can start from the simple design of model and quickly deploy the same in Keras with little delay to achieve the results, which is key for research. Keras gives huge advantage to the beginner developers, researchers and scientists. A person need not worry about low-level computations and directly delve into deep earning deployment using Keras.

Fig. 2 Sequential MLP deep
learning model visualization
plot

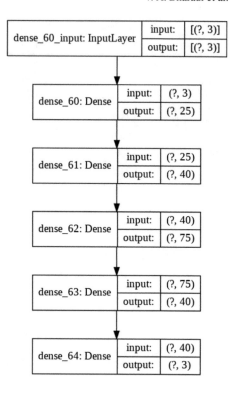

A deep learning model having one input, three hidden layers, and one output layer
is designed in Keras. For the MLP nodes activation function used in rectified linear
(ReLU) activation function. The Keras model summary is given below. The model
is of sequential type.

Once the model is ready, it is compiled and then trained. For training purpose out
of the 1081 student's data, 75% data are used for model training and testing, and
then, the remaining 25% data are used for finding the accuracy. The deep learning
network is trained with 400 epochs and batch size of "5".

In the training phase, the model has reached up to 95% accuracy, but the detailed
analysis will be discussed in the next section.

4.3 Making the Model Robust to the Variations in the Data

One of the most common problem faced by the deep learning researchers is the
overfitting of the model [24]. The overfitting means the model is virtually remem-
bering the training data, and the performance of the model degrades on the real-life
data, beyond the training set. This problem is solved by the regularization techniques. Regularization is a technique in which slight modifications are made to the

learning algorithm so that the model generalizes in a better way. This results in the improvement of the model's performance on the unseen data [25]. To improve the performance of the current model, addition of Gaussian noise and dropout layers is done, and the model is trained and evaluated. The results will be discussed in next section.

4.4 K-fold Cross-Validation

It is a general practice to use k-fold cross-validation for the evaluation of the performance of a classification algorithm. A model's performance is given by the accuracy estimate, and the reliability of the accuracy estimate is indicated by a relatively small variance over the various sets of input data. Several studies therefore recommended to repeatedly perform k-fold cross-validation [26]. Cross-validation (CV) is a technique based on resampling procedure; this is used to evaluate machine learning models on a limited data sample. To implement CV, it is required to keep aside a sample/portion of the data; this sample should not be used to train the model; later, this sample is used for testing/validating.

One variation of k-fold CV is stratified k-fold approach. In this variation of k-fold cross-validation, stratified folds are returned, i.e., the labels are equalized to have a uniform variation. Each set containing approximately the same ratio of target labels as the complete data.

K-fold CV has a single parameter referred as "k" that indicates the number of groups that a given data sample is to be split into, hence, the name k-fold cross-validation. A specific value for k is chosen, e.g., $k = 10$ becoming tenfold cross-validation.

This procedure shuffles the dataset randomly, then splits the dataset in k-folds, and on each such variation, the model is tested, and the final accuracy is averaged, and the variance of accuracy is also calculated. For the evaluation of the model proposed in Fig. 3, k-fold cross-validation with $K = 10$ is performed for both the normal as well as regularized model. The results are given in the next section.

5 Results

The sequential deep learning model with and without regularization is implemented and tested in Keras. To compare the performance of the model, other statistical classifiers are also implemented. The list is as follows:

1. Support vector machines (SVMs) [27]
2. Logistic regression [28]

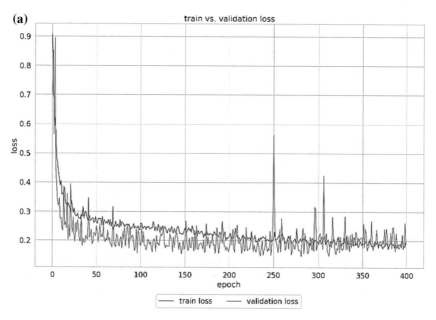

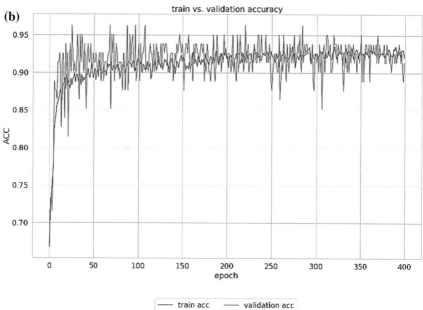

Fig. 3 Deep network training details—**a** Training and validation loss, **b** Training and validation accuracy for the model shown in Fig. 3

3. Naive Bayes classifier [29]
4. K-nearest neighbors [30]
5. Decision tree [31]

Figure 5 shows the final performance of the deep learning and the statistical classifiers implanted as per the discussion in Sect. 4. The classifiers were implemented in Keras on the Google Colab platform [32]. Deep learning classifiers have given the best performance as they have the ability to model complex nonlinear functions. The deep learning classifier has given 91.51% accuracy while classifying slow, fast, and average learners. The regularized deep learning classifier has given 87.08% accuracy. Next best performance is given by logistic regression, KNN classifier, logistic regression, decision tree, and SVM.

The regularization technique and k-fold CV has further strengthened the findings, and the observed performance range is up to 90% for the deep learning ANN classifier. Finally, the best performance is expected from the regularized DNN classifier; the expected range is 87–90% of accuracy. The summary of these results is given in Fig. 4.

Once the classifier is trained and tested, this model can be saved as an h5 file format. These models can be called in Python script to predict the cognitive level of any student provided the required academic details are available. The code and the implementation for the abovementioned research are available at https://doi.org/10.5281/zenodo.4153494 [33].

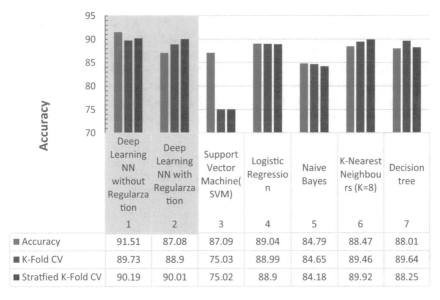

	Deep Learning NN without Regularzation	Deep Learning NN with Regularzation	Support Vector Machine(SVM)	Logistic Regressio n	Naive Bayes	K-Nearest Neighbou rs (K=8)	Decision tree
	1	2	3	4	5	6	7
■ Accuracy	91.51	87.08	87.09	89.04	84.79	88.47	88.01
■ K-Fold CV	89.73	88.9	75.03	88.99	84.65	89.46	89.64
■ Stratfied K-Fold CV	90.19	90.01	75.02	88.9	84.18	89.92	88.25

Fig. 4 Performance comparison of various classifier models implemented in the research

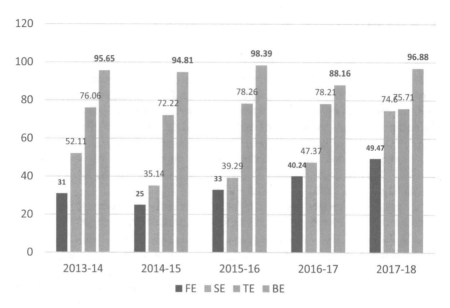

Fig. 5 Improvement in student's performance as they go to higher year of engineering degree

6 Conclusion

In this paper, a method for classification of cognitive levels of the students based on their academic performance. The deep learning-based approach presented here has given accuracy of 91.51%; further, this model was regularized; the accuracy of the regularized model was found to be in the range of 87–90%. K-fold and stratified k-fold cross-validation was also performed on the models proposed. This process will automate the work of teachers to classify the student's cognitive levels, and they can plan their teaching–learning activity accordingly to improve the understanding of the concepts at the students end.

Acknowledgements This work is sponsored by University of Mumbai Minor research Grant Project ID: 1001, Sanctioned in Dec 2019. The hardware for this project work is given by NVIDIA; they have given two Jetson Nano boards (2GB) for the research work.

References

1. Winn AS et al (2019) Applying cognitive learning strategies to enhance learning and retention in clinical teaching settings. MedEdPORTAL J Teach Learn Resour 15:10850. Web
2. Visser L, Korthagen FAJ, Schoonenboom J (2018) Differences in learning characteristics between students with high, average, and low levels of academic procrastination: students' views on factors influencing their learning. Front Psychol 9. Web

3. Rosenblatt F, The perceptron: a probabilistic model for information storage and organization in the brain
4. Mcculloch WS, Pitts W (1990) A logical calculus of the ideas immanent in nervous activity
5. Pham T, Tran T, Phung D, Venkatesh S (2017) Predicting healthcare trajectories from medical records: a deep learning approach. J Biomed Inform 69:218–229. https://doi.org/10.1016/j.jbi.2017.04.001
6. Sehar A (2013) Influence of the constructivist learning approach on students' levels of learning trigonometry and on their attitudes towards mathematics. Hacettepe Üniversitesi Eğitim Fakültesi Dergisi 28(3):219–234
7. Anees S (2017) Analysis of assessment levels of students' learning according to cognitive domain of bloom's taxonomy. Online Submis 1–14. Print
8. Koparan T, Güven B (2015) The effect of project-based learning on students' statistical literacy levels for data representation. Int J Math Educ Sci Technol 46(5):658–686. Web
9. Bouhamed H, Ruichek Y (2018) Deep feedforward neural network learning using local binary patterns histograms for outdoor object categization. Adv Model Anal B 61(3):158–162. https://doi.org/10.18280/ama_b.610309
10. Hinton G et al (2012) Deep neural networks for acoustic modeling in speech recognition: the shared views of four research groups. IEEE Signal Process Mag 29(6):82–97. https://doi.org/10.1109/MSP.2012.2205597
11. Mohamed AR, Dahl GE, Hinton G (2012) Acoustic modeling using deep belief networks. IEEE Trans Audio Speech Lang Process 20(1):14–22. https://doi.org/10.1109/TASL.2011.2109382
12. Cireşan DC, Meier U, Gambardella LM, Schmidhuber J (2010) Deep, big, simple neural nets for handwritten digit recognition. Neural Comput 22(12):3207–3220. https://doi.org/10.1162/NECO_a_00052
13. Yu D, Deng L (2011) Deep learning and its applications to signal and information processing [exploratory DSP]. IEEE Signal Process Mag 28(1):145–154. https://doi.org/10.1109/MSP.2010.939038
14. Bengio Y (2009) Learning deep architectures for AI. Found Trends Mach Learn 2(1):1–27. https://doi.org/10.1561/2200000006
15. Bharadi VA, Mestry HA, Watve A (2019)Biometric authentication as a service (BaaS): a NOSQL database and CUDA based implementation. In: 2019 5th international conference on computing, communication, control and automation (ICCUBEA), Pune, India, pp 1–5. https://doi.org/10.1109/ICCUBEA47591.2019.9129570
16. Bharadi VA, Tolye S (2020) Distributed decomposed data analytics of IoT, SAR and social network data. In: 2020 3rd international conference on communication system, computing and IT applications (CSCITA), Mumbai, India, pp 180–185. https://doi.org/10.1109/CSCITA 47329.2020.9137785
17. Bharadi VA, Meena M (2015) Novel architecture for CBIR SAAS on Azure cloud. In: 2015 international conference on information processing (ICIP), Pune, pp 366–371. https://doi.org/10.1109/INFOP.2015.7489409
18. D'silva GM, Bharadi VA (2015) Modified online signature recognition using software as a service (SaaS) model on public cloud. In: 2015 international conference on information processing (ICIP), Pune, pp 360–365. https://doi.org/10.1109/INFOP.2015.7489408
19. Zhang Z, Shan S, Fang Y, Shao L (2019) Deep learning for pattern recognition. Pattern Recogn Lett. https://doi.org/10.1016/j.patrec.2018.10.028
20. Bouhamed H (2020) COVID-19 deaths previsions with deep learning sequence prediction. Int J Big Data Anal Healthc 5(2):65–77. https://doi.org/10.4018/ijbdah.20200701.oa1
21. Karlik B, Olgac A (2010) Performance analysis of various activation functions in generalized MLP architectures of neural networks. Int J Artif Intell Exp Syst (IJAE) 1(4):111–22. http://www.cscjournals.org/csc/manuscript/Journals/IJAE/volume1/Issue4/IJAE-26.pdf
22. Ramchoun H, Amine M, Idrissi J, Ghanou Y, Ettaouil M (2016) Multilayer perceptron: architecture optimization and training. Int J Interact Multimed Artif Intel 4(1):26. https://doi.org/10.9781/ijimai.2016.415
23. Chollet F (2015) Keras: the python deep learning library. Keras.Io

24. Tao Z, Muzhou H, Chunhui L (2018) Forecasting stock index with multi-objective optimization model based on optimized neural network architecture avoiding overfitting. Comput Sci Inform Syst 15(1):211–36. https://doi.org/10.2298/CSIS170125042T

25. Vasicek D (2019) Artificial intelligence and machine learning: practical aspects of overfitting and regularization. Inf Serv Use 39(4). https://doi.org/10.3233/isu-190059

26. Wong TT, Yeh PY (2020) Reliable accuracy estimates from K-fold cross validation. IEEE Trans Knowl Data Eng 32(8):1586–94. https://doi.org/10.1109/TKDE.2019.2912815

27. Burges CJC (1998) A tutorial on support vector machines for pattern recognition. Data Min Knowl Discov 2(2):121–67. https://doi.org/10.1023/A:1009715923555

28. Sperandei S (2014) Understanding logistic regression analysis. Biochemia Medica 24(1):12–18. https://doi.org/10.11613/BM.2014.003

29. Chen S, Webb GI, Liu L, Ma X (2020) A novel selective Naïve Bayes algorithm. Knowl Based Syst 192. https://doi.org/10.1016/j.knosys.2019.105361

30. Maillo J, Ramírez S, Triguero I, Herrera F (2017) KNN-IS: an iterative spark-based design of the k-nearest neighbors classifier for big data. Knowl-Based Syst 117:3–15. https://doi.org/10.1016/j.knosys.2016.06.012

31. Song YY, Lu Y (2015) Decision tree methods: applications for classification and prediction. Shanghai Arch Psychiatr 27(2):130–35. https://doi.org/10.11919/j.issn.1002-0829.215044

32. Carneiro T, Medeiros Da NóBrega RV, Nepomuceno T, Bian G, De Albuquerque VHC, Filho PPR (2018) Performance analysis of google colaboratory as a tool for accelerating deep learning applications. IEEE Access 6:61677–61685. https://doi.org/10.1109/ACCESS.2018.2874767

33. Bharadi VA, Prasad KK, Mulye YG (2020) Using deep learning techniques for the classification of slow and fast learners (version 1.0). Zenodo. https://doi.org/10.5281/zenodo.4153494

Voting Ensemble-Based Model for Sentiment Classification of Hindi Movie Reviews

Ankita Sharma⑩ and **Udayan Ghose**⑩

Abstract Lately, distinct classifiers have been seen to complement each other in classification performance. This gave rise to the belief of using an ensemble of many simple models instead of using one complex fine-tuned model for classification tasks. Movie review mining or movie review sentiment classification or sentiment analysis (SA) of movie reviews is the process of abstracting the attitude and feelings of the writer from movie reviews text, meaning thereby whether sentiment expressed in written reviews is of positive valence or negative valence. It is known that ensemble models are a promising way of solving various classification problems. These days, researchers have proposed many models for Hindi movie review mining. However, existing research investigations neglect the potency and competence of ensemble models in Hindi movie review sentiment analysis. The voting ensemble is a very popularly used ensemble technique in which many machine learning classifiers (MLCs) can be combined for yielding better predictive performance. This paper proposes a majority voting ensemble model of classifiers for doing binary sentiment classification of Hindi movie reviews. The proposed voting ensemble model has been compared with baseline models. Experimental results on our Hindi movie reviews unfold that our proposed voting ensemble model delivers better acceptable performance when compared to other stated models for sentiment classification on Hindi movie reviews.

Keywords Sentiment classification · Movie reviews · Hindi · Supervised classifiers · Ensemble

A. Sharma (✉) · U. Ghose
University School of Information, Communication and Technology,
Guru Gobind Singh Indraprastha University, Delhi, India
e-mail: Ankitasharma2711@gmail.com

U. Ghose
e-mail: udayan@ipu.ac.in

A. Shukla et al. (eds.), *Computational Intelligence*, Lecture Notes in Electrical Engineering 968, https://doi.org/10.1007/978-981-19-7346-8_40

1   Introduction

There is a saying, "pen is mightier than the sword;" this is very much true in today's contemporary world. With the emergence of Web 3.0, people are posting their opinions over the Internet, and it is becoming common for people to learn what other people are liking or disliking before buying a product or watching movie or Web series, etc. Companies, businesses, political parties, etc., are investing money to make sense of data that is being posted online as online opinions are a good surrogate of word of mouth. Machine learning is the science of getting computers to act by feeding them data and letting them learn on their own [1]. The key to machine learning is data; machines learn just like humans; humans need to collect information and data; similarly, machines must also feed data to learn, make sense of it, and make a decision. Movie review mining or movie text classification categorizes movie review text as positive or negative. Fundamentally, movie text classification using machine learning can be of two types: supervised or unsupervised. Supervised learning is like learning with a teacher; direct feedback is given. In this, the machine learns under guidance; it is called supervised learning. In supervised learning, machines learn by feeding them labeled data and explicitly telling them that this is the input and how the output must look. This deals with the classification and regression problems. Whereas in unsupervised learning, all the data is unlabeled, and the algorithm learns to understand the data and identifies patterns and structures. Here, no direct feedback is given. In this type of learning, the data have not been labeled. The machine itself figures out the dataset given and must find the hidden patterns to make predictions about output [2]. It resolves the clustering and association problems. This paper has been confined to supervised learning algorithms for sentiment classification of movie reviews of a resource-poor language Hindi [3].

The remainder of the paper is organized as follows: Sect. 2 reviews the literature related to movie reviews domain. In Sect. 3, reason about choosing Hindi movie sentiment classification is provided. Section 4 gives details of proposed approach followed for performing binary sentiment classification on Hindi movie reviews. Section 5 discusses experimental results obtained; it also describes and compares performance of our proposed voting ensemble model with baseline classifiers which were applied. In Sect. 6, we have concluded our work and proposed further future research directions.

2   Related Literature

Below is the literature related to SA in the movie review domain.

Authors in [4] have proposed a Senti-lexicon algorithm for performing SA on a dataset of 300 movie reviews retrieved from Twitter. A list of positive, negative, negation words, and emoticons were also considered separately. The accuracy obtained was 70.66% by employing the Senti-lexicon algorithm. Future work includes sarcasm detection, finding time sensitivity, and forged review detection.

In [5], an attempt to predict box office success is made by employing a model that uses decision trees, regression, and neural networks. Binary, categorical, interval-dependent variable was employed, and SEMMA methodology was used. Dataset used was obtained from opus data and was enhanced using data from IMDB. Results show binary variable is giving favorable misclassification errors and is most successful. Other text mining techniques, movies released during the same time will also be considered in the future.

An attempt has been made in [6] to perform binary sentiment classification on Hindi movie reviews. A dataset consisting of Hindi movie reviews was made by collecting reviews from Twitter, Facebook, IMDB, etc. Random forest and support vector machine were applied. Promising results were obtained. Random forest (RF) obtained an accuracy of 91.07%, while support vector machine (SVM) obtained an accuracy of 89.73%. Future work is to include neutral reviews also along with positive and negative reviews for sentiment classification.

Authors in [7] have attempted to predict the success class of Indian movies. The data such as hero, heroine, director, producer, writer, release time, and marketing budget are considered, and appropriate weights are assigned to each component. Neural network (NN) was employed for prediction. 93.3% accuracy was obtained, and results concluded that by increasing the number of hidden layers, the neural network's performance improves, but computation time also increases. In the future, an enhanced equal number of training patterns and unsupervised MLCs will be used.

The point in [8] was to perform movie box office prediction by employing various MLCs like SVM, logistic regression, multi-layer perceptron, random forest, Gaussian Naïve Bayes, AdaBoost, and stochastic gradient descent. Seven hundred fifty-five movie reviews from 2012 to 2015 were considered. Pre-released and post-release features were considered, and tenfold cross-validation was used. Results state that multi-layer perceptron has performed best compared to other applied models and achieved an accuracy of 58.53%. The limitation was movie sequel, and the genre was not considered for this study.

Authors in [9] have attempted to predict net box office success of movies in India by employing linear regression, polynomial regression, KNN, RF, DTs, SVM, Naive Bayes (NB), and NN and did multiclass classification of movies using logistic regression (LR), KNN, RF, DTs, SVM, NB, and NN. A dataset consisting of 250 Bollywood movies was taken. Thirty-two features were used. Results show NN achieves an accuracy of 50% and performs better in terms of net Indian box office collection prediction and multiclass movie classification. Future work is to employ convolution NN to capture people's reaction sentiment while watching movie trailers at home and in theaters.

Our study is different from the other studies as follows. Our study considers the potency and competence of the ensemble method for Hindi movie review sentiment analysis, which researchers otherwise neglect. Most researchers have used the

English review dataset; very few have worked on a resource-scarce Hindi review dataset. As per our knowledge, there are very few studies in which Hindi movie review sentiment classification was dealt with using ensemble learning.

3 Hindi Movie Review Sentiment Classification

Over the past few decades, Bollywood—the Hindi film industry—has evolved tremendously; it has become a multibillion-dollar industry [9]. Bollywood produces films equivalent to the combined production of the largest films producing countries like the USA, Japan, and China [7]. Big money is riding on the Hindi film industry, so the footprint and influence of the Hindi film industry are formidable. There is an imperative need for Hindi movie review mining. It has been observed that significant work has been done for western languages like English. But, very little work has been done for a resource-deficient language like Hindi. Very few resources are available to work with Hindi; even the available resources are in their growing stage and are not standardized. Even there exists no standard movie review dataset to work with. It is also known that people often post critical reviews after watching a movie, or reviewers review the movie and give their views in the form of comments after a few days of movie release on various social networking sites. Reading online movie reviews has become a trend for a personal recommendation. People usually do not prefer to watch a movie having bad or negative reviews. The viewer's greed and reliance on online Hindi movie reviews have attracted researchers to explore this area. Nowadays, with the emergence of UTF-8, a good amount of movie reviews in Hindi are available on the Web have made this an exciting area for data analysis. Through reviews, users quickly decide whether to watch the film or not and book a ticket for films in advance. Hindi movie review mining or Hindi movie text classification categorizes Hindi movie review text as positive or negative. This paper will confine its study to a binary classification of reviews.

4 Proposed Approach Used for Sentiment Classification of Hindi Movie Reviews

In this section, the approach followed and proposed methodology for sentiment classification of Hindi movie reviews is presented [10].

4.1 Dataset Collection

Hindi is a resource-scarce language, and tech advancement for dealing with Hindi text is neoteric. As a result of this, sufficient and annotated datasets are unavailable.

Table 1 Glimpse of Hindi movie reviews

Hindi movie review	Polarity
आंखों देखी दिल को छू जाने वाली फिल्म है।	Positive
बहुत ही निराश करती है 'कबाली'।	Negative
चेन्नई एक्सप्रेस की हर बोगी में मिलता है मनोरंजन।	Positive
कट्टी बट्टी एक उबाऊ फिल्म है और इससे कट्टी करने में ही भलाई है।	Negative
लुटेरा फिल्म में पाखी के किरदार में सोनाक्षी सन्निहा ने वाकई जान डाल दी है।	Positive

The dataset used is primary. Movie reviews in Hindi language were collected from online Web site [11], and they were manually labeled with positive or negative valence. The objective or neutral reviews were not taken into consideration. A total number of 1200 Hindi movie reviews were collected. Glimpse of movie dataset is given in Table 1.

4.2   Preprocessing of Dataset

In this step, movie reviews are transformed into the form for performing efficient classification. All the irrelevant words and symbols will be removed. Since our dataset is primary, hence contains no missing values. For the preprocessing step, Hindi reviews were tokenized, Hindi stop words, non-Hindi text words, numbers, and any special characters were removed.

4.3   Feature Extraction

The words or phrases that help find out the sentiment in movie review text are called features. Term frequency–inverse document frequency (Tf–idf) was used to extract the features from movie review text. The best thing about Tf–idf is that it uses all dataset words as vocabulary. The frequently occurring words in movie reviews will have high Tf, but if it occurs in most documents, its idf gets reduced. This balances out everything. The frequency of non-important words is penalized, and important words phrases receive a boost.

4.4   Dataset Splitting

It has been observed that prediction results depend on train and test spilt values. Supervised MLCs require training data for model training and testing data for model evaluation. If there is less training data, then prediction values will have higher variance values. Therefore, getting the right train test ratio is important. Our dataset is divided into two datasets training and testing datasets, respectively. 75% of movie reviews was used for training, and 25% was used for testing. Train test split can be imported from sklearn.model_selection import train_test_split.

4.5   Brief Description of the Classifiers Used

The following supervised learning classification algorithms were applied to Hindi movie reviews for prognosticating binary sentiment polarity labels in written movie reviews [12, 13].

4.5.1   Decision Tree (DT)

DT is a non-parametric algorithm used for regression and classification problems. The motive is to prognosticate the value of the target class by using rules deduced using data features. DTs require less training time and effort.

4.5.2   Support Vector Machine (SVM)

SVM is a supervised classifier that is discriminative in nature. It classifies the data into two classes by finding the optimal hyperplane using training instances to classify new unseen instances. This algorithm is effectual in high dimensional space and can be imported from Scikit learn library.

4.5.3   Naïve Bayes (NB)

NB is a widely used supervised classification algorithm; it assumes that all variables are not correlated to each other in the dataset. It is based on the Bayes theorem. It performs well in the case of categorical features, as well as with large training instances. NB can be imported using Scikit learn library.

4.5.4   K-Nearest Neighbor (KNN)

KNN is a supervised non-parametric algorithm used for classification and regression problems. The result of this algorithm is class membership. An instance is assigned a membership class common to its K number of nearest neighbors. This lazy algorithm requires no training instances but just the distance calculation to predict the membership class. It can be imported from sklearn.neighbors library.

4.5.5   Logistic Regression (LR)

LR is a linear supervised classification algorithm widely used for binary classification problems. It is straightforward to implement understand and requires fewer training instances. Therefore, it is very effective to train. This algorithm performs reasonably well when the data are linearly separable. It can be imported from sklearn.linear_model import library.

4.5.6   Random Forest (RF)

RF is a supervised algorithm based on the ensemble of many DTs following the bagging technique. Random forest, as the name suggests, comprises of two words random and forest; by forest, it is meant that it is made up of a large number of trees, and by random, it is meant that process of finding the root and splitting feature node will be done randomly. If the number of trees taken is greater in number, the more accurate results are obtained. It can be imported from sklearn.ensemble module.

4.5.7   AdaBoost (AB)

AB can be imported from sklearn.ensemble module. The algorithm initially trains on a weak learner, mainly DTs, after training weak learners, makes predictions on training instances. Following this, the algorithm increases relative weight for misclassified training instances. Again, training is repeated on updated weights. The exact process is repeated until no misclassified training instance is left.

4.6   Evaluation of Metrics

Performance of various baseline models used and proposed voting ensemble technique are evaluated using accuracy, precision, recall, and F1-score. Accuracy is correctly classified movie reviews to the total number of Hindi movie reviews in the dataset. Precision is correctly prognosticated positive movie reviews to a total number of positive reviews in the dataset. The recall is correctly prognosticated positive movie reviews to an actual number of positive movie reviews. F1-score finds the mean value between recall and precision.

5   Results and Evaluation

Bollywood—the Hindi film industry has evolved tremendously; the footprint and influence of the Hindi film industry is formidable. There is an imperative need for Hindi movie review mining. The extraction of opinions from Hindi movie review text plays a significant role in forecasting the success and inclination of people toward a particular movie. Positive reviews signifies that the movie is worth watching. However, negative reviews indicate that the movie is not good enough to watch. In this paper Python, 3.8.0 was used for implementation. Initially, various supervised MLCs such as KNN, SVM, RF, NB, AB, LR, and DT were applied to our Hindi movie reviews to predict the sentiment polarity of Hindi movie reviews as positive or negative.

Results obtained by applying individual classifiers implied that LR has performed the best and has obtained 84% accuracy, followed by SVM and NB, which obtained 83 and 82% accuracy, respectively. Classifiers such as RF and AB obtained the same accuracy of 78%. On the other hand, the DT obtained an accuracy of 74%.

While KNN performed the worst, we randomly experimented with different values of K. We obtained an accuracy of 62% when we randomly took the value of K as 5 and 11. But, the accuracy value improved by 2%, when we applied grid search and obtained an accuracy of 65% with K = 23. We obtained an accuracy of 65%, recall of 65%, precision of 66%, and F-measure of 64%, when applying an optimal value of K as 23 using grid search. In our case, LR, which is a discriminative model, performs the best, while KNN performs the worst.

The obtained results were not very promising. So, instead of applying other classifiers to the Hindi movie reviews, we combined all the classifiers using the concept of ensemble learning. Ensembling is a technique in which multiple MLCs are combined to form one robust model with enhanced performance [12]. MLCs used are base learners or weak learners fitted on training datasets and used to make predictions. Base models can comprise different models or the same models with changed hyperparameters. The final prediction is obtained by combining predictions from all base models. There are many methods to combine predictions to get a final prediction. Some are maximum voting, averaging, and stacking [14]. The most straightforward and widely used ensemble is maximum voting [15]. Suppose there are three base learners, and these base learners independently make a prediction, for instance, for the class label. The final prediction label of the class will be the one that has got the highest weight [16]. In maximum voting, all base learners are being assigned equal weights.

We proposed a voting ensemble of classifiers given in Fig. 1 comprising all classifiers applied above, and the pseudocode algorithm is stated beneath. Hindi movie reviews dataset was randomly divided into a 75:25 train test ratio. KNN, SVM, DT, AB, RF, LR, and NB were applied, and L1, L2, L3, L4, L5, L6, L7, respectively, are the polarity labels predicted for the particular Hindi sentence-level movie review. The final polarity label of Hindi movie review was the polarity label which obtained the maximum vote or more the half of the vote of the applied classifiers. This way, polarity labels were predicted for all Hindi movie reviews. We used an odd number of classifiers to ensure no equal label predictions.

Algorithm: Majority Voting-Based Ensemble of Classifiers.

Prerequisite

Input: Hindi movie reviews set R = (R_1, R_2, R_3, ..., R_{1200});
Polarity label class set C = (C_1, C_2, C_3, ..., C_n);
Base learners set B = (KNN, SVM, DT, AB, RF, LR, NB);

Steps:

1. Divide dataset into train test ratio of 75:25
2. Apply base learners for classification.
3. Generate polarity labels (L1, L2, L3, L4, L5, L6, L7)
4. Each predicted polarity label deemed as votes.
5. If total predicted positive votes > total predicted negative votes
6. Output movie review as positive
7. Else movie review is negative
8. Repeat previous steps for all reviews

Output: Predicted polarity label based on majority voting.

Proposed voting ensemble for predicting the sentiment polarity of Hindi movie reviews as positive or negative obtained cheering results. Our proposed voting ensemble of classifiers obtained an accuracy of 88%, the precision of 89%, recall of 88%, and F1-score of 88%, which was the highest and better than all the individual classifiers employed. This implied that it is good to use voting ensemble while doing binary sentiment classification on Hindi movie reviews. Table 2 is showing

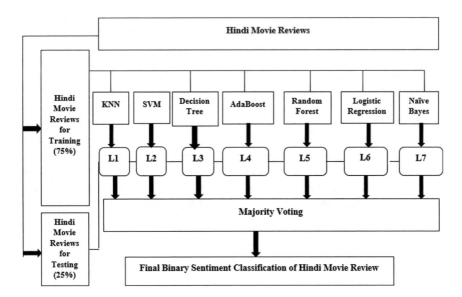

Fig. 1 Proposed voting ensemble of classifiers for binary sentiment classification of Hindi movie reviews

Table 2 Results obtained by all classifiers applied on Hindi movie reviews

Classifier applied	Accuracy (%)	Precision (%)	Recall (%)	F1 score (%)
K-nearest neighbor	65	66	65	64
Decision tree	74	76	74	74
Random forest	78	78	78	77
AdaBoost	78	78	78	77
Naïve Bayes	82	83	83	82
Support vector machine	83	84	83	83
Logistic regression	84	84	84	84
Proposed voting ensemble	*88*	*89*	*88*	*88*

accuracy, precision, recall, and F1-score results of all the applied classifiers along with our proposed voting ensemble of classifiers.

6 Conclusion

It is known that ensemble models are a promising way of solving various classification problems. However, existing research investigations neglect the potency and competence of ensemble models in sentiment analysis of movie reviews in the Hindi language. Over the past few decades, Bollywood—Hindi film industry has evolved tremendously, and there is an imperative need for Hindi movie review mining. It has been observed that very little work has been done for a resource-deficient language Hindi. This paper presents an open problem of binary sentiment classification of Hindi movie reviews. Because of the nature, essence, and structure of Hindi, sentiment analysis of Hindi language text is not effortless. Initially, individual classifiers such as SVM, DT, LR, NB, AB, KNN, and RF were applied, followed by our proposed voting ensemble to Hindi movie reviews. Results obtained from the experiments conducted unfold that our voting ensemble of classifiers based on maximum voting is superior to all individual classifiers and achieved reasonably good performance. Our future direction is to use deep learning models and stacking ensemble techniques for binary sentiment classification of Hindi movie reviews and to use word embedding as a feature extraction technique. Also, we will be increasing the size of our Hindi movie review dataset. Another paper encompassing future direction is planned for further work.

References

1. Sharma A, Ghose U (2021) Lexicon A linguistic approach for sentiment classification. 11th international conference on cloud computing, data science & engineering (confluence). IEEE, India, pp 887–893

2. Sharma A, Ghose U (2020) Sentimental analysis of twitter data with respect to general elections in India. Procedia Comput Sci 173:325–334
3. Jha V et al (2016) Sentiment analysis in a resource scarce language: Hindi. Int J Sci Eng Res 7 (9):968–980
4. Mumtaz D, Ahuja B (2016) Sentiment analysis of movie review data using Senti-lexicon algorithm. In: 2nd international conference on applied and theoretical computing and communication technology (iCATccT), IEEE, pp 592–597
5. Galvao M, Henriques R (2018) Forecasting model of a movie's profitability. In: 13th Iberian conference on information systems and technologies (CISTI), IEEE, pp 1–6
6. Nanda C et al (2018) Sentiment analysis of movie reviews in Hindi language using machine learning. In: International conference on communication and signal processing (ICCSP), IEEE, pp 1069–1072
7. Kaur A, Nidhi Ap (2013) Predicting movie success using neural network. Int J Sci Res (IJSR)
8. Quader N et al (2017) Performance evaluation of seven machine learning classification techniques for movie box office success prediction. In: 3rd international conference on electrical information and communication technology (EICT), IEEE, pp 1–6
9. Kanitkar A (2018) Bollywood movie success prediction using machine learning algorithms. In: 3rd International conference on circuits, control, communication and computing (I4C), IEEE, pp 1–4
10. Shah R et al (2019) Sentiment analysis on Indian indigenous languages: a review on multilingual opinion mining. arXiv preprint arXiv:1911.12848
11. Filmi Beat Hindi Homepage. https://hindi.filmibeat.com/reviews/, Last accessed 1 Jan 2022
12. Ishrat N et al (2021) Use of novel ensemble machine learning approach for social media sentiment analysis. In: Analyzing global social media consumption, IGI Global, pp 16–28
13. Srivasatava S et al (2017) An ensemble-based NLP feature assessment in binary classification. In: International conference on computing, communication and automation (ICCCA), IEEE, pp 345–349
14. Behera R et al (2016) Ensemble based hybrid machine learning approach for sentiment classification—a review. Int J Comput Appl 146(6):31–36
15. Rojarath A et al (2020) Probability-weighted voting ensemble learning for classification model. J Adv Inform Technol 11(4)
16. Sarkar K (2020) Heterogeneous classifier ensemble for sentiment analysis of Bengali and Hindi tweets. Sādhanā 45(1):1–17

Brain Tumor Detection with Artificial Intelligence Method

Shweta Pandav◉ **and S. V. B. Lenina**◉

Abstract Detection of brain tumor is very challenging task in today's medical world. Nowadays, many doctors prefer ready-made methods to diagnose the tumor from given magnetic resonance imaging (MRI) images. Accurate and less time-consuming methods are used frequently to identify it. This paper introduces new techniques for finding and differentiating brain tumors. The new methods include support vector machine (SVM) and convolution neural network (CNN) with Softmax classifier. The SVM uses histogram of oriented gradients (HOGs) as a feature to label the given dataset. Further, the same feature of the test image is verified against labeled sets and classified. The SVM gives accuracy of 87.33%. In CNN, multilayer structure is used. It contains a convolution layer followed by the pooling layer and finally fully connected layer. In CNN, specifically, Alexnet is preferred for classification. In Alexnet, rectified linear unit (ReLU) is used as an activation function. At the output of it, Softmax is used as activation to classify the images. The CNN with Softmax classifier gives improved accuracy of 98.68%. In addition, the performance measures are also found out such as specificity, sensitivity, and precision. The sensitivity with SVM is 93.18%, and with CNN, it increases up to 100%. Comparing both methods with the obtained results, the CNN with Softmax classifier gives better results. Proposed method with CNN produces only one misclassified image as an advantage over the existing method, while SVM gives nine misclassified images. The methods are tested on 253 MRI images which are detected and classified correctly with Matlab.

Keywords SVM · CNN · HOG · MRI

S. Pandav (✉) · S. V. B. Lenina
Shri Guru Gobind Singhji I.E. & T Nanded, Nanded, Maharashtra 431603, India

S. V. B. Lenina
e-mail: lvbirgale@sggs.ac.in

© The Author(s), under exclusive license to Springer Nature Singapore Pte Ltd. 2023   485
A. Shukla et al. (eds.), *Computational Intelligence*, Lecture Notes in Electrical
Engineering 968, https://doi.org/10.1007/978-981-19-7346-8_41

1 Introduction

Tumor is the fastest growing cause of death among humans older than 65 years. Tumor is abnormal growth of tissues. Tumors are checked by taking computer tomography (CT) scan or magnetic resonance imaging (MRI) of affected body parts. Other histopathology methods also detect tumors such as blood tests and biopsy of tumor parts. Tumors are mainly divided into two types benign (non-cancerous) and malignant (cancerous). Malignant tumors spread within a small amount of time in the body and are dangerous for survival of humans. Segmentation of tumor is a challenging task. It may be of any shape, size, and may appear at any location of the brain with different intensities. Early detection of tumors increases the chance of recovery in patients. Timely checking of a tumor is helpful to remove it completely. The radiologists mark the tumor region manually in MRI images. Researchers worked on engineering methods for finding the area of affected regions in brain tumors.

The proposed method uses support vector machine along with feature extraction techniques to find the tumor in brain images. The second technique used is an artificial intelligence algorithm based on CNN to classify cancerous and non-cancerous images. The first method gives more mistakenly classified images, while the second method with Softmax classifier corrected all the incorrectly classified images into the right one. CNN is a very innovative technique in brain tumor segmentation. It gives increased network accuracy.

The paper is arranged as follows: Sect. 2 describes the related work in the field of brain tumor segmentation; Sect. 3 explains methodology used in proposed research; Sect. 4 gives experimental setup; Sect. 5 highlights the results, while Sect. 6 includes conclusion, and at the end, references are added.

2 Related Work

Hondre and Kokare explained identification of tumor in brain by edge detection and watershed segmentation methods. It gives good results, and it is a very basic method [1]. Mahmud et al. proposed K-means and bisecting K-means algorithm [2]. In this paper, results are tested against 100 MRI images where K-means gives accuracy as 78.81% in 1.672 s. Sachdeva et al. worked on genetic algorithms with SVM [3]. They worked with 428 real-time MRI images and identified five different classifiers and got accuracy as 91.7%. Shrinivas and Roy proposed K-means and fuzzy C-means clustering algorithm [4]. It uses data from Web site, and noticed FCM gives better results. El Kaitouni and Tairi proposed comparative study between hidden Markov and deep learning approaches [5]. Out of which, K-means accuracy is 97.17% as compared with deep learning approaches. Haritha in her paper evaluated K-means LBP, morphological operation, edge detection, and its combination on random images. She found K-means with LBP, and edge detection gives satisfactory results [6]. Pavlov et al. worked on BRATs-2018 dataset and proposed machine learning with ResNet and illustrated the number of pixel-wise segmentation get reduced in it [7]. Gargouri et al. worked on mean square error, mean error, and Otsu's multilevel threshold and iterative closest point matching (ICP) as explained in [8]. In this paper, 21 MRI brain images are checked with half symmetry divided and are tested against MSE with ICP and gives better results. Shukla and Kumar Sharma in their research work done computer-based cropping segmentation with anisotropic filter and post-processing with median filter on Kaggle dataset [9]. After this, the author applied erosion followed by dilation and got the highest accuracy with it. Su et al. used the SVM active learning algorithm for automated glioblastoma disease segmentation [10]. With this method, accuracy is 77.7%, but accuracy gets increased as 88.4% with knowledge-based algorithms. Bougacha et al. worked on 50 different real-time images [11]. They proposed K-means with watershed and genetic algorithm as well as optimized FCM, and after comparing all, concluded last method gives best results.

Gupta and et al. in their paper compared five different methods [12]. The five methods include artificial neural network-based Levenberg–Marquardt method, SVM and ANN-based segmentation, local independent projection base segmentation, wavelet and Zernike-based segmentation, and segmentation based on geometric transformation invariant method. Out of which, local independent projection-based segmentation gives best results. Mzoughi et al. proposed histogram-based equalization techniques such as AHE, CLAHE, AIR-AHE, and BPDHE are checked on standard dataset as BRATS-15 [13]. This author calculated performance measures as PSNR and entropy. Out of which, adaptive histogram equalization technique gives better results.

Gupta et al. worked on a selective block approach with local binary pattern [14]. They applied the methods on 100 images from CT scan center and got accuracy as 99.67%.

El-Melegy et al. proposed deep learning with WEKA algorithm in which 20 different classifiers are used [15]. Dice method is used for accuracy checks. The results are obtained on BRATS-16. After completion of the result, a random forest classifier is obtained as the best classifier. Tulsani et al. proposed morphological watershed segmentation, K-means, and FCM methods [16]. Out of which, K-means does segmentation in 0.6416 s. The complete study on supervised and unsupervised classification methods is proposed by Lu et al. compared in [17], in which ANN with GMM gives accurate results. Iriawan and Pravitasari in their research work used K-means cluster, FCM, Gaussian mixture model, Fernandez–Steel skew normal (FSSN) methods for segmenting tumor and noise reduction [18]. In their methods, FCM is used for Gaussian noise reduction, and GMM is proved effective for salt and pepper noise reduction. FSSN is efficient for both the noise reduction.

Afshan et al. proposed histogram thresholding, K-means clustering, FCM, and combination to find the tumor [19]. In which FCM with K-means clustering illustrates the best outcome.

Siar and Teshnehlab proposed CNN with a feature extraction method for tumor identification [20]. It produces accuracy as 98.6% with two misclassified images.

Gargouri et al. worked on Otsu's threshold and entropy-based segmentation as explained in [21]. In this paper, 19 MRI brain images are checked with a proposed method and demonstrated Harvada gives good results.

Laha et al. worked on skull stripping methods with Ostu's global thresholding on three datasets ISBR, LPBA40, and OASIS [22]. The global thresholding is followed by analysis and removal of connected components, and finally, morphological operations give correct brain mask. The overall sensitivity is 94.34% as compared to old skull stripping method. Gurbina et al. worked on brain tumor segmentation based on various wavelets and SVM method [23]. In this work, they found the loss of edges in segmentation is prevented with CWT wavelet as well as SVM with proper training dataset gives better results. Keerthana et al. worked with K-means and SVM algorithm as a hybrid technique to get better results [24]. Shinde et al. in their research work highlights on different techniques to segment the brain tumor [25]. It includes various segmentation, classification, and feature extraction techniques along with data mining algorithm. Suresha et al. in their research work proposed combination of K-means and SVM to detect tumor, and it is found to be effective method [26].

3 Methodology

Various methods are studied in literature surveys; some of the methods are more advanced than proposed one such as ResNet [7]. Although ResNet is a new CNN method, it is a very complex method; it uses more than 20 layers in its architecture and takes more time to train the given data.

The new method uses AlexNet with CNN. It is a very simple method; it uses only eight layers and takes less time to train data. So, this method is preferred for brain tumor detection in the proposed work.

The proposed methods SVM and CNN are explained in further sub-sections.

3.1 Support Vector Machine

The support vector machine (SVM) is trained machine learning algorithm used in classification of images. The SVM algorithm plots each data item as a point in n-dimensional space (where n is the number of features used) in which the feature value is the same as the value of a particular coordinate. Then, with the help of the main plane, the two classes clearly distinguished. SVM produces a clear margin to separate the classes. In high-dimensional spaces, it produces good results. It gives accurate results when the number of sample points is smaller than the number of dimensions. It requires less memory because it uses a small set of training points in support vectors. Proposed method uses histogram of oriented gradients (HOG) as a feature vector to extract it from brain MRI images. The dataset is trained with an SVM classifier. The test images are checked against its own HOG feature with classifiers labeled set and then classified correctly.

3.2 Convolution Neural Network

One of the special deep learning networks is CNN. It has multilayer structure which is explained as

(i) *Convolution layer*: It is the main part of the CNN. The network's computational weight age is taken by this layer. In this layer, multiplication between two matrices is done, where one matrix has values of understandable parameters called a kernel, and the other matrix is the limited portion of the receptive field. If the image has three (RGB) channels, the kernel height and width will be dimensionally small, but the depth extends up to all three channels.

(ii) *Pooling layer*: This layer added to reduce the number of repeated layers to one. It decreases the structural size of the representation. The required quantity of computation and weight is reduced with this layer. Every slice of the image is checked one by one by the pooling method.

(iii) *Fully connected layer*: Every neuron in this layer is completely connected with all neurons in the previous and upcoming layer. The result is obtained with matrix multiplication. The FC layer gives the connection between input and output. Figure 1 shows a block diagram of CNN.

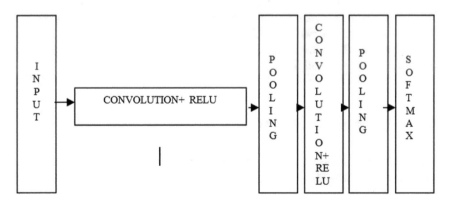

Fig. 1 Block diagram of CNN

4 Experimental Setup

The experimental setup consists of a dataset, its simulation, and performance measures of proposed methods.

4.1 Dataset

The dataset used in this paper includes brain MRI images of 253 patients, including normal and tumorous patients who went to MRI centers. After examination, the 155 patients are found to be having cancerous images, and 97 images are of healthy patients.

4.2 Simulation

The brain MRI images are observed and tested with Matlab 2021 software. The 70 percent of data is used for training, and 30 percent is for testing purposes. The two algorithms proposed in this paper; first is support vector machine, and another is CNN with Alexnet. For training purposes, 178 images and for testing 75 images are used. In SVM, the HOG features are extracted from trained images and tested against labeled data.

In CNN, the Alexnet is used. The input to this classifier is the images of size 227 × 227. Alexnet contains five convolution layers and three fully connected layers. The activation function used in it is ReLU. The activation used in output layer is Softmax. It has a total of 62.3 million neurons. All the results are tested based on accuracy, sensitivity, specificity, and precision.

4.3 Performance Measures

The performance measure criteria used in both the methods are formulated as in Eqs. (1)–(4), respectively.

$$\text{Accuracy} = \frac{(TP + TN)}{(TP + TN + FP + FN)} \tag{1}$$

$$\text{Sensitivity} = \frac{TP}{(TP + FN)} \tag{2}$$

$$\text{Specificity} = \frac{TN}{(TN + FP)} \tag{3}$$

$$\text{Precision} = \frac{\text{TP}}{(\text{TP} + \text{FP})} \qquad (4)$$

where, TP: True Positive

TN: True Negative

FP: False Positive

FN: False Negative.

5 Results

The results are tested on a total 253 patients with cancerous and non-cancerous brain MRI images from standard dataset available on Kaggle Web site. For training, 178 images and for testing 75 images are used. As SVM is used in prior art to detect brain tumors, but none of the existing methods extract HOG as a feature for classification. Alexnet with CNN is preferred over ResNet because of its simplicity and requiring less computational time. This paper proposes two new algorithms used to detect tumors and classify the images into normal and abnormal images. In SVM method, histogram of oriented gradient (HOG) is used as a feature to extract it from brain MRI images. It trains the dataset with an SVM classifier. The test images are checked against image HOG features with classifiers labeled set and then classified correctly.

According to SVM, we get the accuracy 87.83%, sensitivity 93.13%, specificity 82.75%, and precision 89.13%, while with CNN and Alexnet in combination increases accuracy up to 98.68%, sensitivity up to 100%, specificity up to 96.55%, and precision to 97.82%. As per the obtained result, CNN with Softmax layer gives best classification accuracy as 98.68%. The suggested technique uses maximum epoch size as 50 and mini batch size as 300 with a fully connected layer. Table 1 gives the results obtained on test data images.

Figure 2 shows misclassified images with SVM.

With CNN, only, one image is incorrectly classified as shown in Fig. 3.

The network accuracy graph is shown in Fig. 4, and the loss has been shown in Fig. 5, respectively.

Table 1 Results obtained on test data brain images

Methods	Accuracy (%)	Sensitivity (%)	Specificity (%)	Precision (%)	False
SVM (proposed)	87.83	93.18	82.75	89.13	9
CNN + Softmax (proposed)	98.68	100	96.55	97.82	1
CNN + Softmax [19]	98.67	100	94.64	98.26	2

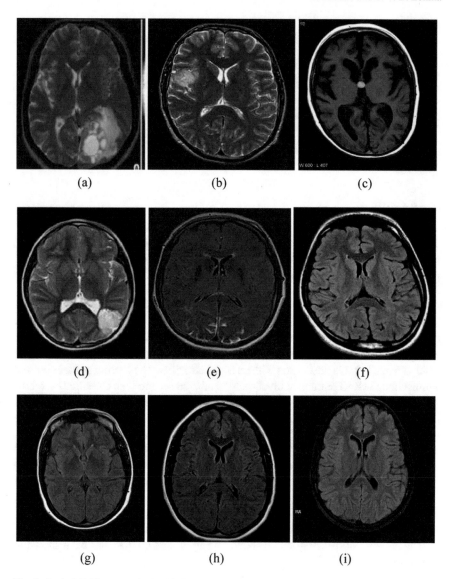

(a)   (b)   (c)

(d)   (e)   (f)

(g)   (h)   (i)

Fig. 2 Brain MRI images misclassified with SVM

6   Conclusion

In this paper, two new techniques are proposed to identify and classify brain cancer images. The SVM with feature extraction gives accuracy of 87.83%, and a total nine images are misclassified. With the new artificial intelligence algorithm that CNN and Softmax in combination gives 98.68% accuracy, and only, one image is misclassified.

Fig. 3 Brain MRI images
misclassified with CNN

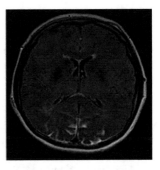

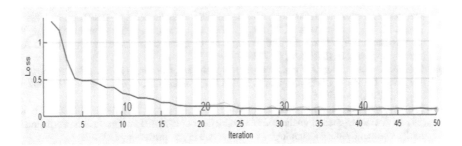

Fig. 4 Testing network accuracy graph with CNN

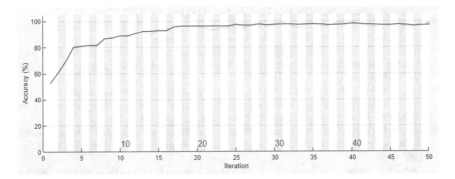

Fig. 5 Testing network loss graph with CNN

The existing papers, with CNN method, give two misclassified images whereas the proposed method reduces the number of misclassified images, and only, one image is misclassified. The comparison between two proposed methods shows CNN gives best results and accuracy because it has a strong classifier with multilayered hyperplanes. From the given dataset of 253 patients, 155 images are classified into tumorous images, and 97 are classified as non-tumorous, and only, 1 is misclassified.

Different performance measures are tested such as accuracy, sensitivity, specificity, and precision. From obtained results, CNN with Softmax proves to be an efficient method to classify tumor images.

References

1. Handore S, Kokare D (2015) Performance analysis of various methods of tumor detection. In: International conference on pervasive computing (ICPC)
2. Mahmud M, Mamun M, Hussain M (2018) Comparative analysis of K-means and bisecting K-means algorithms for brain tumor detection. In: International conference on computer, communication, chemical, material and electronic engineering (IC4ME2)
3. Sachdeva J, Kumar V, Gupta I, Khandelwal N, Ahuja CK (2011) Multiclass brain tumor classification using GA-SVM. In: Developments in e-systems engineering, pp 182–187
4. Srinivas B, Rao GS (2018) Unsupervised learning algorithms for MRI brain tumor segmentation. In: Conference on signal processing and communication engineering systems (SPACES), pp 181–184
5. El Kaitouni SEI, Tairi H (2020) Segmentation of medical images for the extraction of brain tumors: a comparative study between the hidden Markov and deep learning approaches. In: International conference on intelligent systems and computer vision (ISCV)
6. Haritha D (2016) Comparative study on brain tumor detection techniques. In: International conference on signal processing, communication, power and embedded system (SCOPES), pp 1387–1391
7. Pavlov S, Artemov A, Sharaev M, Bernstein A, Burnaev E (2019) Weakly supervised fine tuning approach for brain tumor segmentation problem. In: 18th IEEE international conference on machine learning and applications (ICMLA)
8. Gargouri F, Ben Hamida INA, Chtourou K (2014) Automatic localization methodology dedicated to brain tumors based on ICP matching by using axial MRI symmetry. In: 1st international conference on advanced technologies for signal and image processing (ATSIP), pp 209–212
9. Shukla M, Kumar Sharma A (2020) Comparative study to detect tumor in brain MRI images using clustering algorithms. In: 2nd international conference on innovative mechanisms for industry applications (ICIMIA), pp 773–777
10. Su P, Xue Z, Chi L, Yang J, Wong ST (2012) Support vector machine (SVM) active learning for automated Glioblastoma segmentation. In: 9th IEEE international symposium on biomedical imaging (ISBI), pp 468–473
11. Bougacha JB, Ben Slima M, Ben Hamida A, Ben Mahfoudh K, Kammoun O, Mhiri C et al (2018) Efficient segmentation methods for tumor detection in MRI images. In: 4th international conference on advanced technologies for signal and image processing (ATSIP)
12. Gupta KK, Dhanda N, Kumar U (2018) A comparative study of medical image segmentation techniques for brain tumor detection. In: 4th international conference on computing communication and automation (ICCCA), pp 1–4
13. Mzoughi H, Njeh I, Ben Slima M, Ben Hamida A (2018) Histogram equalization-based techniques for contrast enhancement of MRI brain Glioma tumor images: comparative study. In: 4th international conference on advanced technologies for signal and image processing (ATSIP)
14. Gupta N, Seal A, Bhatele P, Khanna P (2016) Selective block based approach for neoplasm detection from T2-weighted brain MRIs (ICSIP), pp 151–155
15. El-Melegy MT, El-Magd KMA, Ali SA, Hussain KF, Mahdy YB (2018) A comparative study of classification methods for automatic multimodal brain tumor segmentation. In: International conference on innovative trends in computer engineering (ITCE), pp 36–40
16. Tulsani H, Saxena S, Bharadwaj M (2013) Comparative study of techniques for brain tumor segmentation. IMPACT 118–12

17. Lu S, Guo X, Ma T, Yang C, Wang T, Zhou P (2019) Comparative study of supervised and unsupervised classification methods: application to automatic MRI glioma brain tumors segmentation. In: International conference on medical imaging physics and engineering (ICMIPE)
18. Iriawan N, Pravitasari AA, Fithriasari K, Irhamah S, Purnami W, Ferriastuti W (2018) Comparative study of brain tumor segmentation using different segmentation techniques in handling noise. In: International conference on computer engineering, network and intelligent multimedia (CENIM), pp 289–292
19. Afshan N, Qureshi S, Hussain SM (2014) Comparative study of tumor detection algorithms. In: International conference on medical imaging, m-health and emerging communication systems (MedCom), pp 251–255
20. Siar M, Teshnehlab M (2019) Brain tumor detection using deep neural network. In: 9th international conference on computer and knowledge engineering (ICCKE), pp 363–368
21. Gargouri F, Jayaswal S, Khunteta A (2016) A comparative study of otsu and entropy based segmentation approaches for tumour extraction from MRI. In: 2nd international conference on advances in computing, communication, & automation (ICACCA) (Fall)
22. Laha M, Tripathi P, Bag S (2018) A skull stripping from brain MRI using adaptive iterative thresholding and mathematical morphology. In: 4th international conference on recent advances in information technology, RAIT
23. Gurbina M, Lascu M, Lascu D (2019) Tumor detection and classification of MRI brain image using different wavelet transforms and support vector machines. IEEE Trans
24. Keerthana A, Kavin Kumar B, Akshaya KS, Kamalraj S (2021) Brain tumour detection using machine learning algorithm. In: ICNADBE journal of physics: conference series
25. Shinde A, Vengurlekar S (2020) Image mining methodology for detection of brain tumor: a review. In: Proceedings of the fourth international conference on computing methodologies and communication (ICCMC)
26. Suresha D, Jagadisha N, Shrisha HS, Kaushik KS (2020) Detection of brain tumor using image processing. In: Proceedings of the fourth international conference in computing methodologies and communication (ICCMC)

Drowsy Driver Detection Using Galvanic Skin Response

Vivek Jangra⬤, **Ramnaresh Yadav**⬤, **Anjali Kapoor**⬤, **and Pooja Singh**⬤

Abstract Since the past decade driver monitoring and automated driving systems have seen significant innovation with the introduction of artificial intelligence, image recognition systems and improved sensors. The whole world has seen an increased concern over driver safety, and constant efforts are being done to improve the same. Drowsiness on the road is a serious threat to safe driving. One counter measure to this is designing prompt alert systems for the driver. The said alert system must serve as an alarming system at first indication of drowsiness to make the driver entirely conscious and then as a suggestion system to take some rest nearby instead of continued driving. Here we developed the prototype circuit to record an individual's electrodermal activity (EDA) by measuring skin conductance level (SCL) to identify the threshold and pattern of skin conductance for timely alerting drivers transitioning to drowsy states.

Keywords Driver alertness · Drowsy driver · Electrodermal activity · Galvanic skin response · Skin conductance level

1 Introduction

One in twenty-five adult drivers (aged 18 years or older) report falling asleep while driving in a span of thirty days [1]. A lot of accidents were reported due to drowsy driving causing fatalities. Drowsy driving is a prominent threat to road safety. Timely and accurate drowsy driver detection is one of the notable ways to prevent these accidents [2]. EDA reflects the activity of the autonomic nervous system. Several parameters of EDA show potential for disclosing information about the mental state

V. Jangra (✉) · R. Yadav · A. Kapoor
Amity University, Noida, India
e-mail: vjangra@amity.edu

R. Yadav
Amity University, Tashkent, India

P. Singh
GL Bajaj Institute of Technology and Management, Greater Noida, India

497

of the driver. In latest research done on driver distraction and inattention, it was determined that the major effect of the driver's mental attentiveness was found on the galvanic skin response (GSR) of the driver.

1.1 The Nervous System

Our peripheral nervous system consists of the autonomic nervous system. The autonomic nervous system is further divided into two sections, sympathetic nervous system and parasympathetic nervous system. We are concerned with the sympathetic nervous system. Its functioning is majorly involuntary and is controlled by external stimuli or emotional state of an individual. One of its functions is to establish a connection between the mental state of a person and the secretion of a number of glands. These glands include the sweat glands, and this is where one can determine the individual's EDA. EDA refers to the change in the skin's electrical properties on change in the emotional state [3]. One of these electrical parameters is SCL. A reaction to external stimuli or a variation in the emotional state of a person shows a characteristic increase or decrease in the SCL, and this is referred to as the galvanic skin response [4]. The two terms SCL and GSR are often used interchangeably.

1.2 Galvanic Skin Response and Driver Alertness

In an alert state of mind, the sweat glands are more active resulting in an increased perspiration and higher skin conductance (or lower skin resistance). However in a drowsy or inactive state of mind, the sweat gland activity is decreased resulting in lower values of skin conductance (or higher skin resistance) [4]. A driver's state of mind gradually transitions from alert to inactive due to fatigue and results in drowsy driving occurrences. Our main aim is to develop a circuit to measure and test the threshold value of skin conductance below which it can be declared that the individual is shifting to a drowsy state and should take rest instead of continuing to drive. A basic GSR measurement system consists of two electrodes attached to the skin [5], one constant DC supply to the skin, a voltage divider circuit, an amplifier, and a filter as shown in Fig. 1. In Fig. 1, a quantity called the basal skin response (BSR) is taken from a neutral electrode attached to the skin, and the final measurement done is the voltage difference between BSR output and GSR output. The range of values for SCL is typically 2–20 μs [5]. There are various sites on the body where electrodes can be placed for GSR measurement. These include shoulders, wrist, palm, and fingers. In Fig. 2, three such sites on the hand are shown. We used site 2 for our measurements on account of the least muscular activity on this site during driving which will reduce disturbances in measurement [5]. The type of electrodes connected with the skin has a considerable impact on the sensitivity of the input from the skin. We selected Ag/AgCl electrodes for their superior sensitivity in EDA measurements [6].

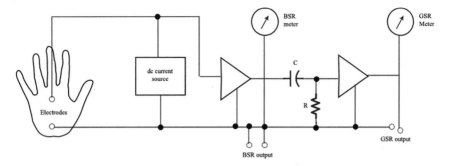

Fig. 1 Basic block diagram of GSR measurement

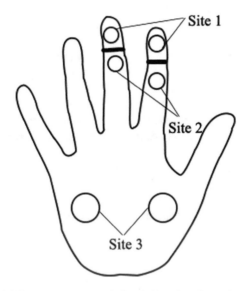

Fig. 2 Three sites of GSR measurement on the hand. Site 1 is volar surface on medial phalanges. Site 2 is volar surface of distal phalanges. Site 3 is thenar and hypothenar eminences

The rest of the work in the paper is arranged as: Sect. 2 presents circuit design configuration and experimental setup. Proposed circuit design is also included in this section. Section 3 shows readings of the proposed circuit. Finally, Sect. 4 describes the conclusion.

2   Circuit Design Configuration and Setup

The experiment setup consists of a subject to whom the Ag/AgCl electrodes were attached, GSR measurement circuit implemented on breadboard, Arduino board to serially transmit the output voltage from the operational amplifier, and a simple excel data acquisition system on a computer for recording and monitoring the change in voltage resulting due to changes in skin conductance level of the subject in real time. An LED has been used as an actuator to indicate whether the individual under monitoring was drowsy or not. For test purposes, the threshold of skin conductance below which the subject is drowsy was selected as 1.61 μS [7, 8].

2.1 Voltage Divider Method for SCL Measurement

In exosmotic measurement with direct current, two electrodes are attached to the subject's skin and connected in series with a reference resistor, as depicted in Fig. 3. Due to the skin's electrical properties, it acts as a resistor in the circuit [9]. A voltage source provides a constant voltage V. The change in the skin conductance is reflected as the change of the voltage through known resistor R. Our output was the measurement of voltage across R from which we calculated the conductance values and plotted them. The recommended limit of current density across skin is 10 μA/cm^2 and for calculation of the known resistance value R, this limit needs to be taken into consideration. By computation method, it was found that when $R = 270$ kW, current density through skin is around 10 μA/cm^2 over the entire range of skin conductance level.

When using direct current method for GSR measurement, the output from the fixed resistor is prone to variations. High frequency noise in the circuit also causes errors in measurement. We employed an active low pass filter of Sallen–Key architecture for its simple design, ease of parameter variation and good performance at low frequencies [10] to minimize the sources of error. As presented in Fig. 4, the low pass filter comprised of resistances R_1, R_2, R_3, R_4, capacitance C_1 and C_2 and an

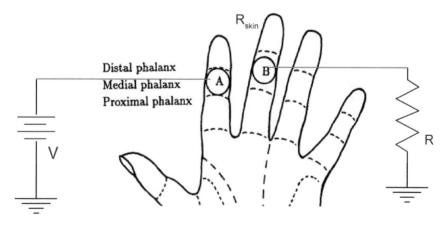

Fig. 3 Equivalent voltage divider circuit

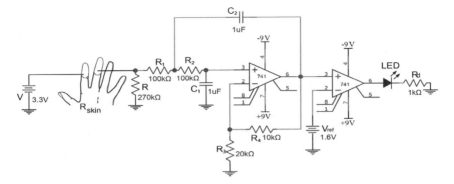

Fig. 4 Circuit diagram of the GSR measurement system

operational amplifier. The cut off frequency was selected to be 1.6 Hz which allows for a dynamic range of SCL to be recorded. The gain factor of the filter was 1.5 which resulted in a sufficient amplification of the output being measured. For a skin conductance of 1.61 μS, a corresponding reference voltage V_{ref} for actuation was calculated to be 1.5 V. If voltage output from the filter fell below V_{ref}, the LED was actuated. This mechanism was performed by a unity gain comparator made from the second operational amplifier as proposed in the circuit in Fig. 4. The resistance R_d was connected to the LED to absorb the incoming excess voltage for damage prevention of the LED.

3   Readings and Inferences

The EDA of 6 subjects was recorded at a sampling frequency of 10 Hz using the circuit designed. From Fig. 4, the voltage calculated from the filter output V_{out} is related to SCL. This was converted into skin conductance (SC) by evaluating the ratio among V_{out} and difference of voltage applied to the skin, i.e., $3.3V$ and V_{out}, multiplied by the fixed resistance R, that is $SCL = \frac{V_{out}}{(3.3V \times 1.5 - V_{out}) \times 270\,k\Omega}$. The chart for one of the subjects taken in the most ideal conditions is described here for evaluation. The mean temperature of the room during the monitoring was 29–32 °C and relative humidity was 72–89%. For uniformity and standardization of data representation, we plotted the z score of SCL values which reflects the distance of the value from the mean of the dataset in terms of its standard deviation. In Figs. 5 and 7, the outcomes of the subject in awake and drowsy state, respectively, are plotted. Contrary to our earlier selection of 1.6 μS for drowsy subjects, the lowest value of skin conductance observed was 4.623 μS. The sections shown in Figs. 6, 8 and 9 exhibits the duration in which the conductance value starts decreasing steadily approaching the threshold for drowsiness in both awake and drowsy scenarios. This steady decrease approximately conforms to a quadratic trend represented with a green and red trend line for awake and drowsy data, respectively. The range of skin conductance level as it decreases in this duration is 5.65–4.623 μS. It was observed that in the active case, certain time intervals values dropped as low as that when the subject was inactive. However, the subject reported being alert in this duration. It is seen in the graph in Fig. 6 that the up and down pattern of the EDA in this case differed from the one in case of the drowsy state of the subject as depicted in Figs. 8 and 9. The values dropped below 5.6 μS for the active case but increased to normal levels much faster in comparison with the case of inactivity. Table 1 lists the slope of the rate of change of these three quadratics and the recovery of zSCL after it, decreased below 0.5 in case of an awake subject is faster by 52.2–95.6% as compared to when the same subject is drowsy. The zSCL plot of the active state is unpredictable than the drowsy state. This is indicative of the level of activity which an active person's mental state goes through while doing the same task as a drowsy person.

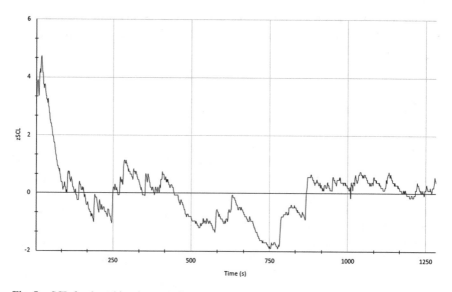

Fig. 5 zSCL for the subject in an awake state

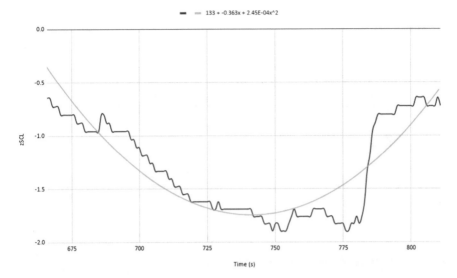

Fig. 6. zSCL section of the subject in an awake state when value falls below threshold

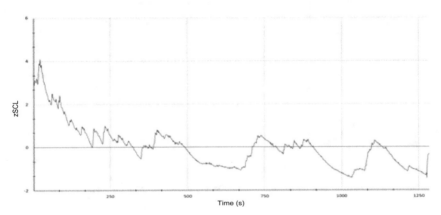

Fig. 7 zSCL for the subject in a naturally occurring drowsy state

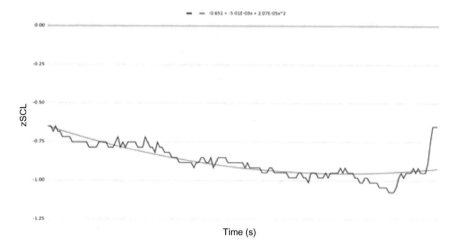

Fig. 8 zSCL section of the subject in drowsy state when value falls below threshold

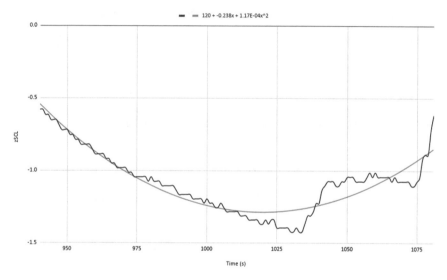

Fig. 9 zSCL section of the subject in drowsy state when value falls below threshold

Table 1 Slopes of the rate of change of quadratic trends in Figs. 6, 8 and 9	Figures	State	Slope of derivative
	Fig. 6	Alert	4.90×10^{-4}
	Fig. 8	Drowsy	0.214×10^{-4}
	Fig. 9	Drowsy	2.34×10^{-4}

4 Conclusion

The prototype circuit presented in this paper recorded trends in EDA of 6 subjects which conform to the normal range and pattern. The circuit designed measured a skin conductance greater by 3 μS than our originally selected threshold value for LED actuation. It was also seen that simply associating a threshold value to drowsiness assessment is insufficient for designing an alert system. The alert system will require a digital signal processing unit which will fire an algorithm at the designated threshold value of SCL to check the subsequent change in SCL and the slope of the change for a fixed time span to finally determine whether the driver is drowsy or not. One important advantage of analyzing the change in SCL response as compared to only analyzing instantaneous values is that relative humidity need not be considered as a factor for calibration of the final device. This will make the system more sophisticated.

References

1. Wheaton AG, Chapman DP, Presley-Cantrell LR, Croft JB, Roehler DR (2013) Drowsy driving-19 states and the district of Columbia, 2009–2010 (Reprinted from MMWR, vol 51, pp 1033–1037, 2013). JAMA-J Am Med Assoc 309(8):760–762
2. Higgins JS, Michael J, Austin R, Akerstedt T, Van Dongen HP, Watson N et al (2017) Asleep at the wheel: a national compendium of efforts to eliminate drowsy driving. Sleep 40(2)
3. Braithwaite JJ, Watson DG, Jones R, Rowe M (2013) A guide for analysing electro-dermal activity (EDA) & skin conductance responses (SCRs) for psychological experiments. Psychophysiology 49(1):1017–1034
4. Cromwell L, Weibell FJ, Pfeiffer EA, Usselman LB (1973) Biomedical instrumentation and measurements. In: Biomedical instrumentation and measurements. Prentice-Hall, Inc., Englewood Cliffs, 457p
5. Dawson ME, Schell AM, Filion DL (2017) The electrodermal system
6. Karki J (2000) Active low-pass filter design. Texas instruments application report
7. Gao ZH, Fan D, Wang D, Zhao H, Zhao K, Chen C (2014) Muscle activity and co-contraction of musculoskeletal model during steering maneuver. Bio-Med Mater Eng 24(6):2697–2706
8. Malathi D, Jayaseeli JD, Madhuri S, Senthilkumar K (2018) Electrodermal activity based wearable device for drowsy drivers. J Phys Conf Ser 1000(1):012048
9. Edelberg R, Greiner T, Burch NR (1960) Some membrane properties of the effector in the galvanic skin response. J Appl Physiol 15(4):691–696
10. Poh MZ, Swenson NC, Picard RW (2010) A wearable sensor for unobtrusive, long-term assessment of electrodermal activity. IEEE Trans Biomed Eng 57(5):1243–1252

A Vocabulary-Based Framework for Sentiment Analysis

Shelley Gupta, Urvashi, and Archana Singh

Abstract The variety of knowledge analysis that is earned from the news reports, user reviews and social media updates or a small blogging website is named opinion mining. Its associated approaches are employed to analyse the sentiments provided as input file. The reviews of people towards events, brands, product, or company are often legendary part of sentiment analysis. The responses of general public are collected and measured temporarily by researchers to perform evaluations. The recognition of sentiment analysis is growing these days since the numbers of views being shared by folks on the small blogging sites are increasing. All the emotions are often categorized into three totally different classes referred to as positive, negative and neutral. We have used lexicon-based approach to determine the scores of sentiment sentences. Python language is employed during this analysis to implement the classification algorithmic programme on the collected knowledge.

Keywords Sentiment analysis · Opinion mining · Text · Vocabulary · Lexicon

1 Introduction

Human activities are influenced by opinions because they are key drivers of human behaviour. When making a decision, we always want to know other people's opinions. Businesses and organizations always want to know what the public thinks of their products and services [1]. A number of online shopping websites, e-commerce websites, social networking websites and other web resources provide platforms for expressing sentiments, which plays a very important role in understanding the sentiments of the online users and customers regarding economic, political, social, etc. issues [2–4]. For three decades, research scholars, intellectuals and professors

S. Gupta (✉) · Urvashi
ABES Engineering College, Uttar Pradesh, Ghaziabad, India
e-mail: shelley.g17@gmail.com; shelley.gupta@abes.ac.in

A. Singh
Amity University, Noida, India
e-mail: asingh27@amity.edu

have been working consistently on sentiment analysis to accomplish these tasks. Sentiment analysis (SA) is a calculation and data analysis of the attitude, feelings, thoughts and conceptions depicted in content towards an entity. It is to identify, extract and model attitudes expressed in texts. Sentiment analysis aids in accomplishing multiple objectives like scrutinizing common people emotions regarding political atmosphere, market perceptions, the quantification of customer gratification, cinematics sales forecasting and many more. Sentiments and reviews evaluations are becoming greatly noticeable due to increased importance of e-commerce, which is an important medium of addressing and analysing sentiments. The users of electronic commerce websites mainly rely on analysis of reviews published by current customers to determine the quality products. However, product developers and service providers consider the re-views posted by customers on the social media very important for improving the performance and standards of their products and services. Customer's opinion is greatly impacted by the reviews of customers given over popular e-commerce websites [5, 6].

2   Literature Review

Conclusion examination manages trademark and grouping feelings or slants that square measure blessing in text. Online media is producing a huge amount of supposition well off records inside such a tweet, standing updates, surveys diary posts and so on. Conclusion examination of this client created information which is very useful in knowing the assessment of the group. AI approaches are frequently utilized for examining conclusions from the content. Some slant examinations measure performed by breaking down the twitter posts with respect to electronic item like PDAs, PCs and so forth utilizing machine learning approach. By acting slant investigation in an unmistakable area, it's feasible to recognize the effect of space information in assessment arrangement. They offered a trade highlight vector for ordering the tweets as certain, pessimistic or nonpartisan and concentrate individuals' sentiment with respect to item.

Another examination attempted to pre-handle the data set, subsequently removed the descriptor from the data set. AI approaches like Naïve Bayes [7], most Entropy and SVM [8] along the edge of the phonetics orientation based generally WordNet [9] that separates equivalents and connection for the substance include. Towards the end, they estimated the presentation of classifier as far as review, exactitude and precision.

A few specialists had partner approach that declare tweets from the Twitter small-scale blogging site exposed to pre-handling and characterized, upheld their enthusiastic substance as sure, negative and unbiased or superfluous and analyses the presentation of arranged ordering calculations upheld their precision and review in such cases. Further, the paper moreover examines the utilizations of this investigation and its constraints. Assortment of AI like Naïve Bayes and Random Forest models performed assessment investigation on item survey data.

Some incorporate this field encased tests with perspective plan on journal posts. One in all the investigates likewise oversees review of viewpoint-based evaluation studying from untagged free-form matter client reviews while not anticipating that customers should answer any requests. The tweet recuperation procedure needs access tokens from the twitter fashioner site, and a dash of code that plays out the action of recouping those tweets. Opinion mining is a delightful examination subject because of knowing an individual's good or negative emotions from the content he/she composes is an essential present data assembling and choosing. For example, inescapable informal organizations' sites, (for example, Twitter, Facebook and YouTube) urge their clients to add to their substance by posting their remarks concerning various parts of life that produce an outsized stream of information. Among the advantages of applying opinion mining on such information are [1–4, 10, 11]:

1. **Online client feedback**: Merchandise and services would like continuous testing and improvement. This method will extremely like client reviews or feedbacks. These reviews can offer indication concerning usefulness of products and services. Creating this method automatic and online can save time and cash to e-commerce websites and improve their client support and relations.
2. **Belief polls**: Opinion mining helps call manufacturers to urge overall read about community and its desires by observation folks reviews on social networks.
3. **Advertising**: Blogs square measure increasing on World Wide Web and anyone will reach to what post on blogs. This could create them helpful to post advertisements. We utilized a mixture of qualitative and quantitative ways to develop, and so by trial and error verify, a gold-standard sentiment lexicon that's particularly linked to small journal like text. They combine these lexical options considerately for five hypothesize rules that demonstrate semantic and syntactic arrangement that persons utilize for addressing or accentuation sentiment magnitude.

   - **Market sentiment**: Large quantity of information is arisen in the international monetary market for each second. It is compulsory for capitalist to stay on prime of such information in order that they'll administer their portfolios productively. The sentiment analysing tools provide a technique of estimating the merchandise sentiments.
   - **Gauging the recognition of the merchandise, books, movies, songs and services** by examining the info produced from completely distant sources such as online feedback, tweets, mannerisms (swipes and likes), etc.

Sentiment analysis make use of various machine learning algorithms as shown in Fig. 1. The sentiment analysis approach mainly uses the bag-of-words (BoW) [12] illustration. BoW technique considers the sovereignty of lexicons but evades the significance of linguistics along with subjective info in the text [13]. Entire words within the text area unit thought of uniformly influential. The BoW illustration is often utilized for sentiment analysis, directing into great spatial property of the feature area. Machine learning algorithms diminish this high-dimensional feature area with the assistance of feature selection approaches that decides solely necessary options

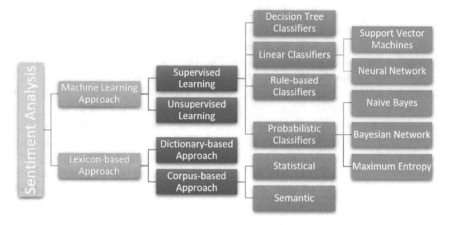

Fig. 1 Techniques of sentiment analysis [15]

by removing the hissing and tangential options. Recently, machine learning-based sentiment analysis approaches area unit encouraging prominence within the field [14].

Use of a vocabulary is during one in everything about 2 primary ways to deal with conclusion investigation, and it includes calculative the feeling from the semantics direction of word or expressions that happen in a book. With this methodology, a wordbook of good and negative words is required, with a good or negative estimation worth dispensed to everything about words. Entirely, unexpected ways to deal with making word references are anticipated, along with manual and programmed approaches. Commonly speaking,in vocabulary-based methodologies, a touch of instant message is drawn as a sack of words [16].

In our work, we've set to utilize a vocabulary-based methodology [17] to maintain a strategic distance from the necessity to get a labelled training set. The most impediment of AI models is their dependence on labelled data. Itis amazingly problematic to affirm that adequate and accurately named information can be acquired. Other than this, the established truth that a dictionary-based methodology can be all the more essentially comprehended and changed by an individual's is considered a huge favourable position for our work. We tend to establish it simpler to get a fitting vocabulary than gather and mark, and applicable corpus gave that the data from online media region unit made by clients from wherever in the world.

3   Proposed Framework

Our approach Fig. 2 determines to influence the benefits of penurious rule-based modelling to build a machine learning-based sentiment analysis framework [8] that:

Fig. 2 Flow chart of
proposed framework

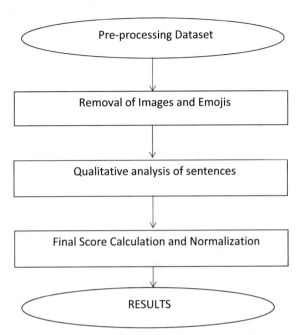

(1) Its endeavour well on social media vogue sentiment sentences, however, pronto
    applicable to numerous domains,
(2) It does not need any coaching knowledge, however, has natural language lexicon
    based on a generalizable, valence-based and man-organized gold customary
    approach,
(3) Adequately speedy for utilization with social media streaming information, and
(4) It has not experienced diminishing velocity attainment.

The approach [17] has started by building a listing and examining of galvanized
well-established sentiment word banks. To this, we tend to later amalgamate various
lexical options prevalent in online opinion sentences of small blogs. It incorporates
the entire list of Western-style emojis and emoticons like ":D" presents a "laugh" and
customarily illustrate positive polarity opinion), online sentiment-related acronyms
and initialisms (e.g. LOL and WTF area unit both sentiment-laden initialisms), and
ordinarily used slang with sentiment worth (e.g. "nah", "meh" and "giggly").

We next dissect a deliberate example of 400 positive and 400 negative online
media text bits from a greater starting arrangement of 10 K irregular posts pulled from
open timetable, which upheld their supposition scores and exploitation the example
slant investigation motor. Example could be a net digging module for Python, and
furthermore, the pattern module could be a phonetic correspondence measure (NLP)
toolbox that use WordNet [9] to achieve conclusion with regards to the English
descriptors utilized in the content.

We next used analysis techniques to spot properties and characteristics of text that
have an effect on the perceived sentiment intensity of text [17].

- Punctuation, explicitly the accentuation (!), will build the extent "The food here is sweet"!!! is a ton of serious than "The food here is acceptable".
- Capitalization, explicitly exploitation ALL-CAPS to pressure a feeling applicable word inside the nearness of option non-uppercase words, will expand the greatness of the opinion force. "The food here is GREAT"! passes on a ton of force than "The food here is pleasant!"
- Degree modifiers (otherwise called intensifiers, promoter words or degree qualifiers) sway sentiment force by either expanding or diminishing the power. For example, sufficient, "The administration here is awesome" is a great deal of intense than "The administration here is acceptable", though "The administration here is possibly acceptable" lessens the power.
- The contrastive combination "yet" signals a move in assumption extremity, with the opinion of the text following the combination being prevailing. "The food here is incredible, anyway the service is frightful" has blended slant, with the last half directing the overall rating.
- Negation flips the polarity of the text. A negated sentence would be "The food here isn't extremely all that great".

4 Result

Table 1 depicts the scoring based on the five conditions mentioned in the proposed framework [17].

Table 1 Scores

Comments	Positive score	Negative score	Neutral score	Overall score
I love the ambience of this place!!!	0.458	0.0	0.542	0.72
We are going to have so much fun here!!!!!	0.358	0.0	0.64	0.66
I LOVE this place	0.622	0.0	0.378	0.71
The food here is AMAZING	0.531	0.0	0.469	0.6739
You look extremely gorgeous in this dress	0.4169	0.0	0.583	0.647
This weather too good for roaming outside	0.326	0.0	0.674	0.44
The ambience is good but food is horrible	0.154	0.374	0.4719	-0.585
I would love to go with you but I do not like that crowd	0.17	0.175	0.655	-0.0168
I do not wish to come here again	0.15	0.374	0.4719	-0.585
I do not like the vibe coming off her	0.170	0.175	65.5	-0.178

- **Punctuation**: The first two comments in the table depict that when punctuations are inserted at the end of the comments, the scoring tends to gravitate towards the positive side more.
- **Capitalization**: When the capitalization is used in the comments, it intensifies the positive or negative aspect of the score.
- **Degree Modifiers**: They are also known as *intensifiers* or *booster words*. They intensify the positivity or negativity of the sentences to a great extent.
- **Contrastive Conjunction**: The use of contrastive conjunction *"but"* flips the polarity of the sentence.
- **Negation**: It catches almost all of the cases where negation *"not"* is used in the sentence.

5 Conclusion

We report the deliberate turn of events and investigation of our methodology. Utilizing a blend of subjective and quantitative ways, we build and through exact perception approve a gold-ordinary rundown of lexical that are explicitly receptive to notion in little blog-like settings. We at that point consolidate these lexical alternatives obligingly for five general principles that exemplify syntactic and syntax shows for communicating and underlining notion power. The outcomes aren't exclusively reassuring they are so very wonderful. Technique acted furthermore as option amazingly considering feeling investigation apparatuses.

6 Future of Opinion Mining

In future, the effect of arranged elective element decrease strategies like idle Dirichlet distribution might be examined. More tests should be led inside the future to pass judgement on a hundred and fifty effect of grouped area and district explicit boundaries. Stretching out notion mining to elective areas may bring about entrancing new outcomes. In future, the use of a great deal of mix of n-grams and have coefficient that offers a higher exactness level than this can be suspected of. The work depleted this examination is simply connected with grouping feeling into two classifications (dual characterization) that is a good classification and bad classification. Inside the future turn of events, a multiple classification of conclusion clustering like positive, negative and independent so on may be taken into thought. During this work, the primary objective is on discovering major points that appear explicitly as expressions of social media. The finding of understood alternatives is left to future work. As troupe learning strategies might want huge amounts of figuring time, and equal registering methods ought to be handle these drawbacks. A huge impediment of outfit learning ways is that the absence of interpretability of the outcomes, and furthermore, the information learned by troupes is inconvenient for people to know.

Along these lines up, the interpretability of outfits is another imperative exploration heading. Forthcoming sentiment burrowing structure might want more extensive and more profound normal and standard sense data bases. This may bring about a superior comprehension of phonetic correspondence assessments and can a ton of quickly overcome any issues between multimodal information and machine processable information. Blending logical contemplation of emotions in with the reasonable designing objectives of breaking down assumptions in semantic correspondence text can bring about a ton of bio-enlivened ways to deal with the arranging of insightful sentiment mining frameworks fit for taking care of phonetics data, making analogies, learning new emotional data, and sleuthing, seeing, and "feeling" feelings [18].

References

1. Liu B (2012) Sentiment analysis and opinion mining. Synth Lect Hum Lang Technol 5(1):1–167
2. Gupta S, Garg O, Mehrotra R, Singh A (2021) Social media anatomy of text and emoji in expressions. In: Smart computing techniques and applications. Springer, Singapore, pp 41–49
3. Mehrotra R, Garg O, Gupta S, Singh A (2022) Opinion mining of pandemic using machine learning. In: Advances in data and information sciences. Springer, Singapore, pp 225–231
4. Gupta S, Singh A, Ranjan J (2022) Online document content and emoji-based classification understanding from normal to pandemic COVID-19. Int J Performability Eng 18(10)
5. Liu B, Zhang L (2012) A survey of opinion mining and sentiment analysis. In: Mining text data. Springer, Boston, pp 415–463
6. Ravi K, Ravi V (2015) A survey on opinion mining and sentiment analysis: tasks, approaches and applications. Knowl-Based Syst 89:14–46
7. Mukherjee S, Sharma N (2012) Intrusion detection using naive Bayes classifier with feature reduction. Procedia Technol 4:119–128
8. Manek AS, Shenoy PD, Mohan MC, Venugopal KR (2017) Aspect term extraction for sentiment analysis in large movie reviews using Gini Index feature selection method and SVM classifier. World Wide Web 20(2):135–154
9. Fellbaum C (2010) WordNet. In: Theory and applications of ontology: computer applications. Springer, Dordrecht, pp 231–243
10. Gupta S, Bisht S, Gupta S (2021) Sentiment analysis of an online sentiment with text and slang using lexicon approach. In: Smart computing techniques and applications. Springer, Singapore, pp 95–105
11. Gupta S, Singh A, Ranjan J (2020) Sentiment analysis: usage of text and emoji for expressing sentiments. In: Advances in data and information sciences. Springer, Singapore, pp 477–486
12. Zhang Y, Jin R, Zhou ZH (2010) Understanding bag-of-words model: a statistical framework. Int J Mach Learn Cybern 1(1–4):43–52
13. Agarwal B, Mittal N (2016) Machine learning approach for sentiment analysis. In: Prominent feature extraction for sentiment analysis. Springer, Cham, pp 21–45
14. Agarwal B, Mittal N (2016) Prominent feature extraction for sentiment analysis. Springer International Publishing, Berlin, pp 21–45
15. Devopedia (2020) Sentiment analysis. Devopedia. Retrieved 11 Sept 2020, from https://dev opedia.org/images/article/105/8215.1532752754.png
16. Rice DR, Zorn C (2013) Corpus-based dictionaries for sentiment analysis of specialized vocabularies. Polit Sci Res Method 1–16

17. Gilbert CHE, Hutto E (2014) Vader: a parsimonious rule-based model for sentiment analysis of social media text. In: Eighth international conference on weblogs and social media (ICWSM-14). Available at 20 Apr 16, vol 81, p 82. http://comp.social.gatech.edu/papers/icwsm14.vader.hutto.pdf
18. Dagar V, Verma A, Govardhan K (2021) Sentiment analysis and sarcasm detection (using emoticons). In: Applications of artificial intelligence for smart technology. IGI Global, pp 164–176

Decentralized Framework to Strengthen DevOps Using Blockchain

Sandip Bankar and Deven Shah

Abstract DevOps is the combination of software development management tools and cultural philosophies and practices that focus on cross-departmental integration and automation. Conflicts between the development and operations teams frequently occur when software development teams are focused on ensuring that consumers receive software. Concerns over privacy and security have also arisen. In the diverse setting of fast software development and deployment, the suggested system DevOp-sChain seeks to foster confidence in individuals who are difficult to trust and resistant to change. It does this by deploying DevOps with real business and technology constraints. Smart contracts and blockchain technology can be used to control how project artifacts move and who can see them. This makes sure that transactions are safe and secure. The proposed system is built on the decentralized Hyperledger Fabric Blockchain. The proposed framework appears to function effectively when the block size, endorsement rules, and blockchain characteristics are adjusted, according to an evaluation of its performance.

Keywords Blockchain · Software development cycle · DevOps · Agile development

1 Introduction

Software development techniques are necessary in order to improve the software development process because software changes quickly, it's hard for organizations working on software projects to choose the right software development method [1]. They have worked on making software practices faster and more reliable. The Waterfall model, was the first software development methodology, was followed by the agile model [2, 3]. Businesses face problems with agile methods when it's hard to

S. Bankar (✉)
SVKM's NMIMS Navi Mumbai, Mumbai, Maharashtra, India
e-mail: sandipbnkr@gmail.com

D. Shah
LR Tiwari College of Engineering, Mumbai, Maharashtra, India

A. Shukla et al. (eds.), *Computational Intelligence*, Lecture Notes in Electrical Engineering 968, https://doi.org/10.1007/978-981-19-7346-8_44

figure out what their customers' changing needs are. DevOps is an IT culture that grew out of the agile method to improve how different stakeholders in software development and operations work together at all stages of the software's lifecycle [4]. But businesses that are usually in a central location may not always be able to use the services. If a document is changed, the most up-to-date version should be easily accessible to everyone who has a stake in the process. Businesses nowadays are being impacted by the DevOps culture, which encourages production servers to release more software often [5]. Security is receiving less attention in this rush to increase the number of releases every day and hour. Security teams use software to keep things safe, but software that is quickly deployed is more likely to have flaws as the number of files increases. Artifacts, configuration files, and code repositories are important parts of a project because they hold important private information. This information needs to be shared regularly with the software development team's stakeholders. It is hard to share and keep track of this data because it needs to be used and updated by the right people. It also helps stakeholders trust each other [6]. So, the current development environment needs new ideas, like DevOps, so that any changes to or access to a file can be tracked and safely stored. To avoid request bottlenecks and keep DevOps processes running smoothly, files need to be stored on a secure platform and in a decentralized way [7]. DevSecOps was made so that security could be built into each step of software development to reduce security holes. For DevSecOps to work well, all the stakeholders in a business need to work together more, even though some stakeholders are still hesitant to trust every other stakeholder or organizational conflicts tend to happen [8]. Given the importance of security and privacy, there may be a need for blockchain-enabled secure, which can provide a history of every access and updates [9]. Taking these challenges into account, Hyperledger Fabric [10] permissioned private blockchain is used, which is a type of blockchain technology.

The rest of this paper is divided into the following five sections. In Sect. 2, the paper talks about the work that goes with it. Section 3 goes into detail about the proposed framework. In Sect. 4, we talk about how it was put into action and what happened with a performance analysis. Section 5 is the last part of the report. It is a summary of the research.

2   Related Work

Although increasing the number of releases through DevOps is getting more and more popular, security is sometimes neglected. Many authors have written on DevOps issues in recent years, but few have attempted to make DevOps operate better. Few authors have talked about the problems with software development practices and pointed out many problems and research gaps [11, 12]. In research paper [12], the authors focused on how to build applications and test them over time, how to make development and testing results more visible and known in Continuous Integration (CI) and how to find mistakes and violations. They have encountered issues with security and scalability pipelines when attempting to increase the dependability and reliability of the deployment process. After continuous testing, researchers have added a security gate to DevOps using the tool Sync to check for dependencies and widely known vulnerabilities in an effort to increase security [13].

In a research paper [12], the results and problems of DevOps are looked at from the point of view of five different kinds of businesses. They said that the adoption and use of DevOps in organizations was not important because it required changes in the industrial, structural, and social sectors. Problems at DevOps include a high demand for information, unclear roles between the development and operations teams, difficulty convincing higher-ups, and a lack of general understanding. Authors Paez et al. have focused on the uniqueness of artifacts, versioning tools, naming conventions for different types of versions, and the ability to track changes to artifacts. Also, source code files, documentation, and artifacts are put under version control to keep a semantic versioning approach [14]. A study on the problems that make it hard for large companies to use DevOps found five key ones. Adapting to DevOps in large enterprises can be hard because of problems with communication and collaboration, resistance to change, the strategic plan staying with senior management, and a lack of information about DevOps [15].

Authors Kamuto et al. tried to improve DevOps practices by adding security layers, but they didn't solve all of DevOps' problems [15]. To improve security, there needs to be a place where changes that aren't supposed to happen or are illegal can be found and fixed [16]. So, the objective of this work is to create an atmosphere to enhance software development speed and efficiency without disturbing regular DevOps practices.

3   System Architecture for the Proposed Framework

The proposed DevOpsChain framework for DevOps software development using blockchain is shown in Fig. 1. Presented framework is the DevOps stakeholders-centric framework, which takes into account the different types of users, such as the Administrator, Director, Project Manager, Developer, and Tester. In every DevOps phase, documents are generated and get stored on Blockchain. To communicate with applications and peers, the DevOpsChain framework employs Certificate Authority (CA) as membership provider services and user interfaces. Peers use consensus mechanisms, they hold smart contracts, blockchains, and state databases.

When a client signs up for the membership system service, the administrator makes sure that the client is a real person. A certification authority is used by the membership service to make key pairs that are used to sign and encrypt the keys for each node. Anyone with a stake in the project can do things like create the project, update files, upload files, and view files [7]. Every peer node that is part of the network has a blockchain called a "ledger" that stores transactions that can't be changed and the chaincode business logic. The client makes a request to create a file, get a file back, or change a document. The endorsing peer then prepares the results by

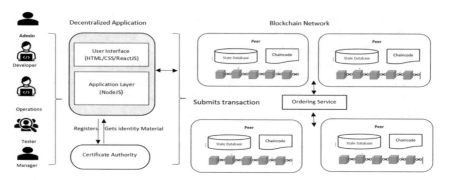

Fig. 1 Architecture of the proposed DevOpsChain system

executing the transaction proposal, and read–write endorsement responses are sent back to the requester client. Then, the client sends the read–write set to the orderer as a transaction response. The orderer runs the ordering service on the inside. The ordering service notifies the individuals who placed the orders to create a block after accepting the transactions that have been accepted. The block is sent to all of the peers as soon as it is created. In this scenario, the blockchain must be used to store all huge data files. However, this significantly slows down the entire blockchain system. So, just the hash of each file is maintained in the blockchain and the InterPlanetary File System (IPFS), a distributed file storage system for huge files, is used.

4 Implementations and Results

This suggested DevOpsChain framework is built utilizing a permissioned blockchain platform that makes use of Hyperledger Fabric and Hyperledger Composer for effective file tracking and storage, as well as greater data privacy and security. This web application's front end is built using HTML-CSS, JavaScript, and ReactJS. The suggested system is divided into two components: continuous operations and continuous development. Proposed system can store software development repositories and reports to blockchain in the complete project cycle. Files accesses and modification history are maintained on immutable storage so that any update or access can be easily tracked and availability of documents is increased.

4.1 Continuous Development

Continuous development covers the DevOps stages: planning, continuous development, and version control. To set up continuous development in Hyperledger, need to set up participants, transactions, assets, and rules for controlling access. In this

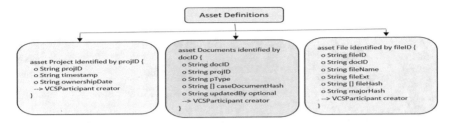

Fig. 2 Assets definitions

case, the logic demanded for a transaction processor function, and code written in JavaScript is used.

In this case, the project artifacts, files, and projects are the assets, and the people who are working on the project are the participants. Figure 2 shows the definitions of the assets used in the model that is being proposed.

A project manager creates the project, adds any user to the project, and updates the files. In the Create User operation, initially, the admin creates any type of user, and then the user gets certificates. Using a certificate, a user can login to the system. When several developers work on the same file, it becomes very critical to manage updates made by developers, and then it needs to be resolved. So, blockchain-based version control is created to maintain versions separately as per the versions. In the continuous development phase of DevOps, this phase is about how version control is used to manage project repository versions. In this system, several branches are created, which are managed by a blockchain-based version control system, so that any update or change is stored in immutable storage.

4.2 Continuous Integration and Delivery

Continuous operations include integration, deployment, and delivery that happen all the time. Once a developer writes or changes code, it needs to be added to the main project build or repository. In this phase, an integration server based on a blockchain is used, as shown in Fig. 3. In this case, a smart contract makes sure that changes to the code will go through the steps for a production release. When a developer submits code to blockchain-based version control many times during the day, chaincode (smart contract) will observe changes in blockchain-based source code management. A dedicated form worker would get the source code whenever the modification was discovered. Then, at that point, the completed form application is sent to the test workers for the User Acceptance Test (UAT).

With permissioned blockchain, all the people who have a stake in the software development process—developers, testers, managers, and clients—are brought together on a single platform. It helps protect the content of the software development process, its updates, and its history so that there isn't a blame game. It also

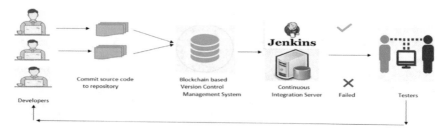

Fig. 3 Architecture for Continuous Integration

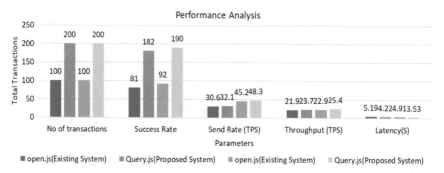

Fig. 4 Results of proposed system

improves the content's integrity without stopping the work of the different DevOps tools. Figure 4 shows how the whole DevOpsChain project development process works. The whole work is made up of development and operations. Developers write code repositories that will be stored in version control systems based on blockchain that can't be changed. Once the code is successfully moved to the main branch, it will be integrated and built by the integration server using blockchain-based version control. It will then be deployed and released to the production server.

4.3 Significances of Proposed System

Proposed DevOpsChain framework uses the blockchain technology to protect DevOps documents without disturbing DevOps phases. In every phase where DevOps generates documents, they will be tracked and stored on decentralized blockchain environments that will create trust and improve the availability of documents. The DevOpsChain framework helps to close the gap between the development and operations teams by using blockchain and DevOps, which are two of the most promising technologies. Here, the new idea is aimed at a lot of important things. It includes pluggable, consensus-based, and secure storage of DevOps documents in the distributed ledger of blockchain technology as a form of secure data access. The integrity of

DevOps documents needs to be protected in these ways. This helps make the documents and their changes more open and accountable [17]. A DevOps document with all the relevant information, including a project ID number, is saved to the blockchain whenever any of the DevOps stages receive new data. These papers were stored using a variety of DevOps technologies before being transferred to the blockchain. Adding security was hard because manual processes and third-party tools were needed.

Significances of Proposed System

- All software development files are maintained on the blockchain so that their access and updates are traced.
- All stakeholders from different DevOps tools that are in different places are brought together on one platform.
- It helps keep DevOps from turning into a game of "who's to blame?" by building trust among all stakeholders.
- It keeps you from having to use security testing tools and do security checks by hand.
- It helps reduce the skills gap by making all the most recent changes available to all stakeholders, which speeds up the software delivery process.
- It helps address performance bottlenecks through distributed configuration management and decentralized log management.

4.4 Performance Evaluation

In this section, blockchain assessment measurements are used to test how well the current system works. Performance is monitored using the Hyperledger Caliper benchmarking framework, and the outcomes are examined using criteria like Success Rate, Send Rate (TPS), Throughput (TPS), and Latency (S) [18, 19]. They loaded it with 100,200 open round transactions and query transactions with three sub-rounds that each had a separate rate control in order to evaluate how the proposed framework would function. Performance is compared using Success Rate Send Rate (TPS), Throughput (TPS), and Latency (S). The Success Rate displays the proportion of transmitted transactions that were processed and added to the blockchain. The Transaction/Read Throughput (TPS) of your blockchain network measures how many transactions (or requests) it can process in a single second. Transaction/Read Latency (s) metrics are based on how long it takes for a client to submit a transaction or query, for it to be processed, and for it to be written to the ledger. The Send Rate (TPS) shows how many transactions were sent to the network successfully. Based on the system with Hyperledger Fabric 1.4 network [7], the proposed system is compared to the system that is already in place.

The graphical representation of Fig. 4 shows proposed method performs better than existing system in terms of transaction Throughput, Success Rate, and Latency when tuned with block size, endorsement policies, and blockchain parameters.

5   Conclusion

This work presents a decentralized blockchain-enabled framework with the aim of not only building trust in DevOps software development but also enhancing the security of software documents. Implementation is done using Hyperledger Fabric and Hyperledger Composer. By adding permissioned blockchain to the DevOps culture, all software documents are now easily accessible, and changes to project files can be tracked, audited, and can't be changed, so there's little to no room for fraud. In turn, this makes an organization more honest and productive. Success Rate, Send Rate (TPS), Throughput (TPS), and Latency are also used to measure performance (S). Tuning the blockchain, block size and endorsement parameters can improve how well the network works. In this paper, blockchain technology is used to improve the DevOps method of software development. In the future, the authors want to keep this idea alive by making it possible for different blockchain platforms to work together.

References

1. Kumar G, Bhatia PK (2014) Comparative analysis of software engineering models from traditional to modern methodologies. IEEE, pp 162–167. ISBN: 978-1-4799-4910-6/14 $31.00
2. Sharma S, Sarkar D, Gupta D (2012) Agile processes and methodologies: a conceptual study. Int J Comput Sci Eng 4(05):892. ISSN: 0975-3397
3. Sureshchandra K, Shrinivasavadhani J (2008) Moving from waterfall to agile. IEEE, pp 97–101. ISBN: 978-0-7695-3321-6
4. Banica L, Radulescu M, Rosca D, Hagiu A (2017) Is DevOps another project management methodology? Inf Econ 21(3/2017):39
5. Roche J (2013) Adopting DevOps practices in quality assurance. Commun ACM 56(11):38–43
6. Ahmed Z, Francis SC (2019) Integrating security with DevSecOps: techniques and challenges. In: International conference on digitization (ICD), pp 178–182
7. Bankar S, Shah D (2020) DevOps project artifacts management using blockchain technology. In: ECAI&ML international conference, pp 115–120. ISBN: 978-1-7369640-5-7
8. Bashir I (2018) Mastering blockchain. PACKT Publishing, Birmingham, pp 10, 12, 16–23, 27–32, 35–38, 44, 359, 362–373
9. Dhillon V, Metcalf D, Hooper M (2016) The hyperledger project. In: Blockchain enabled applications. Springer, pp 139–149
10. Hyperledger composer (2017) https://hyperledger.github.io/composer/v0.19/installing/instal ling-index
11. Shahin M, Ali Babar M, Zhu L (2017) Continuous integration, delivery and deployment: a systematic review on approaches, tools, challenges and practices. IEEE Access
12. Lwakatare LE, Kilamo T, Karvonen T, Sauvola T, Heikkila V, Itkonen J, Kuvaja P, Mikkonen T, Oivo M (2019) DevOps in practice: a multiple case study of five companies. Inf Softw Technol 114:217–230
13. Ur Rahman AA, Williams LA (2016) Software security in DevOps: synthesizing practitioners' perceptions and practices. In: IEEE/ACM international workshop on continuous software evolution and delivery-CSED, pp 70–76
14. Paez N (2018) Versioning strategy for DevOps implementations. In: Congreso Argentino de Ciencias de la Informática y Desarrollos de Investigación (CACIDI), pp 1–6

15. Kamuto MB, Langerman JJ (2017) Factors inhibiting the adoption of DevOps in large organisations: South African context. In: 2017 2nd IEEE international conference on recent trends in electronics, information & communication technology (RTEICT), pp 48–51
16. Agrawal P, Rawat N (2019) DevOps: a new approach to cloud development & testing. In: International conference on issues and challenges in intelligent computing techniques (ICICT), pp 1–4
17. Mukne H, Pai P, Raut S, Ambawade D (2019) Land record management using hyperledger fabric and IPFS. In: 10th international conference on computing, communication and networking technologies (ICCCNT), pp 1–8
18. Hyperledger caliper (2018) HyperledgerCaliper. https://hyperledger.github.io/caliper/v0.4.2/getting-started/#architecture
19. Bankar S, Shah D (2022) Integration of hyperledger fabric blockchain in software development. Int J Eng Res Technol (IJERT) 11(01)

Application of Machine Learning for Analysis of Fruit Defect: A Review

Siddharth Tulli and Yogesh

Abstract Machinery analysis can be very fruitful in agricultural field, as defect prophecy can be done in earlier stages of fruit maturity which can help farmers in taking necessary steps. Even there are some seasonal fruits which are sorted in nonuniform manner (in refer to fruit maturity). So, computerized analysis can help in detecting: accuracy of fruit maturity, defect, quality level and even predictions of a harvest can be made. Increase in machine learning can also increase consumer choices and preferences as well as help farmers in saving time and money. This paper provides the detailed overview of various methodologies used for fruit analysis, like surface color method and feature selection, in which few supplementary models are being used for example: RGB color model with HSI model and many more. At last, in this paper, we will conclude some better methods with algorithms than can show better results.

Keywords Machine learning for fruit defect · Application · Agriculture · Computerized analysis · Methodology

1 Introduction

Since, agriculture plays the very crucial role in economy as it not only offers food and raw material, it also gives employment to various fields, specially, to traders and retailers. And for them, customer's satisfaction is the main key for their business. So, for them, providing qualitied fruits (raw food items) at reasonable or lower rates become essential. But it is tough to check each and every fruit manually and even if they try, then as well the accuracy of fruit quality will lag and rate will also increase.

S. Tulli
Department of Information Technology, Amity School of Engineering and Technology, Amity University Uttar Pradesh, Noida 201313, India

Yogesh (✉)
Department of CSE, Chitkara University Institute of Engineering & Technology, Chitkara University, Rajpura, Punjab, India
e-mail: eceyogesh@gmail.com

© The Author(s), under exclusive license to Springer Nature Singapore Pte Ltd. 2023   527
A. Shukla et al. (eds.), *Computational Intelligence*, Lecture Notes in Electrical Engineering 968, https://doi.org/10.1007/978-981-19-7346-8_45

Hence machinery analysis can be very fruitful in this field. This will not any save time but this will escalate the quality check. Defect prophecy can be done at early stages of fruit maturity because of which farmers can take necessary steps to control the further crop failure; moreover, extra use of fertilizers can be prevented because increase in fertilizers may lead to heavy metal/chemicals accumulation in fruits. There are seasonal fruits which are sorted in nonuniform manner (with respect to fruit maturity), with increase in technology proper segregation can be done. Crop prediction can also be done by some analysis.

With help of machine learning, we can develop a system in which fruit checking and defect analysis can be done automatically. In this, we need to create a database of fruit analysis, and with support of algorithmic expressions, this can be achieved. Texture, size and color are the main parameters for fruit quality identification. There are numerous methodologies which are used for inspection, and their databases are being managed. Further, in this paper, different methodology are being discussed like surface color analysis, feature selection, feature extraction, tomography and automation process. In this paper, apple fruit is being discussed, but these processes can be done for any fruit. At last, paper is concluded with better method.

2   Literature Review

Machine learning can enhance the agricultural field with proper analysis and detect any defect if any. The quality check can be maintained. Even prediction of quality and quantity can be made which will further help farmers to know about the crop quality and could take necessary steps. Therefore machine learning (ML) is essential for improvement of fruit quality and accuracy. There are several methodologies discussed below and also have a very promising accuracy.

Surface Color Analysis: Using this, we can analyze the color and can make predictions. In this, RGB color model which describes perception of color [1] of 3D image [2] is used with HSI which gives the intensity of colors in a graphical manner [3] that can be transferred from RGB color format [4] by formulas as well. After which further 3 steps are there for analyzing: BP network, color histogram, quality evolution [5, 6]. Feature Extraction: In this, PCA was used to evaluate the ripening stage [7]. Feature Selection: In this, generalization of fruits into different classes using multivariate method [8]. This involves REF and SVM classifiers methods [9].

There are various more methods which are conferred below in this paper.

3   Methodology

3.1   Use of Surface Color Analysis

The surface color tells a lot about fruit like maturity and defects. So, to analyze surface color, images of fruit are taken by CCD camera, which is presented in RGB color model, for example, black is represented as $R = G = B = 0$; then the HSI is evaluated after which several three steps are taken.

In RGB color model, different color can be reproduced by additively combining red, green and blue color in different ways. In general sense, RGB color model describes perception of color [1], as shown in Fig. 1.

HSI color model gives the color representation of color intensity in histogram. Hue color attribute describes a pure spectrum of color. The histogram is useful tool to identify the color character, as shown in Fig. 2.

HSI color format can be transferred from RGB color format as follows [4]:

$$I = 1/3(R + G + B) \tag{1}$$

$$S = 1 - 3/(R + G + B)[\min(R, G, B)] \tag{2}$$

$$H = \{\theta, B <= G \,\&\, 360 - \theta B > G\} \tag{3}$$

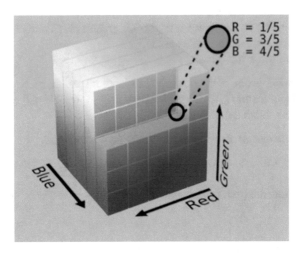

$$R = 1/5$$
$$G = 3/5$$
$$B = 4/5$$

Fig. 1 RGB 3D representation [2]

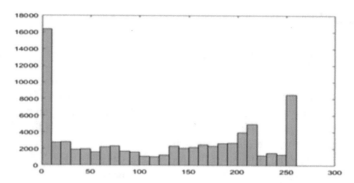

Fig. 2 Histogram of color intensity [6]

Fig. 3 Flowchart

$$\theta = \cos -1\{[1/2[(R - G) + (R - B)]]/[(R - G)2 + (R - B)(G - B)]1/2\} \tag{4}$$

Here, I: intensity, R: red color amount, G: green color amount, B: blue color amount, S: saturation, H: Hue and θ: angle.

Then further 3 steps are there:

Step 1: BP Network is one of the powerful bio-inspired pattern recognitions in many researches and application [5]. This network has self-study and analysis ability which can accurately establish the mapping.

Step 2: Color Histogram shows the color intensity for analysis.

Step 3: Quality Evolution evaluates the fruit quality with surface color and BP network.

Here is the flow diagram for this method (Fig. 3).

3.2 Feature Extraction

In ripening of apple, the skin color as well as some characters are also revolute. A prompt corrective action (PCA) is used to evaluate the classification of ripening

Fig. 4 Scanning of ripening of skin [6]

stages. For this analysis, eight variables are used (RPI, W, L^*, a^*, b^*, E, h^* and C^*), and they are arranged in a matrix form [7].

The skin color change confirms the stages. All sides of sample (apple) are scanned. Change in color is almost similar, hence we can segregate the ranges through CVS in terms of days through analysis that is: Day 1 to 8—E is 5.53 + −0.99 [7]. Similarly, 8–16 is different. The way of ripening skin of apple scanning can be seen from Fig. 4.

3.3 Feature Selection

To generalize apple into different classes, we need a maximize feature separation between the classes. In this case, multivariate method of feature selection by backward elimination is applied [8]. Further, here is the summary of abstract from paper [9]. In this method, recursive elimination feature (REF) technique is used with SVM classifiers. And SVM classifier were used.

1. *Theory of SVM and SVM-REF Technique*: SMV is basically an idea to map data into a high-dimensional space and find optimum separating hyperplane (OSH) with maximal margin. And the RFE method is based upon the value of margin on different set features [9]. The selected featured is performed through backward sequential elimination procedure followed by maximum principal margin.
2. *SVM-Based Classification*: To classify the maturity of classes, they are divided into four different levels, total $2^{4-1} - 1 = 7$ numbers of binary SVM are combined by the maximum hamming distance rule in such a way that each of them aimed at separating into different combination of classes [9]. The code world used for four different classes tells about the formulation, the individual classifiers are trained for two meta class problems, where discrete meta classes, original classes (odd classes) create one class and other (even classes) class establishes the second class. During testing, outputs of classifiers should be −1 or 1.
3. *Training, Testing and SVM Parameters Optimization*: In classification model, it is trained to test using 2 techniques but, in this system, it is alienated into six subclasses with five subclasses used to train the SVM decision surface, the lasting subclass is used for testing. In this cycle of training and testing processes, each cycle will be a rounded off and obtained by means of four measures, i.e., sensitivity (SE), specificity (SP), predictivity (PR) and accuracy (AC), and they

are calculated using following formulas [9].

$$Sensitivity = (TP)/(TP + FN) \tag{5}$$

$$Specificity = (TN)/(TN + FP) \tag{6}$$

$$Predictivity = (TP)/(TP + FP) \tag{7}$$

$$Accuracy = (TP + TN)/(TP + TN + FP + FN) \tag{8}$$

Here, TP: True Positive; TN: True Negative; FP: False Positive; FN: False Negative.

These Coefficient Variable (cv) tests are mainly optimized for SVM classifiers parameters like C and kernel function to obtain average classification performance.

4. *Feature Selection*: After finalization, the feature selection processed with SVM-RFE is computed [9]. All features applied to model in which each iteration of RFE, the worth of ranking score r_q of the remaining $(27 - i)$ features are found, and then the feature with the lowest r_q value is removed. Rank $R = 27 - I$ will be allocated to remove feature. The succeeding cycle begins with $\{27 - (i + 1)\}$ number of features and continues to eliminate the following least important feature according to their ranks [9]. This process will be continued till there only one feature remains. At the end of this process, margin-based general ranking of all the features is attained according to their position on classification.

3.4 Tomography

This is image scanning method which can be done through the practice of any kind of penetrating. This method is used in radiology, archeology, biology and various other areas of science. This method is not too efficient for large scale of inspection but can detect if any infection is there inside the apple or a fruit. This can be used for data analyses only for few segments/ranges of fruit. In Fig. 5, the Tomographic Image is represented.

3.5 Automation Process

This is projected in order to detect the fruit skin defect based on fruits skin defect based on shape and color. This basically removes the background and in presence of shadow by ACM algorithm which will intensifies the speed for segmentation of defects. Instead of using all image pixel, only pixel of fruit shape is utilized. There

Fig. 5 Tomographic Image of apple [6]

are some expressions to achieve the above. This is only an abstract of the paper [10]. In Sect. 3.4, Fig. 6 shows the background elimination [11].

Other model which can be used is artificial neuron network (ANN), this is immensely parallel distributing processing system that is built upon artificial neurons that has convinced performance characteristic resembling to biological neurons of human brain. This network is characterized by its architecture that repeats in which pattern is connected between nodes. This method is used to regulate the connection of weights and the activation function [12].

Flowchart

See Fig. 7.

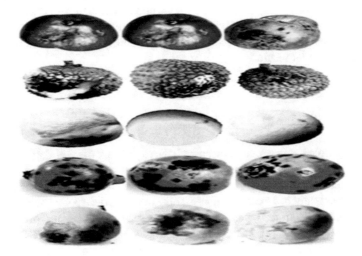

Fig. 6 Background elimination [11]

Fig. 7 Flowchart of steps of functioning

The flowchart tells the detailed manner of working or processes which is followed. Different methodologies which are discussed are used in step of analyzing the images rest everything remains constant. And table provides the detailed overview of all the methodologies like advantages and disadvantages.

4 Result

As several methodologies which have been discussed above have their merits and demerits (which is also discussed in Table 1). So, selection of any method depends on various factors. The core distinguisher of every method is the main reason of its selection. Like in surface color analysis, prediction of fruit quality and defects is based on surface color and intensity of fruit. In feature extraction, prediction of fruit maturity is made to predict the nutrition level or if any defect is there. In feature selection, we generalize fruit into different classes based on common feature and compare them. Tomography is used to scan the internal feature of fruits. And in Automation Processing, with help of algorithms, self-analyze system is made. Hence, the selection of the method for defect identification, evaluation of different factors like investment, scalability of process, profitability of project, etc., is required for getting the better result.

5 Conclusion

Many methodologies have been discussed above with merits and demerits. Every method has a common problem of data management, but in Automation Processes data management processes are done by itself. However, accuracy is the main concern as this process is still in developing phase, compared to surface color analysis, feature extraction and selection. But this (automation process) can be improved if accurate image segregation can be done using high end hardware and develop efficient processes. Moreover, with the help of regular comparison for a limited period of time between the predictive fruit quality and the tomographic results, area of improvement can be identified and improvement in processes can be implemented to enhance the

Table 1 Observation/comparison table of different methodologies

S. No.	Method	Accuracy	Advantages	Disadvantages
1	Surface color analyses	Accuracy is about 95.83% [13]	• Analysis can be made without harming the fruits on large scale • Reduces cost for checking each fruit and save time	• The picture should be clear • Need to invest good sort of money • Huge database needs to be managed which require high storage
2	Feature extraction	92.6% accuracy [7]	• We can know the maturity of fruits • The freshness could be known	• Sharpe picture quality is required to make accurate analysis • Need huge amount of data management system to generalize fruits into different class otherwise process will lag. Moreover, stringent monitoring is required for transportation
3	Feature selection	Accuracy is about 90% [11]	• We can categorize fruits into different classes • Stringent monitoring is not required	• Shape is critical visual feature for image description which cannot be defined precisely because it is hard to measure similarity between shapes • Difficult to tell exact quality of fruit after segregating into different classes based on common features
4	Tomography	• 100% accurate for detecting nonunion [14] and used by scientist for anything to observe	• Can detect defect even from inside • Shows highest accuracy • Can tell all the properties of fruit accurately	• Can harm the fruit with radio rays • Cannot be done for every fruit on large scale, moreover need to do scan manually • This is an old method for fruit detection

(continued)

Table 1 (continued)

S. No.	Method	Accuracy	Advantages	Disadvantages
5	Automation processing	• ANN shows 93.5% of accuracy [12]	• Image extraction from whole can be done automatically • Can check and analyze every fruit directly as human do by ANN • Database is also maintained automatically	• Since this under developed, more work and researches are being going so it not much accurate as others • Development process is quit complex • High storage capacity is required

output or result. Such implementation of self-efficient and improved data structure can lead to error free analysis. In the future, this will not only affluence the procedure of fruit segregation into different classes but also increase the income of farmer. For example, in Indian economy, more than 20% of GDP is dependent on agriculture and allied sector so there is a huge scope for commercialization of this process. There are few giant brands in India who have already taken a step forward in this direction.

With the help of accurate fruit segregation and fruit quality check, the better quality of fruits can be supplied to the end consumer. Consumers will also have a good amount of information concerning the products they buy. They can also have more option to choose the product on basis of process followed, fertilizers and pesticides used while growing the fruits. This in longer run will reduce the usage of chemical-based fertilizers and pesticides and accordingly will improve our environment by reducing different kind of pollutions including soil, water and air pollution.

References

1. Wang Y, Cui Y, Chen S, Zhang P, Huang H, Huang GQ (2010) Study on HSI color model-based fruit quality evaluation. In: 3rd international congress on image and signal process (CISP), vol 6, pp 2677–2680
2. Wikipedia. RGB color space. https://en.wikipedia.org/wiki/RGB_color_space. Accessed 22 July 2021
3. Yogesh, Dubey AK, Arora RR (2018) A comparative approach of segmentation methods using thermal images of apple. In: 7th international conference on reliability, infocom technologies and optimization (ICRITO), AIIT, Amity University Uttar Pradesh, Noida, India, 29–31 Aug 2018
4. Pratt WK (2007) Digital image processing: PIKS scientific inside, 4th edn. Wiley-Interscience, Hoboken, NJ
5. Dayhoff JE (1990) Neural network architectures: an introduction. Van Nostrand Reinhold, New York
6. Yogesh, Dubey AK, Arora RR. Fruit defect prediction model (FDPM) based on three-level validation. J Nondestr Eval. Received: 9 Sept 2020, accepted: 16 May 2021

7. Perez SC, Perez JC, Mendez JVM, Dominguez GC, Santiago RL, Flores MJP, Vazquez IA (2017) Evolution of the ripening stages of apple (golden delicious) by means of computer vision system. IAgrE

8. Bishop CM (1995) Neural networks for pattern recognition. Oxford University Press, New York, NY

9. Nandi CS, Tudu B, Koley C (2014) A machine vision-based maturity prediction system for sorting of harvested mangoes. IEEETIM 63

10. Moradi G, Shamsi M, Sedaghi MH, Alsharif MR (2011) Fruit defect detection from color images using ACM and MFCM algorithms. In: International conference on electronic devices, systems and applications (ICEDSA), 24 Apr 2011

11. Yogesh, Dubey AK, Arora RR (2019) Computer vision based analysis and detection of defects in fruits causes due to nutrients deficiency. Springer Science+Business Media, LLC, part of Springer Nature 2019. Received: 11 Sept 2019, accepted: 25 Nov 2019

12. Shekar R et al. Fruit classification system using computer vision: a review. Int J Trend Res Dev 5(1). ISSN: 2394-9333

13. Yogesh, Dubey AK, Arora RR (2020) Multiclass classification of nutrients deficiency of apple using deep neural network. Springer-Verlag London Ltd, part of Springer Nature 2020. Received: 17 July 2020, accepted: 19 Aug 2020

14. Bhattacharyya T, Bouchard KA, Phadke A, Meigs JB, Kassarjain A, Salamipour H (2014) The accuracy of computed tomography for the diagnosis of tibial nonunion. J Bone Joint Surg Am 96(21):5

COVID-19 Data Clustering Using K-means and Fuzzy c-means Algorithm

Anand Upadhyay, Bipinkumar Yadav, Kirti Singh, and Varun Shukla

Abstract Corona Virus Disease 2019 (COVID-19) is a contagious disease caused by severe acute respiratory symptoms. It has been declared a global pandemic since 2019 by the World Health Organization. Countries are in an authoritarian state of preventing and controlling this pandemic, and the USA is the central hub. The COVID-19 virus has also shown variance. As an outcome of the genetic recombination of genes that arise from coronavirus, their short life span results in mutations that promote new strains. However, the number of individuals who passed their lives is still counted. Additionally, it is crucial to analyze the spread of the virus before it is deferred in the lungs. In this research, the effort has been taken to predict the proliferation of the virus through various chest radiography images by data clustering. In this study, two clustering algorithms, i.e., the K-means algorithm and the Fuzzy c-means algorithm, have been used better to analyze the spread of the virus in the lungs. These algorithms are further being compared and evaluated for the precise result of both models. This study helps to recognize the most suitable clustering model for the COVID-19 prediction and spread of the virus in the lung.

Keywords K-mean · Fuzzy c-mean · Clustering · COVID-19

1 Introduction

Communicable diseases are grounds for numerous microorganisms that impart from one to other individuals by different creatures. The COVID-19 falls under this category where COVID-19 infection spreads from one to another individual through the air; therefore, it is called airborne infection. The COVID-19 is a swift, contagious disease with vast and high danger symptoms with high chances of death among humans. Therefore, it is imperative to detect COVID-19 signs among humans so that infected persons can be quarantined and the spread of infection can be stopped.

A. Upadhyay · B. Yadav · K. Singh (✉) · V. Shukla
Thakur College of Science and Commerce, Thakur Village, Kandivali (East), Mumbai, Maharashtra 400101, India
e-mail: kirticism@gmail.com

© The Author(s), under exclusive license to Springer Nature Singapore Pte Ltd. 2023
A. Shukla et al. (eds.), *Computational Intelligence*, Lecture Notes in Electrical Engineering 968, https://doi.org/10.1007/978-981-19-7346-8_46

Various kits, methods, and procedures are developed and used for detection recognition of COVID-19 infection and symptoms related to COVID-19. Researchers developed different medical-related kits, treatment, medicine and detection mechanisms. In recent studies, the predicting process for the occurrence, proliferation, and change in infectious diseases mainly included regression prediction models to predict contagious diseases. In contrast, the people from various technology developed various instruments and methods with the use of technology, i.e., e-Governance, IoT, machine learning, AI and big data to detect COVID-19, detection of COVID-19 symptoms, implementation of COVID-19 appropriate behaviors to stop the spread of infection, patients related data management, etc. Medical practitioners failed to accurately predict the prognosis of COVID-19 patients upon their admission until later stages of the disease. The course of COVID-19 can take unpredictable turns where a patient's condition can seemingly deteriorate rapidly to a critical state. This particular critical situation even cannot understand by skilled physicians. So, to enhance the understanding of the COVID-19 infection spread, the two folds comparison and prediction methods are implemented in a flow of abstract, introduction, literature review, comparative analysis, methodology, results, conclusion and references. Here is delineated a prediction of COVID-19 using the radiography data. In this research paper, the different clustering techniques, i.e., K-means and Fuzzy c-means algorithms, are used to perform the clustering of radiography images. K-means clustering is one of the splendid clustering algorithms used for pattern identification. In contrast, Fuzzy c-means clustering is a soft clustering approach, where each data point is assigned a likelihood or probability score to belong to that cluster. Further, the performance of both methods is evaluated by various parameters and evaluation techniques.

2   Literature Review

Today, every sector of our daily lives, including health care, business, media, law enforcement agencies and more, are interrelated with technology; choosing the suitable parameters is one of the challenges involved in training a model or the challenge when looking for benefits of using a particular machine learning algorithm to use for prediction [1]. The author Huang et al. has been noticed that skin lesions are side effects of the COVID and not symptoms that are proven to be harmless [2]. Additionally, the use of four most used convolutional neural networks [3, 4], i.e., ResNet18, ResNet50, Squeeze Net and DenseNet-121 [5], has been made, which makes the model decision boundaries overstrained if the training data does not consist of examples we are looking for in the class [6]. Also, Sengupta et al. emphasized clustering analysis by clustering the districts of India in the phase of the Covid-19 outbreak [7]. This draws similarity with Abdullah's et al. and Hutagalung's et al. study where the research discusses recovery and death cases using K-means clustering for province clustering, conversely to mitigate the small number of samples; we can also employ transfer learning, which transfers knowledge extracted by pre-trained models to the model to be trained [8, 9]. Alternatively, tailored deep learning models are suggested to detect pneumonia infection cases; however, CNN cannot encode the position of an object and visual patterns [10]. Clustering algorithms such as the K-means algorithm and Fuzzy c-means algorithm are not only the finest for visual predictions but efficient and straightforward enough to work quickly with big dataset [11]. Our main finding is to compare both clustering algorithms for high clustering performance and the construct robust model for COVID-19 prediction.

3   K-mean and Fuzzy c-mean Clustering

3.1   K-means

K-means is an excellent and most straightforward clustering algorithm, which uses an iterative technique that splits the unlabeled dataset into different clusters in such a way that each dataset belongs to only one set that has similar properties, where k is a number of centroids that need to be created in the process, for example: if $k = 2$, there will be two clusters as shown in Fig. 1 alternately if $k = 3$, there will be three clusters. If a data point is associated with a centroid, it would be assigned as the cluster of that centroid. The purpose of this algorithm is to decrease the sum of the range between the data point and their correspondent clusters.

The K-means algorithm is as follows:

Step-1: Select the clusters k.

Step-2: Select random K points or centroids.

Step-3 Allocate every data point to their nearest centroid, which will form the predefined k clusters.

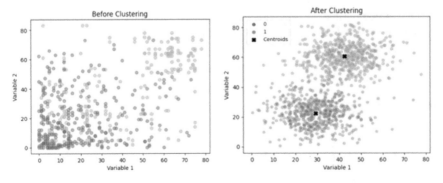

Fig. 1 K-means clustering algorithm

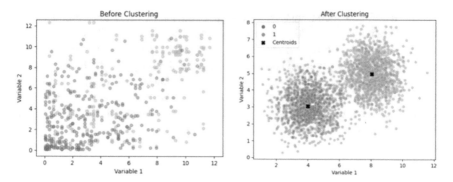

Fig. 2 Fuzzy c-means algorithm

Step-4: Recompute the centroid of newly formed clusters.

Step-5: Repeat steps 3 and 4.

3.2 Fuzzy c-means

Fuzzy c-means (FCM) is the technique of clustering that permits a single block of data to belong to more than one cluster. This technique is frequently utilized in pattern identification; it works by allocating members to every data point correspondent to all centroid based on the range between the cluster and the data point, as shown in Fig. 2. If the data is nearest to the centroid, then its belonging toward the particular centroid is more.

4 Methodology

A methodology is a core component of any research and innovation. Here, the algorithms are executed using Python and Python dependent machine learning packages. The detailed methodology steps are as in Fig. 3.

4.1 Radiology Image Acquired

The radiography image of the chest is freely available online at Kaggle also the dataset is acquired from the year 2019 to 2020 for the COVID-19 prediction. The data is easily available for study and research purposes for various issues related to the COVID-19 pandemic.

Fig. 3 Steps for COVID-19
image clustering

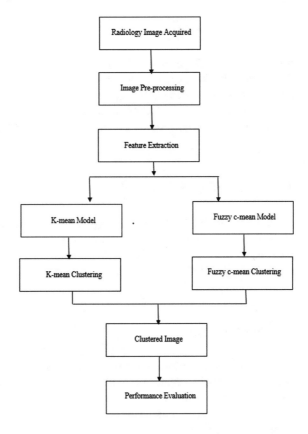

Data Source—https://www.kaggle.com/tawsifurrahman/covid19-radiography-database.

4.2 Image Preprocessing and Characteristic Extraction

Preprocessing is an important fundamental step of any classification and clustering process. In image processing, the image is enhanced and improvised to collect the features properly for appropriate classification or clustering purposes. Here, the histogram equalization method is applied to smoothen the image after pixel-based color information is extracted for clustering purposes.

4.3 Clustering Model Creation and Image Clustering

In this step, the K-means and Fuzzy c-means models are prepared using various Python libraries. Numerous Python libraries are richer with various machine learning algorithms. Once the model is prepared, the images are supplied to the model for clustering purposes. The model clusters the data and shows the clustered values, and those values are once again mapped to image format as clustered images.

4.4 Performance Evaluation

After the clustering and clustered images, another essential part is the performance evaluation of K-means and Fuzzy c-means algorithm. Here, various techniques and approaches are used to specify the performance of the clustering algorithm, which is explained in the result section of the research paper.

5 Results

After the successful implementation of the K-means and Fuzzy c-means algorithm, now it is essential to evaluate the performance of the algorithms in terms of their clustering performance and clustering results for COVID-19 image clustering. Here, various techniques, i.e., Silhouette score, Davis Bouldin score and Calinski Harabasz score, are calculated.

5.1 Silhouette Coefficient Score

It is used to measure the standard of the clustering algorithm, which ranges from -1 to 1. The formula to calculate the value for Silhouette coefficient score is as follows:

$$s(i) = \frac{b(i) - a(i)}{\max(b(i) - a(i))} \tag{1}$$

In Eq. (1), $a =$ average intra-cluster represents the average distance between each data point in the cluster and $b =$ average inter-cluster, which indicates the average distance among all the data points in the cluster.

5.2 Davis Bouldin Score

The score represents the average similarity measure of each cluster with the cluster most similar to it. The minimum value represents the better clustering; the score ranges from 0 to 1. The formula for Davis Bouldin score is as follows:

$$\text{DB} = \frac{1}{n} \sum_{i=1}^{nc} Ri \tag{2}$$

In Eq. (2), $i = \max(Rij)$, $i = 1, 2 \ldots nc$, $j = 1, 2, \ldots nc$ and $i \neq j$.

5.3 Calinski Harabasz Score

The score is also known as Variance Ratio Criterion, where it is a ratio of the sum of between clusters dispersion and inter-cluster dispersion of all the clusters. A higher score indicates better performance.

$$s = \frac{\text{tr}(B_k)}{\text{tr}(W_k)} \times \frac{n_E - k}{k - 1} \tag{3}$$

In Eq. (3), $W_k = \sum_{q=1}^{k} \sum_{x \in C_q} (x - c_q)(x - c_q)^T$ $B_k = \sum_{q=1}^{k} n_q (c_q - c_g)(c_q - c_g)^T$.

Table 1 describes the values of various parameters that are calculated. The above quantitative value table concludes that the performance of the K-means algorithm is more acceptable than the efficiency of the Fuzzy c-means algorithm. Conclusively, K-means obtained a Silhouette score 0.0003 higher than Fuzzy c-means on a scale of -1 to 1. Additionally, Davis Bouldin score for Fuzzy c-means is 0.003 less than K-means on the scale of 0–1. Furthermore, in terms of the Calinski Harabasz score, we see that K-means is more efficient as there is a difference of 635.0996 between it and fuzzy c-means (Figs. 4 and 5).

Table 1 Performance evaluation score of the algorithms

Technique	No. of clusters	Silhouette score	Davis Bouldin score	Calinski Harabasz score
K-means	3	0.6246	0.5375	273,942.3767
Fuzzy C-means	3	0.6243	0.5345	273,307.2771

Fig. 4 K-means clustered image

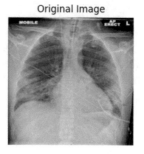
Original Image

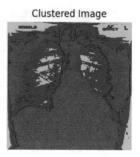
Clustered Image

Fig. 5 Fuzzy c-means clustered image

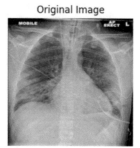
Original Image

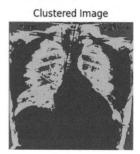
Clustered Image

6 Conclusion

In this paper, a clustering algorithm has been used to cluster COVID-19 radiology images. The suggested methodology and techniques show that the K-means is more efficient than the Fuzzy c-means. In the future, the K-means can be evaluated with some other clustering algorithm, and its performance can be compared with different algorithms too. Instead of only image data, the researchers can include other clinical trial data and clinical diagnosis feature data. Clustering algorithms are only flexible with clear images, so K-means and Fuzzy c-means algorithms fail to detect the blurred, tilted and unsharpened data.

References

1. Kwekha-Rashid AS, Abduljabbar HN, Alhayani B (2021) Coronavirus disease (COVID-19) cases analysis using machine-learning applications. Appl Nanosci 1–13
2. Huang Y, Pinto MD, Borelli JL, Mehrabadi MA, Abrihim H, Dutt N, Lambert N, Nurmi EL, Chakraborty R, Rahmani AM, Downs CA (2021) COVID symptoms, symptom clusters, and predictors for becoming a long-hauler: looking for clarity in the haze of the pandemic
3. Cohen JP, Dao L, Roth K, Morrison P, Bengio Y, Abbasi AF, Shen B, Mahsa HK, Ghassemi M, Li H, Duong T (2020) Predicting covid-19 pneumonia severity on chest X-ray with deep learning. Cureus 12(7)

4. Houssein EH, Abohashima Z, Elhoseny M, Mohamed WM (2021) Hybrid quantum convolutional neural networks model for COVID-19 prediction using chest X-ray images. arXiv preprint arXiv:2102.06535
5. Makris A, Kontopoulos I, Tserpes K (2020) COVID-19 detection from chest X-ray images using deep learning and convolutional neural networks. In: 11th Hellenic conference on artificial intelligence, Sept 2020, pp 60–66
6. Minaee S, Kafieh R, Sonka M, Yazdani S, Soufi GJ (2020) Deep-covid: predicting covid-19 from chest X-ray images using deep transfer learning. Med Image Anal 65:101794
7. Sengupta P, Ganguli B, SenRoy S, Chatterjee A (2021) An analysis of COVID-19 clusters in India. BMC Public Health 21(1):1–21
8. Abdullah D, Susilo S, Ahmar AS, Rusli R, Hidayat R (2021) The application of K-means clustering for province clustering in Indonesia of the risk of the COVID-19 pandemic based on COVID-19 data. Qual Quant 1–9
9. Hutagalung J, Ginantra NLWSR, Bhawika GW, Parwita WGS, Wanto A, Panjaitan PD (2021) COVID-19 cases and deaths in Southeast Asia clustering using K-means algorithm. J Phys Conf Ser 1783(1):012027
10. Hammoudi K, Benhabiles H, Melkemi M, Dornaika F, Arganda-Carreras I, Collard D, Scherpereel A (2021) Deep learning on chest X-ray images to detect and evaluate pneumonia cases at the era of covid-19. J Med Syst 45(7):1–10
11. Fränti P, Sieranoja S (2019) How much can k-means be improved by using better initialization and repeats? Pattern Recogn 93:95–112

Body Temperature and Oxygen Level Detection System

Karan Pathania, Nakshtra Kumar, Pallavi Choudekar, Ruchira, and Kamlesh Pandey

Abstract A device comprising an oximeter and a module for detecting body temperature has been designed so that a person can readily check his or her health in crucial situations. This was accomplished by programming Arduino to output values measured by sensors such as the MAX30102 (Particle Sensor) and GY-906-BCC (Infrared Sensor). We've all been dealing with a global pandemic for the past year. As a result, there have been numerous coronavirus discoveries. The COVID-19 virus primarily affects an individual's respiratory system, lowering the patient's oxygen levels, and it causes a rise in body temperature. This approach can be quite valuable in such situations and can aid in the regular monitoring of an individual's health.

Keywords Oximeter · Body temperature detection · Arduino

1 Introduction

As the whole world is suffering from the pandemic, there is a need for all of us to unite and fight the current situation in any way possible. This virus comes from a large family of coronaviruses, i.e., COVID-19. A person infected by it may develop multiple symptoms. The study has shown it commonly found symptoms in the patients are the difficulty in breathing and rise in temperature of body of an individual. It is seen if a person is not taken care of in such times, it may have fatal results. To avoid such scenarios, patient's health monitoring is very critical at such times.

For health monitoring purposes, a system can be developed to give out the level of oxygen and body temperature without causing any inconvenience to the patient. An Arduino is used for such a system as it is compact, easy to use, and easily programmable with any other component used. It gives out accurate and real-time values measured by the sensors. For this system, MAX30102 (particle sensor) and GY-906-BCC (infrared sensor) can be used. The MAX30102 is a particle sensor that

K. Pathania · N. Kumar · P. Choudekar (✉) · Ruchira · K. Pandey
Amity University Uttar Pradesh, Noida, India
e-mail: pallaveech@gmail.com

© The Author(s), under exclusive license to Springer Nature Singapore Pte Ltd. 2023   549
A. Shukla et al. (eds.), *Computational Intelligence*, Lecture Notes in Electrical
Engineering 968, https://doi.org/10.1007/978-981-19-7346-8_47

detects the pulse by which it calculates the blood concentration in any region of the body. Further, it measures the SpO2 level in the blood used for measuring the oxygen level in the blood. The GY-906-BCC, a non-contact infrared sensor for monitoring anyone's temperature, was employed. These sensors are very easily used with an Arduino and program.

This system is developed which measures the level of oxygen and temperature of body of an individual simultaneously. This is achieved by using MAX301202 and GY-906-BCC sensors together and mounting them on a single platform. Hence, it is proficient of giving out the value all at once, which allows a person to monitor his or her health regularly and easily. This system reduces the scope for errors as it gives out digital values, hence leaving no room for human parallax errors. As we know, the virus spreads vigorously by human contact. It can also help in reducing the spread of the virus in such a scenario of a pandemic.

2 Methodology

It is significant to do research and develop a model to be implemented. Understanding the operation of the many components employed in the system, as well as the circuit constructed for it, is equally critical. Assembling of components has been done according to the circuit. The next step was to program the Arduino to receive the system's output using the Arduino IDE. We tested the very first prototype developed and made changes as per the requirement of the system.

3 Virus Background

In December 2019, this virus was first originated in China and later reached almost every country globally [1]. On January 30, 2020, the IHR emergency committee (International Health Regulations) of WHO announced a "public health emergency of international concern". One month later, seeing the increase of infected countries, WHO declared this outbreak of COVID-19 a pandemic on March 11, 2020.

COVID-19 comes from a family of coronaviruses [2]. The coronavirus is mainly found in animals like camels, cats, cattle, and bats. Most viruses from this family do not spread in humans, but a few, like MERS-CoV, SARS-CoV, and SARS-CoV-2 can infect a human being.

The SARS-CoV-2 genome is very much like the SARS-CoV-1 genome [3]. SARS-CoV is mainly related through the common cold and flu. SARS-CoV and MERS-CoV have evolved, and they are causing severe respiratory problems in humans [4]. Its nucleotide is 89% indistinguishable from bat SARS-COVZXC21 and 82% to human SARS-CoV, which might be behind the explanation of why COVID-19 looks like SARS-COV-2. Viral transmission through contaminated patients has been evolved through human-to-human social contacts and is straightforwardly related to several optional cases.

3.1 How It Targets the Patient?

From the study, it shows that an individual infected with SARS-CoV experiences respiratory, digestive, neurological, and urogenital problems. Taking the respiratory aspect into account, when the infection gets inside the host's body, it [5] diffuses alveolar harm with fluctuating levels of intense exudative highlights, edema and hyaline films, association, and fibrosis. There is an invasion of macrophagic or mixed cells. It also results in big multinuclear cells causing abnormal responsive pneumocytes and damaging the arteries and veins of the respiratory system. These damages in the patient's respiratory tract make it patient difficult to breathe, and the individual feels suffocated. This further results in the deteriorating levels of oxygen in the patient.

Another, major symptom of COVID-19 is the rise in body temperature, resulting in fever. In this way, an individual's body reacts to the insurgence of the virus [6]. Later infection, pyrogens are released into the body on the triggering of the immune response. This is done by the human body in order to recognize these pathogen-associated molecular patterns. This results in an elevation in the body temperature of the host (Fig. 1).

There are five levels of infection when a virus is introduced to the body of an individual. The very first stage is incubation. This is the period when humans get exposed to the virus. Next is the prodromal stage. In this period, humans experience a mild fever. Hence, detecting the rise of temperature in this stage can help a person to

Fig. 1 5-stages of infection

take precautions and start taking care of himself/herself to prevent it from worsening the situation.

4 System Functionality

This oximeter and body temperature measuring system is a small and light device for non-invasively testing physiological parameters. It detects oxygen level and temperature of an individual by just placing their finger on this device. It is not only useful for patients, but it can also be used by the general public to assess his/her health at home only without going out and exposing himself/herself to the outer in a pandemic.

4.1 Components Used

It comprises following components:

- Arduino UNO
- 2 LEDs
- MAX30102 heart rate sensor
- 20 × 4 I2C LCD
- Two 220 Ω resistors
- GY-906-BCC non-contact infrared temperature sensor module.

Figure 2 shows the step-by-step working of the system developed. The first block is the SpO2 level and patient's body temperature. This is measured by the sensors used, i.e., MAX30102 and GY-906-BCC, as indicated in the further stage. Data is collected by sensors and then transferred to the Arduino UNO used. The data received by the Arduino UNO in this stage is the raw data. In the 4th stage, this raw data is then calculated by the Arduino according to the program that we have uploaded to it using the Arduino IDE. After the value is calculated as per the requirements, i.e., we need temperature in *C and oxygen level in percentage, it then transfers these calculated values to a 20 × 4 I2C display. Here, it displays the final values of the oxygen level and body temperature measured by the sensors on the patient.

4.2 How It Functions?

- **Oximeter**: An oximeter works on a very simple principle [7]. The SpO2 (saturation of Oxygen in arteries) level is measured to detect the percentage of hemoglobin. For this purpose, we have used MAX30102 heart rate sensor [8]. The SpO2 level is measured using red and infrared light. The magnitude of reflected red and infrared light from the patient's body is used to do this.

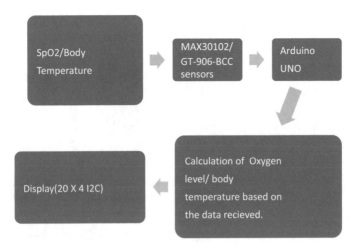

Fig. 2 Block diagram of the system developed

A Photolethylmsography (PPG) signal is sent to the system to calculate the change in the blood volume in the vicinity of the sensor's attachment, i.e., generally a fingertip. This change in the blood volume is due to heart beat of the person. Hence, pulse rate can also be detected. An average pulse rate of the person is between 60 and 100 BPM, varying according to the individual's activity was involved with, i.e., sleeping, running, stress, etc. Therefore, it can be used in measuring oxygen level in the blood. It mainly depends on the concentration of hemoglobin in the blood. The average oxygen level ranges from 90 to 100%. Values varying less than this can indicate that a person is not well.

- **Body Temperature Measuring System:** Infrared sensor measures temperature on the phenomenon that every body or object emits radiation energy. When the IR sensor is subjected to any body or object, it detects the intensity of radiation and hence calculating the object temperature.

IR sensors mainly uses four law of radiations for the temperature measurement [9]:

- Kirchoff's Law: The amount of absorption will be equal to the amount of emission when the object or body is at thermal equilibrium.
- Wien's Displacement Law: As the body temperature or an object increases, the wavelength of the radiation at which the maximum amount of radiation energy was emitted becomes shorter.
- Planck's Law of Black Body: The intensity of the radiation emitted out by each body or object is the function of the object/body temperature.
- Stephan Boltzmann Law: Infrared energy emitted out by any object or body is directly proportional to the object/body temperature (Fig. 3).

Fig. 3 Circuit diagram of
the system

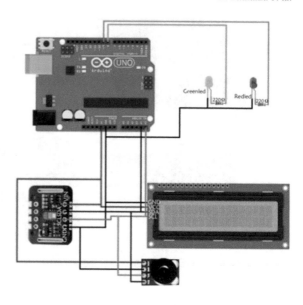

5 Result

- Result of the oximeter developed

  See Table 1.

- Result of body temperature measuring system

  See Table 2.

  Figure 4 is the final system that is developed. It consists of the components that were discussed earlier in this research. Temperature measuring system is programmed such that it shows values in the integer form only hence it displays a round off value. The value displayed by the system is not alarming hence we can see no LED glowing.

Table 1 Oximeter results

S. No.	Respondent	SpO2 level (%)
1	Respondent 1	96.45
2	Respondent 2	93.69
3	Respondent 3	94.23
4	Respondent 4	92.77
5	Respondent 5	94.38

Table 2 Body temperature results

S. No.	Respondent	Body temp. measured by system (*c)	Body temp. measured by digital thermometer (*c)	Deviation (%)
1	Respondent 1	38	37.76	0.006
2	Respondent 2	37	37.42	0.01
3	Respondent 3	37	37.26	0.007
4	Respondent 4	38	37.84	0.004
5	Respondent 5	36	37.11	0.03

Fig. 4 System developed

6 Conclusion

The system developed in this research is capable of meeting the project's requirements. This technology can monitor an individual's blood oxygen level and body temperature. With a few minor errors, the study's conclusion matches the expected outcome. The body temperature system's final output shows values with inaccuracies ranging from 0.004 to 0.01%. The inaccuracies do not alter the final result displayed with a large value, causing it to diverge from the true value. If extreme pressure is applied to the sensors, or if there is a barrier (cloth, glass, and rubber) in between the sensor and the patient's finger, the device may display sudden numbers.

References

1. Pawar PA (2014) Heart rate monitoring system using IR base sensor and Arduino Uno. In: 2014 conference on IT in business, industry and government (CSIBIG). IEEE, pp 1–3
2. Srividya B, Satyanarayana V (2018) Personal lung function monitoring system for asthma patients using internet of things (IOT). Int J Res Electron Comput Eng 6
3. Zhiao Z, Chnaowei, Nakdahira Z (2013) Healthcare application based on internet of things. In: Proceedings of IEEE international conference on technology application, Nov 2013, pp 661–662
4. Islam MM, Rahaman A, Islam MR (2020) Development of smart healthcare monitoring system in IoT environment. SN Comput Sci 1:185. https://doi.org/10.1007/s42979-020-00195-y
5. Rajasekaran A, Indirani G (2018) Real time health monitoring system using Arduino with cloud technology. Asian J Comput Sci Technol 7:29–32. https://doi.org/10.51983/ajcst-2018.7.S1.1810
6. Khan MM (2020) IoT based smart healthcare services for rural unprivileged people in Bangladesh: current situation and challenges. In: 1st international electronic conference on applied science, MDPI, Switzerland, pp 1–6
7. Riazul Islam SM, Kwak D, Humaun Kabir M, Hossain M, Kwak K-S (2015) The internet of things for health care: a comprehensive survey. IEEE Access 3:678–708
8. Carmel-Neiderman NN, Goren I, Wasserstrum Y, Rutenberg TF, Barbarova I, Rapoport A et al (2018) Structured, protocol-based pulse-oximetry measurement improves the evaluation of hypoxemic patients at hospital admission. Isr Med Assoc J 20:147–150
9. Krishnan DSR, Gupta SC, Choudhury T (2018) An IoT based patient health monitoring system. In: 2018 international conference on advances in computing and communication engineering (ICACCE), pp 01–07. https://doi.org/10.1109/ICACCE.2018.8441708

Generating ISL Using Audio Speech

Devesh Tulsian, Pratibha Sharma, Nancy, and Purushottam Sharma

Abstract The communication gap between other people and deaf people is bigger than ever. The main aim of our project is to help people with hearing disabilities understand the spoken content using an interface that accepts audio/text as input and converts them to corresponding sign language for deaf people. The audio input will be given by a person with limited knowledge and is therefore not able to communicate with deaf people. The system automatically uses various artificial intelligence and natural language processing to produce the corresponding sign language gesture. The system uses speech to text API, lemmatization for reducing the given form of word or words making sure of the root word to be recognized, tokenization for a thorough understanding of the word or phrase entered, parts of speech tagging for the understanding of the context according to the grammar rules, and parsing to produce the best possible output.

Keywords Indian sign language · Natural language processing · Artificial intelligence · Lemmatization · Tokenization · Parts of speech tagging

1 Introduction

A sign language is a language in which communication is done through various hand gestures and body language. Sign languages have been around since the very beginning of time. In the ancient era, people used to communicate only through sign language. Slowly and gradually the evolution of spoken language has happened and now we live in a world where communication is mostly done through spoken language. When we think of expressing ourselves, the first mode we go for is our speech but what if the other person just cannot get what we speak. What if all our efforts of expressing ourselves are not exactly reaching its destination. The statistics say, in approximately 50 lakhs people suffer from hearing disabilities, partial or whole [1]. The communication gap between the people who have limited knowledge

D. Tulsian · P. Sharma · Nancy · P. Sharma (✉)
Amity University Uttar Pradesh, Noida, India
e-mail: psharma5@amity.edu

© The Author(s), under exclusive license to Springer Nature Singapore Pte Ltd. 2023   557
A. Shukla et al. (eds.), *Computational Intelligence*, Lecture Notes in Electrical
Engineering 968, https://doi.org/10.1007/978-981-19-7346-8_48

about sign language and the people who have enough knowledge to communicate in sign language is increasing day by day which makes it even harder to explain why it is required as it does not strike a healthy common person as to how the other person must be having a hard time understanding them. This is therefore leading to miscommunication or no communication with deaf people at all. People with hearing disorders find it very hard to grasp what the other person is trying to say, and therefore, they are forced to face various difficulties in doing various day to day activities, so why leave out such a large ratio of the country who can never be neglected due to such trivial issues, why not take them along when they would understand what is being conveyed through a real-time sign-language visually presented to them. There are approximately 300 certified ISL interpreters in India [1], and therefore, there is an increasing need for a solution that can help people interpret ISL.

2 Indian Sign Language

Sign language can be defined as the means of communication with the use of body movements where hands, arms, and expression play a vital role. The sign language have historically been into practice before the speech itself. It may be roughly expressed as bare scowl (grimace), shrugs, and pointing. Alternatively, you can use a combination of subtle nuances of coded manual signals, enhance with facial expressions, and in some cases complement with words spelled in the manual alphabets. It can be used to bridge the gap where voice communication is difficult, such as between speakers of mutually incomprehensible languages or when one or more communicators is deaf.

There are a lot of spoken languages in the globe, such as English, French, and Urdu, among others. Similarly, hearing-impaired people employ a variety of sign languages and phrases around the world. American Sign Language (ASL) is used in the US, British Sign Language (BSL) is used in the UK, and Indian Sign Language (ISL) is used in India to convey thoughts and communicate with one another.

2.1 History of Indian Sign Language

Language is a versatile school that includes an arbitrary set of codes that make up concepts. It began as a means of easing intrapersonal transactions and has easing intrapersonal transactions and has evolved since into an important tool for the notion of a machine's organized illustration of its ideas. Contact and motion are said to have dominated interaction among monkeys, the closest relatives of humans, and spoken communication among prehistoric paleontological organisms. Because of the convenience of production, price of transaction, and gain over area and time, language, as an arbitrary machine of verbal exchange, gradually followed sounds as the primary detail of the transaction, thus contriving to make a human speech, related to aural-oral transactions, the number one shape of language, and primary mode of verbal exchange.

Institute focused on ISL coaching and study was advocated by Indian deaf in the 2000s. The Eleventh Five Year Plan (2007–2012) noted what the people with hearing impairments needed were surprisingly ignored and proposed the establishment of a signal language studies and education center, as well as the sale and training of instructors and interpreters [1].

As a result, in 2011, the Social Justice and Empowerment Department approved the established mission of the Indian Sign Language Research and Training Centre (ISLRTC) as an independent center for the Indira Gandhi National Open University (IGNOU) in Delhi [1].

2.2 Grammar of ISL

Indian Sign Language, like other languages, has its unique grammar. It is unaffected by the spoken language, whether English or Hindi. The manual depiction of spoken English or spoken Hindi is not the same as sign language. It has some specific characteristics, such as:

1. Each number is represented with a hand gesture this appropriate for that number. For example, the sign for 45 will be the symbol of four followed by the symbol for five.
2. The signs for "male/man" and "female/woman" come before the indications for familial relationships. The interrogative sentences with questions like WHAT, WHERE, and so on are represented by asking questions at the end of the sentences.
3. Non-manual movements such as mouth patterns, mouth gestures, facial expressions, body postures, head position, and eye gazing make up the ISL. The word order in ISL is essentially Subject-Object-Verb (unlike English which is Subject-Verb-Object).

3 Literature Survey

The authors have proposed a system that converts Indian Sign Language to natural language using translating gloves. The gloves are powered by Arduino UNO, and then using various sensors, the glove can detect various signs gestured by a person wearing gloves [2].

In the paper "A translator for American Sign Language to text and speech," a system to convert American Sign Language to speech or text is proposed. The proposed system takes input from the user's webcam/camera and gives an output in the natural language. The system has achieved 98.7% efficiency [3].

Using MATLAB, this double-handed Indian Sign Language to speech converter con verts the sign language to the natural language. The system uses a minimum eigenvalue algorithm to detect the various gestures in the input and then uses text to speech synthesis to convert the detected gesture to speech [4]. Following the previous approach, this system uses HSV technology to detect various gestures in the input to produce the output [5].

This is one of the few systems which gives a two-way communication medium for deaf people. This system uses neural networks and a hidden Markov model to generate the sign language. The main drawback of this system is that the output given by the system is very animated which may be difficult for some people to understand and therefore, cannot be implemented in the real world [6].

This is one of the first systems which converts English to Indian Sign Language but as mentioned by the authors in the paper, the system has very few words in its dictionary and is therefore not able to generate sign language for every word entered as input. This problem is resolved in our system as we split the non-existing word into letters and gives the output in the form of every letter. This at least gives output to the user rather than skipping the word completely [7] (Table 1).

4 Proposed Work

The proposed system will first take the audio as input from an external or in-built microphone and will convert it into English text. The English text obtained will be further tokenized into words using natural language processing. After this, source language will be converted to phrase structure trees by a parser using transfer-based translation. Further, ISL grammar rules will be used to convert the phrase structure tree to an ISL modified tree which represents the structure and grammar of ISL. In this step, the accuracy of our system will be improved using lemmatization and parts of speech tagging. The words which do not have their sign language videos in the database are further split into letters, and then, the sign language of the letters is shown one by one. This increases the efficiency of the overall system as it does not skip any word and the whole message is fully conveyed in sign language. The sign language videos are self-recorded by all the authors.

5 Comparison with Existing Models

Most of the existing models convert the input from the user to American Sign Language or British Sign Language. There are very a small number of models which successfully convert text or audio input from the user to Indian Sign Language. The models which do convert the input to ISL, sometimes skip the word/words which are not in their video/sign dictionary. This model not only does not skip any word

Table 1 Literature survey

Source	Objective	Methodology	Conclusion
[2]	To develop a smart glove that uses various sensors to detect various sign language gestured by the person wearing the gloves	Arduino UNO and various sensors to sense the movement	The authors would like to enhance the project into virtual reality using a centralized IoT hub
[3]	To detect static ASL sign signals and then convert them into speech	ADA boost and Haar-Like classifiers	20 frames per second were captured through video input to increase the accuracy of the system
[4]	A double hand sign recognition system was introduced in this paper	The system utilizes a minimum eigenvalue algorithm to detect and analyze the video input	The proposed system was successfully able to detect the various signs, and then, an audio output was given by using text to speech algorithms
[5]	An Android app is proposed which recognizes various hand gestures and then gives an output in the natural language	The system uses the HUE color model for hand tracking, further, MATLAB is used for pattern recognition	The system detected that using the traditional RGB color model the accuracy of the segmentation was getting reduced. Further work is required to increase the accuracy in various lighting conditions
[6]	A unique 2-way communication model has been proposed by the authors	The system utilizes neural networks and the hidden Markov model to produce the satisfactory output	The system uses American Sign Language as its database. The mobile application does its job in establishing a 2-way communication between a deaf and a normal person
[7]	The authors have suggested a model which gives an animated SIGML output and takes English audio as input	The system utilizes HamNoSys and NLP to give the preferred output	The animated person also lips the input text so that hard to hear person can also make out what the animation is trying to sign
[8]	Authors have only proposed a system that converts a speech input to an ISL output	They propose to use artificial intelligence and natural language processing in the system	They wish to grow their ISl dictionary so that they can increase the performance of the model
[9]	The proposed system takes hand gestures as input and provides English text as output	The system uses an RGB color model to detect the various gestures. Authors have also utilized various pattern recognition algorithms to detect the sign	The system only detects alphabets and numbers, and the authors wish to expand the model so that it also recognizes words and sentences

(continued)

Table 1 (continued)

Source	Objective	Methodology	Conclusion
[10]	The system is yet another ISL to English translation	Authors have used OH and PCA analysis which have given them better results	The system also recognizes 2 hand gestures, but the authors wish to use HMM and SVM models for larger datasets

but also tokenizes the word into letters if the sign language video for that word is not present in the database. This ensures that the communication between 2 people is completed successfully.

This model can further be improved by adding features like multiple words/phrase/sentence identification. This would allow the model to show a sign language video of the multiple words/phrases/sentences in one go if the video for the same is present in the database. This would ensure smoother and faster communication.

6 System Architecture

See Fig. 1.

7 Natural Language Processing (NLP)

Natural language processing is when a computer uses artificial intelligence to interpret the spoken speech or the entered text like how human beings interpret various languages. Natural language processing uses various machine learning and deep learning tools to convert the natural language or the spoken language into a machine language [11]. The various steps included in the full process of natural language processing are:

7.1 Tokenization

Tokenization is splitting up a text into words or sentences called tokens. Tokens help the computer to understand the meaning of the text entered. This a very key step in natural language processing as it has many uses such as:

- Counting of words/sentence in the text.
- Counting the frequency of a particular word/sentence/phrase.

Fig. 1 System architecture

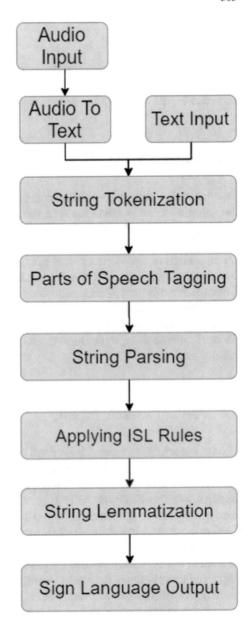

- There is various type of tokenization.
- Whitespace: It splits up the text at each whitespace encountered. Especially, in a language like English in which meaningful words are separated by whitespaces.
- Dictionary Based: In this, a dictionary is created beforehand then various texts are split with respect to entries in the dictionary. If a token is not found in the dictionary, then special techniques are used to handle the situation.

7.2 Parsing

Parsing in natural language processing is a process through which syntactic meaning is extracted from every token created after tokenization. It uses various grammar rules and syntax analysis to determine the meaningfulness of each token.

7.3 Lemmatization

Lemmatization is the process of reducing the submitted form of a word or words by making sure the root word of the submitted word belongs to the language. In lemmatization, the root word is known as "lemma." A lemma is a canonical lexicon or the citation form of a group of words. Lemmatization converts the word to its meaningful base form considering its context. The reason that lemmatization is preferred more over stemming is that it considers the morphological analysis of the words. Lemmatization is used when a valid word is required which exists in the dictionary to understand its actual meaning [12]. For example, plays, playing, and played are all forms of the word "play," thus "play" is the lemma or the root form of all these words. To use lemmatization, Python has the NLTK library which provides the WordNet lemmatize function.

7.4 Parts of Speech Tagging

Parts of speech have emerged from English language grammar. In English grammar, we have different constructs like nouns, adjectives, adverbs, and prepositions which are collectively known as parts of speech. Some parts of English grammar are also required when we deal with NLP. Thus, whenever we use NLP, we need to use the concept of parts of speech to understand the context of the sentence. In a similar fashion to parts of speech in English, we use POS tagging which is also known as parts of speech tagging in NLP. POS tagging refers to the conversion of the sentence to a list of tuples. Each tuple mainly has two parts, i.e., word and its corresponding tag which is collectively written as (word, tag). Tag helps us to identify whether the word is an adverb, adjective, verb, noun, etc.

In our proposed system, to further improve the process of lemmatization, we have assigned parts of speech tagging to every word. This improves the accuracy of word to root word conversion.

Table 2 Showing different non-manual component

Semantic relation	Syntactic category	Example
Synonymy (similar)	Noun, verb, adj, and adv	Pipe, tube rise, ascend sad, unhappy, rapidly, and speedily
Antonymy (opposite)	Noun, verb, adj, and adv	Wet, dry, rapidly, slowly top, bottom, rise, and fall
Hyponymy (subordinate)	Noun	Sugar maple, maple, tree, and plant
Meronymy (part)	Noun	Brim, hat, gin, martini ship, and fleet
Toponymy (manner)	Verb	March, walk, whisper, and speak
Entailment	Verb	Drive, ride, divorce, and marry
Derivation	Adj and adv	Magnetic, magnetism, simple, and simple

7.5 Natural Language Generation

NLG uses artificial intelligence to generate written or spoken words of the human language. It uses various data analysis algorithms to convert data into a human language. The main use of NLG is in chatbots and artificial homes such as Alexa/Siri/Google Home.

7.6 Semantic Relationship Detection

Primarily, relationship detection refers to understanding two objects and pictures the relationship between the two in the image. And semantic relationships are referred to as the interrelations that existed between the meanings of the phrases, words, or sentences while speaking. Understanding the absolute nature of the semantic relationship between the set of words given is significant sets a firm ground for the other applications of NPL such as question–answering, speech recognition, and summarization.

Table 2 has been referenced from [13].

8 HamNoSys Symbol

Hamburg Notation System was first developed in 1985, by a group of hearing and deaf people. Unlike the other scripts for writing sign languages, HamNoSys isn't meant as a realistic writing tool for day to day communication. Rather, its motive is like that of the International Phonetic Alphabet, that's used to transcribe the sounds of any spoken language in a constant way. As the larger number of possible parameter versions did not allow for a well-known alphabet acquainted to the users, newly

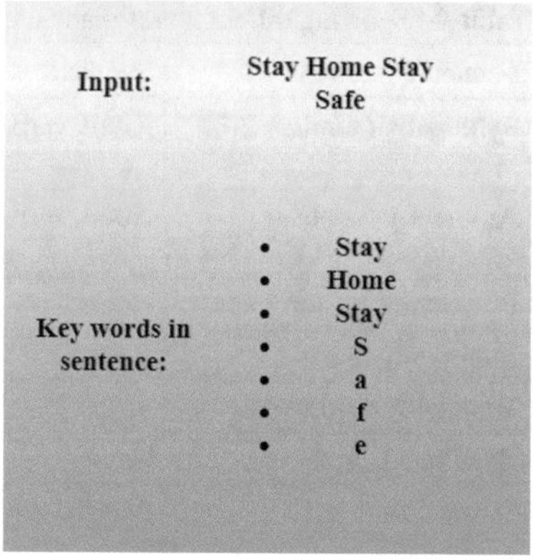

D. Tulsian et al.

Fig. 2 Example 1

Input:	Stay Home Stay Safe
Key words in sentence:	• Stay • Home • Stay • S • a • f • e

created glyphs must be designed in a manner that facilitates memorizing or maybe deducting the meaning of the symbols wherever possible.

As it appeared apparently, given the state of the art in sign language research, a notation system might now no longer be able to address all factors of signal formation description for all signal languages proper from the beginning, HamNoSys must allow each for a standard evolution and specializations. New versions of the system should no longer render antique transcriptions invalid.

9 Implementation

In Fig. 1, the user has entered "Stay Home Stay Safe" as the input then the keywords in the sentence are {Stay, Home, Stay, Safe}. The sign language video for "Safe" is not present in the database so it is further tokenized. Therefore, the final list of keywords becomes {Stay, Home, Stay, S, a, f, e}.

In Fig. 2, the user has entered "I wish a happy birthday" as the input. The model uses various natural language algorithms and applies the Indian Sign Language grammar rules and turns the sentence into "Me wish you happy birthday".

Here are some screenshots of the actual web page showing the video output for "Stay Home Stay Safe".

10 Conclusion

In our research paper, we have presented an English to Indian Sign language translation system which shows how using natural language processing (NLP), we can erase the communication barrier in society especially for deaf and dumb people. The system first performs the audio to text conversion module followed by text to ISL rules-based con-version. Then using NLP, we have integrated lemmatization with

Fig. 3 Example 2

Fig. 4 Implementation of the word 'stay'

Fig. 5 Implementation of the word 'home'

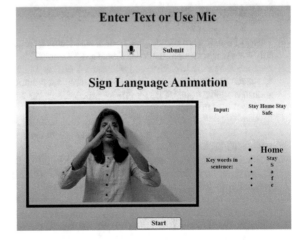

Fig. 6 Implementation of
the letter 'r'

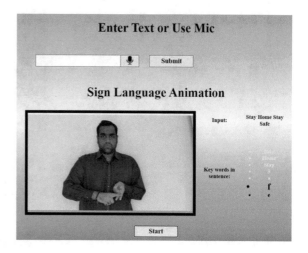

parts of speech tagging to increase the system's ability to convert ISL words to their corresponding root word. At last, our personalized recorded video output is shown according to ISL grammar rules. This system can be integrated into many areas like schools, airports, railway stations, and hospitals which could be a boon for Indian society.

The future scope of our project is to enhance our database by adding more recorded videos of ISL common phrases and to include the reverse system of this project, i.e., sign language to text/audio which will help deaf people in communicating their thoughts with other people. In this way, we will create a 2-way communication system that could prove to be an asset for the Indian society soon.

References

1. History | Indian Sign Language Research and Training Center (ISLRTC). Government of India. http://www.islrtc.nic.in/history-0. Accessed 17 Feb 2022
2. Heera SY, Murthy MK, Sravanti VS, Salvi S (2017) Talking hands—an Indian sign language to speech translating gloves. In: IEEE international conference on innovative mechanisms for industry applications, ICIMIA 2017—proceedings, pp 746–751. https://doi.org/10.1109/ICIMIA.2017.7975564
3. Truong VNT, Yang CK, Tran QV (2016) A translator for American sign language to text and speech. In: 2016 IEEE 5th global conference on consumer electronics, GCCE 2016. https://doi.org/10.1109/GCCE.2016.7800427
4. Dutta KK, Kumar SRK, Kumar AGS, Arokia SSB (2016) Double handed Indian sign language to speech and text. In: Proceedings of 2015 3rd international conference on image information processing, ICIIP 2015, pp 374–377. https://doi.org/10.1109/ICIIP.2015.7414799
5. RVS Technical Campus, IEEE Electron Devices Society, Institute of Electrical and Electronics Engineers. In: Proceedings of the international conference on electronics, communication and aerospace technology (ICECA 2017), 20–22 Apr 2017
6. Priya L, Sathya A, Raja SKS (2020) Indian and English language to sign language translator—an automated portable two way communicator for bridging normal and deprived

ones. In: ICPECTS 2020—IEEE 2nd international conference on power, energy, control and transmission systems, proceedings. https://doi.org/10.1109/ICPECTS49113.2020.9336983

7. Sharma DS, Sangal R, Singh AK et al (2016) Automatic translation of English text to Indian sign language synthetic animations. NLP Association of India, pp 144–153
8. Jadhav K, Gangdhar S, Ghanekar V (2021) Speech to ISL (Indian sign language) translator. Int Res J Eng Technol
9. Institute of Electrical and Electronics Engineers (2012) Proceedings on 2012 1st international conference on emerging technology trends in electronics, communication and networking (ET2ECN), 19–21 Dec 2012. IEEE
10. Tripathi K, Nandi NBGC (2015) Continuous Indian sign language gesture recognition and sentence formation. Procedia Comput Sci 523–531
11. Gelbukh A (2005) Since 1997 he is the Head of Natural Language Processing Laboratory of the Center for Computing Research of the National Polytechnic Institute, Mexico
12. Socher R, Bengio Y, Manning CD (2012) Deep learning for NLP (without magic). In: Association for computational linguistics
13. WordNet: a lexical database for English George A. Miller

Classification of Patient's Heartbeat Obtained by ECG Using Active Learning

Neha Shukla, Anand Pandey, and A. P. Shukla

Abstract While physicians can correctly detect various anomalies in heartbeat signals by using electrocardiograms (ECGs) of different individuals, supervised machine learning has mixed results. The diversity of activities performed by patients is often severe class disproportion, and the cost of obtaining and mapping them to label data for individual patients makes the issue more complicated. We generate synthetic biological data using SimGANs and solve these issues by performing patient-adaptive and task-adaptive heartbeat categorization via active learning. Our approach performed much better than the previous methods of heartbeat classification when evaluated on a benchmark database of ECG recordings on the two main classification tasks specified by the medical sciences. Furthermore, our method needed almost 90% less patient-customized training data than the techniques we evaluated.

Keywords ECG · Active learning · SimGANs · Machine learning · Deep learning

1 Introduction

In India, one of the major causes of death is a heart attack or artery disease. During the COVID-19 pandemic, we saw a significant increase in heart stroke cases of patients who have recovered but due to some side effects their arterial walls became thick and they suffered heart attack [1]. It is also caused by the deposit of fat, cholesterol in the arteries. Early estimation could help the individual and society when inexpensive and simple treatment plans such as diet or drugs could be followed to reduce the

N. Shukla (✉) · A. Pandey
CSE Department, SRM Institute of Science and Technology, Delhi-NCR, Meerut Road, Modinagar, India
e-mail: ns3951@srmist.edu.in

A. Pandey
e-mail: anandp@srmist.edu.in

N. Shukla · A. P. Shukla
CSE Department, KIET Group of Institutions, Delhi-NCR, Ghaziabad, India
e-mail: ap.shukla@kiet.edu

© The Author(s), under exclusive license to Springer Nature Singapore Pte Ltd. 2023 571
A. Shukla et al. (eds.), *Computational Intelligence*, Lecture Notes in Electrical Engineering 968, https://doi.org/10.1007/978-981-19-7346-8_49

damages caused to heart tissues before it becomes incurable. WHO report indicates that CVDs cause 31% of global deaths, out of which at least three-quarters of deaths occur in low or medium-income countries. One of the primary reasons behind this is the lack of primary healthcare support and the inaccessible on-demand health monitoring infrastructure [2]. Electrocardiogram (ECG) is considered one of the essential attributes for continuous health monitoring required for identifying those at serious risk of future cardiovascular events or death [3].

For a single patient, an electrocardiogram (ECG) may capture over 100,000 heart-beats in 24 h [4]. For a doctor, it is unlikely to examine all of them. So, an automatic technique to analyse these long-term ECG signals is required to assist doctors in gaining a better understanding of a patient's physiological condition and risk of poor cardiovascular events [5, 6]. Labelling the various kinds of heartbeats is often an essential step in such a study. This labelling simplifies an ECG to a collection of symbols that may be transferred from patient to patient [7]. Over a dozen, distinct kinds of heartbeats may be identified in ECG records by trained doctors. Researchers have had mixed results utilizing supervised machine learning methods to achieve the same results [8–11]. As the physiological systems generate data in a variety of ways, the classifier performance is compromised when an unseen or unexpected data is fed to it for prediction. This is mainly due to the variety in the shape and timing of the ECG signals that are generated by damaged cardiovascular systems. As a result, global classifiers are notoriously inaccurate, and as a result, are seldom utilized for real-time applications [12]. Hu et al. [13] were among the first to develop a patient-adaptable automated ECG heartbeat classifier, which could easily classify ventricular ectopic beats (VEBs) and non-VEBs. They [13] combined a global with local classifier and trained it on the patient's ECG record for first 5 min, resulting in a hybrid method. De Chazal and Reilly [14] added patient-specific expert information to improve the performance of a global classifier. The first 500 tagged beats of each record were used to train their local classifier [15]. Ince et al. [16] created a patient-adaptable classification method utilizing artificial neural networks (ANN) more recently by include the initial five minutes of every recording that is present in the training set.

Hand-coded rule-based systems for heartbeat categorization have shown some success [17]. Premature ventricular contraction (PVC) is a particularly hazardous form of ectopic heartbeat that Hamilton devised a rule-based method for identifying [12]. Rule-based algorithms, although fairly accurate, are over fitted; hence, they can only be utilized for a particular classification job. And, in order to be helpful in real world, a classifier must be able to adapt not just to unseen data, but also to novel classification problems, since the classification job at hand may differ depending on the condition of patient or even the clinician. Because the area of ECG research is always developing, signal analysis techniques should be able to change as well. We demonstrate how active learning may be effectively used to both patient-adaptive and task-adaptive heartbeat categorization issues in this article.

Our approach was created keeping clinical context in conscious: it needs zero labelled data at first, and does not need any customized parameters, and performs well on an unbalanced data set. Our approach outperforms current framework machine learning algorithms which are used for heartbeat classification and requires significantly a smaller number of training data when applied to data from the MIT-BIH Arrhythmia Database. Furthermore, our algorithm beats a rule-based system for detecting a certain kind of irregular beat. Finally, we describe the results of the categorization technique in a prospective study involving two cardiologists.

Electrocardiography (ECG) and photoplethysmography (PPG) are most commonly used methods for determining heart activity [3]. These two kinds of signals are critical for medical applications as they capture both the basic mean heart rate (HR) and more specific information such as heart rate variability (HRV). However, skin-contact ECG/BVP sensors are used to detect these signals, which may be uncomfortable and unpleasant for long-term monitoring. Remote photoplethysmography (rPPG), which aims to monitor heart activity remotely and without any touch, has been quickly evolving in the recent years to address this issue [9]. The possible changes in the patient's skin's surface body are measured by an ECG, which captures the patient's cardiac electrical activity [18]. In most healthy individuals, the ECG starts with a P-wave, then a QRS complex, and finally, a T-wave, as measured from Lead II [12, 19]. Cardiac irregularities may cause the heart's normal sinus rhythm to be disrupted, and they can range from benign to life threatening, depending on the kind and frequency of the abnormalities [20].

2 Background Study

Previous methods mostly include conventional machine learning approaches as described below:

K-Nearest Neighbour (KNN): The KNN is a supervised learning algorithm in which we have to establish a learning function between the input and output features [21]. KNN can be used to solve both regression problem in which the output is a discreet value and classification problem in which we divide the new instances in different classes [22]. KNN works on the principal of similarity in which distance is calculated between a new data feature and previously classified features. The various distance metrics available for KNN algorithm are:

$$\text{Euclidean Distance: } ED(i, j) = \sqrt{\left|x_{i1} - x_{j1}\right|^2 + \left|x_{i2} - x_{j2}\right|^2 + \cdots + \left|x_{ip} - x_{jp}\right|^2}$$

It is the measurement of distance between two points i and j on a Cartesian plane.

$$\text{Manhattan Distance: } distance(i, j) = \left|x_{i1} - x_{j1}\right| + \left|x_{i2} - x_{j2}\right| + \cdots + \left|x_{ip} - x_{jp}\right|$$

This distance is measured between two points i and j along axis at right angle.

Logistic Regression: The logistic regression works by using a sigmoid activation function. The output is generally a probabilistic value which shows the confidence in the assumed hypothesis [23]. Logistic regression is further classified into binary logistic regression where the output lies in either of two classes, multinomial logistic regression where the output can lie in three or more classes but the data is unordered and ordinal logistic regression where the data can lie in three or more classes and order is associated with that data.

Random Forest Classifier: It is a meta-estimator that uses technique of ensemble learning to fit multiple decision tree into it which is further used to classify data in different classes [23]. It uses averaging to improve the predictive accuracy and hence controls the over fitting problem which is generally encountered by decision tree. The features are always randomly permuted at each split.

Support Vector Machine (SVM): These methods can be used for classification, regression, and outlier detection. It is generally used as it shows high efficiency in higher dimensional space. It is a versatile algorithm as different kernel functions can be specified for decision functions [23]. It works by constructing a hyper plane or set of hyper planes in a higher or infinite dimensional space which is used to solve problems of regression, outlier analysis and classification.

XGBoost Classifier: It uses a more iterative approach, but uses ensemble learning in the background. It combines different models and trains them in succession as the error done by previous model is corrected by the next model [23]. When there is no further correction needed, the process is stopped.

3 Methods

The two major components of our heartbeat categorization system are described in this section. We start with the feature extraction procedure and then move on to the classification approach.

3.1 Feature Extraction

Unnecessary number of parameters becomes an overhead while training a model, so it is better to remove parameters having high correlation. This technique is known as feature extraction. It is performed using various feature extraction mechanisms like PCA, LDA or genetic algorithms. It limits the dimension of huge data set to feasible number of parameters for training using some clustering technique.

We pre-process and segment the ECG before obtaining feature vectors. The R-peaks of each heartbeat were detected using PhysioNet's automated R-peak detector [24]. We then used the technique outlined in [25] to eliminate baseline drift from the signals. The data was then segmented into individual heartbeats based on prede-termined periods before and after the R-peak, ensuring that each beat had the same amount of samples.

Total 67 features are used denoted as $\vec{x}$ where first 60 $\{x_1, x_2, x_3 \ldots x_{60}\}$ features are using a Daubechies 2 wavelet, which are retrieved from the ultimate 5 levels of a 6 level wavelet decomposition. The next three x_{61}, x_{62}, x_{63} are the beat's adjusted energy in various parts, and x_{64}, x_{65}, x_{66} represent the average RR interval, as well as the before and post RR interrupts were standardized by a local mean. The x_{67} is the geometrical gap between the actual beat and the median beat on the record which is recalculated on every 500 heartbeats.

3.2 Classification

We aim to create a patient-adaptable heartbeat classifier technique that could be used to solve a variety of binary heartbeat classification issues. The classifier was created for use in a clinical environment, when doctors do not have much time to identify beats, much alone adjust classifier parameters. As a result, it was critical that the technique only required a few cardiologist-labelled heartbeats but that no user-defined variables be used. Through examination of these objectives, leads us to create the technique, which incorporates several concepts from the researchers [12, 14, 25–28].

Many suggested SVM active learning methods require that one begins with a set of labelled data or that the first training examples are chosen at random, as in [27]. We begin with a pool of data that is entirely unlabelled in our applications. Moreover, since there is frequently a imbalance in the class (for example, a very large number of beats recordings include fewer than a dozen of PVCs), selecting less or even modest quantity of randomized samples is rare to provide representative samples of a record. The first query selection is critical. If just one class of beats is searched, the algorithm may come to a halt. More generally, the binary job has no bearing on the selection of the initial set of queries, thus the debut inquiry should include additional instances from every beat classes of the record. In order to rapidly find depiction samples from every class, we utilize clustering. Prior settling on hierarchical clustering, we tested with several clustering methods.

Hierarchical clustering beats other common clustering methods, such as k-means, on aggregate. This, we think, is due to the fact that simply changing the linking criteria, hierarchical clustering may generate a number of distinct clusters. In order to overcome the intra-patient variance seen in ECG data, we decided to utilize two complimentary linking criteria. The average linkage is the first statistic. The average gap between two clusters, x and y, is defined as the mean gap between all pairs of objects in y and z by average linkage. This connection tends to combine clusters with small variations and is biased towards generating clusters with comparable variances. Ward's linkage [29], stated in Eq. 1, is the second linkage criteria.

$$d(y, z) = \text{ss}(yz) - [\text{ss}(y) + \text{ss}(z)] \tag{1}$$

When y and z are merged, $\text{ss}(yz)$ is the intra-cluster sum of squares for the resultant cluster. The intra-cluster sum of squares, $\text{ss}(y)$, is calculated as the sum of squares of the distances among all cluster objects and the cluster centroid.

We train a linear SVM and apply it to all of the data after the first queries have been tagged. As most heartbeat data sets are roughly linearly separable and linear SVMs need minimal tuning parameters, we employ them. Then, on or within the margin of the SVM, we re-cluster the data, increasing the maximum number of clusters for each iteration. The beat from each cluster that is closest to the SVM decision border is then queried.

As previously stated, our system would come to a stop if there was no unlabelled data on or within the boundary. However, certain recordings, such as those containing fusion beats (a mix of regular and aberrant rhythms), are exceptions. Many beats may fall inside the SVM's margins, resulting in a clinician labelling hundreds of beats that contribute little information. When more training data has little to no impact on the answer, one should intuitively cease searching. As a result, the algorithm stops when the difference in the margin between repetitions is less than ε'.

We utilized $\text{SVM}_{\text{light}}$ [28] to train the linear SVM at each iteration of our method, which we developed in MATLAB. Throughout all of the trials, we kept the linear SVM's cost parameter, $C = 100$, constant. Based on prior cross-validation tests, this value was chosen. The stopping precision ε' was maintained at $\varepsilon' = 10^{-3}$. Typical ECG recordings include 2–5 classes of beats, but they may contain more; we conservatively choose $k = 10$ based on this a priori information. Throughout all of the tests, this value remained consistent.

We conducted a series of tests using data from various patients and for various classification tasks to see whether our suggested method for heartbeat classification was useful. First, we compare the performance of a classifier created using our method against that of two previously published classifiers. Then, using the same pre-processing and characteristics as our proposed active learning approach, we create our own passive learning classifier to directly evaluate the effect active learning has on the categorization of heartbeats. Finally, we put our approach to the test with real cardiologists.

We provide classification results in terms of sensitivity (S), specificity (SF) and positive predictive value (PoPrV) in our tests (PoPrV). The F-score is used to assess overall performance. In medicine and information retrieval, the F-score is a widely recognized performance assessment metric in which one data class (typically the positive class) is more significant than the other [14]. We choose this metric because there is a lot of class imbalance in the issue of heartbeat classification, thus the S (aka recall) and PoPrV (aka accuracy) are more significant than the SF.

4 Classifier Performance

The MIT-BIH Arrhythmia Database (MITDB) [24], which is a widely utilized benchmark database of forty-eight half-hour ECG recordings that have been sampled at 360 Hz from 47 distinct individuals, was utilized to assess performance. From a pool of 4000 recordings, 23 of these tracks, designated 100–124, were chosen at random. The continuing 25 records, numbered from 200 to 234, were chosen as they include uncommon clinical trials that would not have been examined if all forty-eight records had been randomly picked. There are about 109,000 cardiologist-labelled heartbeats in the database. Each beat is assigned to one of sixteen distinct classes. In some ways, the MITDB's data is too excellent. It was recorded at 360 Hz, which is greater than the usual sample rate for Holter monitors used to capture most long-term clinical data. The data set was resampled and the pre-processed ECG signal was represented at 128 Hz to mimic this kind of data. The association for the Advancement of Medical Instrumentation (AAMI) has suggested two major categorization tasks: identifying ventricular ectopic beats (VEBs) and detecting supraventricular ectopic beats (SVEBs). Other studies looking on patient-adaptive heartbeat classification have focused on these two objectives. Ince et al. [16] and De Chazal and Reilly [14] recently developed techniques that integrate global and patient-specific data. Ince et al. trained a global classifier on 245 hand-selected beats from the MITDB, then modified it by training it on labelled data from the first five minutes of each test record. Table 1 gives the results of their testing on 44 of the 48 recordings from the MITDB (all records with timed beats were removed). De Chazal and colleagues trained their global classifier on all of the data from 22 patients in the MITDB, then modified it by training on labelled data for the first 500 beats of each test record. Figure 1 also includes their findings of testing on 22 entries from the MITDB that are distinct from the ones used in the global training set.

As given in Table 1, the technique presented here outperforms the methods described in [14, 16] for each job. Our approach utilized 45 labelled beats per record on average to categorize VEBs versus non-VEBs (compared to approximately 350 beats for [16] and 500 beats for [14]). Our approach required significantly fewer labelled beats for the job of identifying SVEBs. Because the class imbalance issue is more severe, and supraventricular beats are tougher to differentiate from regular sinus rhythm beats, identifying SVEBs is much more challenging than detecting VEBs.

Table 1 Comparison of our methods with other classifiers

Classifier	Ventricular ectopic beats				Supraventricular ectopic beats			
	S	SF	PoPrV	F-score	S	SF	PoPrV	F-score
De Chazal et al.	92.6	98.3	95.2	93.9	86.4	94.0	43.0	57.4
Ince et al.	82.7	97.4	86.4	84.5	64.5	97.0	54.6	77.5
Proposal 1	98.2	99.4	99.0	98.8	87.3	100	99.2	92.9
Proposal 2	99.4	99.9	99.1	99.2	89.0	100	99.8	94.1

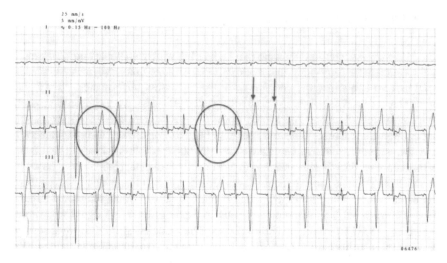

Fig. 1 Classifier using active learning

4.1 Active Learning

The way we have designed our feature vector and the criteria of selection of training data set leads to the performance gap between our model and the other models as given in Table 1. To test this idea, we conducted a study that explicitly contrasts the effects of actively versus passively choosing the training set, while keeping all other variables constant (e.g. identical pre-processing, identical feature vectors, etc.). We compare a VEB versus non-VEB classifier trained on the first 500 beats of each record to a linear SVM classifier developed on the first 500 beats of each sample using our method for each of the 48 recordings in the MITDB. We keep track of the number of queries performed for each patient, as well as the performance of each classifier. These findings demonstrate that active learning may significantly decrease the labour cost of generating highly accurate classifiers. Despite not having any global training data, the passive method performed better than [16] and nearly as well as [14], indicating that our feature vector offers some benefit.

4.2 Clinical Experiments

We conducted an experiment with two cardiologists using data from another cohort of patients hospitalized with NSTEACS to see whether our method could be used in a real-world situation. Unlike those in the MITDB, the ECG tracings in this database are not very clean, including a significant amount of noise and many artefacts. This makes them more reflective of the data that a clinical algorithm is likely to encounter. We looked at four records selected at random from a group of patients who had at least one episode of ventricular tachycardia in the seven days after randomization. We examine the first half-hour of each record, yielding a test set of 8230 heartbeats.

We utilized a somewhat different halting criteria established before in these trials. Each cardiologist was given an ECG plot of the heartbeat to be classified and the beats around it, similar to the one shown in Fig. 1, as our system selected beats to be labelled. The cardiologist was then instructed to identify it using the following key: 1 = obviously non-PVC, 2 = ambiguous non-PVC, 3 = ambiguous PVC, 4 = plainly PVC. Because the cardiologists disagreed on how to name certain beats, one was asked to identify an average of 15 beats each record, while the other was instructed to label approximately 20 beats per record. Each cardiologist required approximately 90 s per record to complete the procedure.

5   Conclusion and Future Work

We proposed a methodology that can automatically categorizing activity in ECG recordings that might be used in clinical settings. The customizable patient variations in the anatomy and synchronization features of the ECG generated by damaged cardiovascular systems, as well as the diversity in the classification assignments that a physician may wish to undertake, make the issue difficult to solve. We present a technique for performing patient-adaptive and task-adaptive heartbeat categorization using active learning to solve these issues. Primary classification assignments of VEBs and SVEBs as specified by AAMI, our approach outperformed other previously published algorithms when evaluated on the most commonly used benchmark database of cardiologist annotated ECG recordings. Furthermore, our approach needed almost 90% less training data than the systems it was compared against. For a third frequent classification job, we demonstrated that our approach outperforms a SOTA hand-coded techniques.

We performed short research with two cardiologists to see whether our technique might be used in practice. With minimal training, the cardiologists were able to use our method to perform classification assignment and found the result to be satisfactory. These early findings are very promising, and they indicate that active learning may be utilized in a clinical environment to not only decrease labour costs but also to enhance performance. There is, of course, still space for development. We utilized the same input parameters in all of the trials; fine-tuning these factors may enhance the findings. However, parameter tweaking is difficult in a clinical environment; therefore, further research into automatically parameter hyper-tuning is required. Examining on early tests, we think that improving performance while reducing the overall number of needed labels may be accomplished by first learning the optimum number of required clusters for each record. By beginning with a global classifier and then modifying it via active learning, it may be feasible to further decrease the amount of expert work needed.

References

1. Kiranyaz S, Ince T, Gabbouj M (2015) Real-time patient-specific ECG classification by 1-D convolutional neural networks. IEEE Trans Biomed Eng 63(3):664–675
2. Niu X, Han H, Shan S, Chen X (2018) SynRhythm: learning a deep heart rate estimator from general to specific. In: Proceedings of international conference on pattern recognition, vol 2018, no i, Aug 2018, pp 3580–3585. https://doi.org/10.1109/ICPR.2018.8546321
3. Weimann K, Conrad TOF (2021) Transfer learning for ECG classification. Sci Rep 11(1):1–12
4. Gupta V, Saxena NK, Kanungo A, Gupta A, Kumar P et al (2022) A review of different ECG classification/detection techniques for improved medical applications. Int J Syst Assur Eng Manag 1–15
5. Syed Z, Guttag J, Stultz C (2007) Clustering and symbolic analysis of cardiovascular signals: discovery and visualization of medically relevant patterns in long-term data using limited prior knowledge. EURASIP J Adv Signal Process 2007. https://doi.org/10.1155/2007/67938
6. Exner DV et al (2007) Noninvasive risk assessment early after a myocardial infarction. The REFINE study. J Am Coll Cardiol 50(24):2275–2284. https://doi.org/10.1016/j.jacc.2007.08.042
7. McDuff D (2018) Deep super resolution for recovering physiological information from videos. In: IEEE conference on computer vision and pattern recognition workshops, vol 2018, June 2018, pp 1448–1455. https://doi.org/10.1109/CVPRW.2018.00185
8. Chaichulee S et al (2017) Multi-task convolutional neural network for patient detection and skin segmentation in continuous non-contact vital sign monitoring. In: Proceedings of 12th IEEE international conference on automatic face & gesture recognition, FG 2017 and 1st international workshop on adaptive shot learning for gesture understanding and production, ASL4GUP 2017, biometrics in the wild, Bwild 2017, Heteroge, pp 266–272. https://doi.org/10.1109/FG.2017.41
9. Yu Z, Li X, Zhao G (2020) Remote photoplethysmograph signal measurement from facial videos using spatio-temporal networks. In: 30th British machine vision conference 2019, BMVC 2019
10. Tiwari AK, Shukla N (2021) Brain tumor segmentation using CNN. In: Recent trends in communication and electronics: proceedings of the international conference on recent trends in communication and electronics (ICCE-2020), Ghaziabad, India, 28–29 Nov 2020, p 411
11. Karim AM (2022) Effective classification of ECG signals using enhanced convolutional neural network in IoT. arXiv preprint arXiv:2202.05154
12. Hamilton P (2002) Open source ECG analysis. Comput Cardiol 29:101–104
13. Hu YH, Palreddy S, Tompkins WJ (1997) A patient-adaptable ECG beat classifier using a mixture of experts approach. IEEE Trans Biomed Eng 44(9):891–900. https://doi.org/10.1109/10.623058
14. De Chazal P, Reilly RB (2006) A patient-adapting heartbeat classifier using ECG morphology and heartbeat interval features. IEEE Trans Biomed Eng 53(12):2535–2543. https://doi.org/10.1109/TBME.2006.883802
15. De Haan G, Jeanne V (2013) Robust pulse-rate from chrominance-based rPPG, pp 1–9
16. Ince T, Kiranyaz S, Gabbou M (2009) A generic and robust system for automated patient-specific classification of ECG signals. IEEE Trans Biomed Eng 56(5):1415–1426. https://doi.org/10.1109/TBME.2009.2013934
17. Wang T, Lu C, Sun Y, Yang M, Liu C, Ou C (2021) Automatic ECG classification using continuous wavelet transform and convolutional neural network. Entropy 23(1):119
18. Wiens J, Guttag JV (2010) Active learning applied to patient-adaptive heartbeat classification. In: Advances in neural information processing systems 23: 24th annual conference on neural information processing systems 2010, NIPS 2010, pp 1–9
19. Sokolova M, Japkowicz N, Szpakowicz S (2006) AI 2006: advances in artificial intelligence. In: 19th Australian joint conference on artificial intelligence, Hobart, Australia, 4–8 Dec 2006. Proceedings, pp 1015–1021. [Online]. Available: https://doi.org/10.1007/11941439_114

20. Kirkwood L (2015) Chimerica. Chimerica. https://doi.org/10.5040/9781784600266.00090006
21. Tiwari AK, Sharma S, Kumar D. Hybrid approach for brain tumor segmentation
22. Gupta V, Mittal M, Mittal V, Gupta A (2022) An efficient AR modelling-based electrocardiogram signal analysis for health informatics. Int J Med Eng Inform 14(1):74–89
23. Bahrami M, Forouzanfar M (2022) Sleep apnea detection from single-lead ECG: a comprehensive analysis of machine learning and deep learning algorithms. IEEE Trans Instrum Meas
24. Goldberger AL et al (2000) PhysioBank, PhysioToolkit, and PhysioNet: components of a new research resource for complex physiologic signals. Circulation 101(23). https://doi.org/10.1161/01.cir.101.23.e215
25. Sternickel K (2002) Automatic pattern recognition in ECG time series. Comput Methods Programs Biomed 68(2):109–115. https://doi.org/10.1016/S0169-2607(01)00168-7
26. Dasgupta S, Hsu D (2008) Hierarchical sampling for active learning. In: Proceedings of 25th international conference on machine learning, pp 208–215. https://doi.org/10.1145/1390156.1390183
27. Buff RI (2009) The deported. Am Q 61(2):417–421. https://doi.org/10.1353/aq.0.0077
28. Kamvar SD, Klein D, Manning CD (2002) Interpreting and extending classical agglomerative clustering algorithms using a model-based approach. In: Proceedings of 19th international conference on machine learning, no 1, p 8
29. Ward JH (1963) Hierarchical grouping to optimize an objective function. J Am Stat Assoc 58(301):236–244. https://doi.org/10.1080/01621459.1963.10500845

Career Path Prediction System Using Supervised Learning Based on Users' Profile

Hrugved Kolhe, Ruchi Chaturvedi, Shruti Chandore, Gopal Sakarkar, and Gopal Sharma

Abstract In the past few years, technology has changed drastically and due to COVID-19 pandemic, people spend more time on screen. The use of social media platforms has also been increased and this affects the human mind and decision taking ability. Online career counseling is largely supported these days and hence this paper proposes an online career prediction system using supervised machine learning based on the user's profile. This research attempted to develop a model for the user which predicts the career path in a precise manner and gives actionable feedback and career recommendations to encourage them to make significant career judgments.

Keywords Career prediction · Online career counseling · Career guidance · Machine learning · Supervised machine learning · Decision tree · SVM · Random forest

H. Kolhe (✉) · R. Chaturvedi · S. Chandore
Department of Artificial Intelligence, G H Raisoni College of Engineering, Nagpur, India
e-mail: geeky.hrugved143@gmail.com

R. Chaturvedi
e-mail: rch432@gmail.com

S. Chandore
e-mail: shrutichandore86@gmail.com

G. Sakarkar
Associate Professor, D Y Patil Institute of Masters of Computer Application and Management, Pune, India
e-mail: g.sakarkar@gmail.com

G. Sharma
Vice President of Technology, MyCaptain, Bangalore, India
e-mail: gopal.gy.sharma@gmail.com

© The Author(s), under exclusive license to Springer Nature Singapore Pte Ltd. 2023
A. Shukla et al. (eds.), *Computational Intelligence*, Lecture Notes in Electrical Engineering 968, https://doi.org/10.1007/978-981-19-7346-8_50

583

1   Introduction

1.1   *Predictive Analytics*

A prediction system is a model that is used to predict or forecast future events or outcomes, based on statistical analysis [1]. Predictive analytics involve predicting the future trends and outcomes. The two approaches that are used for predictive analytics are machine learning techniques and regression techniques [2]. Common machine learning algorithms types are supervised learning, semi-supervised learning, unsupervised learning, reinforcement learning, transduction and learning to learn [3]. However, this paper uses supervised machine learning algorithms for predicting the next step of career. Insurance companies use working professional's data with the help of a third-party and identifies the type of insurance those professionals will be interested in with the help of predictive analytics [4]. Banks use the same technique to identify fraudulent and make customers alert about frauds [5]. It has applications in sales as well. For example, pharmaceutical companies use predictive analytics to prodigy sales in particular areas and become alert for expiry of that medicine [6].

1.2   *Supervised Machine Learning*

Supervised learning is defined as learning with labeled data. In this, a machine is trained to predict a target variable based on observations [7]. Supervised learning algorithms are very well-known machine learning algorithms and are widely used in classification and regression tasks. Predictive models having the labeled dataset use the supervised learning approach on a large scale as this machine learning technique is very efficient in classifying the dataset in the required classes based on the classification algorithm used [8]. Complex trees, boosted trees, random forests, nearest neighbors, neural networks, support vector classifiers, local kernel-weighted methods are some of the prominent algorithms used in the predictive models as these algorithms support better data exploration and data handling [9, 10]. When there is a case where data is not labeled correctly which is a very common case when we collect the data from the Internet and the most data available on the Internet is incompletely labeled due to human cost, for this problem, we have deep learning techniques [11, 12], we also use the semi-supervised learning approach in which who basic assumptions are used that are clustering assumption and manifold assumption. Both assumptions are used for the data distribution, which is very helpful in the case where data is incompletely labeled [13]. In this study, we have used the XGBoost classifier, decision tree, random forest and support vector classifier for the prediction of the model, then we have compared the scores of the applied algorithms and selected the most suitable algorithm for the dataset, and the output of the prediction is portrayed by deploying our model in a GUI. The work presented in this literature paves the way for designing and implementing the career prediction model in multiple fields. This will definitely guide a large group of students in choosing the appropriate career path.

1.3 Definitions

Students are required to identify their capabilities and interests so that they can pursue a career that will be beneficial for them to grow and learn. Hence, knowing capabilities and interest can improve performance and can motivate students to give themselves a right direction with a target of achieving a career [14].

Career prediction is a technique that can predict if a person is fit for a particular career or not. It helps in identifying the innate skills of a person and identifies his or her interest in a particular field [15]. Learning about careers is hard. It is hard to find out for which field a person is satisfied. It is hard to find out what careers people will prefer in the future. To prepare for the future, one needs to know where they are headed. Recently, technology has changed drastically [16]. In the past few years, one could not realize the changing trend so quickly, but today, the world has a variety of tools to predict career development. In this paper, an online career prediction system using Machine Learning [17].

1.4 Objective and Challenges

In this work, it is found that users are not able to choose a suitable career path for them due to problems like absence of guidance, unfamiliarity with the emerging career options with technology, lack of information about several skills and the talent required in the workplace. Most importantly, they do not know about their strengths and weaknesses [18]. Peer and parents pressure restrict them to choose the career of their own choice. The purpose of this research is to understand the user's personality, skills, interests, academic and non-academic performances and not only recommend best-suited courses according to their performance but also recommend career path and career domains that should not be chosen. The objective of the recommender system is to help the users with their decision-making skill and suggest users with the recommendation of online courses, making counseling groups for the same skilled users [19, 20].

2 Literature Review

Choosing a career right after 10th grade or 12th grade is very common in the Indian education path. Students often get confused while choosing the educational stream if their goals are not clear, some of them choose the career path under the parent's pressure or under the influence of their peers. Competition in today's world has grown significantly, so this might be a serious issue for the students with no career target. To solve this problem and help the students who find it difficult to make a career decision, multiple theories have been published by the researchers. Recently, machine learning has made it a lot easier to predict a career using different kinds of recommendation algorithms [21]. But the problem with most recommendation systems is that they use only the traditional questionnaire assessment parameters to assess the candidate, while the fact that the daily routine and the behavioral data can have a significant impact on the performance of the candidate, might help the recommendation system to make the right prediction which significantly increases the chances of success. Therefore, the behavioral data such as preferences, promptness, unique habits and mental state combined with the assessment statistics will help the recommendation system perform better than the systems using only the traditional assessment statistics [22, 23].

A lot of research has been done in the deep learning field using convolutional neural networks where a large dataset in the form of resumes was dimensionally reduced using the principal component analysis, and this data was fed to the multi-layer convolutional neural network [24]; in this study, NLP was also used as the data in the resume consist of text as well as the numerical and this was done using the word embedding technique. Using deep neural classifiers with NLP can boost the performance of the model by a significant amount. Using deep neural networks helps to optimize the value of prediction by backpropagation as this technique is efficient in handling the error by resetting the weights between neurons [25].

Social media handles used by the student have a great influence in predicting what are the actual interests of the student. As almost every student nowadays broadcasts their lives online which creates their digital fingerprint [26], this can help in predicting the occupation matching the student's personality considering the fact that one can excel in the field where he/she is comfortable and interested to work. Twitter and LinkedIn can be some of the most important social media platforms to access the personality of the student as this can help the prediction system in analyzing the student's fields of interest, what domain or technology is the student curious about and many such factors.

3 Proposed Methodology

3.1 Offline Career Counseling

Career counseling is a procedure that enables you to recognize your strengths and weaknesses as well as helps you to take important decisions to achieve the right career choice [27]. For accurate career prediction, many parameters are needed like users' academic scores in different subjects, specializations, logical capabilities, knowledge of programming and subjective details like interests, skills, involvement in different

Fig. 1 Career counseling steps

competitions, workshops, certifications, reading books and many more [28]. As all these factors play an important role in deciding growth toward a career area, all these things are taken into consideration by career counselors during offline career counseling [29].

The method of career counseling is differentiated into five main stages. Each of these stages is carried out sequentially to get accurate and beneficial results. The initial stage is to select the counselors of your choice. A good relationship with a counselor plays an important role in deciding the best career path. The second stage is to gather information so that counselors get to know more about the user and their preferences. Counseling can be done in multiple ways like asking several questions and filling forms. The third stage is the exploration stage which is an analytical stage. This analytical stage includes users' social behavior, values, personality, abilities, background and some tests like psychometric assessments and aptitude tests. With this piece of information, counselors can make better decisions regarding users' career paths. Correct counseling will explore which career path is right for a better future. The next stage is to create an action plan according to users' preferences to achieve your goals. This plan of action allows you to decide the career paths that suit users. The final stage is to implement your action plan under the direction of your counselor [30, 31] (Fig. 1).

3.2 Online Career Counseling

Online career counseling methods are emerging in the pandemic era of COVID-19. Due to the COVID-19 situation, all the educational institutions are asked to go with an E-learning pattern for students [32]. Hence, online learning is now part of the education system. Everyone, whether they are students or professionals who want to switch their career, wants to find a job suited to their personality and guarantees their professional and personal growth. To find a suitable career option, students or professionals who want to take their next career step need to approach a professional career counselor. In the previous years, to resolve such career-related problems, people used to approach a career counselor. In the recent years, online career counseling platforms have been created and now it is being done with the help of machine learning. Some people are not necessarily good at deciding what to do with the rest of their lives. Online career counseling saves time and effort and works the same as an offline career counseling system, but the suggestions are software-based. Online career guidance is a modern, fast and effective tool for career development, which helps professionals to find new jobs, find the right jobs for talents or find better jobs for their current positions.

The main advantage of online counseling is the user is comfortable with the questions asked and they can honestly assess themselves. Users do not hesitate to give an honest review for themselves that increases the accuracy of predicting the right career for the user. Hence, it decreases the chances of choosing the wrong career path [33].

There are many online platforms available for online career counseling. In general, online career counseling, the system assesses the user with 40–50 questions that include the user's interests, personality assessment, academic and non-academic questions. There are a lot of factors that can affect one's career. Hence, researches are going on to discover user's behavior toward several things including social media likes, dislikes, activeness on social and professional platforms.

3.3 Social Media: An Approach

Due to the increase in technology, social media platforms are used worldwide over the past few years. And, due to COVID-19 pandemic, about 50–70% usage of the Internet has been increased and 50% of the time has been spent on social media by the people [34].

With this fact, it has been observed that user's behavior and activity on social media also affect the important career-related decisions as according to a survey, social media bridges the communication gap between the provider and the needy and also increases employee's knowledge about different domains [35].

This can be a great source of information that can serve as features in a dataset and can increase the model's accuracy of a career prediction system. However, at the time of research, this information has not been used for preparing the dataset.

3.4 Dataset

Due to the unavailability of the required dataset, this research has been done on the manually created dataset. The inspiration of the data features was taken from the forms which the multinational companies release for hiring the candidates.

The data created, consists of a total 6800 rows and 20 columns (features). The dataset includes academic and non-academic areas which are numeric (like 'logical quotient rating,' 'hackathons,' 'coding skills rating,' 'public speaking points') and categorical columns (like, 'extra-courses did,' 'certifications') that also include text.

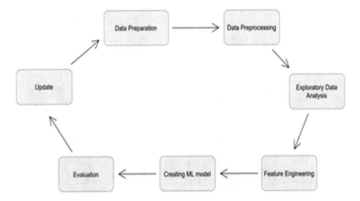

Fig. 2 Flowchart of whole process

The targeted column suggests job roles ('job role suggestion') according to the features in the dataset.

3.5 Implementation of Algorithm

After creating the dataset, exploratory data analysis (EDA) has been done to explore the created dataset through visualizing it. After this, feature engineering has been done to make the dataset ready for the model. Several techniques like binary encoding, dummy variables creation have been applied for encoding the dataset into numeric form (Fig. 2).

Following is the correlation matrix which describes the relation among logical quotient rating, hackathons, coding skills and public speaking points (Fig. 3).

Since the model is for multiclass classification dataset, the proposed model has been tested with support vector machine, decision tree and random forest, out of which decision tree outperformed the others. Hence, the model uses the decision tree algorithm for predictions.

4 Result

4.1 GUI Outputs

The project has used *Streamlit Cloud Service* for building an interactive user interface for testing and demo purposes (Fig. 4).

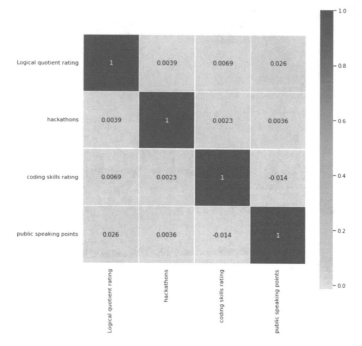

Fig. 3 Flowchart of whole process

Fig. 4 GUI snapshot 1

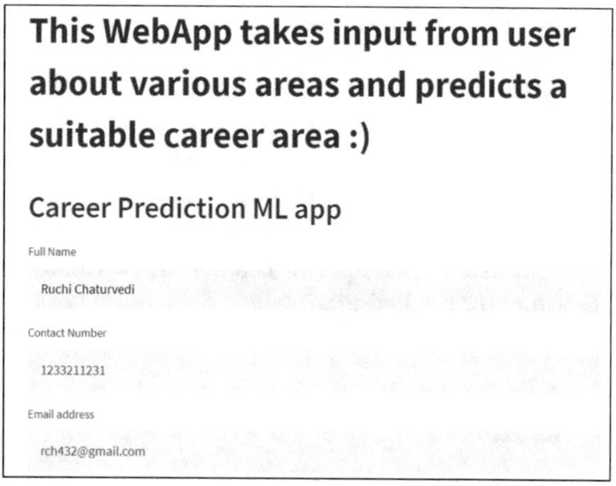

This is the user interface that has been created with the help of *Streamlit Cloud Service*. Firstly, it asks for personal details of the user, like name, email address and contact number. Then, it asks for various academic details and non-academic details of the user (Figs. 5, 6 and 7).

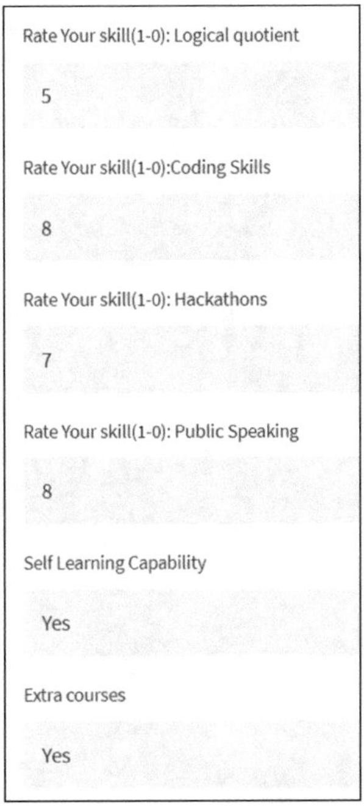

Fig. 5 GUI snapshot 2

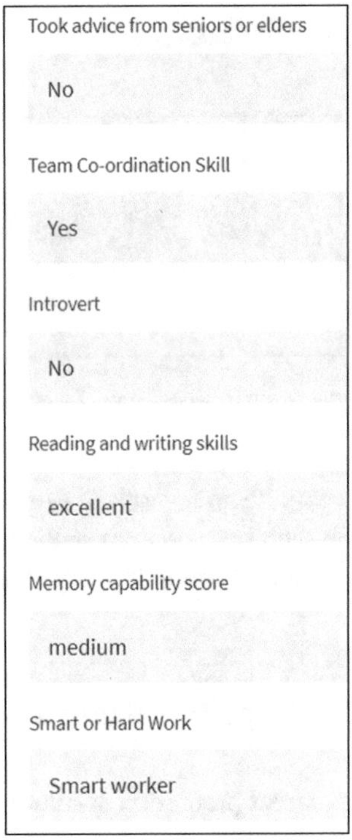

Rate Your skill(1-0): Logical quotient

5

Rate Your skill(1-0):Coding Skills

8

Rate Your skill(1-0): Hackathons

7

Rate Your skill(1-0): Public Speaking

8

Self Learning Capability

Yes

Extra courses

Yes

Fig. 6 GUI snapshot 3

Took advice from seniors or elders

No

Team Co-ordination Skill

Yes

Introvert

No

Reading and writing skills

excellent

Memory capability score

medium

Smart or Hard Work

Smart worker

Fig. 7 GUI snapshot 4

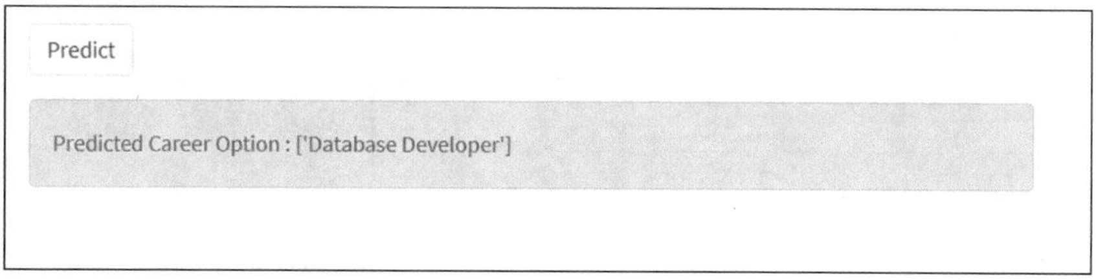

Fig. 8 GUI snapshot for predicted output

As it can be seen in Fig. 8, it asks the user to submit the form and the output can be seen in the output section.

Here is the link for the WebApp—*Career Path Prediction.*

5   Conclusion and Future Work

A career prediction system is a tool that predicts users' career and helps the user to make decisions concerning their life and will also help the organizations to provide better career counseling services. It is based on the data provided by the user. The model asks some questions related to users' profession, what they are doing currently, education, age, gender, academic and non-academic performance, etc. The career prediction system is easy to use. It asks users to give information about their current profession. It asks to give information about the profession they want to have in the future, information about the education that they have. And in output, it predicts the most favorable career decision. Hence, it helps the user to use their most favorable career and also saves them from making wrong decisions.

The scope of this research is vast and depends on the data related to a user. More the data the more will be the accuracy. The future scope of this project is to develop website and mobile application that works on the principle of artificial intelligence. This research can also include the prediction of those career areas that should not be chosen by the user. This can save user's time, effort and can reduce dilemmas about the next right career step.

References

1. Cranmer S, Desmarais B (2017) What can we learn from predictive modeling? Polit Anal 25(2):145–166. https://doi.org/10.1017/pan.2017.3
2. Nassif AB, Azzeh M, Banitaan S, Neagu D (2016) Guest editorial: special issue on predictive analytics using machine learning. Neural Comput Appl 27(8):2153–2155. https://doi.org/10.1007/s00521-016-2327-3
3. Oladipupo T (2010) Types of machine learning algorithms. In: New advances in machine learning. https://doi.org/10.5772/9385
4. Nyce C, Cpcu A (2007) Predictive analytics white paper. American Institute for CPCU. Insurance Institute of America, pp 9–10
5. Kumar V, Garg ML (2018) Predictive analytics: a review of trends and techniques. Int J Comput Appl 182(1):31–37
6. Eckerson WW (2007) Predictive analytics. Extending the value of your data warehousing investment. TDWI Best Pract Rep 1:1–36
7. Kolla N, Giridhar Kumar M (2019) Supervised learning algorithms of machine learning: prediction of brand loyalty. Int J Innov Technol Explor Eng (IJITEE) 8:11
8. Niculescu-Mizil A, Caruana R (2005) Predicting good probabilities with supervised learning. In: Proceedings of the 22nd international conference on machine learning—ICML'05. https://doi.org/10.1145/1102351.1102430
9. Apley DW, Zhu J (2020) Visualizing the effects of predictor variables in black box supervised learning models. J R Stat Soc Ser B (Stat Methodol) 82(4):1059–1086. https://doi.org/10.1111/rssb.12377
10. Henri G, Lu N (2019) A supervised machine learning approach to control energy storage devices. IEEE Trans Smart Grid 1. https://doi.org/10.1109/tsg.2019.2892586
11. Goodfellow I, Bengio Y, Courville A (2016) Deep learning. MIT Press, Cambridge
12. Mathur R, Sakarkar G, Kalbande K, Mathur R, Kolhe H, Rathi H (2023) Orthopantomogram (OPG) image analysis using bounding box algorithm. In: Asari VK, Singh V, Rajasekaran R, Patel RB (eds) Computational methods and data engineering. Lecture notes on data engineering and communications technologies, vol 139. Springer, Singapore. https://doi.org/10.1007/978-981-19-3015-7_5

13. Zhou Z-H (2017) A brief introduction to weakly supervised learning. Natl Sci Rev 5(1):44–53. https://doi.org/10.1093/nsr/nwx106
14. Campagni R, Merlini D, Sprugnoli R, Verri MC (2015) Data mining models for student careers. Expert Syst Appl 42(13):5508–5521. https://doi.org/10.1016/j.eswa.2015.02.052
15. Heppner MJ, Paul Heppner P (2003) Identifying process variables in career counseling: a research agenda. J Vocat Behav 62(3):429–452
16. Li L, Jing H, Tong H, Yang J, He Q, Chen B-C (2017) NEMO. In: Proceedings of the 26th international conference on world wide web companion—WWW'17 companion. https://doi.org/10.1145/3041021.3054200
17. Vidyapriya C, Vishhnuvardhan RC. Student career prediction
18. Roy KS et al (2018) Student career prediction using advanced machine learning techniques. Int J Eng Technol 7:26
19. Heap B et al (2014) Combining career progression and profile matching in a job recommender system. In: Pacific Rim international conference on artificial intelligence. Springer, Cham
20. Qu H et al (2016) What is my next job: predicting the company size and position in career changes. In: 2016 IEEE Trustcom/BigDataSE/ISPA. IEEE
21. Sripath Roy K, Roopkanth K, Uday Teja V, Bhavana V, Priyanka J (2018) Student career prediction using advanced machine learning techniques. Int J Eng Technol 7(2.20):26. https://doi.org/10.14419/ijet.v7i2.20.11738
22. Alalwan N, Al-Rahmi WM, Alfarraj O, Alzahrani A, Yahaya N, Al-Rahmi AM (2019) Integrated three theories to develop a model of factors affecting students' academic performance in higher education. IEEE Access 7:98725–98742. https://doi.org/10.1109/access.2019.2928142
23. Nie M, Xiong Z, Zhong R, Deng W, Yang G (2020) Career choice prediction based on campus big data—mining the potential behavior of college students. Appl Sci 10(8):2841. https://doi.org/10.3390/app10082841
24. He M, Shen D, Zhu Y, He R, Wang T, Zhang Z (2019) Career trajectory prediction based on CNN. In: 2019 IEEE international conference on service operations and logistics, and informatics (SOLI). https://doi.org/10.1109/soli48380.2019.8955009
25. Harrouk AI, Barbar AM (2018) A psycholinguistic approach to career selection using NLP with deep neural network classifiers. In: 2018 IEEE international multidisciplinary conference on engineering technology (IMCET). https://doi.org/10.1109/imcet.2018.8603068
26. Kern ML, McCarthy PX, Chakrabarty D, Rizoiu M-A (2019) Social media-predicted personality traits and values can help match people to their ideal jobs. Proc Natl Acad Sci. https://doi.org/10.1073/pnas.1917942116
27. Milot-Lapointe F, Savard R, Le Corff Y (2019) Effect of individual career counseling on psychological distress: impact of career intervention components, working alliance, and career indecision. Int J Educ Vocat Guid. https://doi.org/10.1007/s10775-019-09402-6
28. Obeid C, Lahoud I, El Khoury H, Champin P-A (2018) Ontology-based recommender system in higher education. In: Companion of the web conference 2018 on the web conference 2018—WWW'18. https://doi.org/10.1145/3184558.3191533
29. Rangnekar RH et al (2018) Career prediction model using data mining and linear classification. In: 2018 fourth international conference on computing communication control and automation (ICCUBEA). IEEE
30. Hirschi A, Froidevaux A (2020) Career counselling. In: Gunz H, Lazarova M, Mayrhofer W (eds) Routledge companion to career studies. Routledge, London, pp 331–345. https://doi.org/10.4324/9781315674704
31. Hirschi A, Froidevaux A (2019) Career counseling. https://doi.org/10.4324/9781315674704-20
32. Kumar A (2021) Impact of Covid19 on education system. In: International J Eng Res Technol (IJERT) 10(06)
33. Vignesh S, Shivani Priyanka C, Shree Manju H, Mythili K (2021) An intelligent career guidance system using machine learning. In: 2021 7th international conference on advanced computing and communication systems (ICACCS). https://doi.org/10.1109/icaccs51430.2021.9441978
34. Pandya A, Lodha P (2021) Social connectedness, excessive screen time during COVID-19 and mental health: a review of current evidence. Front Hum Dyn 3:684137. https://doi.org/10.3389/fhumd.2021.684137

35. Babu S, Hareendrakumar VR, Subramoniam S (2020) Impact of social media on work performance at a technopark in India. Metamorphosis J Manag Res 19(1):59–71. https://doi.org/10.1177/0972622520962949

An Improved Technique for Risk Prediction of Polycystic Ovary Syndrome (PCOS) Using Feature Selection and Machine Learning

Nitisha Aggarwal⊙, Unmesh Shukla⊙, Geetika Jain Saxena⊙, Manish Kumar, Anil Singh Bafila⊙, Sanjeev Singh⊙, and Amit Pundir⊙

Abstract Polycystic ovary syndrome (PCOS) is an endocrine disorder that affects more than five million women globally in their childbearing age. The study suggests that accurate and specific machine learning models in conjunction with relevant feature selection methods can play an essential role in detecting PCOS. The statistical feature selection algorithms such as Chi-Square, ANOVA, and Mutual Information identify insignificant features from the data. The present research revealed that the random forest classifier achieved 93.52% accuracy on a feature set suggested by the ANOVA test. The results indicated no significant decline in accuracy, and other parameters like $F1$-score and specificity improved with a substantial reduction in computational time. The model's effectiveness is measured by the AUC that varies between 0.82 and 0.98; the higher the value, the better the model's classification ability. The paper reports improved model performance by suggesting methods to increase AUC, improve recall and specificity. The improved performance of the proposed machine learning model shall help optimize and scale data-driven diagnosis of PCOS at a higher rate and enable better decision-making.

Keywords ANOVA · Chi-Square · Feature selection · Machine learning · Polycystic ovary syndrome

N. Aggarwal · U. Shukla · A. S. Bafila · S. Singh
Institute of Informatics and Communication, University of Delhi South Campus, Delhi, India

G. J. Saxena · A. Pundir (✉)
Department of Electronics, Maharaja Agrasen College, University of Delhi, Delhi, India
e-mail: amitpundir.du@gmail.com

M. Kumar
School of Computer and Information Sciences, Indira Gandhi National Open University, Delhi, India

1 Introduction

Polycystic ovary syndrome (PCOS) is perhaps the most common endocrine disorder encompassing various other health issues in women. PCOS is evidenced by the presence of 20 or more follicles on one or both ovaries. While this symptom is detected in ultrasound, there are cases where this diagnostic technique is not recommended, like in women younger than 20 years of age [1]. The diagnosis is challenging because the symptoms are often nonspecific. For example, irregular menstruation cycle or hair loss, or acne, by themselves, maybe the signs of many other disorders and may not be the explicit biomarkers or identifiers of PCOS. It, therefore, becomes imperative for researchers to find a new diagnostic technique that accurately, precisely, and efficiently explain a subject's symptoms and signs of PCOS. If not detected early and treated adequately in the worst-case scenario, PCOS may lead to various complications such as diabetes, infertility, and uterine cancer. Healthcare practitioners use hormone blood tests and signs and symptoms of excess male hormones to diagnose PCOS. Anti-Müllerian Hormone (AMH) is one such biomarker for PCOS, with higher levels indicating this syndrome's existence [2, 3]. Still, women with the syndrome may have different symptoms and multiple imbalances; it is crucial to consider other parameters and investigate their relationships and levels of significance to get an accurate diagnosis quickly using the latest machine learning (ML) techniques. Today, PCOS is found in 6–10% of the female population and is a cause of up to 30% of infertility, with no cure, and minimal research has been done that covers the PCOS spectrum in its entirety [4].

The present investigation uses the Kaggle dataset [5], collected from 10 different hospitals across Kerala, India, containing 541 records with 44 physical and clinical features. The ML requires preprocessed data for training and testing models. In addition, in ML models having high dimensionality [6], the visualization and analysis of features need large memory and high computational power. The feature selection process can address such problems by identifying redundant features that are not considered during data processing. As a result, it enhances model performance with reduced processing time [7–11]. Few such filter-based feature selection techniques like Chi-Square, ANOVA, and Mutual Information (MI) are easy to use, fast, and effective. They select input variables having a strong relationship with the target variable [12]. In contrast, the wrapper feature selection techniques train the classifier on different subsets of features and measure the model's performance. The filter method is computationally less complex, faster than the wrapper method, and the feature set selected is general and can be applied to any model. In this study, filter-based methods are used for feature selection. In addition, ML predictive models such as logistic regression (LR), support vector machine (SVM), Naive Bayes (NB), decision tree (DT), random forest (RF), extended gradient boosting with random forest (XGBRF), and CatBoost are trained with different parameters to classify the presence of PCOS in a patient.

2 Literature Survey

The PCOS is characterized by signs and symptoms, which generally do not manifest immediately or are often misdiagnosed as ordinary conditions of pubertal development. To detect, decipher, and predict the existence of PCOS, researchers have proposed various ML-based techniques. Cheng and Mahalingaiah [13] developed an ML-based classification system to classify 2000 randomly selected, manually labeled ultrasound images with an accuracy of 80.84%. Dewi and Wisesty [14] used feature extraction and competitive neural network (CNN) to achieve 80.84% accuracy on ultrasound images. In another work on ultrasound images by Sumathi [15], the authors extracted the features using the watershed algorithm, performed classification by CNN, and achieved 85% accuracy. Usage of canny edge detection for the classification of follicles attained an accuracy of 88% in a study by Padmapriya and Kesavamurthy [16]. Maheswari et al. [17] analyzed ultrasound images and proposed a novel approach, *furious flies*, for feature identification. Metabolic features and biomarkers are also used to diagnose PCOS, as Mehrotra et al. [18] reported. In another study by Denny et al. [19], features were extracted by principal component analysis (PCA) and classified using NB, LR, KNN, classification and regression trees (CART), RF, and SVM. Results obtained using RF classifier were best for PCOS prediction with an accuracy of 89.02%. Bharati et al. [20] developed a univariate feature selection algorithm model to find the best features to predict PCOS using several classifiers such as gradient boosting, RF, LR, hybrid random forest, and logistic regression (RFLR). RFLR method achieved the best test accuracy of 91.01%. Nandipati et al. [21] suggested a method using correlation matrix and recursive feature elimination method reporting RF as the best suitable classifier, which yielded an accuracy of 93.12%. Inan et al. [22] report resampling techniques such as synthetic minority oversampling techniques (SMOTE) and edited nearest neighbor (ENN) to select features using statistical correlation methods—ANOVA and Chi-Square test. Classification done on selected features by XGBoost achieved a cross-validation score of 96.03% in detecting patients not having PCOS. Higher-level AMH [23] combined with other features also plays an essential role in diagnosing PCOS.

3 Methodology

ML techniques were used to design the model and identify the best approach to diagnosing PCOS using the Kaggle dataset [5]. The dataset of physical and clinical features for predicting PCOS contains 541 records with 44 features, including a PCOS indicator label having a value of either 1 or 0. During the preprocessing of the dataset, three records having missing values were discarded. In addition, features like the *Serial number* of the patient and *Patient file number*, which did not have any significance in the proposed study, were not considered. Standard scaler was used to standardize the data to have a mean value of 0 and a standard deviation of 1 for comparing features having different measurement units. For selecting significant features, statistical feature selection methods Chi-Square test [24], ANOVA [25], and MI [26] were used, and score matrices for each feature with the target were determined. These methods are filter-based methods, which are computationally less expensive in terms of time and space. PCOS indicator label is categorical by nature, and these methods are suitable for categorical output variables. The grid search algorithm for hyper-tuning of parameters was used for each classifier. Feature selection and model training were done on 70% of the data, and 30% of the data was used to evaluate the performance metric.

4 Results and Discussion

4.1 Results Using Preprocessed Dataset with Full Features

In this experiment, all the features were used for training and evaluating the model. The CatBoost classifier yielded the best performance on the test dataset with 94.44% accuracy, 93% precision, 95% specificity, 94% sensitivity, 94% $F1$-score, and 0.87 MCC taking a processing time of 393 ms. An accuracy of 93.52% and area under the curve (AUC) of 0.98 was achieved using the RF classifier with a processing time of 21.5 ms, significantly lesser than CatBoost classifier, saving a significant amount of computational time without compromising in terms of performance. In Fig. 1a, the receiver operating characteristic (ROC) curves show the trade-off between sensitivity and specificity. The figure shows that CatBoost and RF perform best among classifiers as they generate curves closer to the top-left corner, indicating better performance. The performance metrics of the classifiers with optimized hyperparameters are shown in Fig. 1b.

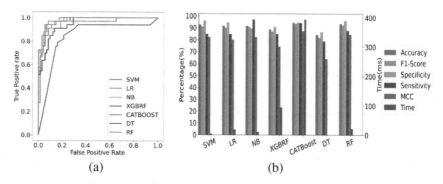

Fig. 1 Result with preprocessed full features dataset. **a** ROC curves for the pre-processed full features dataset for best-performing classifiers. **b** Best-performing classifiers on preprocessed full features dataset

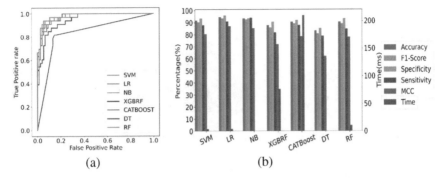

Fig. 2 Result with selected features using Chi-Square. **a** ROC curves for the selected features (no. of features = 11, alpha = 0.01) for best-performing classifiers. **b** Best-performing classifiers on selected feature set

4.2 Result Using Chi-Square Feature Selection Method

The Chi-Square feature selection method with an alpha value of 0.01 suggested 11 significant features. The classifiers were trained and tested on the selected feature set. LR yielded the best results considering overall performance metrics and processing time with an accuracy of 92.59%, AUC of 0.97, a sensitivity of 91%, and a high $F1$-score. The area under the ROC curve, Fig. 2a for the SVM classifier was 98%, suggesting that the model works best with the reduced features set, saving significant computational cost. The comparison of performance metrics of the classifiers with optimized hyperparameters is shown in Fig. 2b. The processing time also reduces in almost all classifiers with reduced features, as evident in Figs. 1b and 2b.

4.3 Results Using Mutual Information (MI) Feature Selection Method

Different classifiers were trained and tested using the features selected by the MI algorithm. The classifier's performance (ROC Curves in Fig. 3a) is best with the set of 17 features. The random forest classifier obtained the highest accuracy of 92.57% in 19.7 ms processing time, attaining an MCC of 0.85, and AUC of 0.98. Follicle number in both ovaries, physical fitness (skin darkening, hair growth, weight gain), menstruation cycle length, and hormone profile are the top selected features by the MI algorithm. The features are mostly the same as obtained by the Chi-Square method for accurate PCOS prediction as given in Table 1. It is visible in Fig. 3b that SVM, LR, and NB have less processing times than the other classifiers with comparable accuracies and other performance parameters.

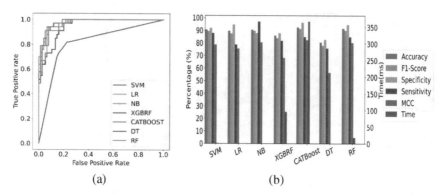

Fig. 3 Result with features selected by MI algorithm (17 features). **a** ROC curves for selected features for best-performing classifiers. **b** Performance of classifiers on the selected feature set

Table 1 Features selected by Chi-Square, MI, and ANOVA

Chi-Square	MI	ANOVA
11 attributes	17 attributes	11 attributes
Follicle number (right)	Follicle number (right)	Follicle number (right)
Follicle number (left)	Follicle number (left)	Follicle number (left)
Fast food	Skin darkening	Skin darkening
Pimples	Facial hair growth	Facial hair growth
Hairfall	PRL (ng/mL)	Weight gain
Skin darkening	Menstruation cycle length	Menstruation cycle regular
Facial hair growth	Fast food	Fast food
Weight gain	BMI	Pimples
Menstruation cycle length	Weight gain	AMH
Menstruation cycle regular	Menstruation cycle regular	Weight
Hip (inch)	Waist (inch)	BMI
	AMH (ng/mL)	
	Avg. F size (L) (mm)	
	Weight	
	Hairfall	
	Blood group	
	RBS (mg/dL)	

4.4 Results Using ANOVA F-Test Feature Selection Method

The best performance metrics were obtained with a set of 11 features suggested by ANOVA. With a reduced feature set, the best classification accuracy of 93.52% was obtained with the RF classifier. The values of MCC and AUC for RF were 0.85 and 0.98, respectively. The ROC curves in Fig. 4a show the performance of various classifiers. The performance evaluation of different classifiers with hyperparameters optimization with 11 features selected using ANOVA has been shown in Fig. 4b.

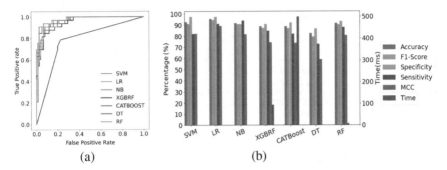

Fig. 4 Result with selected features by ANOVA (11 features). **a** ROC curves of features selected for best-performing classifiers. **b** Performance of classifiers on selected feature set

5　Discussion

Statistical feature selection algorithms were applied to the full feature dataset for selecting the most significant features to predict PCOS. Using a dataset with reduced features, the accuracy and other performance parameters obtained were similar to those attained with the full feature dataset. The downsizing of features saves a lot of machine time and other complexities. Most of the features selected by different feature selection methods in the present study are similar as reported in Table 1, and the literature also shows [27] that if the follicle is more than 12–20 per unit area in an ovary and is detectable in a radio scan, it can be diagnosed as PCOS. Other symptoms like irregular menstruation, pimples, weight gain, high anti-Müllerian hormone (AMH), skin darkness, and hair loss are also significant for the diagnosis. The comparative revealed that SVM (linear) and RF classifiers performed better with all the feature selection algorithms. For the full-featured pre-processed dataset, the CatBoost classifier performed best with an accuracy of 94.44%. However, it also took the highest processing time of 393 ms. With reduced features, the results obtained for all the classifiers with the Chi-Square algorithm were comparable to the full-featured dataset, although the dimensions were approximately one-third of the full-featured set. With feature set selected using MI and ANOVA, the SVM, LR, and RF yielded better performances. The performance in terms of accuracy was comparable but the computational time reduced significantly. As in proposed approach, holdout approach has been used, and in [28], hold out result has considered for comparison. The accuracy of proposed approach is better in comparison with previous research as given in Table 2.

Table 2 Comparison of the result of the proposed approach with earlier approaches

Classifier	Technique	Number of selected features	Accuracy (%)	References
RF	Principal component analysis (PCA)	23	89.02	[19]
Hybrid random forest and logistic regression (RFLR)	Univariate feature selection	10	91.01	[20]
RF	Synthetic minority oversampling technique	41	93.12	[21]
LR	Correlation	10	92	[29]
RF	ANOVA F-value with outlier removal (holdout)	10	85	[28]
RF	ANOVA	11	93.52	This study

6 Conclusion

An efficient methodology for a highly accurate and effective method of assessing the presence of PCOS in a patient, based on statistical feature reduction and machine learning algorithms, is proposed. The study was performed using the Kaggle PCOS dataset [5] using feature selection method Chi-Square, ANOVA, and MI to predict PCOS. The selected features were processed with classifiers such as SVM, NB, LR, DT, RF, XGBRF, and CatBoost classifiers. Performance evaluation metrics like accuracy, F1-score, MCC, and AUC were calculated to assess the overall efficiency of the models for effectively ascertaining the risk of PCOS with least computational time. Each model has been evaluated, and the experimental results show AUC of the models ranging from 0.82 to 0.98. LR yielded the best results on reduced features dataset derived using Chi-Square algorithm with an F1-score of 93%. The RF classifier performed the best in most cases and outperformed all other models in conjunction with the ANOVA feature selector with an accuracy of 93.52% and a significant reduction in computation time. All feature sets show that follicle no. is a distinguishable feature for the diagnosis of PCOS. Other physical parameters like excessive hair growth on skin, baldness, darkening of skin, and body measurements have strong correlations with the outcome. The improved performance of the model offers promising applications in healthcare services as the extracted prominent features give strong evidence of the presence of PCOS and may help in early and speedy risk assessment.

Acknowledgements The authors would like to thank Institution of Eminence (IoE), University of Delhi for the research grant under FRP scheme.

References

1. McCartney CR, Marshall JC (2016) Polycystic ovary syndrome. N Engl J Med 375(1):54–64
2. Aziz M, Sidelmann JJ, Faber J, Wissing ML, Naver KV, Mikkelsen AL (2015) Polycystic ovary syndrome: cardiovascular risk factors according to specific phenotypes. Acta Obstet Gynecol Scand 94(10):1082–1089
3. Barber TM, Wass JA, McCarthy MI, Franks S (2007) Metabolic characteristics of women with polycystic ovaries and oligo-amenorrhoea but normal androgen levels: implications for the management of polycystic ovary syndrome. Clin Endocrinol (Oxf) 66(4):513–517
4. Barthelmess EK, Naz RK (2014) Polycystic ovary syndrome: current status and future perspective. Front Biosci (Elite Ed) 6:104–119
5. Kottarathil P (2020) Polycystic ovary syndrome (PCOS)—version 3. https://www.kaggle.com/prasoonkottarathil/polycystic-ovary-syndrome-pcos
6. Domingos P (2012) A few useful things to know about machine learning. Commun ACM 55(10):78
7. Wettschereck D, Dietterich TG (1995) Mach Learn 19(1):5–27
8. Wettschereck D, Aha DW, Mohri T (1997) Artif Intell Rev 11:273
9. Yang M, Nataliani Y (2018) A feature-reduction fuzzy clustering algorithm based on feature-weighted entropy. IEEE Trans Fuzzy Syst 26(2):817–835
10. Chen R, Sun N, Chen X, Yang M, Wu Q (2018) Supervised feature selection with a stratified feature weighting method. IEEE Access 6:15087–15098
11. Imani M, Ghassemian H (2015) Feature extraction using weighted training samples. IEEE Geosci Remote Sens Lett 12(7):1387–1391
12. Liu H, Motoda H (1998) Feature extraction, construction, and selection: a data mining perspective. Springer Science-Business Media, LLC, New York
13. Cheng JJ, Mahalingaiah S (2019) Data mining polycystic ovary morphology in electronic medical record ultrasound reports. Fertil Res Pract 5:13
14. Dewi RM, Wisesty UN (2018) Classification of polycystic ovary based on ultrasound images using competitive neural network. J Phys Conf Ser 971(1):012005
15. Sumathi M (2021) Study and detection of PCOS-related diseases using CNN. Conf Ser Mater Sci Eng 1070:01206
16. Padmapriya B, Kesavamurthy T (2015) Diagnostic tool for PCOS classification. In: Goh J, Lim C (eds) 7th WACBE world congress on bioengineering 2015, vol 52. Springer International, pp 182–185
17. Maheswari K, Baranidharan T, Karthik S (2021) Modelling of F3I based feature selection approach for PCOS classification and prediction. J Ambient Intell Human Comput 12:1349–1362
18. Mehrotra P, Chatterjee J, Chakraborty C, Ghoshdastidar B, Ghoshdastidar S (2011) Automated screening of polycystic ovary syndrome using machine learning techniques. In: 2011 annual IEEE India conference, Hyderabad, pp 1–5
19. Denny A, Raj A, Ashok A, Ram CM, George R (2019) i-HOPE: detection and prediction system for polycystic ovary syndrome (PCOS) using machine learning techniques. In: IEEE region 10 conference (TENCON), pp 673–678
20. Bharati S, Podder P, Mondal MRH (2020) Diagnosis of polycystic ovary syndrome using machine learning algorithms. In: IEEE region 10 symposium (TENSYMP), pp 1486–1489
21. Nandipati SCR, XinYing C, Wah KK (2020) Polycystic ovarian syndrome (PCOS) classification and feature selection by machine learning techniques. Appl Math Comput Intell 9:65–74
22. Inan MSK, Ulfath RE, Alam FI, Bappee FK, Hasan R (2021) Improved sampling and feature selection to support extreme gradient boosting for PCOS diagnosis. In: IEEE 11th annual computing and communication workshop and conference (CCWC), pp 1046–1050

23. Dewailly D, Lujan ME, Carmina E, Cedars MI, Laven J, Norman RJ, Escobar-Morreale HF (2013) Definition and significance of polycystic ovarian morphology: a task force report from the androgen excess and polycystic ovary syndrome society. Hum Reprod Update 20(3):334–352

24. Franke TM, Ho T, Christie CA (2012) The chi-square test: often used and more often misinterpreted. Am J Eval 33(3):448–458

25. Scheffe H (1959) The analysis of variance. Wiley, New York

26. Hutter M (2002) Distribution of mutual information. Adv Neural Inf Process Syst 1:399–406

27. Lawrence MJ, Eramian MG, Pierson RA, Neufeld E (2007) Computer-assisted detection of polycystic ovary morphology in ultrasound images. In: Fourth Canadian conference on computer and robot vision, pp 105–112

28. Neto C, Silva M, Fernandes M, Ferreira D, Machado J (2021) Prediction models for polycystic ovary syndrome using data mining. In: Antipova T (eds) Advances in digital science. ICADS 2021. Advances in intelligent systems and computing, vol 1352. Springer, Cham

29. Tanwani N (2020) Detecting PCOS using machine learning. Int J Modern Trends Eng Sci (IJMTES) 7(1):1–20

Fog-Enabled Framework for Patient Health-Monitoring Systems Using Internet of Things and Wireless Body Area Networks

Ankush Kadu and Manwinder Singh

Abstract Wireless body area networks (WBANs) have a significant role in the automation of remote patient monitoring systems (via the Internet) for large hospitals, potentially reducing paramedic staff responsibilities. These systems, on the contrary, create a large volume of sensed data, necessitating time-bounded services, dependability, data preparation, and effective communication technology. One of the acceptable choices to improve patient monitoring systems is the Internet of things (IoT) with the notion of fog computing. This paper focuses on the patient monitoring system's needs before proposing and implementing a hierarchical layer-based IoT architecture incorporating WBANs, fog computing, and cloud services. The tested architecture uses an embedded system and an open-source prototyping platform. Results are tracked, saved, and evaluated after testing. The findings demonstrate that this architecture meets the stringent criteria of medical applications by delivering reliable communication.

Keywords Fog computing · WBAN · IoT · Patient health monitoring systems · Wearable devices

1 Introduction

Early detection and treatment are essential for everyone to remain fit and solid. A fast-paced lifestyle, the consumption of unhealthy foods, and high-pressure jobs cause different illnesses. The IoT is receiving much attention in today's medical services environment. In general, an IoT-based framework can connect numerous products and sensors via the Internet. Each linked gadget has its unique identity, allowing it to communicate information without requiring human intervention. In terms of efficacy, rationality, and accessibility, the Internet of things (IoT) integration

A. Kadu · M. Singh (✉)
School of Electrical and Electronics Engineering, Lovely Professional University, Phagwara, Punjab, India
e-mail: manwinder.25231@lpu.co.in

© The Author(s), under exclusive license to Springer Nature Singapore Pte Ltd. 2023
A. Shukla et al. (eds.), *Computational Intelligence*, Lecture Notes in Electrical Engineering 968, https://doi.org/10.1007/978-981-19-7346-8_52

in the medical services sector have resulted in a revolutionary advance in a health-monitoring system. The organization of IoT has become a lot easier because of the availability of free integrated development environment (IDE) and programming development kits (SDK) programming [1, 2].

The IoT enabled remote health monitoring, commonly known as wireless body area sensor networks (WBASNs), but it is first necessary to comprehend WBASNs. A WBASN comprises sensor nodes with low power, miniaturization, wearability, and lightweight properties. These sensor nodes are reliable and can monitor continuously with little memory [3, 4]. These wearable sensor nodes continually monitor physiological data and transfer it to a coordinator node (PDA or any other appropriate node) for preprocessing utilizing RF signals before memory fills (radio frequency). The coordinator node is connected to the human body or close to the body. The central server (CS) is used to maintain track of data for various patients. The WBASN architecture has three sections: on-body wearable sensors, coordinator, and control system. Communication occurs at two levels: between the sensor node and the coordinator and between the coordinator and the CS. The central pushing element for WBASN development is user requirements. Usability, security, privacy, interoperability, availability, and safety are just a few of the criteria. WBASN employs physiological, environmental, and bio-kinetics sensors to meet these needs. These sensors, such as EMG, ECG, temperature, humidity, blood pressure, blood glucose, and motion sensors, are now commercially accessible. The industry was forced to explore possible applications because of the range of WBASN sensors (medical and non-medical applications). Wearable WBAN (wearable health monitoring, asthma and sleep staging, etc.), implant WBAN (cancer detection, etc.), and remote control WBAN (ambient assisted living (AAL)), patient monitoring and tele-medicine system [5, 6], etc. are examples of medical applications. In contrast, non-medical applications include entertainment and video streaming apps. Because communication between sensor nodes includes channel access methods, choosing a suitable and efficient MAC protocol is critical [2, 7].

The IoTs are a forward-thinking concept for communication advancements. The IoT is a collection of entities with various components such as sensors, hardware, and programming. These actual implanted objects can communicate with others like articles, much like the Internet. Identify each transmitting object by its physical characteristics (UID or MAC or IP, and so on). By 2020, IoTs would consist of 50 billion things [4]. Remote health-monitoring territories use the IoT concept, known as WBASNs. However, it is required first to understand WBASNs. Sensor hubs make up a WBASN, which indicates qualities such as low force, scaled-down, wearable, and lightweight, among others. These sensor hubs are robust and well equipped for continuous monitoring with limited memory [3, 4]. These wearable sensor hubs continuously monitor physiological data and send it to a facilitator hub (PDA or another suitable hub) for preprocessing using RF signals before memory fills (radio frequency). IoT-based medical care is few persistent illnesses and continuous area monitoring [4]. Continuing advancements spurred the modernization of the applied sciences in data and correspondence innovation. The wireless body area network (WBAN) [1] enhanced the applications in medical care by reducing the number of distant sensors and other electronic devices. The extra options created by technological advancements have also lowered healthcare costs and treatment delays. WBANs are brighter, more minor, have a shorter battery life, have higher quality of service (QoS) requirements, and handle diverse organization traffic [8]. The Web of things (IoT) is a unique innovation that connects any item to a company, and this approach is ideal for WBAN engineering of medical care administrations [9, 10]. The constant advancement of innovation that prepares for the expansion of relationships over the Internet and the development of the capacity to deal with data has created more significant chances for the global health business, specifically telemedicine. Data sharing, information analysis, the IoT, wearables, cloud innovation, and mechanical technology are all on the rise as development drivers for the next decade. With these perspectives pointing to responses to the massive volume of information used in medical care, the requirement for predictable exactness in complex methodology, and the growing demand in medical care administration, it is clear that artificial intelligence (AI) plays a prominent role in technological activity and application. Computerizing clinic logistics is needed to increase productivity in scheduling time and transmitting medical care demands and actions [1, 4].

Fog computing distributes the load of a primary network's information exchange and administration. It improves the cloud administrators' performance by providing more detailed information. Fog computing is a virtualized platform that provides processing, programming interface, systems management, and capacity, among other services. The business administration facilitates communication between an IoT and cloud connection. Fog computing has important when applications and administrations are sent for a more excellent range in a circulating environment. The device can detect a person's heartbeat and provide a hard or soft rest. The most modern cells nowadays come with built-in sensors, such as the Samsung Note 4 with heartbeat sensors, a gyro meter, and an accelerometer integrated into PDAs. Fog processing is an all-encompassing aspect of distributed computing in which both registration phases have comparable admissions, which benefits fog computing by reducing cloud worker idleness. Distributed computing creates a complete package that helps customers but also has flaws. One of the advantages of the IoT is the ability to quickly analyze enormous volumes created to process, research, and manage efficiently in various applications. Innovative care frameworks are critical in the fast-paced movement of human life. When fog processing and IoT are relevant to clinical sector applications, it increases the competence of the activities in such medical care frameworks [11, 12].

Users need to monitor their health parameters regularly to reduce the risk of various ailments, which is impossible to manually without visiting hospitals and clinics. As a result, a fog-enabled IoT-based health-monitoring system is required to provide real-time healthcare services to remote sites. The proposed fog-enabled architecture promises lower latency, faster reaction time, increased security, and lower power consumption than existing cloud-based solutions. The planned IoT framework also enhances inhabitants' overall quality of life.

2 Related Work

Both AI and telemedicine's adaptability gave unlimited opportunities for advancement [1]. Kallipolitis et al. [8] present the plan and usage of a feeling examination module incorporated in a current telemedicine stage. The recent examination expects to deliver a novel engineering (SENET) [9], which depends on AI methods and comprises three principle layers. Medical care habitats bring the calculations near information sources [10]. A structure introduced [3] utilizes fog computing alongside IoT, and AI gives a superior and more intelligent medical services insight. The author presents a deal with coordinating electroencephalography [4] based AI components in our e-Health IoT framework using the TensorFlow open-source stage. The author offers a consistent home telemonitoring framework [6] for constant respiratory patients by using the 5G network. Ashapkina et al. have considered an errand [5] to create quantitative measurements for programmed acknowledgment of activity types in the framework for far-off checking of the actual restoration measure. A progressive structure of the framework under investigation is proposed [7] based on the OSI reference model and the degrees of information. They actualized a framework [2] for the continual transmission and show of details from numerous subtle frameworks and approved that there were no issues related to sending and accepting information. The AIR CARDIO venture aims to assess [11] the effect, productivity and viability of a home telemonitoring framework for youngsters with inherent coronary illness. Creator uses IoT to build up the availability between the apparatuses [13], the client, and their organization. Gračanin et al. [14] investigate how innovation is used to give well-being administrations while securing and saving protection.

A devoted clinical decision support system (DSS) and a clinical specialist UI upheld the framework [15] in the computerized location of anomalies/illness weakening. The creators set an objective to dissect [12] the utilization of the accomplishments of current the chance of deciding approaches to improve the nature of telemedicine benefits. Author research work [16] depends on the exploration subject of the insight IoT telemedicine medical services for the old living alone framework. An e-Health model [17] for taking the patient's systolic, diastolic, and pulse communicates the information to a Bluetooth preparing card and sends them progressively to the distant worker through the phone organization. Shahiduzzaman et al. propose [18] a cloud-network edge engineering to improve fall identification, anticipation and security comprising the clinical cloud, edge organic organizations gadgets, such as an intelligent protective cap. An epic IoT framework is proposed [19] with the help of oxygen immersion (SpO_2) estimation sensor, temperature sensor, blood pressure sensor, Bluetooth, Arduino, and APP innovations or strategies. Tripathy et al. present [20] a savvy customer hardware answer for encouraging protected and steady opening after lifting stay-at-home limitations. Zhao et al. present [21] a wearable framework utilizing accelerometers and AI for programmed observing of fetal development.

3 Material and Method

As shown in Fig. 1, the proposed system accomplishes WBANs and depends on IoT devices and wearable or embedded WBANs.

In this framework, WBSNs data collection and transmission to them toward their programmable devices, which can encourage interpreting the information received. The machines are customized by various strategies in which the diagnostic framework is viewed. These devices are programmable and can also be associated with the Internet. Unlike sensor networks, they have no energy impediments and are associated directly with the power. Intelligent techniques and computational hardware utilize to end this. Besides, to accomplish appropriate analysis models, doctors screen the frameworks.

All in all, the proposed design uses supervised machine learning techniques. The programmable devices likewise report anonym events to the connected experts. Primarily framework is categorized into three layers described below:

3.1 Sensor Layer

It includes mobile wireless body sensors node, which mainly depends on diseases. It consists of various wearable sensors. A WBAN is a collection of miniaturized wireless nodes coordinating with each other and including sensors and actuators to supervise the human body's functions and environment. Body area network consists of nodes, each acquiring a specific biological parameter from the human body, processing it and communicating it to the fog nodes. Wireless sensor nodes worn by

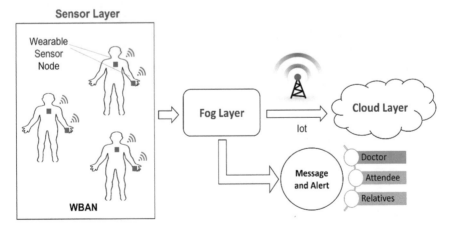

Fig. 1 Proposed framework

the patient enable him to carry out his routine work while being monitored. Short-range wireless communication technologies practice Zigbee and Bluetooth, which utilize license-free ISM bands (2.4–2.4835 GHz) to implement WBAN. Also, Wi-Fi or using IoT will be an option for long-range distances. These standard technologies define a specific protocol to implement the network functionalities. However, other non-standard wireless communication technologies utilize the same frequency spectrum to do the same operations.

3.2 Fog Layer

It includes fog nodes capable of processing and analyzing data, which helps diagnose or predict the nature of the disease. The analytics system will develop under the supervision of experts, which generates alerts messages and data for doctors or attendants who take care of the patient. Fog computing analyses the data and aggregate the resultant factors controlled by a sensor layer device. It also shares the underlying network load in data processing and management. Most importantly, this layer is represented as the server to distribute the process across devices, i.e., fog node. Data classification using an analytics system allows predicting the value of a categorical variable by constructing a model based on one numerical and categorical variable.

3.3 Cloud Layer

It belongs to cloud storage, where data such as medical health records are made available for further use. It also helps to create monitoring system applications for real-time data for doctors, patients, or attendants. Data processed and managed by fog nodes will receive at the cloud layer for application or use case scenarios for UI-based data software or app. Our primary purpose is to study the recognition system based on machine learning techniques, as such awareness is beneficial for assisting the medical staff better in interpreting patient state and vital signs measurements.

4 Results and Discussion

4.1 Experimental Setup

The proposed architecture framework uses sensor, fog, and cloud layer components. Sensors such as breathing, body temperature, and heart rate sensor to monitor parameters in the sensor layer, which is attached to the Node MCU open-source platform. The sensor nodes are mounted on a shield linked to the microcontroller. The Arduino

IDE is used to develop a embedded C program for reading sensor data. This data sent wirelessly using IEEE 802.11 Wi-Fi. The coordinator is attached to the Fog server, which runs on a Raspberry Pi through a Wi-Fi network. For processing, the coordinator delivers data (sensor data) to the fog server. The fog server first processes sensor data and temporarily stores it in a database before continuing to transfer data to the Ubidots Cloud account using the MQTT protocol. A dashboard with a user interface (UI) shows data on a mobile phone, tablet, or monitoring screen.

4.2 Experimental Results

Wearable sensor parameters are monitored, deliberated, and analyzed to assess the system's efficacy in various human activities. The values of each sensor parameter are observed for two sensor nodes connected to healthy people, and the distributions were statistically significant on the cloud. As per the anomaly found concerning the threshold assigned, an alert will be sent to the doctor or caretaker.

Figure 2 shows the respiratory rate, body temperature, air quality, SpO_2, and heart rate person under various test environments extracted from the prototype device in real-time recorded for 30 events at an interval of 1. It has been noted that when a person is healthy, then in the case of monitoring respiratory rate, its value is discovered to be 1breaths0 breath per min; in other cases, it is unhealthy. In the case of body temperature, the healthy person's temperature value is 36.5–38.5 °C, and in the case of heart rate measurement, the practical person's weight is 60–100 beats per min. Similarly, for SpO_2, a healthy person's value is discovered to be above 95%. For false case determination, the air quality index is measured, and it is below 100 PPM for a healthy environment should be below. In the case of an unhealthy person, the anomaly or abnormality is measured value based on a threshold, and the alert signal is generated and referred to the doctor, patient, or caretaker as per the priority given to individuals.

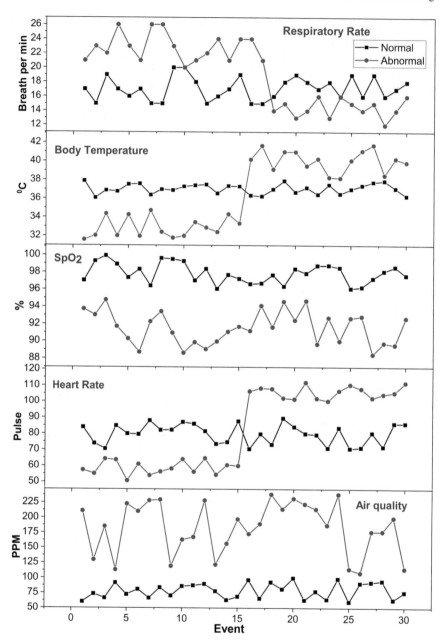

Fig. 2 Analysis of normal and abnormal range for respiration, body temperature, SpO$_2$, heart rate, and air quality (top to bottom)

5 Conclusion

WBANs deliver novel possibilities for various healthcare applications, such as activity identification and patient monitoring. To fully utilize the network in a remote setting, it is vital to summarize their needs and combine them with a practical computation and communication architecture. We adopted the notion of fog computing for computing, which works as an intermediary layer between the network and perception layer of IoTs. Besides, the fog server includes data acquisition and processing, storage, and transmission capabilities. Overall, the system architecture is based on IoTs, with WBAN sending data to the fog server; after that, the data is processed and sent to the network layer. The network layer uses a trustworthy model to offer effective routing. The application layer receives the data, which then sends it to the appropriate interface. The suggested architecture is prototyped utilizing the Node MCU open-source prototyping platform. The findings demonstrate that architecture helps to minimize WBAN load and deliver time-bounded services. We want to expand the system in the future by adding more sensors on a big scale. More technological enhancements and clinical data collection are required to determine whether the telemedicine system can be implemented in practice. However, this proposed methodology has several flaws that can be rectified in the future. One is the security and privacy of data created by multiple layers.

References

1. Pacis DMM, Subido EDC, Bugtai NT (2018) Trends in telemedicine utilizing artificial intelligence. AIP Conf Proc 1933. https://doi.org/10.1063/1.5023979
2. Donati M et al (2020) Improving care model for congenital heart diseases in paediatric patients using home telemonitoring of vital signs via biomedical sensors. In: IEEE medical measurements & applications, MeMeA 2020—conference proceedings. https://doi.org/10.1109/MeMeA49120.2020.9137163
3. Pap IA, Oniga S, Alexan A (2020) Machine learning EEG data analysis for eHealth IoT system. In: 2020 22nd IEEE international conference on automation, quality and testing, robotics—THETA, AQTR 2020—proceedings, pp 20–23. https://doi.org/10.1109/AQTR49680.2020.9129966
4. Angelucci A, Kuller D, Aliverti A (2020) A home telemedicine system for continuous respiratory monitoring. IEEE J Biomed Health Inform 2194(c):1. https://doi.org/10.1109/jbhi.2020.3012621
5. Ashapkina MS, Alpatov AV, Sablina VA, Kolpakov AV (2019) Metric for exercise recognition for telemedicine systems. In: 2019 8th Mediterranean conference on embedded computing MECO 2019—proceedings, June 2019, pp 1–4. https://doi.org/10.1109/MECO.2019.8760024
6. Buldakova T, Krivosheeva D, Suyatinov S (2019) Hierarchical model of the network interaction representation in the telemedicine system. In: Proceedings—2019 21st international conference "complex systems: control and modeling problems", CSCMP 2019, vol 2019, Sept 2019, pp 379–383. https://doi.org/10.1109/CSCMP45713.2019.8976743
7. Choi A, Noh S, Shin H (2020) Internet-based unobtrusive tele-monitoring system for sleep and respiration. IEEE Access 8:76700–76707. https://doi.org/10.1109/ACCESS.2020.2989336

8. Kallipolitis A, Galliakis M, Menychtas A, Maglogiannis I (2020) Affective analysis of patients in homecare video-assisted telemedicine using computational intelligence. Neural Comput Appl 0123456789. https://doi.org/10.1007/s00521-020-05203-z

9. Mani N, Singh A, Nimmagadda SL (2020) An IoT guided healthcare monitoring system for managing real-time notifications by fog computing services. Procedia Comput Sci 167(2019):850–859. https://doi.org/10.1016/j.procs.2020.03.424

10. Banerjee A, Mohanta BK, Panda SS, Jena D, Sobhanayak S (2020) A secure IoT-fog enabled smart decision making system using machine learning for intensive care unit. In: 2020 international conference on artificial intelligence and signal processing, AISP 2020, pp 2–7. https://doi.org/10.1109/AISP48273.2020.9073062

11. Ganesh D, Seshadri G, Sokkanarayanan S, Rajan S, Sathiyanarayanan M (2019) IoT-based Google duplex artificial intelligence solution for elderly care. In: 2019 international conference on contemporary computing and informatics (IC3I), Singapore, Singapore, pp 234–240. https://doi.org/10.1109/IC3I46837.2019.9055551

12. Liau JC, Ho CY (2019) Intelligence IoT (internal of things) telemedicine health care space system for the elderly living alone. In: Proceedings of the 2019 IEEE Eurasia conference on biomedical engineering, healthcare and sustainability, ECBIOS 2019, pp 13–14. https://doi.org/10.1109/ECBIOS.2019.8807821

13. Kaimakamis E et al (2019) Applying translational medicine by using the welcome remote monitoring system on patients with COPD and comorbidities. In: 2019 IEEE EMBS international conference on biomedical and health informatics, BHI 2019—proceedings, pp 1–4. https://doi.org/10.1109/BHI.2019.8834464

14. Gračanin D, Benjamin Knapp R, Martin TL, Parker S (2019) Smart virtual care centers in the context of performance and privacy. In: ConTEL 2019—15th international conference on telecommunications proceedings, pp 1–8. https://doi.org/10.1109/ConTEL.2019.8848553

15. Kolisnyk K, Deineko D, Sokol T, Kutsevlyak S, Avrunin O (2019) Application of modern internet technologies in telemedicine screening of patient conditions. In: 2019 IEEE international scientific-practical conference problems of infocommunications, science and technology (PIC S&T 2019)—proceedings, pp 459–464. https://doi.org/10.1109/PICST47496.2019.9061252

16. Lopez A, Jimenez Y, Bareno R, Balamba B, Sacristan J (2019) E-health system for the monitoring, transmission and storage of the arterial pressure of chronic-hypertensive patients. In: 2019 Congreso Internacional de Innovacion y Tendencias en Ingenieria, CONIITI 2019—conference proceedings. https://doi.org/10.1109/CONIITI48476.2019.8960803

17. Swamy TJ, Murthy TN (2019) eSmart: an IoT based intelligent health monitoring and management system for mankind. In: 2019 international conference on computer communication and informatics, ICCCI 2019, pp 1–5. https://doi.org/10.1109/ICCCI.2019.8821845

18. Shahiduzzaman KM, Hei X, Guo C, Cheng W (2019) Enhancing fall detection for elderly with smart helmet in a cloud-network-edge architecture. In: 2019 IEEE international conference on consumer electronics—Taiwan, ICCE-TW 2019, pp 1–2. https://doi.org/10.1109/ICCE-TW46550.2019.8991972

19. Zhang K, Ling W (2020) Health monitoring of human multiple physiological parameters based on wireless remote medical system. IEEE Access 8:71146–71159. https://doi.org/10.1109/ACCESS.2020.2987058

20. Tripathy AK, Mohapatra AG, Mohanty SP, Kougianos E, Joshi AM, Das G (2020) EasyBand: a wearable for safety-aware mobility during pandemic outbreak. IEEE Consum Electron Mag 2248(c):10–14. https://doi.org/10.1109/MCE.2020.2992034

21. Zhao X, Zeng X, Koehl L, Tartare G, De Jonckheere J, Song K (2019) An IoT-based wearable system using accelerometers and machine learning for fetal movement monitoring. In: Proceedings—2019 IEEE international conference on industrial cyber physical systems, ICPS 2019, pp 299–304. https://doi.org/10.1109/ICPHYS.2019.8780301

Robust Control of Proton Exchange Membrane Fuel Cell (PEMFC) System

Gunjan Taneja, Vijay Kumar Tayal, and Kamlesh Pandey

Abstract The need for energy and its demand is increasing at a high rate in today's world. There is a need to shift toward a cleaner option to save the environment. PEMFC is one of the best choices to achieve energy demand without polluting the surroundings. Uncertainties in PEMFC can be determined using robust control techniques. Three types of control schemes like H infinity, disk margin method, PI method, and loop shaping are used and compared to determine parameters. The MATLAB output results have been compared in terms of performance evaluation parameters viz. rise time, overshoot, peak time, and settling time for H infinity and disk margin method. From the comparison of simulation results, this is observed that disk margin gives far better results and steady-state error is reduced to a large extent. Thus, the disk margin method is much superior to H infinity method. On comparing frequency response for PI and loop-shaping method, parameters like phase margin and gain margin are checked, and it is observed that the PI control scheme gives far better results.

Keywords Robust controller · Uncertainty · Transfer function · Control schemes · MATLAB

1 Introduction

The device which involves the conversion of chemical energy into electrical energy using an electrochemical reaction is referred to as a fuel cell. It involves the use of electrolytes one as fuel and the other as an oxidant for its working. In today's

G. Taneja (✉) · V. K. Tayal (✉) · K. Pandey
Department of Electrical and Electronics Engineering, Amity University Uttar Pradesh, Noida, India
e-mail: gunjan.taneja1998@gmail.com

V. K. Tayal
e-mail: vktayal@amity.edu

K. Pandey
e-mail: kpandey@amity.edu

© The Author(s), under exclusive license to Springer Nature Singapore Pte Ltd. 2023
A. Shukla et al. (eds.), *Computational Intelligence*, Lecture Notes in Electrical Engineering 968, https://doi.org/10.1007/978-981-19-7346-8_53

era, the demand for energy requirement has been increased to a large extent. Fuel cell contributes toward a clean environment and the production of energy. The most prominent type of fuel cell is proton exchange membrane fuel cell (PEMFC). It was first used in NASA program in 1960. It has various advantages like high efficiency, zero-emission, and operations at low temperature [1]. Nonlinear characteristics are present in proton fuel cell processes and their working. It is sensitive to the environment in which it is working and uncontrolled conditions may result in its degradation. So there is a need to have control schemes to prevent damage, decrease the voltage in stack and oxygen starvation, and improve reliability and increase their lifetime [2]. Noise, interference signals, disturbances, and uncertainty also exist in these fuel cells.

The volumetric approach using lumped parameters for the nonlinear model was described by Pukrushpan et al. [3]. A three-step validation method for a different type of fuel cell was given by Min et al. [4]. Niknezhadi et al. [5] worked on LQR techniques. Biagiola worked on the gray box model technique for determining uncertain parameters [6]. Figueroa and Biagiola [7] determines uncertainty using robust techniques. Kim proposed optimization techniques for determining parameters [8]. Mukhtar et al. worked on PID-based optimal controller [9]. Hybrid renewable energy for enhancement of performance was developed by Arora et al. [10]. Fuzzy-based method for integration of the fuel cell is discussed in [11]. Thermal management system was developed by Zhao et al. [12]. The dynamic model was described by Yang and Chen [13].

In the proposed work, robust control scheme is used to determine the uncertainty in transfer function for MIMO PEMFC system. Three types of robust control schemes are used. H infinity control method is used at first to determine the value of norm ad frequency. The disk margin method is then used by considering different transfer functions and compared on basis of phase margin, frequency values. Lastly, the loop control scheme is used and is compared with the PI control scheme on basis of phase margin values. This paper is organized as follows—the model of the PEMFC system is discussed in Sect. 2. Various schemes are then described for determining uncertainties in the transfer function. They are followed by the output and plots obtained using different controllers. Lastly, the conclusion is added followed by the references used.

2 MIMO System of PEMFC

The nonlinear model of PEMFC was transformed to the linear system [14]. The equations for same are given as follows:

```
A = [−10.85 −7.61 0 7.65, −28.69 −32.32 0 28.94, 0 0 −7.329 −
0.04092, 19.84 19.84 89.85 −21.47]
B = [0 −289.9; 0 0; 365.7 0; 0 0]
C = [0 0 0 1; 0 0 1.644*10−5 −2.988*10−7]
```

Input for the system is compressor voltage (V) as $u1$ and stack current (A) as $u2$. Output for the system is pressure (Pa) as $y1$ and outflow rate of a compressor (kg/s) as $y2$. The final transfer function is given as:

$$G(s) = [203.165e - 0.0085s/0.3742s + 1, \ 0.0008724/0.1465s + 1; \ -35.523e$$
$$-0.0085s/0.3742s + 1, \ 0.0001061e - 0.0085s/0.3742s + 1].$$

2.1 Control Schemes

Robust control is a technique that focuses on dealing with uncertainty. It is constructed in such a way that it may be determined within a specific set of uncertain parameters and disturbances. In the presence of mistakes, it strives to achieve the best performance and stability in the system. It primarily operates on unidentified plants with unidentified dynamics and characteristics. Figure 1 depicts the concept of uncertainty. Figure 1 depicts the control loop. Uncertainty enters the system from three directions. There is uncertainty in the plant model. There are also disturbances. From the sensor side, noise entered the system. The nature of this uncertainty can be additive or multiplicative. Because designers have little control over uncertainties, the computer control system is segregated from the plant.

H Infinity Control Scheme. In control theory, H (or "H infinity") methods are used to synthesize controllers with assured performance for stabilization. H methods have an advantage over conventional control approaches in that they can be used to solve problems involving multivariate systems with channel cross-coupling.

Disk Margin Control Scheme. The stability of a closed-loop system against gain or phase fluctuations in the open-loop response is measured by disk margins. The software models work on multiplicative uncertainty on the open-loop transfer function in disk-based margin calculations. The disk margin is a measure that indicates how much uncertainty the loop can withstand before becoming unstable.

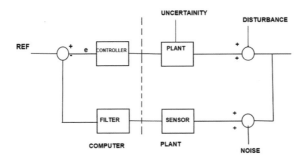

Fig. 1 Robust control system

Loop-Shaping Method. In this method, the shape of the open-loop response is desired is provided by the user, and the loop-shaping function constructs a controller that approximates that shape in loop-shaping controller synthesis.

PI Control Scheme. It consists of the proportional error signal along with the integral error signal. With the presence of a proportional controller, it provides fast action and there will be no steady-state error because of the use of the integral controller.

3 Simulation Results

Four different controllers are compared based on different parameters. For H infinity and disk margin method, step response is obtained and are compared on basis of parameters like rise time, settling time, overshoot, and peak time as in Table 2. PI and loop-shaping methods are compared on basis of phase margin and gain margin values. Frequency response and step response are obtained for the H infinity method as in Fig. 2. Table 1 describes the value of H infinity parameters like h infinity norm and frequency (Fig. 3). Figures 4, 5, 6, 7, 8, 9, 10, 11 and 12 gives different step responses for open loop for four transfer functions, Bode response for four transfer functions and step response obtained using disk margin method. PI and loop response are shown in Figs. 13 and 14.

The value of the H infinity norm for the H infinity controller is 201.4454; frequency is found to be 0 as in Table 1. On basis of the time response shown in Table 2, the rise time for the H infinity controller is 0.8221 s and 0.3923 s for the disk margin method. Settling time for the H infinity controller and disk margin method is 1.4639 s,

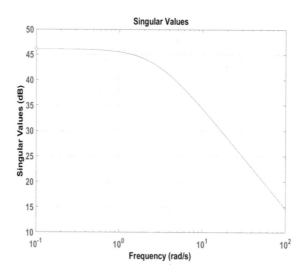

Fig. 2 Frequency response using H infinity controller

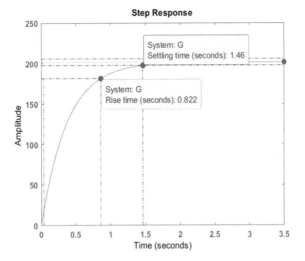

Fig. 3 Step response for the system using H infinity controller

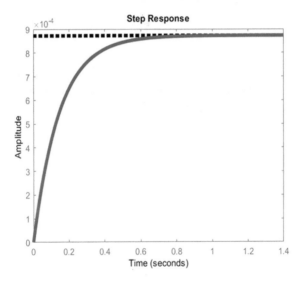

Fig. 4 Step response for 1st T.F open loop without any controller

0.6986, respectively. Overshoot is obtained as 0 in both cases. H infinity has a peak time of about 201.4401 s and disk margin method is 1.8832 s. Frequency response in Table 3 gives the value of phase margin as −0.5153 for PI controller and 90.2858 for loop-shaping method. The gain margin is calculated as [0 Inf] for PI controller and infinite for loop-shaping method. Thus, in this way, comparison is done using different controllers on basis of different parameters.

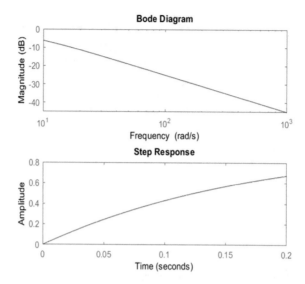

Fig. 5 Bode plot for 1st T.F using disk margin method

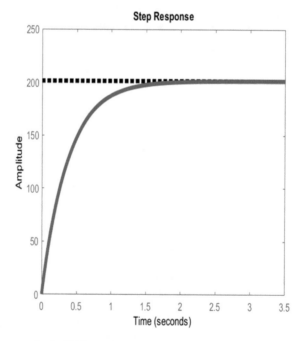

Fig. 6 Step response for 2nd T.F open loop without any controller

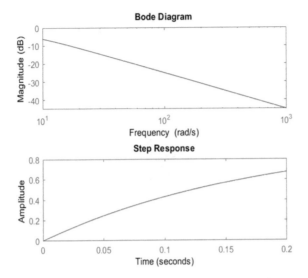

Fig. 7 Bode plot for 2nd T.F using disk margin method

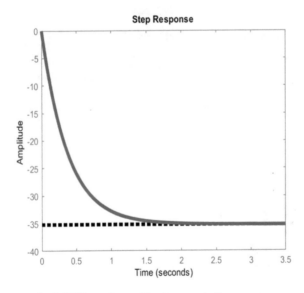

Fig. 8 Step response for 3rd T.F open loop without any controller

4 Conclusion

With the depletion of fossil fuels, alternative energy options are becoming more important. These options require considerable experimental and typical effort to realize new energy resources. PEMFC modeling and simulation enhancements have

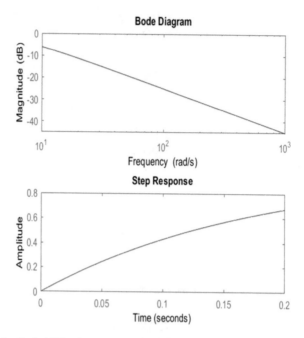

Fig. 9 Bode plot for 3rd T.F using disk margin method

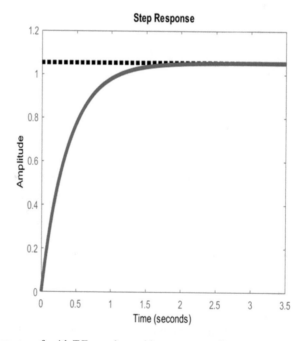

Fig. 10 Step response for 4th T.F open loop without any controller

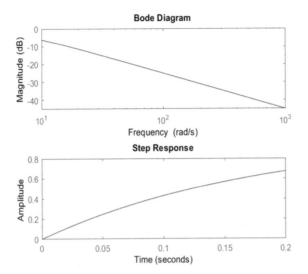

Fig. 11 Bode plot for 4th T.F using disk margin method

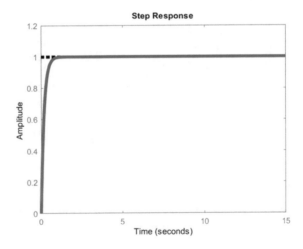

Fig. 12 Step response for the system using disk margin method

been devised to find new alternative remedies. Therefore, PEM fuel cells can be investigated to improve PEM fuel cell modeling and simulation. Uncertainty needs to be traced to ensure the proper working of the PEMFC system. Using various schemes uncertain parameters in transfer function is determined. According to the output in Table 2, it is observed that all parameters like rise time, settling time, overshoot, and peak time are lower using the disk margin method. So it is considered to be the feasible method. On comparing other parameters like phase margin and gain margin in frequency response, PI controller is considered to be better in response.

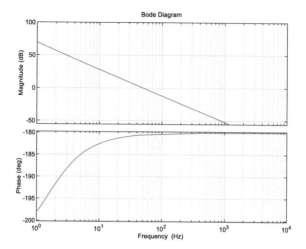

Fig. 13 Bode plot for system using PI controller

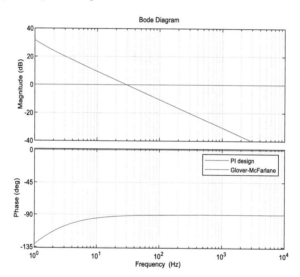

Fig. 14 Comparative plot for PI and loop method

Table 1 H infinity parameters

S. No.	Parameters	H infinity controller
1	Ninf (h infinity norm)	201.4454
2	Fpeak (frequency)	0

Table 2 Comparative analysis of H infinity and disk margin method

Characteristic	H infinity controller	Disk margin
Rise time	0.8221	0.3923
Settling time	1.4639	0.6986
Overshoot	0	0
Peak time	201.4401	1.8832

Table 3 Comparative analysis of PI and loop-shaping method

S. No.	Parameters	PI	Loop shaping
1	Phase margin	−0.5153	90.2858
2	Gain margin	[0 Inf]	Inf

Thus on basis of step response, disk margin is the best method, and on basis of frequency response, PI controller is feasible.

References

1. Han IS, Park SK, Chung CB (2016) Modeling and operation optimization of a proton exchange membrane fuel cell system for maximum efficiency. Energy Convers Manage 113:52–65
2. Fang C, Li J, Xu L, Ouyang M, Hu J, Cheng S (2015) Model-based fuel pressure regulation algorithm for a hydrogen-injected PEM fuel cell engine. Int J Hydrogen Energy 40(43)
3. Pukrushpan JT, Stefanopoulou AG, Peng H (2004) Control of fuel cell power systems: principles, modeling, analysis, and feedback design. Springer, London
4. Min CH, He YL, Liu XL et al (2006) Parameter sensitivity examination and discussion of PEM fuel cell simulation model validation: part II: results of sensitivity analysis and validation of the model. J Power Sources 160(1):374–385
5. Niknezhadi A, Allúe-Fantova M, Kunusch C, Ocampo-Martìnez C (2011) Design and implementation of LQR/LQG strategies for oxygen stoichiometry control in PEM fuel cells based systems. J Power Sources 196(9):4277–4282
6. Biagiola SI, Schmidt C, Figueroa JL (2014) Model uncertainty estimation of a solid oxide fuel cell using a Volterra-type model. J Frankl Inst 351:4183–4197
7. Figueroa JL, Biagiola SI (2013) Modeling and uncertainties characterization for robust control. J Process Control 23:415–428
8. Kim K, von Spakovsky MR, Wang M, Nelson DJ (2012) Dynamic optimization under uncertainty of the synthesis/design and operation/control of a proton exchange membrane fuel cell system. J Power Sources 205:252–263
9. Mukhtar A, Tayal VK, Singh H (2019) PSO optimized PID controller design for the process liquid level control. In: 2019 3rd international conference on recent developments in control, automation & power engineering (RD CAPE), pp 590–593. https://doi.org/10.1109/RDCAPE 47089.2019.8979108
10. Arora R, Tayal VK, Singh HP, Singh S (2020) PSO optimized PID controller design for performance enhancement of hybrid renewable energy system. In: 2020 IEEE 9th power India international conference (PIICON), pp 1–5. https://doi.org/10.1109/PIICON49524.2020.911 3046

11. Khubchandani V, Pandey K, Tayal VK, Sinha SK (2016) PEM fuel cell integration with using Fuzzy PID technique. In: 2016 IEEE 1st international conference on power electronics, intelligent control and energy systems (ICPEICES), pp 1–4
12. Zhao X, Li Y, Liu Z, Li Q, Chen W (2015) Thermal management system modeling of a water-cooled proton exchange membrane fuel cell. Int J Hydrogen Energy 40:3048–3056
13. Yang CW, Chen YS (2014) A mathematical model to study the performance of a proton exchange membrane fuel cell in a dead-ended anode mode. Appl Energy 130:113–121
14. Divi S, Sonawane S, Das S (2018) Uncertainty analysis of transfer function of proton exchange membrane fuel cell and design of PI/PID controller for supply manifold pressure control. Indian Chem Eng 61:1–15

Popularity-Based BERT for Product Recommendation

Srushti Gajjar⬡ **and Mrugendra Rehevar**⬡

Abstract Daily, a large number of ratings and reviews are shared on various social media platforms and websites. Manufacturers wish to create a user recommender system that can discover possible target customers fast and effectively. Different recommendation models are used by online e-commerce platforms like Amazon and Flipkart to deliver different choices to different users. We gathered data of user reviews on Amazon electronic devices for this study. As a result, this research uses review data with ranking (rating) labels on Amazon's electronic items to fine-tune and carry out fine-tuning the bidirectional encoder representations from transformers (BERT) model. We have built a recommendation system that allows recommending products to users/customers related to past product ratings. This study proposes a popularity-based recommender for recommending electronic devices/products to relevant potential consumers. To predict ratings, the proposed recommender system uses fine-tuned BERT. The popularity-based approach looks for products that are currently trending or popular among users, then recommends them to clients right away. In this study, we have recommended popular products to the users/customers which are frequently rated/reviewed by different users.

Keywords Recommendation system · Fine-tuning BERT · Popularity based

1 Introduction

Data has been the most important aspect in everything; however, the amount of data available today is growing at an exponential rate. Despite being the world's

S. Gajjar (✉)
Assistant Professor, Charotar University of Science and Technology, Changa, Anand, Gujarat 388 421, India
e-mail: srushtigajjar.cse@charusat.ac.in

M. Rehevar
Charotar University of Science and Technology, Changa, Anand, Gujarat 388 421, India

second-largest client hub, in June 2015, India had supposed to 354 million users of the Internet, with 500 million anticipated in 2016. Despite the fact that web-based business penetration is modest in comparison with commercial regions such as the United States which was 266 million, 84% and France which was 54 million, 81%, it is experiencing exponential growth, with over half a million new people joining per month [1].

The conventional database management system cannot handle these many datasets efficiently. Database systems cannot store and process datasets in the form of semi-structured and unstructured data such as image, audio, video, JSON documents, wet logs, and search trends, and thus the concept of Big Data came into the picture [1]. According to IBM, "every day, Internet users generate 2.5 quintillion bytes of data enough to account for 90% of the information on the planet today," in the last two years alone [1] (Fig. 1).

The information system for recommendation systems is a well-organized method of checking that filters data based on user preferences or activity. At the outset of the suggestion process, the recommendation system relies on the user's raw data, and then the system observes the customer's activities before offering personalized suggestions [2].

The system may suggest products, such as books, sports, publications, job openings, movies, and advertisements. Netflix uses a recommender algorithm to suggest movies and web series to its members. Similarly, YouTube suggests videos based on viewers and likes. There are numerous instances of widely used recommenders nowadays.

The requirement of the recommendation system is that users benefit from being able to find items that are of interest to them, assist item providers in getting their

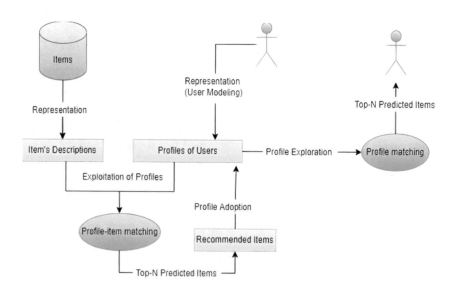

Fig. 1 General concept of recommender system

products to the correct people, users will be able to identify products that are most relevant to them, personalized content, assist websites in increasing user engagement.

A recommendation system provides a collection of tools designed to help the user make decisions. Based on the user's activities, these programs provide the user with data recommendations. Facebook, Google, and Netflix are just a few of the applications that use these recommender systems [2].

Content-based, collaborative, personalized, and hybrid recommender systems are types of recommender systems.

1.1 Content-Based Recommendation System

As an outcome, personalized access to information and user modeling are becoming increasingly critical: Users expect personalized service while filtering through vast amounts of available info likes and dislikes, and tastes [3].

Recommenders guide users to useful or interesting things from a vast selection in a personalized manner. Recommenders generate a list of goods based on the choices of a customer. At Amazon.com, prediction systems are utilized to tailor the online store to each customer's specific demands, likewise, programming labels have been shown to software engineers and baby objects have been shown to new mothers [3].

A content-based recommender determines the number of documents and/or content of products that have previously been rated by a client. The system then creates a model of that person's choices relying on the features of the items that the user has previously ranked [4].

The framework is a structured representation of a customer's tastes that is used to recommend new products to individuals. The basic idea underlying the recommendation process is to match the elements of a content item to the traits of a user account. As an outcome, a relevance evaluation represents the user's level of interest in that thing. Whenever a profile accurately reflects a user's interests, the information access procedure becomes much more efficient. The content analyzer process, which often takes methods from information retrieval systems. It builds a structured item representation by extracting features from unstructured text, which are stored in the repository as objects. In building and maintaining the active's profile, item responses are captured in a certain manner and kept in the review database. These reactions, referred to as annotation or feedback, are used in conjunction with item descriptions to build a model that can predict the true importance of freshly presented things. For recording user input, there are two methods that can be used. This strategy is known as "explicit feedback" when a system necessitates the user's explicit evaluation of elements; the other way is known as "implicit feedback." Which refers to a technique in which feedback is produced without requiring active user involvement, observing, and evaluating the user's behaviors [3].

Advantages of a content-based system are

- User Independence: To establish an active user's profile, content-based models rely solely on the active user's opinions/ratings.
- Transparency—Explicitly listing content features that prompted an object in the list of suggestions could provide info about how the recommender system actually works. These factors should be taken into account while determining whether or not to believe a recommendation.
- New item—Make suggestions for products that have not been reviewed by anyone yet which is done by content-based recommenders [3].

1.2 Collaborative Filtering Based

User behavior information in the form of preferences, feedback, ratings, comments, and actions is collected and analyzed through collaborative filtering systems. Then, it uses commonalities across numerous users or things to predict missing ratings and, as a response, make relevant suggestions relying on this data [5].

Approaches of collaborative filtering generate customer item recommendations based on ratings, such as purchases, without any need for information about the users or the items [5]. If person P1 likes item 1, item 2, item 4, and person P2 likes item 1, item 4, item-5, and what happens if person P3 likes item 1? As an outcome, there is a possibility that person P3 will like item 4 because we already know that person P1, P2, and P3 like items from the first two claims. Figure 2 shows content-based and collaborative filtering recommendations.

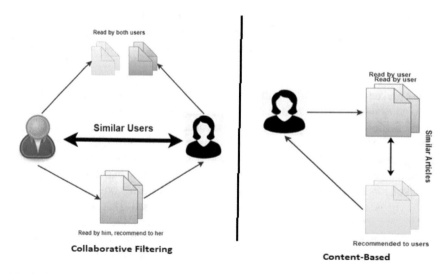

Fig. 2 Collaborative filtering versus content-based filtering

1.3 Hybrid

The hybrid method combines to get the best advantage and improve outcomes while decreasing the worries and obstacles in terms of challenges of these applications, the two advised content-based and collaborative filtering methodologies [5].

There are many approaches for hybrid approaches [6].

- Weighted numerically aggregated every recommended component that received a different score from the system.
- In feature combination to develop recommendation system features, several knowledge sources are integrated.
- In switching, the system presents the user with a number of various recommendation items and selects the most appropriate one based on the user's preferences.

In mixed, the system suggests to the user a combination of objects at the same time. Using content-based or collaborative filtering, great performance and alleviation of issues and challenges are achieved by combining these various approaches.

2   Related Work

Researchers in research paper [7] proposed genetic algorithms which learning personal preferences from customers using algorithms like genetic algorithms which are evolutionary algorithms [8, 9] is one of the most widely used methods. For vending machines [10] which are autonomous, a product recommendation model has been developed. A hybrid recommendation strategy for product items based on a genetic algorithm and user ratings [11] are two examples. To acquire personal preferences from customers, one of the most extensively used approaches is to employ evolutionary algorithms such as the genetic algorithm (GA) [3, 4]. In this research, they offer a product recommender system that uses a genetic algorithm to determine the optimum combination of products to suggest to clients. The study was based on a recording studio power unit recommendation [7]. The genetic algorithm is an essential method for finding answers to optimization issues. Because the solution may be described simply in a bit of array with the number "1" signifying the desired products and "0" or else, the genetic algorithms were picked to fix this problem [7]. The uniform crossover strategy is applied to reproduce the new population. For maintaining genetic variety and preventing premature population convergence, the mutation is a crucial component in the genetic algorithm.

Figure 3 shows the results, which suggest that collaborative filtering using a neighborhood-based strategy is well investigated. A noticeable rise in the year 2012 can be seen in the graph, indicating a current trend in this field of study.

The aim of self-supervised learning [13–15] is to train a network with an auxiliary purpose of automatically extracting ground-truth samples from raw data. The system

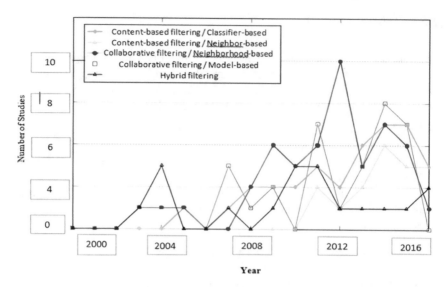

Fig. 3 Timeline of the classification of the studies [12]

generates training signals directly from the correlation of raw data that would be used to train the set. The correlation data gathered by self-supervised learning can then be found in a range of other tasks with ease [16]. Language modeling [13, 15] is a prominent self-supervised natural language processing goal in which the model improves its ability to anticipate future words or/and sentences based on prior sequences. Self-supervised learning has a branch called mutual information maximization [14, 17, 18]. It is based on the InfoMax concept [18] and has achieved significant advances in areas including audio processing [19], natural language understanding [17], and computer vision [14].

In research paper [13], authors used the approach that to improve the original model, they have incorporated many intricately developed self-supervised learning objectives. To generate such aims, they make use of effective correlation signals reflected in the input's intrinsic qualities. They consider the data at many levels of granularity for their work, including attribute, item, segment, and sequence, which represent several points of view on the input. By recording the multi-view correlation, they unite these self-supervised learning aims with the previously defined pretraining strategy in language modeling [13]. They constructed the basic structure for a sequential recommendation model by layering the embeddings, self-attention modules, and predictions layer.

In research paper [20], researches used the BERT pretraining model, the BiLSTM bidirectional long-short memory network, and customer reviews to suggest an upgraded collaborative filtering model based on ConvMF. In comparison with the feature extraction model based on bag-of-words representation, the ConvMF approach uses CNN to extract hidden features from item description information, which is significantly better. The model does, however, have three flaws. First, CNN's local field of view approach can only gather context information from a limited distance, resulting in severe context leakage between long-distance phrases. Second, the typical static word embedding strategy neglects the fact that semantics are context-dependent. Furthermore, in the modeling process, only the item description information and the scoring matrix are used; customer opinions are avoided.

3 Proposed Approach

In this paper, we have used the BERT model for the product recommendation system.

Natural language processing is an artificial intelligence sub-discipline. The research aids in the development of systems that can examine and mimic natural language understanding.

Text machine reading, voice classification, automated speech generation, and machine translation are all examples of automated text or voice generation of natural language processing that require a model. Despite starting using a model that has previously been trained on another problem, rather than having to start from scratch to address a similar experience in terms of the problem. There are several different types of pretrained models: ELMO [21] and GPT [22] from left to right it will look, whereas BERT looks at the text from both the left and right sides. BERT is a feature-based and fine-tuning model that has been pretrained. BERT is powered by a transformer, which is made up of two parts: an encoder and a decoder. BERT merely used the portion encoder and skipped over the decoder.

Pretraining and fine-tuning are the two stages of the BERT technique. BERT is pretrained on two unsupervised tasks: masked LM and next line predictions. BERT's strength is his ability to pay attention to himself from both the left and right sides. This feature can be used by BERT to anticipate the masked word. In conclusion, before a BERT model predicts these masked tokens, certain input tokens must be masked randomly; this type is called Masked LM. Previous models of BERT are BIOBERT, Clinical BERT, AraBERT, SCIBERT, etc.

3.1 Popularity-Based Recommendation System

Data gathering, BERT fine-tuning, and popularity-based recommendation are the three stages of the proposed model. The data collected is split into two pieces. We need labeled data to fine-tune the BERT model. As a result, we use review data with one feature rating out. For recommendations based on review data, the fine-tuned BERT model's anticipated rating value is being used (Fig. 4).

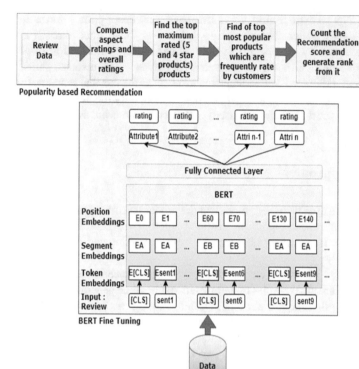

Fig. 4 Popularity-based recommendation model

3.2 Data Collection

Reviews of 4-star and 5-star electrical devices were collected on the Amazon website in order to construct the popularity-based recommendation model. Amazon is an e-commerce, digital streaming, AI, and other technologies are the focus of this multi-national technology corporation. It is the world's largest marketplace. Electronic devices have ratings. The total amount of data collected is 386580. The total number of users 318,173 and total number of products are 24,440.

3.3 Fine-Tuning BERT Model

Transformers' bidirectional encoder representations is a huge neural network with hundreds of millions of parameters. When you train a BERT model from the scratch on a small dataset, it will result in overfitting. As a consequence, starting with a pretrained BERT model that's been trained on a huge dataset is preferred. Model fine-tuning is the process of enhancing the model's training using our tiny sample.

3.4 Popularity-Based Recommendation Process

A popularity-based recommendation system is a form of recommendation system that works on the basis of popularity or anything that's currently popular. These algorithms look for products that are currently trending or popular among consumers and then immediately recommend them (Figs. 5 and 6).

According to the joint plot, popular products (those with higher ratings) are rated more frequently. To increase people's engagement, we can start by recommending them using a popularity-based system and then gradually graduate them to a collaborative system once we have a sufficient number of data points to provide personalized recommendations.

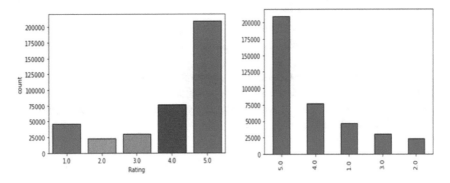

Fig. 5 Ratings to products by users and ratings to products by users in descending order

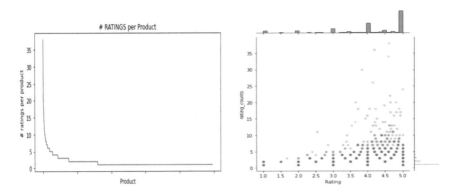

Fig. 6 Ratings per product and ratings count

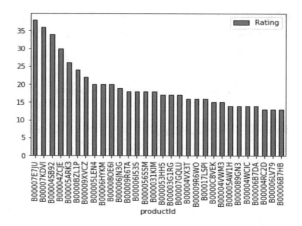

Fig. 7 Top 30 most popular products

The product list of the top 30 most popular products is shown in the bar graph below (Fig. 7).

Figure 8 is a user-selected list, as this is a popularity-based recommendation approach that does not take into account the user. The same products will be recommended.

Root-mean-square error (RMSE) is a commonly used metric for comparing predicted and observed values (sample or population values) by a model or estimator. The RMSE value is 1.11 for the popularity-based model. For our dataset, it is 1.2532474701604854.

4 Conclusion

We introduced a popularity-based recommender system in this study, which uses a fine-tuned BERT model to recommend suitable target customers for electronic equipment. To match the recommendations in the product domain, we gathered Amazon's electronic device overall rating data, which is the world's largest Internet corporation by revenue and the US's second-largest private employer. A popularity-based approach extracts patterns from the previous rating statistics of other comparable users and/or items to anticipate a user's preference for a product. The review data is analyzed using fine-tuned BERT to predict the overall rating in this study. Then, using the anticipated rating values, we obtained user ratings for products, then the top most popular products, and we sorted the products by recommendation score, established a recommendation rank based on score, and suggested/recommended the top high-scoring products to various customers.

```
+-----------------+---------------+--------+------+
|      userId     |   productId   | score  | rank |
+-----------------+---------------+--------+------+
|  AKM1MP6P00YPR  |  B0000010N6   |  5.0   |  1   |
|  AKM1MP6P00YPR  |  B000006705G  |  5.0   |  2   |
|  AKM1MP6P00YPR  |  B000006708H  |  5.0   |  3   |
|  AKM1MP6P00YPR  |  B00009R91K   |  5.0   |  4   |
|  AKM1MP6P00YPR  |  B0000A2U4Y   |  5.0   |  5   |
|  A2CX7LUOHB2NDG |  B0000010N6   |  5.0   |  1   |
|  A2CX7LUOHB2NDG |  B000006705G  |  5.0   |  2   |
|  A2CX7LUOHB2NDG |  B000006708H  |  5.0   |  3   |
|  A2CX7LUOHB2NDG |  B00009R91K   |  5.0   |  4   |
|  A2CX7LUOHB2NDG |  B0000A2U4Y   |  5.0   |  5   |
|  A2NWSAGRHCP8N5 |  B0000010N6   |  5.0   |  1   |
|  A2NWSAGRHCP8N5 |  B000006705G  |  5.0   |  2   |
|  A2NWSAGRHCP8N5 |  B000006708H  |  5.0   |  3   |
|  A2NWSAGRHCP8N5 |  B00009R91K   |  5.0   |  4   |
|  A2NWSAGRHCP8N5 |  B0000A2U4Y   |  5.0   |  5   |
|  A2WNBOD3WNDNKT |  B0000010N6   |  5.0   |  1   |
|  A2WNBOD3WNDNKT |  B000006705G  |  5.0   |  2   |
|  A2WNBOD3WNDNKT |  B000006708H  |  5.0   |  3   |
|  A2WNBOD3WNDNKT |  B00009R91K   |  5.0   |  4   |
|  A2WNBOD3WNDNKT |  B0000A2U4Y   |  5.0   |  5   |
|  A1GI0U4ZRJA8WN |  B0000010N6   |  5.0   |  1   |
|  A1GI0U4ZRJA8WN |  B000006705G  |  5.0   |  2   |
|  A1GI0U4ZRJA8WN |  B000006708H  |  5.0   |  3   |
|  A1GI0U4ZRJA8WN |  B00009R91K   |  5.0   |  4   |
|  A1GI0U4ZRJA8WN |  B0000A2U4Y   |  5.0   |  5   |
+-----------------+---------------+--------+------+
```

Fig. 8 Top 5 recommendations for specific user

References

1. Das D, Sahoo L, Datta S (2017) A survey on recommendation system. Int J Comput Appl 160(7):6–10
2. Amara S, Subramanian RR (2020) Collaborating personalized recommender sys-tem and content-based recommender system using TextCorpus. In: 2020 6th International conference on advanced computing and communication systems (ICACCS). Coimbatore, India, Mar 2020, pp 105–109
3. Lops P, de Gemmis M, Semeraro G (2011) Content-based recommender systems: state of the art and trends. In: Ricci F, Rokach L, Shapira B, Kantor PB (eds) Recommender systems handbook. Springer US, Boston, MA, pp 73–105
4. Mladenic D (1999) Text-learning and related intelligent agents: a survey. IEEE Intell Syst 14(4):44–54
5. Khusro S, Ali Z, Ullah I (2016) Recommender systems: issues, challenges, and research opportunities. In: Kim KJ, Joukov N (eds) Information science and applications (ICISA) 2016, vol. 376. Springer, Singapore, pp 1179–1189

6. Patel YG, Patel VP (2015) A survey on various techniques of recommendation system in web mining. Int J Eng Dev Res 3(4)
7. Janjarassuk U, Puengrusme S (2019) Product recommendation based on genetic algorithm. In: 2019 5th international conference on engineering, Applied sciences and technology (ICEAST). Luang Prabang, Laos, Jul 2019, pp 1–4
8. Hwang C-S, Su Y-C, Tseng K-C (2010) Using genetic algorithms for personalized recommendation. In: Pan J-S, Chen S-M, Nguyen NT (eds) Computational collective intelligence technologies and applications, vol. 6422. Springer, Berlin, pp 104–112
9. Wang Z, Yu X, Feng N, Wang Z (2014) An improved collaborative movie recommendation system using computational intelligence. J Vis Lang Comput 25(6):667–675
10. Lin F-C, Yu H-W, Hsu C-H, Weng T-C (2011) Recommendation system for localized products in vending machines. Expert Syst Appl 38(8):9129–9138
11. Gao L, Li C (2008) Hybrid personalized recommended model based on genetic algorithm. In: 2008 4th international conference on wireless communications, networking and mobile computing. Dalian, China, Oct 2008, pp 1–4
12. Portugal I, Alencar P, Cowan D (2018) The use of machine learning algorithms in recommender systems: a systematic review. Expert Syst Appl 97:205–227
13. Devlin J, Chang M-W, Lee K, Toutanova K (2021) BERT: pre-training of deep bidirectional transformers for language understanding. ArXiv181004805 Cs, May 2019, Accessed: 28 Nov 2021
14. Hjelm RD et al (2021) Learning deep representations by mutual information estimation and maximization. ArXiv180806670 Cs Stat, Feb 2019, Accessed: 28 Nov 2021
15. Mikolov T, Sutskever I, Chen K, Corrado G, Dean J (2021) Distributed representations of words and phrases and their compositionality. ArXiv13104546 Cs Stat, Oct 2013, Accessed: 28 Nov 2021
16. Zhou K et al (2020) S3-Rec: self-supervised learning for sequential recommendation with mutual information maximization. In: Proceedings of the 29th ACM international conference on information & knowledge management, virtual event Ire-land, Oct 2020, pp 1893–1902
17. Kong L, de M. d'Autume C, Ling W, Yu L, Dai Z, Yogatama D (2021) A mutual information maximization perspective of language representation learning. ArXiv191008350 Cs, Nov 2019, Accessed: 28 Nov 2021
18. Linsker R (1988) Self-organization in a perceptual network. Computer 21(3):105–117
19. van den Oord A, Li Y, Vinyals O (2021) Representation learning with contrastive predictive coding. ArXiv180703748 Cs Stat, Jan 2019, Accessed: 28 Nov 2021
20. Wan F et al (2020) Agricultural product recommendation model based on BMF. Appl Math Nonlinear Sci 5(2):415–424
21. Peters ME, Neumann M, Zettlemoyer L, Yih W (2021) Dissecting contextual word embeddings: architecture and representation. ArXiv180808949 Cs, Sep 2018, Accessed: 28 Nov 2021
22. Radford A, Narasimhan K, Saliman ST, Sutskever I (2018) Improving language understanding with unsupervised learning

A Novel Segmentation-Free Approach for Handwritten Sentence Recognition

M. Chethan, R. Anirudh, M. Kalis Rani, and Sudeepa Roy Dey

Abstract This is the era of digitization where everything must be available in digital format for easy processing of data. Through advancements in areas of text processing using improvised and futuristic ML and DL algorithms, digitization of handwritten text occupies an unparalleled position. This paper proposes a segmentation-free approach to convert a handwritten sentence to digital text. It uses convolutional neural networks (CNNs), recurrent neural networks (RNNs) and connectionist temporal classifier (CTC) function. The novel approach adopted is a segmentation-free approach which overcomes the increase in training time and character redundancy. The model achieves an accuracy of ~86% and is able to perform handwritten text recognition on sentences which include alphabets (both upper and lower case), digits and commonly used special symbols. Further, a UI has been added to the model for ease of use.

Keywords OCR · Handwriting identification · CTC · HTR

1 Introduction

Handwriting recognition is basically of two kinds (two approaches), offline and online recognition. Online recognition is based on stylus trajectory, i.e., recognition is real time. Recognition starts once the user starts to write the text on a text pad using a stylus. The other one is offline recognition [1]. Here, the complete text is

M. Chethan (✉) · R. Anirudh · M. K. Rani
PES University, Electronic City, Bangalore, Karnataka 560020, India
e-mail: chethanm@pesu.pes.edu

R. Anirudh
e-mail: anirudhr@pesu.pes.edu

M. K. Rani
e-mail: kalisranim@pesu.pes.edu

S. R. Dey
Deparment of CSE, PES University, Electronic City, Bangalore, Karnataka 560020, India
e-mail: sudeepar@pes.edu

© The Author(s), under exclusive license to Springer Nature Singapore Pte Ltd. 2023   641
A. Shukla et al. (eds.), *Computational Intelligence*, Lecture Notes in Electrical
Engineering 968, https://doi.org/10.1007/978-981-19-7346-8_55

scanned and given as an input to the system (typically a png/jpg file), and then the text is recognized. It is basically performed after the writing is complete. This paper focusses on offline recognition.

The most traditional technique followed in offline handwriting recognition is the segmentation approach [2, 3]. It comes with its own drawbacks. Two of the main short comings of segmentation model is time and redundancy. It takes much time to annotate the dataset on character level. It becomes difficult to segment cursive characters unless techniques like de-slantation are used. Also if the characters are written wide, then it leads to the presence of duplicate characters in the digital text. Now, to deal with these drawbacks, this paper comes up with the segmentation-free approach where individual characters are recognized without segmentation.

The segmentation-free approach maps the image of handwritten text (Matrix) to a character sequence of N length (100 characters here). So, discrete characters are not segmented in preprocessing stage instead they are passed as a whole to the recognizer. This technique is achieved by using the connectionist temporal classifier(CTC) function which maps the time-steps (individual characters) from the RNN to digital characters using probability sums. The important design of the model is this CTC function which counters the drawbacks of segmentation model. It comes with a feature called pseudo-spaces where redundant characters are eliminated even if handwritten texts are widely written. Notably, since the discrete handwritten alphabets are not segmented, both the train and test times are less compared to that of the segmentation model.

The input to the model requires a completely noiseless text with clear white background to achieve best possible and accurate results. The input image gets into various levels of CNNs [2], RNNs and CTC function which is explained thoroughly in the methodology section. This model can identify all the alphabets, both uppercase and lower case, digits (0–9) [4] and commonly used special symbols like full stop, comma, single and double quotations.

2 Methodology Adopted

2.1 Dataset Description

The IAM database file is used to construct the model. It is the popularly used dataset for training models on handwriting recognition. It has wide variety of sub-datasets ranging from images that contain letters to complete sentences. The lines subset of IAM dataset is used here to build the model. It contains 13353 images of handwritten lines (with a total of 115320 words) written across 1539 pages. The database also contains a text file that has every handwritten line with its ground text and with appropriate mapping to their original handwritten images.

The grounds behind using this dataset are the following:

1. Wide styles of handwriting
2. Contains digits and special symbols [2]
3. Minimal preprocessing required.

2.2 Model Description

This image (Fig. 1) explains the high-level design of the model. The model contains 3 important implementations, namely the convolutional neural networks [2, 5], the RNNs and a CTC function (both the greedy decoder and a loss function). This model is trained on variety of characters that include, but not limited to, uppercase alphabets, lowercase alphabets, digits and special symbols.

Convolution Neural Networks (CNNs) A CNN [6] is a type of artificial neural network that is widely used in image and visual domains. It mostly deals with edge detection (feature extractions) from the input images to give rise to further conclusions.

There are a total sum of 7 layers of CNN [5] (Fig. 2) in this model, where in each layer, a fixed size of Kernel is used to detect edges in the input image. The output of CNN's 7th layer is given as an input to the 1st RNN layer [7] .

Recurrent Neural Networks (RNNs) There are 2 layers of RNN (bi-directional LSTM implementation) in the model. The reason behind using LSTM model is we need to propagate the information of handwritten lines through larger distances. A bi-directional RNN is implemented to get two variables, namely, forward and backward and concatenate them to achieve better learning rate. After concatenation, the output is given as an input to the CTC function. The output of RNN has input image converted into horizontal positions called the time-steps (Fig. 3).

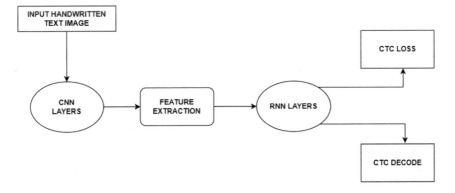

Fig. 1 Proposed methodology which contains CNNs, RNNs and CTC function

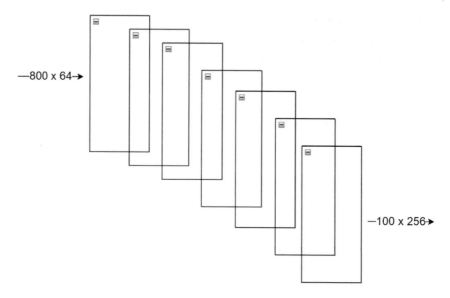

Fig. 2 Dimension of input prior to and the yield after passing through 7 layers of CNN

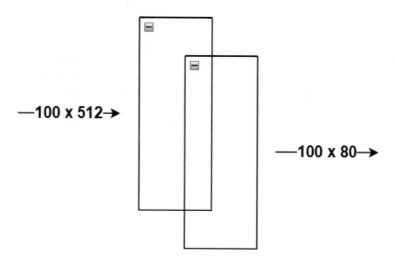

Fig. 3 Dimension of input before and the yield after propagating through 2 layers of bi-directional LSTM RNN

Connectionist Temporal Classifier (CTC) This is where the heart of the model lies. Entire recognition takes place here. It uses CTC greedy decode function to decode the actual text. The handwritten character will be mapped to the digital character based on character scores [6]. Now, during training, both the ground text and handwritten text are given as input to the CTC function. It starts recognizing characters and checks its loss value. It continues to perform the recognition until there is no change in loss/ there is minimal loss (i.e., there are no fixed number of epochs). During testing, only the handwritten text is given as the input to the CTC function, which based on its learning decodes the digital text. Redundancies are reduced using a special technique called pseudo-spaces (pseudo-blanks).

2.3 Experimental Setup

Image Acquisition The headmost step of the making of handwriting recognition model is image acquisition. The IAM dataset which contains huge number
   of handwritten images (over 13 thousand) is used to build(train) the model. About 80% of the images are used for training, while 20% is kept aside for testing.

Preprocessing Since most of the images of IAM dataset are already at preprocessed state, with minimal/no noise, very less preprocessing is used. However, there are 2 important initials that were done at this stage:

A. Damaged images are detected and are replaced with blank images [7].
B. Data augmentation snippet is added here where different styles of same handwriting are created and are given to the model to increase data variety (optional).

Feature Extraction The salient step of the model is this one. One important thing to be noted here is that we will not be wasting time in segmenting/splitting each character individually, rather we will detect edges of the complete word/sentence using CNNs and will be fed into next layer. More layers of CNN are required here since we require a certainly appearing edges for the next stage. The output of CNN is further passed into 2 layers of RNN. LSTM implementation of RNNs (bi-directional) is used. The output of the RNN is given to CTC function (after the creation of time-steps).

Creation of Time-steps This is the step that counters the segmentation model. Time-series data implementation is used to group text into fixed size character chunks called time-steps or time-stamps. Here, we do not really worry about the individual positions of characters since each individual character will be taken care by CTC greedy decoder separately. Now, we have a word/sentence divided into time-steps which is given as an input to CTC decoder.

Generation of Final Text Finally, the CTC function (greedy decoder) starts decoding the individual characters and gives us the final character sequence. Add-ons like autocorrect can be used to further process the obtained digital text.
   During training, both the handwritten text and ground text are given as an input to the CTC function. Hence, it decodes the possible result and estimates the loss. This continues until there is minimal loss/no change in loss value, and the epoch for that particular iteration stops there. The snapshot of the model is saved once the

entire training is completed. A training accuracy of 91.746% was observed once the snapshot is taken. During testing, only the handwritten text is given as an input to the CTC function. Now, it just returns the decoded text as an output.

3   Results

The data in Table 1 explains briefly on the upshots of the ML model obtained after testing and validation. Training accuracy of 91.746% and test accuracy of 85.224% were achieved with an 80:20 split of the original dataset. Along with this, the model was tested on around 800 images of handwritten sentences written exclusively for testing and an accuracy of 82.885% was achieved (Fig. 4).

Table 1 Model results

Dataset	Accuracy (%)
Train-set	91.746
Test-set	85.224
Validation-set	82.885

Fig. 4 Layout of the user interface for this model

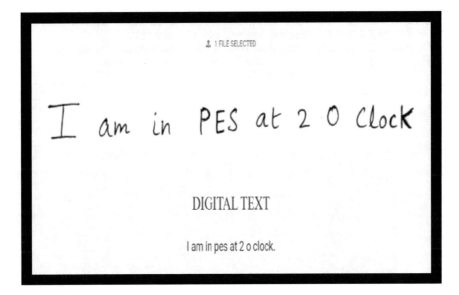

Fig. 5 Example that includes alphanumeric text

4 Discussion and Conclusion

It can be clearly noted that the model is not over-fitted since the training accuracy is greater than the test accuracy. Also, there were different styles of handwriting ranging from simple to slanting to cursive including special symbols and digits that were included in testing and the system could recognize all of them successfully with the above stated accuracy.

This implementation gives the conversion of handwritten sentence to digital text through a UI using the segmentation-free approach. The image with no noise and well segmented with the text is given to the system, and the digital text is obtained on the screen. The accuracy of the obtained text depends heavily on the individual letter clarity in the image as well as the skew in handwriting, i.e., lesser the skew higher the recognition accuracy.

Also there are advancements and algorithms that will further improvise the model to achieve better results like de-slantation, semantic analysis, etc. (Fig. 5).

References

1. Mainkar V, Jyothi M, Katkar A (2020) Handwritten character recognition to obtain editable text—Prof Vaibhav. Ms Poonam R Pednekar (2020)
2. Rao PS, Rao PS (2020) Handwriting recognition—"Offline" approach Upadhyay BD. Handwritten Character Recognition Using Ma
3. Zhang C (2020) Handwritten digit recognition based on convolutional neural network

4. Zhang C et al (2020) Handwritten digit recognition based on convolutional neural network
5. Yang J et al (2020) Handwritten text recognition using faster R-CNN
6. Upadhyay BD et al (2020) Handwritten character recognition using machine learning
7. Pham H (2020) Robust handwriting recognition with limited and noisy data

Decision-Making in Mask Disposal Techniques Using Soft Set Theory

Rashmi Singh, Karuna Khurana, and Pritisha Khandelwal

Abstract In 2003, Maji, Biswas, and Roy developed a method for applying soft set theory to a decision-making problem using Pawlak's rough set approach. Further, research proved that Maji's soft set reductions were inaccurate in 2005, leading to the development of a new method by Chen et al. This article applies soft theory to waste management and disposal decision-making problems. The excessive masks discarded during the COVID-19 era, in particular, must be managed effectively, and the current paper provides a method for better decision-making of the same. The algorithms used are first to compute the reductions and then the reduct soft set is used to choose the ideal objects for decision problems, and then the choice value is calculated. Predefined parameters are sometimes not enough to make precise decisions to solve general or real-time issues. Therefore, additional parameters are added into the existing set, either as a new parameter or generated by the handling of existing ones.

Keywords Soft sets · Waste disposal · Reduct table

1 Introduction

Making a decision is an actual cognitive process that requires data analysis by employing numerous characteristics and features to discover the optimal option based on decision-makers preferences. Dealing with ambiguity and vagueness in real-world decision-making situations demands the application of mathematical procedures in the great majority of instances. This ambiguity and vagueness cannot be dealt with using conventional mathematical tools but with theories such as probability theory, theory of fuzzy sets [1], theory of intuitionistic fuzzy sets [2], theory of vague sets [3], theory of interval mathematics [4], and theory of rough sets [5]. When applied to decision-making processes, soft set theory offers a dependable approach for dealing with uncertainty. In recent years, several different approaches for demonstrating the

R. Singh (✉) · K. Khurana · P. Khandelwal
Amity Institute of Applied Sciences, Amity University Uttar Pradesh, Noida, Uttar Pradesh, India
e-mail: rsingh7@amity.edu

© The Author(s), under exclusive license to Springer Nature Singapore Pte Ltd. 2023   649
A. Shukla et al. (eds.), *Computational Intelligence*, Lecture Notes in Electrical
Engineering 968, https://doi.org/10.1007/978-981-19-7346-8_56

efficacy and efficiency of the soft set theory, which was established by Molodtsov [6], have been offered.

Illustration of the use of soft sets in medical diagnostics, artificial intelligence, and decision-making may be seen in [7, 8]. While studying soft set applications, several new concepts were established to facilitate the job, including tabular representation of a soft set, parameterization, reduct [9], and the properties, operations, types, and structures of a soft set [10].

In recent years, as a result of the pandemic response, a substantial amount of waste has been collected, necessitating the use of an appropriate disposal approach. Dong et al. [11] examined critical parameters influencing the environmental efficiency of waste-to-energy systems such as pyrolysis, gasification, and incineration. In addition to standard waste disposal procedures, Emmanuel [12] presented non-incineration medical waste management solutions. Gupta et al. [13] developed a concise framework for determining the most appropriate technology for various applications. Nema and Ganesh Prasad [14] researched plasma pyrolysis using simulated hospital waste. Rubin and Burhan [15] compiled information about non-combustion technologies to remove persistent organic pollutants. Singh and Umrao [16, 17] proposed the idea of the nearness of finite order, S_n—merotopy in soft set theory. Again, in [18], they presented heminearness spaces in soft set theory.

We discovered and studied the advantages and disadvantages of the many biomedical waste disposal techniques now accessible, as well as a method for picking the optimal one based on parameter and attribute reduction. To choose between them, we turned the problem into a mathematical one with parameters and then utilized the soft sets approach, Chen et al. [9] suggested. We are implementing an enhanced soft set-in response to Chen's decision-making. As a result, this new approach is proven to be more precise than the prior one.

2 Preliminaries

2.1 Waste Survey of India

The biomedical waste rules [19, 20], 1998, and subsequent changes were enacted to regulate biomedical waste management. Numerous healthcare facilities were found to be improperly classifying garbage. If infectious items are found, they must be burned or deposited in an autoclave. This is a difficult task for the majority of medical facilities. Due to improper sorting or negligence, biomedical waste is frequently thrown in the water or landfills and eventually washes up on the shore. This is detrimental to both animals and humans.

New legislation affects how health practitioners dispose of medical waste. Numerous commercial enterprises have ensured proper biomedical waste disposal to obtain accreditation from organizations such as ISO, NABH, and JCI.

Even though India has a range of disposal options, the situation is dire, with the bulk of them being detrimental rather than constructive. "Since March 26, 2020, the National Green Tribunal (NGT) has issued more than 200 licenses for common biomedical waste treatment and disposal facilities (CBWTDF) or common treatment facilities (CTF) around the country. On the other hand, incorrect sorting is caused by a lack of expertise. General trash is coupled with biomedical waste in Tier 2 and Tier 3 cities" [20].

As seen above, the times are critical. People are not concerned about the side effects of what is happening,

To get an idea of how much extra waste is being generated daily due to COVID-19, let's look at some data as on 10/05/21 as around this week India was at the peak of its COVID-19 cases and with the increases in cases was a proportional increase in the amount of waste being generated.

Country—India.

Population (2020)—1,391,531,080.

COVID cases—22,658,234.

Urban population—35.

A formula was given by [21] to get an idea for the waste being generated by disposable masks daily.

$$DFM = P \times UP \times FMAR \times (FMGP/10,000)$$

where

DFM is the daily face mask usage (per pieces), UP is the urban population in percentage, FMAR is the face masks acceptance rate, which may be taken as 80%, FMGP is an assumption that each person uses 1 mask per day, P is the population, FMAR = 80, FMGP = 2, DFM = 1,391,531,080 × 35 × 80 × (2/10,000) = 779,257,404.

The mass of discarded facemask was determined using the assumption (3 g/facemask).

Discarded facemask (tons/day) = 3 × 779,257,404 = 2,337,772,212 g/day = 2576.9527516 tons/day.

Also, a formula was given by [22] to get an idea for the waste being generated by hospitals daily.

Waste generated by hospital = no. of people infected x waste produced per bed. Let waste produced per day per bed = 3.95 kg/bed/day

$$MW = NCC \times MWGR/1,000$$

where
MW = Medical waste (tons/day).
NCC = Number of COVID-19 cases= 22,658,234.
MWGR = rate at which medical waste is generated, that is, 3.95 kg/bed/day.

MW = 22,658,234 × 3.95/1,000 = 89,500.0243 tons/day.
Total daily facemask used (pcs)—779,257,404.
Abandoned facemasks—2576.9527516 (tons/day).
Medical waste—89,500.0243 (tons/day).

Numerous infectious biomedical wastes are being produced during this pandemic. The wellspring of infectious wastes is not produced from the medical health facilities including emergency clinics and worldly clinical offices yet additionally produced by family and isolated center facilities. Home quarantine and isolation are basic practices in numerous nations to manage associated patients with COVID-19. Home quarantine is being utilized for patients with some indications or explorers for at any rate fourteen days.

2.2 Definition

We will present the fundamentals of soft sets and biomedical waste in this section.

Let $Ű$ represents the universal set and $\acute{E}$ is the set of parameters. $\overline{E} = \{ \overset{''}{\acute{o}} \forall i =$
$1, 2, \ldots\}$, $\mathring{A}$ is the subset of $\acute{E}$, $\dashv$ is a mapping of $\acute{E}$ into set of all the subsets of set $Ű$ and $P(Ű)$ is the power set of $Ű$.

Soft Sets. The soft set given by $(\dashv, \overline{E})/\dashv_{\overline{E}}$ over $Ű$ is a parameterized family of the subsets of the universal set $\ddot{\upsilon}$ iff, $\dashv$ is a mapping of $\acute{E}$ into the set of all subsets of the universal set,

$$\dashv : \overset{'}{\overline{E}} \to P(\overset{''}{U}).$$

For all, $\acute{o} \in \acute{E}$, $(\acute{o})$ represents $\acute{o}$ an approximate element of $(\dashv, \acute{E})$.

A soft set $(\dashv, \acute{E})$ over $Ű$ can also be represented a set of ordered pairs,

i.e., $\dashv_{\overline{E}} = \left\{ \left(\overset{''}{\acute{o}}, \dashv(\overset{''}{\acute{o}}) \right) : \overset{''}{\acute{o}} \in \overset{'}{\overline{E}} \right\}$.

Indiscernibility relation. Let $\mathring{A}$ and B be subsets of set of parameters. Assume that $B \subset \mathring{A}$, we associate a binary relation $IND(B)$, known as indiscernibility relation, which is defined as

$$\text{IND}(B) = \left\{ (x, y) \in \overset{''}{U} \times \overset{''}{U} : a(x) = a(y), \forall a \in B \right\}$$

Obviously, IND (B) is .an equivalence relation and $\text{IND}(B) = \cap\{\text{IND}(a)\}$, $a \in B$.

Suppose $V_a = \left\{ E_a^1, E_a^2, \ldots, E_a^{n(a)} \right\}$. 'Define $F_a : V_a \to P\left(\overset{''}{U} \right)$ as

$$F_a(E_a^i) = \left\{ x \epsilon \overset{''}{U} : a(x) = E_a^i \right\},$$

then (F_a, V_a) represents a soft set.

Suppose $A = (a_1, a_2, .., a_m)$, then $S = (\overset{''}{U}, A)$ can be expressed as a soft set.
$(F, Y_{a1} \times Y_{a2} \times .....Y_{am}) = (F_{a1}, V_{a1}) \cap (F_{a2}, V_{a2}) \cap ... \cap (F_{am}, V_{am})$ For every $(p_1, p_2, .., p_m) \epsilon Y_{a1} \times Y_{a2} \times .....Y_{am}$.

$F(p_1, p_2, ..., p_m) = F_{a1}(p_1) \cap F_{a2}(p_2) \cap ... \cap F_{am}(p_m)$ nonempty sets of $F(p_1, p_2, ..., p_m)$ form the collection of the equivalence classes of IND $\overset{\circ}{A}$. As a result, an information representation scheme can be expressed using the soft set.

Core. Assume that R is a set of equivalence relations and suppose that belongs to R. Then, $\overset{\circ}{A}$ is said to be dispensable in R if IND (R) = IND (R − $\{\overset{\circ}{A}\}$); *else*, $\overset{\circ}{A}$ *is indispensable in* R.

The collection R is independent if all $\overset{\circ}{A}$ in R is indispensable in R, else R is dependent. $D \subset G$ is a reduction of G if D is independent as well as IND(D) = IND(G).

The set of all indispensable relations in G will be referred to as the core of G, and it will be denoted by CORE(G). Therefore, CORE(G) = $\cap$RED(G), RED(G) denotes the family of all G reductions.

Soft set parameterization reduction. Let $\overset{'}{U} = \{ \epsilon_1, \epsilon_2, \epsilon_3, \epsilon_4, \epsilon_5 ... \epsilon_m \}$ and $\overset{\acute{}}{E}$ is the set of parameters, i.e.,

$\overset{''}{U} = \overset{''}{0}_1, \overset{''}{0}_2, \overset{''}{0}_3, \overset{''}{0}_4, \overset{''}{0}_5 .....\overset{''}{0}_m$, where $\mathcal{F}_{\underline{\overline{E}}}$ is the soft set.

Define, $\mathcal{F}_{\underline{\overline{E}}} (\epsilon_{ij}) = \Sigma\epsilon_{ij}$, ϵ_{ij}'s are the items in the table of $\mathcal{F}_{\underline{\overline{E}}}$.

Also, $S_{\underline{\overline{E}}}'$ is the collection of objects in $\overset{'}{U}$ that take the maximum value of $\mathcal{F}_{\underline{\overline{E}}}$. For every $\overset{\circ}{A} \subset \overset{\acute{}}{E}$ if $S_{\underline{\overline{E}} - \overset{\circ}{A}}' = S_{\underline{\overline{E}}}'$ then $\overset{\circ}{A}$ is known as dispensable set-in $\overset{\acute{}}{E}$, indispensable in $\overset{\acute{}}{E}$ otherwise.

$\overset{\acute{}}{E}$ is independent if all $\overset{\circ}{A} \subset \overset{\acute{}}{E}$, is indispensable in $\overset{\acute{}}{E}$, $\overset{\acute{}}{E}$ is dependent otherwise.

$G \subset \overset{\acute{}}{E}$ is said to be a reduction of $\overset{\acute{}}{E}$ if G is independent and $S_p' = S_{\underline{\overline{E}}}'$

For $\overset{''}{0} \epsilon \overset{\acute{}}{E}$, $S_{\underline{\overline{E}}}' - \left(\overset{''}{0}\right) \neq S_{\underline{\overline{E}}}'$, it is possible, there exists an $\overset{''}{0}' s\epsilon \overset{\acute{}}{E}$ such that $S_{\underline{\overline{E}}}' - \{\overset{''}{0}, \overset{''}{0}'\} = S_{\underline{\overline{E}}}'$.

3 Selecting Best Technique Using Soft Set Theory

3.1 Methodology

3.2 Analysis

Data set used for selection of the best method of disposal.
  Biomedical waste disposal methods.

- Incineration technique
- Thermal processes

  – Low
  – Medium
  – High

- Chemical-based

  – Chlorine-based technologies
  – Non-chlorine-based technologies

- Lynntech's another technology using ozone for decontamination
- The electron beam technology
- Solar disinfection
- Landfill

Parameters

- $\alpha_1 =$ low cost (low capital cost, low operation cost)
- $\alpha_2 =$ no waste remnant
- $\alpha_3 =$ no toxic emissions
- $\alpha_4 =$ no health effects
- $\alpha_5 =$ safety
- $\alpha_6 =$ no infrastructure requirement
- $\alpha_7 =$ easy to use
- $\alpha_8 =$ reliable treatment
- $\alpha_9 =$ efficient treatment
- $\alpha_{10} =$ no odor

Table 1 Representation of soft sets

U	β_1	β_2	β_3	β_4	β_5	β_6	β_7	β_8	β_9	β_{10}
α_1	1	0	0	0	0	1	1	1	0	0
α_2	0	1	1	1	1	0	0	1	0	0
α_3	1	1	0	0	0	0	1	1	1	0
α_4	0	1	1	1	0	0	0	1	1	1
α_5	1	1	1	1	1	0	0	0	0	1
α_6	1	0	0	0	0	1	1	0	0	0
α_7	0	1	1	1	1	0	1	0	0	1

3.3 Tabular Presentation of a Soft Set

- $\alpha_1 =$ Incineration technique ($\beta_1, \beta_6, \beta_7, \beta_8$)
- $\alpha_2 =$ Thermal processes ($\beta_2, \beta_3, \beta_4, \beta_5, \beta_8$)

  - Low-microwaves and autoclaves
  - Medium-reverse polymerization and thermal de-polymerization
  - High-pyrolysis, dry heat treatment

- $\alpha_3 =$ Chemical-based ($\beta_1, \beta_2, \beta_7, \beta_8, \beta_9$)

  - Chlorine-based technologies
  - Non-chlorine-based technologies
  - Ethylene oxide treatment

- $\alpha_4 =$ Using ozone for decontamination ($\beta_2, \beta_3, \beta_4, \beta_8, \beta_9, \beta_{10}$)
- $\alpha_5 =$ The electron beam technology ($\beta_1, \beta_2, \beta_3, \beta_4, \beta_5, \beta_{10}$)
- $\alpha_6 =$ Landfill ($\beta_1, \beta_6, \beta_7$)
- $\alpha_7 =$ Solar disinfection ($\beta_2, \beta_3, \beta_4, \beta_5, \beta_7, \beta_{10}$).

Soft Set Reduct Table

Consider the soft set, $(F, E).(F, P)$ is a soft subset of (F, E) as $P \subset E$ (Tables 1 and 2).

Find Q, i.e., the reduction of P, then the soft set (F, Q) is said to be the reduct soft set of a soft set (F, P).

Choice Value of β_y

The choice value of an object $\beta_y \in \check{U}$ is C_y, given by.

$C_y = \sum_j \beta_{ij}$, where β_{ij} are the items in the table of the reduct soft set.

Table 2 Tabular format of a soft set (F, E)

U	β_1	β_2	β_3	β_4	β_5	β_6	β_7	β_8	β_9	β_{10}
α_1	1	0	0	0	0	1	1	1	0	0
α_2	0	1	1	1	1	0	0	1	0	0
α_3	1	1	0	0	0	0	1	1	1	0
α_4	0	1	1	1	0	0	0	1	1	1
α_5	1	1	1	1	1	0	0	0	0	1
α_6	1	0	0	0	0	1	1	0	0	0
α_7	0	1	1	1	1	0	1	0	0	1

Algorithm for Selection of the best disposal method

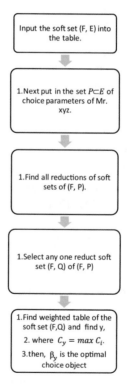

Input the soft set (F, E) into the table.

1. Next put in the set $P \subset E$ of choice parameters of Mr. xyz.

1. Find all reductions of soft sets of (F, P).

1. Select any one reduct soft set (F, Q) of (F, P)

1. Find weighted table of the soft set (F,Q) and find y,
2. where $C_y = max\ C_i$.
3. then, β_y is the optimal choice object

Suppose R_E. is the indiscernibility relation induced by $E = \{\alpha_1, \alpha_2, \alpha_3, \alpha_4, \alpha_6, \alpha_7, \alpha_8, \alpha_9, \alpha_{10}\}$, then the partition defined by R_E is $(\{\beta_1\}\{\beta_2\}\{\beta_3\}\{\beta_4\}\{\beta_5\}\{\beta_6\}\{\beta_7\})$.

Consider the soft set (F, E). For any $P \subset E$, it is sure that (F, P) is a soft subset of (F, E).

Table 3 Tabular format of a soft set (F, P)

U	β_1	β_2	β_3	β_4	β_5	β_7	β_9	Choice value
α_1	1	0	0	0	0	1	0	2
α_2	0	1	1	1	1	0	0	4
α_3	1	1	0	0	0	1	1	4
α_4	0	1	1	1	0	0	1	4
α_5	1	1	1	1	1	0	0	5
α_6	1	0	0	0	0	1	0	2
α_7	0	1	1	1	1	1	0	5

Table 4 Initial reduct table (calculating the choice values)

U	β_2	β_3	β_4	β_5	β_7	β_9	Choice value
α_1	0	0	0	0	1	0	1
α_2	1	1	1	1	0	0	4
α_3	1	0	0	0	1	1	2
α_4	1	1	1	0	0	1	4
α_5	1	1	1	1	0	0	4
α_6	0	0	0	0	1	0	1
α_7	1	1	1	1	1	0	5

Suppose Rp is the indiscernibility relation induced by $P = \{\alpha_1, \alpha_2, \alpha_3, \alpha_4, \alpha_5, \alpha_7, \alpha_9\}$ (Table 3).

Then, the partition defined by Rp is $(\{\beta_1, \beta_6\}\{\beta_2\}\{\beta_3\}\{\beta_4\}\{\beta_5\}\{\beta_7\})$.

Here, max $C_i = C_7$.

Thus, Mr. xyz can either use β_5 or β_7.

Reduct soft sets of (F, P).

Choose $Q = \{\alpha_2, \alpha_3, \alpha_4, \alpha_5, \alpha_7, \alpha_9\}$ (Table 4).

Here, max $C_i = C_7$.

Thus, Mr. xyz can use β_7.

Choose $Q = \{\alpha_1, \alpha_3, \alpha_4, \alpha_5, \alpha_7, \alpha_9\}$ (Table 5).

Here, max $C_i = C_5$ andC_7.

Thus, Mr. xyz can either use β_5 or β_7.

Choose $Q = \{\alpha_1, \alpha_2, \alpha_4, \alpha_5, \alpha_7, \alpha_9\}$ (Table 6).

Here, max $C_i = C_3, C_5$ andC_7. Thus, Mr. xyz can either use β_3, β_5 or β_7.

Choose $Q = \{\alpha_1, \alpha_2, \alpha_3, \alpha_5, \alpha_7, \alpha_9\}$ (Table 7).

Here, max $C_i = C_3, C_5$ andC_7.

Thus, Mr. xyz can either use β_3, β_5. or β_7.

Choose $Q = \{\alpha_1, \alpha_2, \alpha_3, \alpha_4, \alpha_7, \alpha_9\}$ (Table 8).

Here, max $C_i = C_3, C_4, C_5$ andC_7.

Thus, Mr. xyz can either use $\beta_3, \beta_4, \beta_5$ or β_7.

Table 5 Second reduct table with changed parameter choices (calculating the choice values)

U	β_1	β_3	β_4	β_5	β_7	β_9	Choice value
α_1	1	0	0	0	1	0	2
α_2	0	1	1	1	0	0	3
α_3	1	0	0	0	1	1	3
α_4	0	1	1	0	0	1	3
α_5	1	1	1	1	0	0	4
α_6	1	0	0	0	1	0	2
α_7	0	1	1	1	1	0	4

Table 6 Third reduct table with choice values

U	β_1	β_2	β_4	β_5	β_7	β_9	Choice value
α_1	1	0	0	0	1	0	2
α_2	0	1	1	1	0	0	3
α_3	1	1	0	0	1	1	4
α_4	0	1	1	0	0	1	3
α_5	1	1	1	1	0	0	4
α_6	1	0	0	0	1	0	2
α_7	0	1	1	1	1	0	4

Table 7 Reduct soft set can be represented as (calculating the choice values)

U	β_1	β_2	β_3	β_5	β_7	β_9	Choice value
α_1	1	0	0	0	1	0	2
α_2	0	1	1	1	0	0	3
α_3	1	1	0	0	1	1	4
α_4	0	1	1	0	0	1	3
α_5	1	1	1	1	0	0	4
α_6	1	0	0	0	1	0	2
α_7	0	1	1	1	1	0	4

Table 8 The final reduct table leading to our decision (calculating the choice values)

U	β_1	β_2	β_3	β_4	β_7	β_9	Choice value
α_1	1	0	0	0	1	0	2
α_2	0	1	1	1	0	0	3
α_3	1	1	0	0	1	1	4
α_4	0	1	1	1	0	1	4
α_5	1	1	1	1	0	0	4
α_6	1	0	0	0	1	0	2
α_7	0	1	1	1	1	0	4

Choose $Q = \{\alpha_1, \alpha_2, \alpha_3, \alpha_4, \alpha_5, \alpha_9\}$ (Table 9).

Here, max $C_i = C_5$.

Thus, Mr. xyz can use β_5.

Choose $Q = \{\alpha_1, \alpha_2, \alpha_3, \alpha_4, \alpha_5, \alpha_7\}$ (Table 10).

Here, max $C_i = C_5$ and C_7.

Thus, Mr. xyz can either use β_5 or β_7.

If we remove any subset of P that includes at least one of $\alpha_1, \alpha_3, \alpha_4$ and α_5, then the optimal choice objects will be changed. For example, if we delete $\{\alpha_1, \alpha_2\}$ from P, then the optimal choice objects will be $\{\beta_7\}$ (Table 11).

Table 9 Reduct soft set can be represented as (calculating the choice values)

U	β_1	β_2	β_3	β_4	β_5	β_9	Choice value
α_1	1	0	0	0	0	0	1
α_2	0	1	1	1	1	0	4
α_3	1	1	0	0	0	1	3
α_4	0	1	1	1	0	1	4
α_5	1	1	1	1	1	0	5
α_6	1	0	0	0	0	0	1
α_7	0	1	1	1	1	0	4

Table 10 Reduct soft set can be represented as (calculating the choice values)

U	β_1	β_2	β_3	β_4	β_5	β_7	Choice value
α_1	1	0	0	0	0	1	2
α_2	0	1	1	1	1	0	4
α_3	1	1	0	0	0	1	3
α_4	0	1	1	1	0	0	3
α_5	1	1	1	1	1	0	5
α_6	1	0	0	0	0	1	2
α_7	0	1	1	1	1	1	5

Table 11 Final reduct

U	β_3	β_4	β_5	β_7	β_9	Choice value
α_1	0	0	0	1	0	1
α_2	1	1	1	0	0	3
α_3	0	0	0	1	1	2
α_4	1	1	0	0	1	3
α_5	1	1	1	0	0	3
α_6	0	0	0	1	0	1
α_7	1	1	1	1	0	4

Here, max $C_i = C_7$.

Thus, Mr. xyz can choose β_7.

4 Conclusion

When we analyze the set P and obtain Rp, the indiscernibility relation imposed by P, and then find Rp induced by all the Qs, the subset of P constructed so that a different parameter is eliminated each time, we notice that Rp varies for the deletion of four parameters. Particularly here, the soft set has four reduced parameters. These are the top method selection criteria. If these parameters are not the attributes reduction of P and the attributes reduction displayed is not the parameter reduction of P, it is not the lowest parameter set to keep optimal choice objects. We can swap choice and weight. These values may be determined by the importance of each system parameter and where to apply this strategy. Changing the mask's weight may affect the optimum disposal method. The method yields good results with minimal calculation and time.

References

1. Zadeh LA (1965) Fuzzy sets. Inf Control 8:338–353
2. Atanassov K (1986) Intuitionistic fuzzy sets. Fuzzy Sets Syst 20:87–96
3. Gau WL, Buehrer DJ (1993) Vague sets. IEEE Trans. System Man Cybernet 23(2):610–614
4. Gorzalzany MB (1987) A method of inference in approximate reasoning based on interval-valued fuzzy sets. Fuzzy Sets Syst 21:1–17
5. Pawlak Z (1982) Rough sets. Int J Inf Comput Sci 11:341–356
6. Molodtsov D (1999) Soft set theory-first results. Comput Math Appl 37(4/5):19–31
7. Maji PK, Biswas R, Roy AR (2003) Soft set theory. Comput Math Appl 45:555–562
8. Maji PK, Biswas R, Roy AR (2003) An application of soft sets in a decision-making problem. Comput Math Appl 45:555–562
9. Chen D, Tsang ECC, Yeung DS, Wang X (2005) The parameterization reduction of soft sets and its applications. Comput Math Appl 49:757–763
10. Chen D-G, Tsang ECC, Yeung DS (2003) Some notes on the parameterization reduction of soft sets. In: Proceedings of the 2003 international conference on machine learning and cybernetics (IEEE Cat. No.03EX693). Xi'an, China, 3, pp 1442–1445
11. Dong J, Tang Y, Nzihou A, Chi Y (2019) Key factors influencing the environmental performance of pyrolysis, gasification and incineration Waste-to-Energy technologies. Energy Convers Manag 196, 497–512 (2019)
12. Emmanuel J (2001) Health care without harm. Non-Incineration medical waste treatment technologies. Washington, D.C.
13. Gupta VK, Ali I, Saleh TA, Nayak A, Agarwal S (2012) Chemical treatment technologies for waste-water recycling—An overview. RSC Adv 2:6380–6388
14. Nema S, Ganeshprasad K (2002) Plasma pyrolysis of medical waste. Curr Sci 86:271–278
15. Rubin E, Burhan Y (2006) Noncombustion technologies for remediation of persistent organic pollutants in stockpiles and soil. Remediat. J 16:23–42
16. Singh R, Umrao AK (2019) On Finite order nearness in soft set theory. WSEAS Trans Math 18:118–122

17. Singh R, Chauhan R (2019) On Soft heminearness spaces, emerging trends in mathematical sciences and its applications. AIP Conf Proc 2061, 020016-1–020016-5
18. Singh R, Pandey JT, Umrao AK (2020) k, t, d-Proximities in rough set. WSEAS Trans Math 19:498–502
19. Bio-Medical Waste Management Rules (2016) Published in the Gazette of India, Extraordinary, Part II, Section 3, Sub-Section (i), Government of India Ministry of Environment, Forest and Climate Change. Notification; New Delhi, the 28th Mar 2016
20. The Gazette of India Biomedical Wastes (Management and Handling) Rules (1998) Ministry of Environment and Forests, Government of India, India; Notification Dated; 20th July 1998
21. Sangkham S (2020) Face mask and medical waste disposal during the novel COVID-19 pandemic in Asia, Case Stud Chem Environ Eng 2
22. Capoor MR, Bhowmik KT (2017) Current perspectives on biomedical waste management, Rules, conventions and treatment technologies. Indian J Med Microbiol 35:157–164

Computational Analysis of PID and PSO-PID Optimization for MIMO Process Control System

Hafiz Shaikh, Neelima Kulkarni, and Mayuresh Bakshi

Abstract Water level control in MIMO coupled interactive tank seems to be an important aspect of application and the development of process control techniques. Monotonous and precise level measurement can avoid such tragedies. Process control problem has been defined in this paper. PSO-PID controller has been designed as a solution to the water level process control problem. Identification of mathematical model methodology is more beneficial for water level control in coupled tank MIMO system. Observing open loop reaction of the approach can be the first step of system identification. System identification may be used for scrutinizing the actual parameters of the interacting tanks. State-space analysis of coupled tanks is explained in detail along with its transformation into transfer function. In this paper, inherent constraints necessary to do the calculation are discussed. MATLAB is utilized for monitoring the responses. Observations from the PID controller articulate that there is a need for a better controller to enhance the performance. PSO algorithm used for optimization is a novel structure that improves PID performance. The optimization tools used in this paper gives a promising result and a good insight about selection of appropriate one. A detailed explanation of optimizing PID parameters using a PSO-PID controller is specified. Computational analysis along with its discussion is presented in this paper.

Keywords MIMO process control · Mathematical model · PSO-PID · Computational analysis

1 Introduction

Pharmaceutical process industries require water level control to be very precise. Water supplied to the reactors and boilers should be exact. Conventional controllers

H. Shaikh (✉) · N. Kulkarni
Electrical Engineering Department, PES Modern College of Engineering, Pune 411005, India

M. Bakshi
Electrical Engineering Department, Vishwakarma Institute of Information and Technology, Pune 411048, India

© The Author(s), under exclusive license to Springer Nature Singapore Pte Ltd. 2023
A. Shukla et al. (eds.), *Computational Intelligence*, Lecture Notes in Electrical Engineering 968, https://doi.org/10.1007/978-981-19-7346-8_57

like PID do fulfil the need, but the manufacturer always must compromise over the performance index. There different methods to improve controller parameters like fuzzy logic controller fractional order PID controller. These controllers have improved the performance up to certain extent but there is still capability to optimize it.

Pharmaceutical process industries where pH value needs to be controlled consists of two tanks (acid and base) interlinked and passing to the third tank which discharges the final effluent. Particle swarm optimization has very strong potential to obtain desired response and enhance performance index. PSO has been used as an optimization tool for optimizing PID controller for many applications. Implementation of PSO-PID for water level control of coupled tank interacting system brings novelty in the research domain. PSO is a recent metaheuristic technique [1]. PSO make use of cluster communication to contribute personal information about the best possible position and direction to reach their food.

Water level control in coupled tank at different set values is possible by assessing the error response and correcting it to an optimized value. PID controller uses the traditional method to minimize the error by manipulating the three parameters. The analysis says that desired responses are not attained by using this controller. PSO comes into action for improving performance of PID. Error signal currently enhances the objective function of PSO. At every iteration, particle will find the optimized values of PID to minimize the objective function. Depending upon the inhabitants range and quantity of maximum iterations, the pseudo code will work out and a promising output is obtained.

2   Coupled Tank System

Coupled tank interacting expresses an organization that has two analogous or disparate tanks connected along with each other through a connection between them [2]. It has two separate input and outputs. Water is provided to the input of tanks through two pumps. Valve is responsible for controlling output from the tank. The schematic of the system is exemplified in Fig. 1.

Shape of the two tanks connected with each other is rectangular. Specifications of the tanks are as given in Table 1.

Tank one and two have two water inflows $Q1$ and $Q2$, respectively. Water level height for tank 1 and 2 is given by $h1$ and $h2$, correspondingly. $R2$ and $R3$ represent the discharge coefficients of the outlet valves for tanks 1 and 2. The flow of water between the two coupled tanks is given by discharge coefficient $R1$. The coupled tanks perform in an interacting manner because of the link connected between them. Valve $R1$ is responsible for the control of interacting link.

The real photo of coupled tank water level control system is displayed in Fig. 2. The system has three tanks straddling on the panel, nevertheless for the investigation, two tanks at lower position are used which are coupled together. Bubbler method is employed to measure water level in the tank. In this method, an air bubble is

Fig. 1 Schematic
illustration of coupled tank

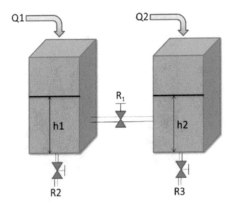

Table 1 Physical
significance of coupled tank

Parameter	Quantity
Tank capacity (volume)	6 lit
Length of tank	0.13 m
Width of tank	0.13 m
Height of tank	0.35 m
Hydraulic resistance of valve (fully open)	15552 s/m^2

inserted inside the water and the pressure of that air bubble on the surface of the water is calibrated in terms of voltage ranging from 0 to 2.5 V. Air filter regulators are mounted to maintain air pressure up to 1 psi.

3 Mathematical Model of Tank

The literature [1] and [3] contain mathematical modelling of a two-tank interactive fluid height monitoring system. In a coupled tank interacting system, determining, and controlling water level has various complications. By constructing a proper controller, a mathematical model aids in controlling the water level in these tanks. The changing aspects for the coupled tank MIMO interacting system can be assessed by evaluating the differential equations for both the tanks [4].

Bernoulli's mass balancing theory makes it easier to figure out the system's differential equation. The essential idea behind this theory is that input minus outflow during a short time interval 'dt' equals the amount of water held in the tank. Differential equations for the tanks and mathematical model are derived using the flowchart shown in the Fig. 3, according to the preceding theory.

State-space model depicted in the flowchart shown above can be transformed to transfer function so as to get four transfer functions namely $G11$, $G12$, $G21$ and $G22$. Figure 4 shows a block diagram interpretation of the system.

Fig. 2 Real photo of coupled tank water level control system

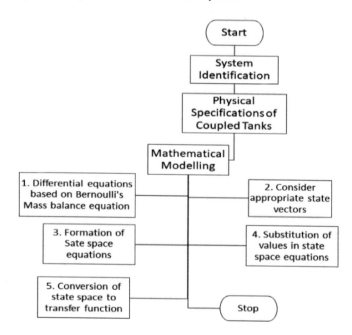

Fig. 3 Flowchart for mathematical modelling of coupled tank water level control system

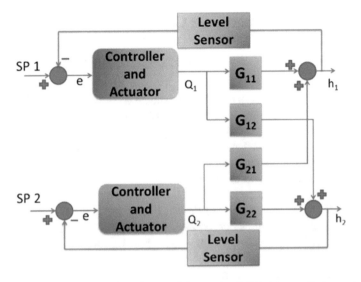

Fig. 4 Block diagram representation of two-tank interactive fluid height monitoring system

Table 2 Physical significance of coupled tanks

Constraints	Value	Unit
a_1	0.0169	m^2
a_2	0.0169	m^2
R_1	7776	s/m^2
R_2	15,552	s/m^2
R_3	15,552	s/m^2

The constraints and their values utilized to find the transfer matrix in the preceding state-space equations are listed in Table 2.

Programming and MATLAB instructions can be used to obtain the coupled tank's transfer function from the state-space equation (Table 3).

Table 3 Transfer function of MIMO interacting tanks system

Constraints	Value
G_{11}	$\dfrac{59.17s+0.6754}{s^2+0.02283s+7.238e^{-5}}$
G_{12}	$\dfrac{0.4503}{s^2+0.02283s+7.238e^{-5}}$
G_{21}	$\dfrac{0.4503}{s^2+0.02283s+7.238e^{-5}}$
G_{22}	$\dfrac{59.17s+0.6754}{s^2+0.02283s+7.238e^{-5}}$

4   Controller Performance for Coupled Tank

4.1   Coupled Tank Response for PID Controller

A classical PID type controller is obvious choice for a controller, and it plays an important role in maintaining an exact level of tank by operating the system in a closed-loop paradigm [5]. The design methodology is a revolutionary strategy that may be implemented in MATLAB Simulink. Two PID controllers are required which will independently control the water level in two tanks. Figure 5 depicts the simulation of a connected tank system in closed loop.

The water level in the connected tanks is controlled by a PID controller based on set points. The SP of tank one is 0.125 m, while tank 2 is 0.1 m. SP minus actual water level coming from the feedback is given as input to the controller. There is a difference in the PID controlled parameters for both the tanks.

Figure 6 shows the values of PID controller parameters after utilizing the auto tuning function. Conventional transfer function-based technique is exploited for adjusting the PID controller. For the coupled tanks, both PID controllers get tuned differently due to the interacting effect.

After tuning the PID, the rise and clearing periods are minimized. Overshoot has been reduced from 10.3–6.51%, as seen in Fig. 6. Manipulation of the PID parameters

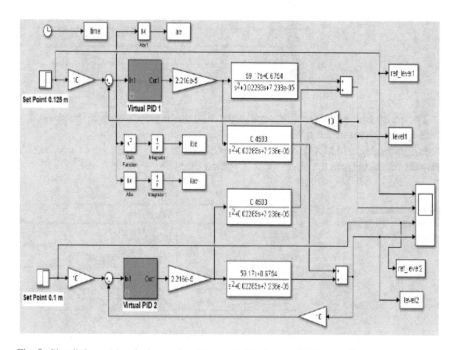

Fig. 5 Simulink model to find water level in coupled tank using PID controller

Controller Parameters		
	Tuned	Block
P	1.5192	1.223
I	0.02151	0.029
D	-30.5986	-3.05
N	0.04965	0.022

Performance and Robustness		
	Tuned	Block
Rise time	58.3 seconds	69.9 seconds
Settling time	270 seconds	237 seconds
Overshoot	6.51 %	10.3 %
Peak	1.07	1.1
Gain margin	Inf dB @ Inf rad/s	Inf dB @ NaN rad/s
Phase margin	61.7 deg @ 0.0234 rad/s	64.6 deg @ 0.0217 rad/s
Closed-loop stability	Stable	Stable

Fig. 6 Constraints tuned by PID controller

can reduce overshoot even further, although this has an impact on rise and settling time. Even if the system's closed-loop performance is stable, some performance attributes must be sacrificed when utilizing a PID control. This means that the same PID parameter will not be able to manage water level at different set points properly.

Figure 7 indicates that there is an overshoot present in the response. In addition, there is some undershoot in the response. In the system, there is no steady-state error. The transient and oscillations in the response are caused by integral action. Derivative activity should be reduced to make the response swift and create a steady state.

5 Particle Swarm Optimization

5.1 Particle Swarm Optimization: Concept

Particle swarm optimization or PSO is an intelligent optimization algorithm it belongs to a class of optimization algorithm called as metaheuristic. PSO is based on a paradigm of swarm intelligence, and it is inspired by social behaviour of animals like fish and bird. For particle 'i', the position of particle is denoted by 'x_i'. This is a vector and it's a member of search space denoted by 'x'. To distinguish between time steps, we add a time index to this position and denoted by $x_i(t)$. Index of particle is

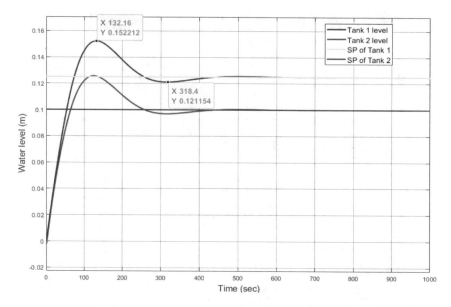

Fig. 7 Water level response of coupled tanks using PID controller

given by 'i' and 'x' is vector of position. In addition to the position, we have a velocity for every particle, which is denoted by '$v_i(t)$'. Velocity describes the movement of particle i in the sense of direction and in the sense of distance and step size. In addition to position and velocity, every particle has a memory of its own best position, best experience denoted by personal best (Pbest). Global best (Gbest) or common best amongst the member of swarm which is the best experience of all particles in the swarm. On every iteration of PSO, position and velocity of every particle is updated according to this simple mechanism. Algorithm of PSO is explained in the flowchart shown in Fig. 8.

5.2 Particle Swarm Optimization: Parameters and Values

Initial parameters of PSO include number of variables, upper bound value, lower bound value, defining objective function, settling number of particles, maximum iteration value, maximum and minimum values of weight and acceleration coefficient. All the values used in the PSO algorithm are shown in Table 4.

The algorithm for PSO is explained as a flowchart in Fig. 9. PID controller parameters when optimized with PSO needs particles to be decided first. Around 30 particles for each parameter are selected. There are three variables to be manipulated to optimize the objective function. The objective function to be optimized here is integral square error (ISE). Integral square error needs to be minimized to attain the preferred

Fig. 8 Flowchart of PSO

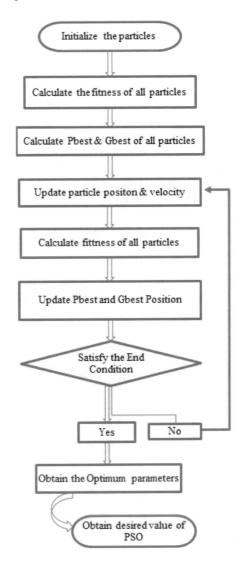

objective. ISE is selected because it reduces the minimal valued error and focus on it.

The upper bound and lower bound values of PSO parameters are decided after experimentation of the system using PID controller only.

Table 4 PSO parameters and their values

PSO parameters	Values (before optimization to both PID controllers)	Values (before optimization to both PID controllers)
Number of variables	3	6
Upper bound	[2.5 0.014 0]	[3.6 0.023 0.00035 1.2 0.007 0.003]
Lower bound	[0 0 −3.5]	[0 0 −3.5 0 0 −0.556]
Objective function	ISE	ISE
Number of particles	30	30
Maximum iteration value	20	5
Maximum value of weight	0.9	0.9
Minimum value of weight	0.2	0.2
Acceleration coefficient	2	2

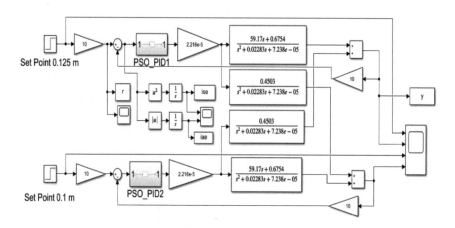

Fig. 9 Simulink model of PSO-PID controller for coupled tank system

5.3 Implementation of PSO-PID Using MATLAB Simulink

Coupled tank water level control using PSO-PID controller is simulated in the Simulink window shown in Fig. 9. ISE and IAE are selected as performance index for optimization purpose. The algorithm for PSO is embedded in PSO_PID block which gets executed after running the model. The simulation runs for 20 number of iterations and then it stops. Reference input and water level output signals are sent to the workspace as time series variables '*r*' and '*y*', respectively. Integral square error (ISE) is calculated in the algorithm and set as the objective function. Pbest and Gbest continuously get updated according to the present value of ISE reported. PID parameters get tuned automatically till 20 iterations get completed.

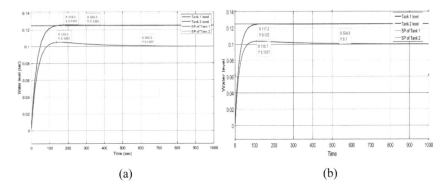

(a)   (b)

Fig. 10 Response using PSO-PID controller, **a** before optimization and **b** after optimization

Water level response of coupled tanks using PSO-PID controller is shown in the Fig. 10a. Rise time is very less as compared to the response using PID controller. For tank 1, the water level perfectly settles without any overshoot, but for tank 2, it can be observed that there is some overshoot in the response. This overshoot can be minimized further by increasing the number of weights and increasing the number of iterations. Increasing number of iterations will, however, result in making the system slow. After certain number of experimentations 20 iterations were found to be optimal for the system. Interaction effect is seen in the response of tank 2.

Improvement in the response can be seen in Fig. 10b, and it is the consequence of optimization in PSO algorithm. Number of iterations are also reduced as a result. PID range for both controllers should be properly implemented to get best response. Water level in tank 2 shows some overshoot but it is little bit reduced as compared to the response shown in Fig. 10a.

Figure 11a shows how the number of weights decided for ISE reduces as per the rise in iteration number. Reduction in ISE will result in better settlement of water level response. Before actual optimization of PSO with actual PID values, the relation is nonlinear, and number of iterations required are more. It is reflected that without PSO the value of ISE does not actually change as desired.

The PID values for both controllers should vary according to the set input. PSO can be perfectly optimized when we give PID range for both the controllers accordingly. Figure 11b shows the response of number of iterations versus weights after optimization. Within five iterations, it gives almost linear response.

6 Performance Analysis and Discussion

Performance evaluation of coupled tank interacting system with PID and PSO-PID controllers is depicted in Table 5. For closed loop with PID controller, the water level achieves the required set point with certain overshoot of 21.76% and 25.64% for tank

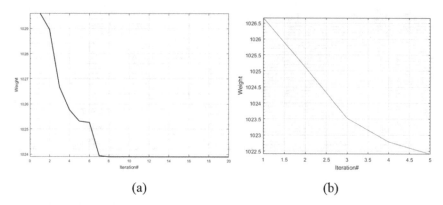

(a)   (b)

Fig. 11 Response of number of iterations versus weights, **a** before and **b** after optimization

one and tank two, respectively, while an undershoot of 3.2% and 3%, respectively, is present in the response. Settling time for attaining the SP for tank one and tank two is 443.4 s and 441.4 s, respectively.

In case of PSO-PID controller, we almost get the promising results are observed. The response is without any overshoot and undershoots. Settling time is also reduced significantly. PSO-PID controller for tank 1 achieves the desired set level of 0.125 m without any peak overshoot and undershoot in 176.3 s. PSO-PID controller for both the tanks possess no offset in the response. The response is smooth and better as compared with PID controller. 20 iteration policies for manipulating the PID parameters prove effective to deliver acceptable implementation from the system, but still there is scope for improvement.

After further optimization in the PSO algorithm, the iterations are reduced to five and there is very good response as seen in Fig. 11b. Settling time for tank 1 is 117.2 s and for tank 2 it is 534 s. % of overshoot is also reduced a lot.

Graphical representation of these performance indices can be observed in Fig. 12a and b. Absolute error depicts the exact peak position of error acting in the system. It also gives us an opportunity to tune the PID at that particular position. IAE determines area beneath the curve of the absolute error. Integral square error (ISE) will castigate the bigger error values than the minute errors. Square of bigger error value will be

Table 5 Comparative analysis of performance parameters

PSO parameters	PID		PSO-PID (before optimization)		PSO-PID (after optimization)	
	Tank 1	Tank 2	Tank 1	Tank 2	Tank 1	Tank 2
% Overshoot	21.76	25.64	0	3.684	0	0.5
% Undershoot	3.2	3	0	0	0	0
Settling time (sec)	443.4	441.4	176.3	569	117.2	534
Offset (m)	0	0	0	0	0	0

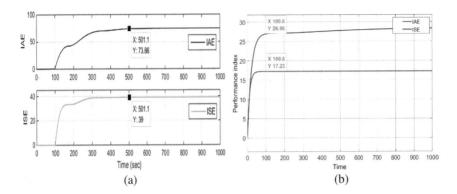

Fig. 12 Graphical representation of IAE and ISE for PID (**a**) and PSO (**b**)

much huge. ISE gives us information about where the bigger error occurs and thus furnish us an opportunity to tune the PID accordingly.

In Fig. 12a, IAE and ISE for PID controller is shown. It can be observed that values of IAE and ISE very high. This indicates that water level does not reach the desired value in optimum time and also contains error. The response is not smooth and has few disturbances at initial state. There is also dead zone nonlinearity present.

Figure 12b shows the response of IAE and ISE when PID controller is optimized with the help of PSO. Reduction in IAE and ISE signifies that water level will reach its desired value without much time lag. Relative evaluation of the performance indices is shown in Table 6. ISE and IAE values are reduced when PSO-PID controller is used. Any feedback control system has the objective of minimizing error, hence a performance measure in terms of IAE and ISE is defined to keep track of the errors at all times from zero to infinity and to minimize them continually. Minimization of performance measure will ensure the minimization of error. Because the error might be negative, these performance measures are usually represented in terms of either absolute value of error or square error. The ISE criterion is also used to generate performance measures for optimum control problems such as linear quadratic regulators and linear tracking control.

Table 6 Comparative analysis of performance indices

Performance index	PID controller values	PSO-PID controller values
Integral absolute error (IAE)	73.86	26.96
Integral square error (ISE)	39	17.23

7 Conclusion

The initial stage in translating a physical system into its mathematical model is system identification. The plant's transfer function is retrieved using a differential equation approach based on the mass balance equation, followed by state-space analysis. MATLAB has shown to be a useful tool for designing and analysing physical systems. The responses of the coupled tanks with PID and PSO-PID controllers are presented in the paper. The purpose of a PID controller gives the system a decent response that can still be optimized. ISE definitely verifies to be perfect performance index to minimize the error present in the water level response. Figure 7 depicts performance and resilience, implying that performance must be sacrificed even after tweaking the PID controller. Intelligent control approaches and optimization tools can be utilized to provide accurate and precise outputs, according to dynamic responses.

When a PID controller is used, the functioning of the system is overridden by the PSO-PID controller. Overshoot and undershoot appear to have been eradicated entirely from the response. The time it takes to settle is also reduced. ISE implicitly considers errors close to zero to be progressively less important. IAE has an issue with the derivative at the origin. Both ISE and IAE additionally trade-off between the error magnitude and its settling time.

References

1. Gaing ZL (2004) A particle swarm optimization approach for optimum design of PID controller in AVR system. IEEE Trans Energy Convers 19(2)
2. Sharma M, Verma P, Mathew L. Design an intelligent controller for a process control system. In:2016 International conference on innovation and challenges in cyber security (ICICCS-INBUSH), 978-1-5090-2084-3/16/$/31.00©2016 IEEE
3. Jaisue R, Chaoraingern J, Tipsuwanporn V, Numsomran A, Pholkeaw P (2012) A design of fuzzy PID controller based on ARM7TDMI for coupled-tank Process. In: 12th International conference on control, automation and systems
4. Sindhwani N, Bhamrah MS (2017) An optimal scheduling and routing under adaptive spectrum-matching framework for MIMO systems. Int J Electron 104(7):1238–1253
5. Olivas EL, Castillo O, Valdez F, Soria J (2013) Ant colony optimization for membership function design for a water tank fuzzy logic controller. In: 2013 IEEE workshop on hybrid intelligent models and applications (HIMA)

Comparative Soil Parameters Anatomization Using ML to Estimate Fertility in Kitchen Garden

Kushagra Kaushik, Pooja Gupta, and Jitendra Singh Jadon

Abstract Farming is a vital constituent of human essence. It is one of India's major employment generator. Agricultural sector employs nearly 50% of India's workforce. It serves as the foundation of our economy. Crop yield is influenced by a variety of factors. Soil majorly affects crop productivity. Improving strategies for predicting agricultural production in various climatic circumstances will assist Gardners and many interested parties in making preferable agronomic and cropping choices. Crop growth estimation is predicting a crop's production depending on the accumulated data with elements such as temperature, moisture, pH, precipitation, and crop name. It provides us with the knowledge of the best cropping selections that can be done in the kitchen gardens or farmlands, given the meteorological circumstances. In this research, machine learning-based comparable examination of soil parameters was conducted to forecast the fertility and crop output. The self-accumulated database has 1800 case entries and six parameters which are pH, temperature, moisture, yield, daylight, and soil texture. By using different ML algorithms, decision tree algorithm predicted with the highest accuracy.

Keywords Kitchen gardening · Agricultural vegetation · Soil parameters · Machine learning · Data mining · Artificial neural networks · Crop yield prediction · Recommendation of the best crop

1 Introduction

India's major occupational sector is agriculture. Nowadays, people are moving towards miniature farming techniques like kitchen gardening, for a good amount of nutritional and healthy vegetables. Crop yield is pretentious by a variety of factors

K. Kaushik
Department of Electronics and Communication, Amity School of Engineering and Technology, AUUP, Noida, India

P. Gupta · J. S. Jadon (✉)
Department of Electronics and Telecommunication, Amity School of Engineering and Technology, AUUP, Noida, India
e-mail: jsjadon@amity.edu

© The Author(s), under exclusive license to Springer Nature Singapore Pte Ltd. 2023
A. Shukla et al. (eds.), *Computational Intelligence*, Lecture Notes in Electrical Engineering 968, https://doi.org/10.1007/978-981-19-7346-8_58

including geological characteristics of the site (such as groundwater, hilly region, or declinations) as well as climatic factors (like moisture content, precipitation, temperature, and sunlight) [1, 2]. Soil being the most crucial variable affecting crop yield. India has a variety of soil types. Specific geographical factors, like average elevation, climatic condition, rainfall, etc., influence the physicochemical attributes of soil [3, 4]. Generally, the four soil types seen in Indian Territory are black, desert, red, and alluvial.

Agricultural planning, i.e. the accessibility of correct and prompt details of various parameters like physical characteristics of soil, meteorological factors, fertilizers consumption, and pesticides consumption, etc., helps the gardeners and farmers to coin the suitable possibilities for crop selection [5]. If the circumstances allow, it would help them attain maximum crop yield while utilizing fewer natural assets, it also aims to bring down the losses because of unfavourable situations [6].

Every day, novel hybrid species are developed. Apparently, these varieties lack the requisite minerals found in naturally grown crops. These man-made methods can also harm the soil. This has a knock-on effect on the environment [7]. Nevertheless, if crop producers have reliable information about crop productivity, the deficit could be effectively diminished. Annually, India tends to lose between 16 and 20% of the overall yield [8]. Crop yield estimation is critical under such circumstances [9, 10].

Nowadays, machine learning is widely used in farming. Heretofore, a data mining-based methodology was proposed that applies various techniques to various crops for plant semantic segmentation distribution and regression models could be used in prediction [11].

ML has been used to describe harvest modelling and evaluation, cultivation practices, and crop supply–demand matching [12, 13]. ML approaches are often used to diagnose yield loss, manage soil, manage water, and anticipate livestock [14].

In the proposed work, crop growth anatomization has been conducted on the authentic data accumulated by employing algorithms like random forest, linear regression, Naïve Bayes, decision trees, logistic regression, support vector machines, Naïve Bayes, KNN, and ANN. Geographic parameters examined for crop growth projection were humidity, temperature, pH, soil texture, daylight for sandy loam, red, black, and alluvial. The research was executed on five crops—tomato, lemon, potato, bottle gourd, and chile.

2 Literature Survey

Artificial neural networks were utilized to estimate Maharashtra rice harvest yield [12]. This was accomplished using a multilayer perceptron model. The suggested model was prepared by utilizing linear regression by utilizing neural networks and Adam as the optimizer. This showed that in order to project the precise information, linear methodology will be insufficient because of the soil factors' acuity, and sophistication. Majority of the researchers' algorithms do not apply one strategy for crop growth estimation, in which all the elements impacting crop growth are examined in one go [13]. From this conclusion, it was suggested that a big dataset should be used. In other study, for predictions, ML models like linear regression, XGBoost, artificial neural network, random forest, and logistic regressions, were employed where precipitation, thermal reading of soil and place, were taken into account [15].

The random forest regressor is shown to be the most effective. For growth estimation, K-mean and customized KNN algorithms were used, with analysis showing that customized KNN provides the greatest possible accurateness [16, 17]. Crop production prediction was accomplished by applying ANN, and it was determined that nonlinear methodologies seem to be particularly appropriate in estimating the cultivation of crops and growth [18]. Yield forecasting by remote sensing technique is widely employed. The major measure of yield projection is the normalized difference vegetation index (NDVI) [19].

As a result of this literature review, the KNN method, the Naïve Bayes method, and the decision tree classifiers were employed in this study as they are less prone to noise distortions, also they function efficiently with large number of class labels in data.

3 Methodology

3.1 Dataset Accumulation

The datasets put forward in this work were acquired by analysing five crops—chile, lemon, potato, tomato, lemon, and bottle gourd. These vegetative-crops were chosen since India is among the world's leading producers and consumers of these crops. Alluvial, black soil, sandy loam, and red soils were used to sow the seeds. The soil was checked on a regular basis. There are 1800 data-points in the dataset, with six parameters: soil texture, yield, temperature, pH, moisture, and daylight. For 45 days, readings of these facets were taken each day, at an interval of three hours. The readings accumulated from the dataset are encapsulated in Table 1.

3.2 Data Processing

The datasets acquired were unprocessed and contained redundant information. There were certain data values missing which were eliminated for reliable prediction.

Table 1 Outcome of the datasets

Crop name	Type of soil	Temperature	pH	Moisture	Sunlight
Tomato	Black, red, alluvial	35–38 °C	6.0–7.0	Good or average level of humidity	Sufficient quantity of daylight during the day
Potato	Red, alluvial, loamy	15–20 °C	4.8–5.8	A good amount of moisture	Adequate sunlight
Chile	Black, alluvial	28–32 °C	6.0–6.8	Adequate moisture	Adequate sunlight
Lemon	Black, loamy	25–30 °C	6.0–7.0	A good amount of moisture	Adequate sunlight
Bottle gourd	Alluvial	28–35 °C	6.7–7.2	Adequate moisture	Adequate sunlight

3.3 Python Application

ML models like support vector machine, random forest, Naïve Bayes algorithm, decision tree, and ANN were applied on the dataset using Python. For ease of prediction, the classifications of good, average, and bad had been transformed to decimal format. The Naive Bayes model utilizes the Bayes classifier. The Gini index was used to help in classification in the decision trees method.

3.4 Accuracy

The datasets were splitted into five batch, with each fold's average accuracy determined. Mean accuracy of these five batches was used to achieve the ultimate accuracy.

3.5 Visualization of Data

The datasets were visualized by Matplotlib, a library in Python. The data-points in characters were assigned with numeric values for the analysis of data.

4 Results

Machine learning algorithms were used to forecast the production of five crops: chile, lemon, tomato, potato, and bottle gourd in sandy loam soil, alluvial soil, red soil, and black soil. Classifications were used to make the forecast, with class 1 representing a low yield, class 2 representing a large yield, and class 3 representing an average yield. The Python programme Matplotlib was used to generate the simulated results for the proposed task.

The histogram in Fig. 1 depicts the various moisture levels and their frequency of recurrence in the dataset that was gathered 0 indicates extreme dryness, 1 indicates dryness, 2 indicates normalcy, and 3 indicates wetness. Figure 2 shows the recurrence of multiple temperature values in the gathered dataset.

Figure 3 displays the various pH values and the number of repetitions with which they occur in the acquired dataset. The appropriate pH, moisture, and temperature range were estimated using charts made using data visualization tools. From the graphs, it is evident that normal is the case. The optimal moisture, pH, and temperature

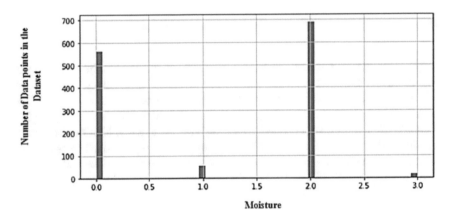

Fig. 1 Graphical representation of moisture values versus data-points in the datasets

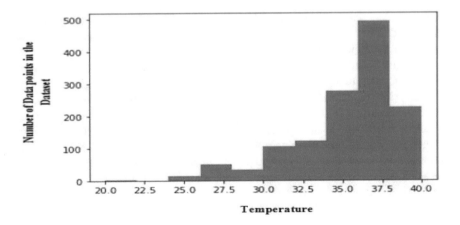

Fig. 2 Graphical representation of temperature values versus data-points in the datasets

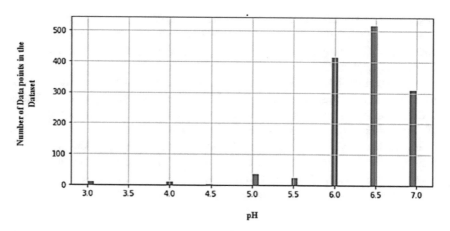

Fig. 3 Histogram presenting PH values versus data-points in the datasets

ranges were two bars tension, 6–7, and 30–35 °C, respectively. The database was bifurcated into five batches for all five methods, and the mean accuracy of every batch was deduced. The precision of the support vector machine algorithm was 93%. The Gini index was used using Naive Bayes to forecast the yield, with an accuracy of 80%. 93.005%. Random forest predicted the yield giving accuracy of 83.33%. Crop yield was predicted with a precision of 93.05% using logistic regression. The crop yield, which forecasted highest accuracy of 94.44% was by the decision trees algorithm.

5 Conclusion and Future Scope

Crop yield prediction was performed on a dataset that was self-obtained in this study. The suggested research shows that decision tree and logistic regression model can be utilized to build relationships among the examined crops and numerous physical variables. After the dataset has been validated, the accuracy of each model is determined. Higher accuracy means that the model worked better on the collected data. When compared to logistic regression, decision trees had 1.435% higher accuracy. Unlike prior studies in this field, the random forest algorithm predicted with the lowest accuracy. It is also possible to figure out the ideal crop for each soil under different situations. To improve model accuracy, the datasets may be enlarged to encompass various climatic conditions and crops. It can also be used to disseminate information via websites and apps to people all around the country. The anticipated research will assist kitchen gardeners in spotting pests and diseases.

References

1. Kumar YJN, Spandana V, Vaishnavi VS, Neha K, Devi VGRR (2020) Supervised machine learning approach for crop yield prediction in agriculture sector. In: 2020 5th International conference on communication and electronics systems (ICCES), pp 736–741. https://doi.org/10.1109/ICCES48766.2020.9137868
2. Gandge Y, Sandhya (2017) A study on various data mining techniques for crop yield prediction. In: 2017 International conference on electrical, electronics, communication, computer and optimization techniques (ICEECCOT). https://doi.org/10.1109/ICEECCOT.2017.8284541
3. Nigam A, Garg S, Agrawal A, Agrawal P (2019) Crop yield prediction using machine learning algorithms. Fifth international conference on image information processing (ICIIP) 2019:125–130. https://doi.org/10.1109/ICIIP47207.2019.8985951
4. Paul M, Vishwakarma SK, Verma A (2015) Analysis of Soil behaviour and prediction of crop yield using data mining approach. International conference on computational intelligence and communication networks (CICN) 2015:766–771. https://doi.org/10.1109/CICN.2015.156
5. Sujatha R, Isakki P (2001) A study on crop yield forecasting using classification techniques. In: Liu J, Goering C, Tian L (eds) A neural network for setting target corn yields. Trans ASAE 44(3):705–713
6. Shastry A, Sanjay HA, Hegde M (2015) A parameter-based ANFIS model for crop yield prediction. In: 2015 IEEE international advance computing conference (IACC). https://doi.org/10.1109/IADCC.2015.7154708
7. Pan G, Li FM, Sun GJ (2007) Digital camera based measurement of crop cover for wheat yield prediction
8. Kale SS, Patil PS (2019) A machine learning approach to predict crop yield and success rate. In: *2019* IEEE Pune section international conference (Pune), MIT World Peace University, Dec 18–20
9. Meeradevi HS (2019) Design and implementation of mobile application for crop yield prediction using machine learning. In: 2019 Global conference for advancement in technology (GCAT), Oct 18–20
10. Bang S, Bishnoi R, Chauhan AS, Dixit AK, Chawla I (2019) Fuzzy logic based crop yield prediction using temperature and rainfall parameters predicted through ARMA, SARIMA, and ARMAX models. In: 2019 Twelfth international conference on contemporary computing (IC3), pp 1–6. https://doi.org/10.1109/IC3.2019.8844901
11. Elavarasan D, Vincent PMD (2020) Crop yield prediction using deep reinforcement learning model for sustainable Agrarian applications. IEEE Access 8:86886–86901. https://doi.org/10.1109/ACCESS.2020.2992480
12. Gandhi N, Armstrong LJ, Petkar O, Tripathy AK (2016) Rice crop yield prediction in India using support vector machines. In: 2016 13th International joint conference on computer science and software engineering (JCSSE), pp 1–5. https://doi.org/10.1109/JCSSE.2016.7748856
13. Terliksiz AS, Altilar DT (2019) Use of deep neural networks for crop yield prediction: a case study of Soybean yield in Lauderdale County, Alabama, USA. In: 2019 IEEE 8th International conference on Agro-geoinformatics. https://doi.org/10.1109/Agro-Geoinformatics.2019.8820257
14. Medar R, RajpurohitVS, Shweta S (2019) Crop yield prediction using machine learning techniques. In: 2019 IEEE 5th international conference for convergence in technology (I2CT), pp 1–5. https://doi.org/10.1109/I2CT45611.2019.9033611
15. Guerif M, Launay M, Duke C (2000) Remote sensing as a tool enabling the spatial use of crop models for crop diagnosis and yield prediction. In: IGARSS 2000. IEEE 2000 international geoscience and remote sensing symposium. Taking the Pulse of the Planet: the role of remote sensing in managing the environment. Proceedings (Cat. No.00CH37120), vol 4, pp 1477–1479. https://doi.org/10.1109/IGARSS.2000.857245

16. Gandge Y, Sandhya (2017) A study on various data mining techniques for crop yield prediction. In: 2017 International conference on electrical, electronics, communication, computer, and optimization techniques (ICEECCOT), pp 420–423. https://doi.org/10.1109/ICEECCOT.2017. 8284541

17. Suresh A, Ganesh Kumar P, Ramalatha M (2018) Prediction of major crop yields of Tamilnadu using K-means and Modified KNN. In: 2018 3rd International conference on communication and electronics systems (ICCES), pp 88–93. https://doi.org/10.1109/CESYS.2018.8723956

18. Rehman TU, Mahmud S, Chang YK, Jin J, Shin J (2019) Current and future applications of statistical machine learning algorithms for agricultural machine vision systems. Comput Electron Agricult 156:585–605

19. Jeevan Nagendra Kumar Y, Spandana V, Vaishnavi VS, Neha K, Devi VGRR (2020) Supervised machine learning approach for prediction in agriculture sector. In: Proceedings of the fifth international conference on communication and electronics systems (ICCES 2020), IEEE Conference Record # 48766; IEEE Xplore ISBN: 978-7281-5371-1

Development of Submarine Simulation for Assessment of Cognitive Skills

Chirag Singh, Anushiv Shukla, Apoorva Murjani, and Dhiraj Pandey

Abstract In today's ever-changing world, migration from legacy processes to modern processes is one of the popular and efficient trends. Cognitive assessment is used as a screening test for job-seeking candidates by many corporations and it is also required for detecting major health diseases such as Alzheimer's disease. Cognitive assessment using a pen-paper test requires participants to imagine the situations given in the test, which leads to inaccurate results defeating the purpose of the whole process. We have tried to resolve this problem by using a more user-friendly and interactive process to judge various cognitive skills by using a simulation method. Simulation can be effectively used as an alternative to currently existing methods, which can be less accurate and lack real-world context. Using a submarine simulation to assess different cognitive skills can provide a much better alternative to a pen-paper test as it offers a more immersive experience and leaves fewer scope for imagination which leads to accurate cognitive skills evaluation comparatively. This submarine simulation after some modification can also be used as a screening or training tool for our upcoming or existing naval officers which will help our country to save large amounts of resources in the form of time and money.

Keywords Cognitive assessment · Simulation · Submarine simulation · Cognitive skills

C. Singh (✉) · A. Shukla · A. Murjani · D. Pandey
JSS Academy of Technical Education, Noida, India
e-mail: chirag123singh@gmail.com

A. Shukla
e-mail: anushivshukla1999@gmail.com

A. Murjani
e-mail: apoorvamurjani@gmail.com

D. Pandey
e-mail: dhip2@yahoo.co.in; dhirajpandey@jssaten.ac.in

1   Introduction

The art of gathering information and making sense of that gained information is referred to as cognition. Cognitive ability tests are becoming a more common component of applicant screening exams, not only to assist in discovering eligible individuals but also to speed up and simplify the hiring process. That is why, when evaluating external job applications, 76% of businesses with more than 100 workers utilize various types of cognitive screening exams. It is also frequently used to detect minor cognitive impairment (MCI). It is recommended by the Alzheimer's Association for everyone who has any concerns about their memory should take up a cognitive assessment. The most common types of cognitive tests are the Montreal Cognitive Assessment (MoCA), Mini-Mental-State-Examination (MMSE), and Mini-Cog. Assessing cognitive abilities using these tests often leads to unreliable and inaccurate results as they are lengthy, and a participant often loses interest in between. This simulation will be able to give more precise and dependable results by making the whole process more interactive and participant oriented. Complex problems of various industries can be analyzed and solved by using simulation modeling. It provides a clear and precise understanding of the problem and a way of analysis that is easy to understand, communicate, and verify. Simulation software provides an excellent methodological route for instructors in training processes as students can use realistic scenarios during training and learning sessions [1].

Simulation tools with technologies like virtual reality (VR) have been demonstrated to increase awareness and the capability of workers when compared to traditional learning methods [2]. Cognition assessment using a submarine simulation is more beneficial because the submarine simulation provides a challenging environment to explore and test out different scenarios and, in this way, many cognitive abilities can be assessed effectively [3]. This simulation can work as a more effective and engaging cognitive assessment process than a simple pen-paper test. The text-based technique can be viewed as presenting multiple-choice queries, which may limit the types of SA information that can be asked [4]. Participants taking the test will be more emerged in the whole process and will make real-time choices that will be more in line with their actual cognitive abilities. This research work has further been distributed into six sections, namely Related Work, Implementation, Results, and Conclusion.

2   Related Works

In this section, the previous research works related to the assessment of various cognitive skills discussed in this paper will be outlined. This section has been divided into the following subsections, with each sub-section discussing the work done in the analysis of these cognitive skills.

2.1 Situation Awareness

Situation awareness (SA) is defined by Nguyen et al. [5] as the perception of environmental elements and events concerning time or space, the comprehension of their meaning, and the projection of their future status.

Freeze-probe technique

Nguyen et al. [5] have described these techniques in their paper. The simulator is randomly frozen. Participants must answer the questions during this frozen period. Based on this concept, we have designed our classification task. To calculate a SA score, the participant will answer some SA queries based on the present state of the simulated environment. The situation awareness global assessment technique (SAGAT) [6] and SA control room inventory (SACRI) in Hogg et al. [7] are the majorly used techniques in this category.

Real-time probe technique

Experts create queries either during the task or before and administer them at appropriate points during the participant's performance. SPAM used by Durso et al. [8] is a typical real-time probe technique. The situation awareness rating technique (SART) [9] and situation awareness rating scales technique (SARS) [10] are majorly used post-trial self-rating techniques. Situation awareness behavioral rating scale (SABARS) [11] is a common observer assessment approach that uses a five-point rating scale. We have decided to go with the combination of the freeze-probe technique and real-time probe technique as it comfortably fits in our simulated environment.

2.2 Attention

Numerous methodological approaches have been developed to investigate the many dimensions of attention. Michael and Greher [12] in their paper mentioned how Digit Span, Virtual Search and Attention Test (VSAT), and Search and Cancelation of Ascending Numbers (SCAN) are associated with each other.

3 Implementation

In this simulation work, an effort has been made to assess the cognitive skills of the participant as well as making the participant experience pleasant using an interactive storyline and a series of tasks involving different and challenging activities simultaneously [13–15]. The simulator is made using Unity3D and C. The process of assessing the cognitive skills is divided into four tasks, namely classification task, torpedo avoidance task, access code and information sending task, and close quarter maneuvering task. Each task assesses the participant by calculating a score and by that score, the final report for various cognitive skills is calculated. The final score obtained in each task is then allotted to the cognitive skill associated with that task. The environment involves a set of monitors placed on a rectangular table which gives it a view of the control room of a submarine. Throughout the simulation, brief and descriptive messages in the form of alerts and warnings are presented to help the participant in transition from one task to another.

3.1 Classification Task

Each task is performed sequentially starting from the classification task in which a radar display is shown with moving circular-colored objects, which are considered as movement of vessels in this simulation [16]. After watching the radar for 30 s, the participant must answer some questions which are presented on the monitor to the left of the radar and all the questions are related to what was previously shown on the radar. The movement of objects is according to the script. The cognitive skills assessed via this task are situation awareness, simultaneous attention, category formation, and response time.

3.2 Torpedo Avoidance Task

After the incomplete classification of nearby vessels by the user as friendly, merchant, and enemy, the user moves to the next task where surprisingly torpedoes have been fired on the submarine by the enemy submarine, and to evade the torpedo, the participant must operate the submarine such that torpedoes do not collide with the submarine and simultaneously he/she must answer some questions. The duration of the task is fixed and is set to 60 s to operate the submarine using WASD keys and clicking on options of the questions presented on the bottom of the screen. Once looked, the movement of torpedoes is designed in such a way that they will follow the target wherever it will move, thus, making them a tricky thing to escape. It will be challenging for the participants to operate the submarine and answer the questions simultaneously. The cognitive skills assessed via this task are speed of information processing, sustained attention, motor and memory skills.

3.3 Access Code and Information Sending Task

Now the submarine is safe from torpedoes, but the participant is required to update the navy base about the situation, an access code is shown in the beginning to the participant while entering the simulated control room and now the participant must recall that access code and enter it in a keypad by clicking on the buttons of the keypad. The participant is not allowed to go through if the correct code is not entered. If he/she can successfully recall the code and enter it, maximum points are given, with each incorrectly entered code, points will be deducted. If the participant chooses to take a hint, he/she can use it to decode the access code. Also, there is an option to see the complete code. After entering the right access code, a questionnaire is presented, and each correct answer will result in 1 point. After the questionnaire is completed, the message conveying the situational update is conveyed to the navy base. The cognitive skills assessed via this task are working memory, pattern recognition, visual and spatial processing.

3.4 Close Quarter Maneuvering

After getting attacked by the torpedoes, the unidentified contact in the first task turns out to be an enemy which has deployed some mines in the water, and now the participant must destroy them such that it does not harm the submarine. In this time-sensitive task, the participant must try to tap maximum tiles and as task time increases the speed of incoming tiles also increases. The duration of this task is 30 s. The cognitive skills assessed via this task are response inhibition and sustained attention.

4 Results

For our test, we selected five random people of age group from 20–25, out of those five, three were male and two were female. All the participants were taken voluntarily. No participant has any prior information about any part of the simulation. 30 min before the simulation, participants were briefed about the details and story of the simulation. An instruction manual was also designed for every participant to read before starting the simulation. The average time to complete the whole simulation assessing 10 different cognitive skills was 14 min which was 4–5 min more than the common cognitive tests, but the number of cognitive skills assessed in our simulation was more diverse in nature. Two reports were generated after the completion of the simulation for every participant, which were the score of different tasks and the score of different cognitive skills. Figure 1 shows the score of participants two in different cognitive skills.

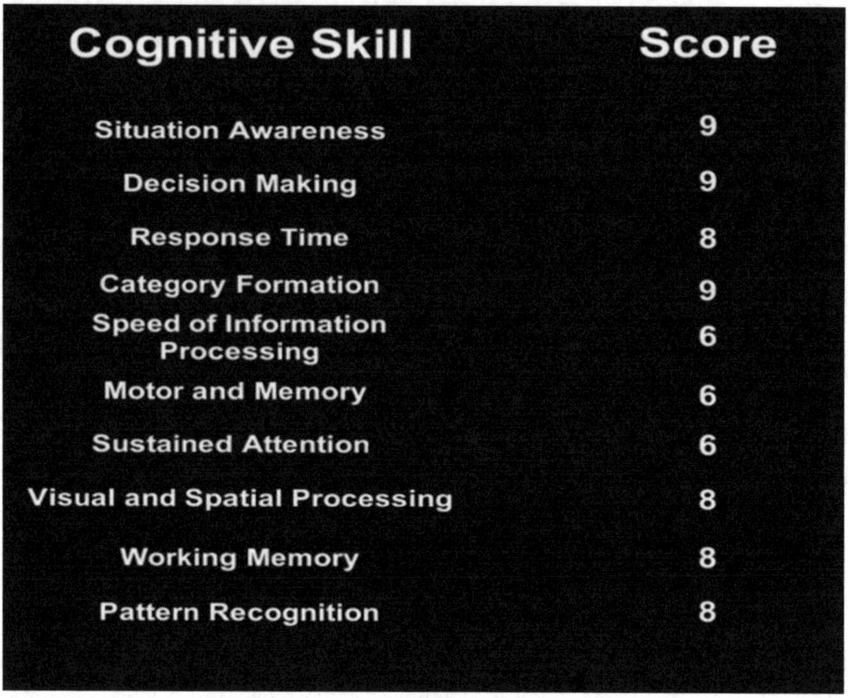

Cognitive Skill	Score
Situation Awareness	9
Decision Making	9
Response Time	8
Category Formation	9
Speed of Information Processing	6
Motor and Memory	6
Sustained Attention	6
Visual and Spatial Processing	8
Working Memory	8
Pattern Recognition	8

Fig. 1 Screenshot of simulation result

5   Conclusion

Every industry is investing its resources in simulating its legacy processes due to various advantages that come with simulations. Cognitive skills can be assessed successfully and effectively using this simulation. There is a massive need for this type of advanced process to assess cognitive skills as the applications of cognitive tests span over different fields including job recruitment, health sector, and screening individuals. This method of assessment will provide us with major long-run benefits in terms of time and money and will also result in more accurate test results and indulging test experience. The real challenge lies in designing different and efficient levels to assess more advanced cognitive skills so that the purpose of switching to simulation gets fulfilled. The same simulation can be used by completely different industries by using the same environment and changing the number and types of tasks. There is a massive shortage in the number of submarine simulations available in our country. A huge amount of money and time is spent to train our naval officers. It is high time for India to investigate a simulated environment as an alternative to train its officers.

This simulation could work as a starting point in building a more complete virtual submarine to start training Indian naval officers more efficiently and cost-effectively.

This simulation can also work as a testing environment for shortlisting naval students. Many resources are required from the selection of the appropriate candidates to training them and finally turning them into naval officers. Simulation training has been going on for a long time in countries like the United States of America, Australia, and China. Simulations are not limited to only training of naval officers, but they can also be used in tests to determine people having the neck for the Indian Navy. The submarine simulation will cut off the resource requirement exponentially and will help our country in making the whole process more efficient and faster.

Acknowledgements The concept of this work was supported by Institute of Nuclear Medicine and Allied Sciences Lab (INMAS), Defense Research and Development Organization (DRDO), Ministry of Defense, India.

References

1. Campos N, Nogal M, Caliz C, Juan A (2020) Simulation-based education involving online and on-campus models in different European universities. Int J Educ Technol High Educ 17
2. Bergamo PAS, Streng ES, de Carvalho MA, Rosenkranz J, Ghorbani Y (2021) Simulation-based training and learning: a review on technology-enhanced education for the minerals industry. Miner Eng 175:107272, ISSN 0892-6875
3. Loft S, Bowden V, Braithwaite J, Morrell D, Huf S, Durso F (2014) Situation awareness measures for simulated submarine track management. Human Factors
4. Strybel TZ, Chiappe D, Vu KPL, Mira-montes A, Battiste H, Battiste V (2016) A comparison of methods for assessing situation awareness in current day and future air traffic management operations: graphics-based versus text-based online probe systems
5. Nguyen T, Lim C, Nguyen ND, Gordon-Brown L, Nahavandi S (2019) A review of situation awareness assessment approaches in aviation environments. IEEE Syst J
6. Endsley MR (1995) Measurement of situation awareness in dynamic systems. Human Factors
7. Hogg D, Folles K, Volden F, Torralba B (2007) Development of a situation awareness measure to evaluate advanced alarm systems in nuclear power plant control rooms. Ergonomics
8. Durso FT, Hackworth CA, Truitt TR, Crutchfield J, Nikolic D, Manning CA (2000) Situation awareness as a predictor of performance for En route air traffic controllers
9. Taylor RM (1990) Situational awareness rating technique (SART): the development of a tool for aircrew systems design. AGARD
10. Waag WL, Houck MR (1994) Tools for assessing situational awareness in an operational fighter environment. Aviat Space Environ Med 65(5, Sect 2, Suppl):A13–A19
11. Matthews M, Beal S (2002) Assessing situation awareness in field training exercises
12. Michael R, Greher BA (2000) Measuring attention: an evaluation of the search and cancellation of ascending numbers (SCAN) and the short form of the test of attentional and interpersonal style (TALS)
13. Roberts A, Stanton N (2018) Macro cognition in submarine command and control: a comparison of three simulated operational scenarios. J Appl Res Memory Cogn
14. Roberts APJ, Stanton NA, Fay D (2017) Land Ahoy! Understanding submarine command and control during the completion of inshore operations. Human Factors
15. Roberts A, Stanton N, Fay D (2015) The command team experimental test-bed stage 1: design and build of a submarine command room simulator. Proc Manuf 3:2800–2807
16. Loft S, Sadler A, Braithwaite J, Huf S (2015) The chronic detrimental impact of interruptions in a simulated submarine track management task. Hum Factors

Road Lane Line and Object Detection Using Computer Vision

Akshit Sharma, Tejas Vir, Shantanu Ohri, and Sunil Kumar Chowdhary

Abstract For humans making, the car travel along the road lane line is easy as us humans can perceive the lanes as one of your common tasks, but when it comes to autonomous vehicles it is usually not the same case. For an autonomous, the main challenge is dealing with different varieties of roads, this paper discuses Indian roads specifically as on these roads sometimes the lane lines are not easily visible or they are not even marked. This paper discuses about the vision-based faction of algorithms that uses camera, it comprises of camera calibration using chessboard images to correct the distortion. Use of bird-eye view and conversion to different formats like HLS/RGB/HSV/YCrCb/Lab formats also using thresholding the warped images by trial-and-error thresholding, plotting Sobel-gradient with appropriate channels and in the end performing the lane pipelining to detect the lane lines in a video. For object detection, we have also use YOLO model. These techniques helped us to detect the lane lines. We used OpenCV computer vision framework that works upon the above functionalities using Python to obtain the results and to compare them.

Keywords Driving assistance · Road · Safety · Images · Lane detection · Intelligence · Calibration

1 Introduction

Autonomous driving generally refers to self-driven vehicles. It is a vehicle equipped for detecting its current circumstance and working without human association. Carrying out this process is a very crucial task when it comes to any autonomous vehicle, and this process may also be known as advanced driver-assistance system (ADAS) lane detection (it tells if the vehicle is in lane or not and in between objects). Another very crucial task is detecting the objects nearby so as to minimize the chance of collision. Detecting objects in this project is done by using YOLOv3, and it helps to detect what kind of object is in the frame and marks it in the form of numbering.

A. Sharma · T. Vir · S. Ohri · S. K. Chowdhary (✉)
Amity University Uttar Pradesh, Noida, India
e-mail: skchowdhary@amity.edu

© The Author(s), under exclusive license to Springer Nature Singapore Pte Ltd. 2023   693
A. Shukla et al. (eds.), *Computational Intelligence*, Lecture Notes in Electrical Engineering 968, https://doi.org/10.1007/978-981-19-7346-8_60

2 Related Work

To minimize the mis-happenings, the object detection plays a vital role in notifying the system that any object which comes into the frame and alerts the driver of the situation. Due to this sole reason, the industry has a very wide outlook to overcome these challenges. There are many sensor-based approaches which use LIDARs, RADARs, or even ultrasonic sensors for lane and object detection, respectively, by extracting features from point cloud and place reliance on the well-known road lane structure [1].

A typical lane detection using OpenCV follows the steps like read and decode the video files into frames then converting the selected frames into gray-scale followed by reducing the noise using the Gaussian-Blur. Then, using the Canny algorithm, the detection of the lane lines is done, in this project, we have not used Canny algorithm instead we used bird-eye view and then converting the un-warped images into RGB/HSV/HLS/YCrCb/Lab format, out of all the chosen formats only the S and L channel from HLS, Y and Cr channel from YCrCb color space look promising and are able to identify the lane lines easily. We chose these color channels because after combining they were easily able to detect the lane lines and were almost free from noise. Despite having various pro, these methods have short outcomings and are limited to certain features. This is the reason why most of the lane detection techniques [2–5] use deep learning convolution neural networks as it is more reliable and gives higher precision. The major problems with these techniques are that they are not officially verified, and many car companies are not in favor of using this technique as their main tool for self-driving.

3 Proposed Method

Libraries used: numpy, cv2, matplotlib.plyplot, moviepy, queue, pytube, math, datetime, collections, scipy.stats, scipy.optimise, and yolo_model.

So, in this paper, we have divided the workflow in two parts:

1. **Lane detection on static images.**
2. **Lane detection on video file.**

3.1 Lane Detection on Static Images

Firstly, we will discuss about part 1 that is lane detection on static images. The workflow is as follows:

Input Chess Board Images

After importing all the important libraries on our notebook, we need to collect all the chessboard images and save them on our device and upload one by one on our notebook.

We use these chessboard images as a metric for checking or fixing the camera calibration because patterns in the chessboard images which are black and white can be distinct and one can easily detect in the following image.

Camera Calibration

Camera calibration step, the camera is calibrated by and checked by plotting corners of the chessboard images. Here, the chessboard images that we have uploaded on out notebook are distorted images.

After, calibrating our camera we made our chessboard images undistorted with the help of OpenCV module (Fig. 1).

Input Test Images

After the calibrating our camera, we are going to first prepare test images that we will be using in our project. So, we prepared around seven images here six of those were images of Indian roads and one was of an USA road.

Distortion Correction

After preparing our test images, images that we collected originally were all distorted images and after collecting them and uploading them on our notebook with the help of OpenCV module we undistorted them.

Figure 2 here represents both distorted images of roads and undistorted images that we converted using OpenCV.

Conversion to Bird-Eye View

After undistorting the image, we transform them into a bird-eye view or top-eye view. We did this to get a better view of the lanes we transform the test images to a bird-eye view.

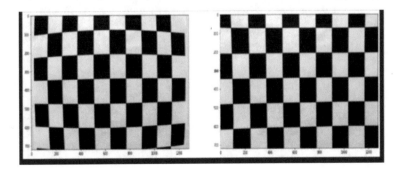

Fig. 1 Plotted corners on chessboard images

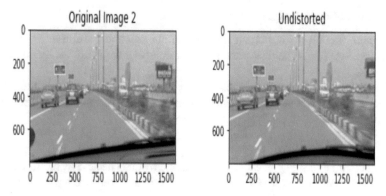

Fig. 2 Original images and undistorted images

Figure 3 shows the original image and bird-eye view of the image, and we could clearly see that lane lines are clearly visible here.

Conversion from Original to RGB/HLS/HSV/YCrCb Lab Format

After converting our original image to bird-eye view, we are going to use these converted bird-eye view images and convert them into the following RGB/HLS/HSV/YCrCb lab formats.

Figure 4 here represents the RGB/HLS/HSV/YCrCb lab formats using the OpenCV module and set the color space as gray.

Y/CR/L/S Conversion to Binary Image

As images were clear in Y, Cr, L, and S channels so we used these channels and decided to convert them into a binary image.

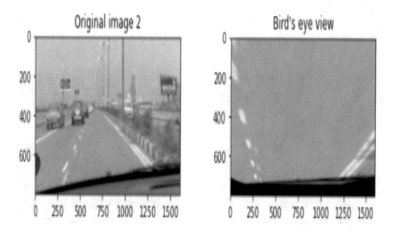

Fig. 3 Original images and bird-eye view images

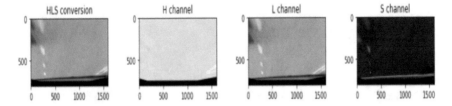

Fig. 4 HLS lab conversion of bird-eye view images

Binary image is an image where pixels are stored which can only have around two colors, it usually consists of black of white (Fig. 5).

Thresholding of Warped Images

After converting to binary image, we are going to use our warped images that we converted earlier and threshold them on three thresholding values which are (0.5, 1.8), (0.7, 1.3), and (0.2, 1.1).

Threshold is basically used to differentiate and image into many classes of pixels (Fig. 6).

Sobel-Gradient Image Conversion and Pixel Conversion

After converting them into threshold images, we will be converting them into Sobel-gradients.

After converting them into Sobel-gradient, we will be combining every image that we have used to detect the lane as shown in Fig. 7 after combining them we will be plotting a graph to determine position of the lanes that we found out.

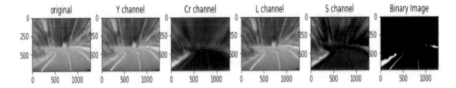

Fig. 5 Y/Cr/L/S conversion to binary image

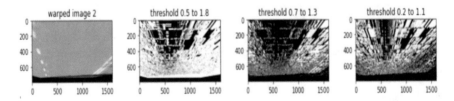

Fig. 6 Thresholding images of warped images

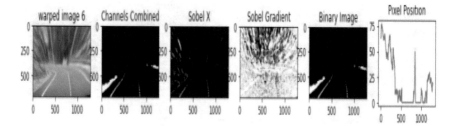

Fig. 7 Combined images, pixel position, and Sobel-gradient

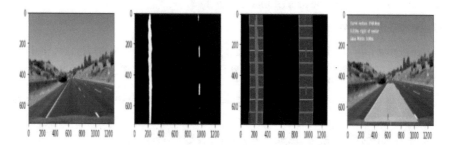

Fig. 8 Original images, combined images, drawing-lanes, and output images

Fitting Lines and Highlighting the Lanes Detected

After drawing the pixel graph using matplotlib library, we must fit lines and plot the lanes that were detected on pixel graph. We check the pixel position graph and on the combined image we made rectangular blocks if left side of the lane was detected it was colored red and when left side was detected inside the rectangular boxes it was colored blue and lanes were plotted how they were detected on the same image as a reference after getting the lane lines we plot them on the original image using OpenCV and give out our output as shown in Fig. 8.

3.2 Lane Detection on Video File

This part is the most crucial part of the proposed design which is the video pipelining, which will give us the main output of detection with objects and the lane lines.

Input Video File

The first is to input video file, as we are using Google Colab. We first uploaded the test video on YouTube then using the YouTube link we imported the test video onto the Colab workspace.

Detecting Lane

The initial step is lane recognition. All the other things we recognize from vehicles to zebra lines exists in setting to the paths on which we are driving. We are worried about the guard distance to the vehicle before us in our path.

There are a couple of features of paths that we can exploit. Paths lines are equal; they are white or yellow. Generally, they are consistent have a standard width all through the street. For a windshield-mounted camera, they will normally begin uniformly dispersed with regards to the picture outline. From one edge to another, the recognized bends will be constant.

Perspective Transform

The point which limits the amount of opposite distance toward this multitude of lines is our vanishing point. We utilize numerical build to back it out.

$$\text{self}.\text{vanish\_pt} = \text{np}.\text{matmul}(\text{np}.\text{linalg}.\text{inv}(\text{Lhs}), \text{Rhs}).\text{reshape}(1)$$

We therefore utilize this vanishing point to make a bunch of source focuses to guide to the objective focuses (corners of the top-view picture). We have observed setting the top-view size as 360×360 px is by all accounts adequate for the remainder of the interaction regardless of whether the objective picture is 720 px high.

$$\text{source\_pts} = \text{np}.\text{array}([\text{p1}, \text{p2}, \text{p3}, \text{p4}], \text{dtype} = \text{np}.\text{float32})$$

Creating Mask out of Perspective Image

The perspective image (three channel RGB) is not usable yet. We need to go it to a mask from which we can extricate the path data out. This is the most precarious advance in the whole interaction. The vehicle may be driven in conditions like: dawn, early afternoon, night, concealed, inside a woodland on a roadway, etc. A portion of these progressions is steady and some as a structure/over-span shadow is very abrupt. While we can disengage the white paths in the RGB shading space (255, 255, 255) yellow is somewhat precarious. Moving to the HLS shading space is a smidgen more reasonable, and we will utilize the HLS changed over picture as the beginning stage for separating the mask.

We begin by lower and upper thresholding of yellow and white masks, this is done because the lighting conditions never remain the same sometimes it is going to be full of light, sometimes it is going to be shady or less lighting so in order to ease the process we have to update the thresholds frequently. The most sensitive factor is the lower yellow HLS threshold, if we were to have picked up the wrong number, then it will not work so a lot of hit and trials were done is order to get the right number.

Detecting Lane Starting and Width

We presently have the masks gotten from the top-view. We need to begin extricating the path data out of it. As an initial step, we want to decide the beginning of the left and right paths. Tops on the histogram of the masked matrix give us simply that. A few times the path may bend left/right so it is more reasonable to utilize lower segment near the vehicle to process the histogram [2].

On most streets, the paths are of a standard width of 3.5–3.75 m. We can utilize this data to plan the pixel organize framework in the top-view picture to this present reality facilitate framework the picture addresses.

Sweeping Window

We have recognized the path begins in the last step. We need to extricate the pixels containing the path lines in this step. The initial step is to make a horizontal pass and get the pixels limited by every one of the left and the right square. Accordingly, we decide the midpoint of these pixels to decide the horizontal path of the window in the following stage. We continue to rehash the means to extricate the pixels for the following line until we have covered the whole picture (Fig. 9).

Now we faced some difficulties while performing this step that sometimes in the window there can be a few or sometimes a large number of pixels. Now the main dilemma arises that how to figure out the x-position on the window. So, now in each of these cases we have to dismiss the data in the window as unusable and placed in a rough approximation of the following window x-position. So, we designed an alternative way to perform window sweeping by using an 'else' condition.

In case the adjoining line has been filled we can balance its situation by the path width and proceed. In any case, on the off chance that an adequate number of lines have been prefilled we can utilize the overall bend to assess the following position. On the off chance that that does not work out, we can utilize the positions got in the past edge to proceed. Assuming that nothing functions as a last response, we can proceed vertically.

Finding Line Center

Now from the previous step, we were able to gather the left and right coordinates of the lane, now we just had to fit a bend onto these points. This step was supposed to be a pretty plain and simple task but the program picked up a noise by the thresholding.

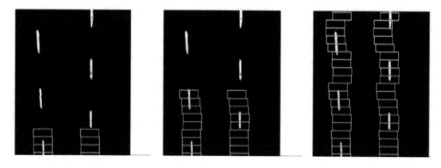

Fig. 9 Lane hotspots

Table 1 Dataset comparison

Method	mAp-50	Time (ms)
[G]FPN FRCN	**59.1**	172
RetinaNet-50-500	50.9	73
YOLOv3-320	51.5	**22**
YOLOv3-416	55.3	29

Now to resolve this situation, we used the centroid of the points so that we can take an estimation of the points on the bend.

Lane Switching and computing the offset

While driving it's natural to change lanes so as to avoid collision, so the program will switch the lane when needed by the driver. For every one of the lanes, we have figured out the lane start position. The midpoint of which provides us with the focal point of the path. The focal point of the camera in the top-view gives the area of the focal point of the vehicle. We can interpret it from the pixel arrange framework to real-world offset by utilizing the proportion we had processed beforehand.

When the vehicle does a lane switch, we offset everything by one path width. We save the old points for the path line, normal to both the paths, and reset the other.

Identifying the Objects on the Lane

There are many ways for detecting objects, but for this project we have used YOLO: real-time object detection. Specifically, we have used YOLOv3 because it is balancing in terms of accuracy and computation cost [6, 7] (Table 1).

4   Result

See Fig. 10.

5   Conclusion

In this paper, we proposed a way to detect different types of Indian roads lanes. The main challenge for detecting the road lane lines will be the: Too much or too less lighting, non-visible lane markings. We have also taken help of YOLOv3 in order to detect the objects that come in the frame. Lane detection in foreign roads is easier as compared to Indian roads because of the more visible lane lines. Our method can detect the blurred lane lines with more efficiency and can detect the curves more accurately with less visible lines for simultaneous detection for road boundaries as well.

Fig. 10 Locating cars and marking on window (top left)

References

1. Yalcin O, Sayar A, Arar OF, Akpinar S, Kosunalp S (2013) Approaches of road boundary and obstacle detection using LIDAR. IFAC Proc Vol 1(PART 1):211–215. https://doi.org/10.3182/20130916-2-tr-4042.00025
2. Kang S, Han DS (2017) Traffic lane estimation using road width information. In: IEEE international conference on consumer electronics—Berlin, ICCE-Berlin, vol 2017, pp 53–54. https://doi.org/10.1109/ICCE-Berlin.2017.8210588
3. Almeida T, Santos V, Lourenço B (2020) Scalable ROS-based architecture to merge multi-source lane detection algorithms. Adv Intell Syst Comput AISC 1092:242–254. https://doi.org/10.1007/978-3-030-35990-4_20
4. Komori H, Onoguchi K (2020) Lane detection based on object detection and image-to-image translation. In: Proceedings—international conference on pattern recognition, pp 1075–1082. https://doi.org/10.1109/ICPR48806.2021.9412400
5. Kim JG, Yoo JH (2019) HW implementation of real-time road & lane detection in FPGA-based stereo camera. In: 2019 IEEE international conference on big data and smart computing BigComp 2019—Proceedings. https://doi.org/10.1109/BIGCOMP.2019.8679333
6. Ying C, Dinghui W (2007) Multi-sensor-based lane detection and object tracking method. In: Proceedings 26th Chinese control conference CCC 2007, (1), pp 137–140. https://doi.org/10.1109/CHICC.2006.4347050
7. Stevic S, Dragojevic M, Krunic M, Cetic N (2020) Vision-based extrapolation of road lane lines in controlled conditions. In: 2020 Zooming innovation in consumer technologies conference ZINC 2020, pp 174–177. https://doi.org/10.1109/ZINC50678.2020.9161779

Evaluation of Support Vector Machine and Binary Convolutional Neural Network for Automatic Medicinal Plant Species Identification

Sachin S. Bhat, Alaka Ananth, Anup S. Shetty, Deepak Nayak, Prasad J. Shettigar, and Sagar Shetty

Abstract Enormous amount of diversified plant species are available in India. Recognition and classification of these species have become a major challenge and an important research field. Though different parts of plants can be used in identifying their genre, leaf is most useful and effective method in classification. Machine learning brings an ideal way to automate this system. A separate dataset is built by collecting 20 different leaf samples available mainly in Southern India. More than 20,000 such samples are collected to build this dataset. Here, we used two different machine learning models namely support vector machine and binary convolutional neural network. These algorithms gave a promising results of 79% and 89.5%, respectively. Various analytical methods are used to evaluate the performance of these models.

Keywords Plant classification · Neural network · Support vector machine · Leaf dataset

1 Introduction

In India, the usage of medicinal plants is an important source of food, medicine, fragrance or odour, flavours, food colour, etc. Preserving such plants is a part of biodiversity and plant genetic resources. The medicinal plants in India are around 8000, and 95% are collected from the forest area either by destructive means or by deforestation. The widespread growth nature of these leaf species are spread over

S. S. Bhat (✉) · A. S. Shetty · D. Nayak · P. J. Shettigar · S. Shetty
Shri Madhwa Vadiraja Institute of Technology and Management, Bantakal, India
e-mail: sachinbhat88@gmail.com

A. Ananth
NMAM Institute of Technology, Nitte, India

© The Author(s), under exclusive license to Springer Nature Singapore Pte Ltd. 2023
A. Shukla et al. (eds.), *Computational Intelligence*, Lecture Notes in Electrical Engineering 968, https://doi.org/10.1007/978-981-19-7346-8_61

many areas of the country. To be generic 41% herbs including grasses, 26% trees, 17% shrubs, 16% climbers are seen across the country. But with the increasing population and extension of urban areas, the extinct situation is a threat for the plant generic resources. The geographical distribution of vegetation area with unique medicinal and herbal plant cultivation is important in original system of medicine.

The different plant parts like roots, bark, fruits, seeds and leaves are collected from the natural habitats. The extinction of forest and wild habitats is a menace for the geographical area changes which leads to the destruction of the greenery region. The present generation is unaware of the medicinal benefits of rich Indian herbal plants. Many plants can be taken as raw food for getting antioxidants, vitamins and minerals. Also some leaves are more useful in improved vision, skin care, bronchial problems, digestion and reduced risk of many diseases like cholesterol level, stomach acid, vomiting and joint pain, etc. These leaves are more with nutritional values and also with vitamin A, C.

Leaf recognition and identification are very useful for plant health prediction and yield estimation and to know the herbal values of it. There are about 500,000 plant species in the world in which manual recognition is always a time consuming and inefficient. The identification of plant species is an important research area and its automatic leaf-based recognition system requires focus on design and development of user friendly identification mechanisms that would assist recognition process. The real fact is due to the recent urbanization and biodiversity loss, herbal plant species has to be preserved and understood by the professionals such as environmental protectors, foresters, agronomists, etc. The botanist conservatively does the plant classification for their floral parts, fruits and leaves. As the flowers and fruits are not suitable for perfect identification, leaves are considered as the best choice as they are available in abundance and for longer duration. The leaves are virtually 2D in shape and can be collected at all seasons. Leaves play a crucial role in medicine, industry, foods, etc. The skill of recognizing these leaves is for conservation of endangered species and rehabilitation of destroyed forestry. It is necessary and yet valuable to develop a computerized automated system for various leaf species identification and classification to help the researchers and public to identify easily.

The huge amount of leaf species is discriminated based on their unique physical shapes through geometric features and venation architectures. This feature has paved the way to devise machine learning methods for perfect identification of leaf species. We have utilized both machine learning and deep learning-based algorithms to recognize and classify 20 different plant species predominantly grown in south India region.

Remaining part of the paper is organized as follows. Section 2 provides an overview of the concepts as well as the preliminaries need to understand the rest of the sections. We propose the methodology based on two algorithms in Sect. 3. Analysed results are presented in Sect. 4. We conclude in Sect. 5.

2 Background

Kataoka et al. [1] implemented segmenting and leaf classification of tomato plants in uncontrolled environment using supervised and unsupervised learning. A self-organizing map (SOM) neural network is applied for the training stage to group colours from a set of images containing vegetation. This work has provided better segmentation rate in uncontrolled environments than the segmentation rate obtained by a colour index technique by using a Bayesian classifier using the two histogram models. Ahonen and Pietikäinen in [2] and Mouine et al. [3] proposed an original method for plant species recognition based on leaf observation. It uses TSL, SC2 and multi triangular approaches of a scanned image by a digital camera with untextured background. Though the leaf colour is not sufficiently discriminant to be used alone in a plant identification task. Hence, Bai and Ren [4] used SC2 descriptor with a more accurate description of the leaf margin and leaf boundary.

Automated leaf image detection literature involves statistical feature matching approaches [5] for appropriate edge detection. More semantic edge boundaries shall be identified using [6] which is learned over very large datasets [7]. Colour and shape feature analysis has been extensively applied over leaf detection literature [8]. Active polygons [9] and active contours [10] are noteworthy to mention. Histograms [11] are widely used for background image separation. For faster detection, leaves required to have a plain white background. Overlapping leaves are also dealt with in the literature [12].

The similar work based on the supervised learning using SVM extracts the feature for image classification. The feature vector contains 100 plant images and obtained the classification accuracy to 78%.

Other features like leaf tip [13], leaf base [14] and leaf petiole [15] are also considered for leaf image classification. Texture analysis was combined with shape above margin and base for better classification [16]. Venation of leaves [17] was also analysed. Extensive research on applying deep learning for automated plant species identification is found in the last decade. AlexNet, ResNet, DenseNet, SqueezeNet and other CNN architectures have transformed automatic leaf classification research into remarkable dimensions.

A consolidated plant datasets available mainly for Indian habitat is given in Table 1.

3 Methodology

3.1 Dataset Description

This dataset includes the medicinal plants mainly used in the Western Ghat part of the Indian subcontinent. Twenty different widely available plants are considered while building the dataset. Each class contained an average of 1000 leaf samples.

Table 1 Datasets available for plant classification

Reference	Dataset name	Species	Size	Algorithm	Accuracy (%)
[18]	Western Ghat dataset	48	50,000	CNN	93
[19]	Medicinal leaf	40	1500	ANN	98.35
[20]	Ayurleaf	40	2400	AlexNet	98.46
[21]	Deepherb	40	2515	VGG16, VGG19	97.5
[22]	Indigenous	100	64,000	VGG16, VGG19	97.8 and 97.6
[23]	MedicinalPlant	10	1054	SVM	96.11
[24]	Medicinal plant	6	300	Multilayer perception	99.01
[25]	Ayur Bharat	10	10,000	ResNet10	96.53
[26]	Leaf	20	1000	YOLOV2	96

After acquisition through a digital camera, all images are saved in the size of 228 × 228 pixels. Proprocessing is done by applying Gaussian filter which considerably removes the noise present in the image. Each class is annotated with appropriate botanical annotation. Entire dataset is split in a ratio of 80:20:20 into training, test and validation samples. Count of individual samples for the 20 different plant species are given in Fig. 1.

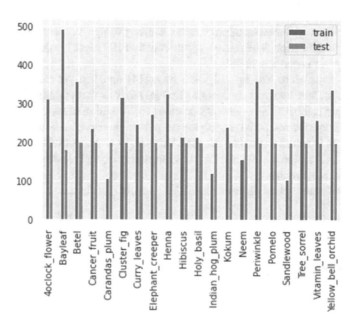

Fig. 1 Description of dataset with train and test split

3.2 Architecture

The basic idea is to classify the plant species without edge detection and segmentation. Hence, we have employed two types of machine learning techniques for classification. They are given as follows:

Support vector machines (SVMs) are basically a supervised learning algorithm. SVM without edge detection involves a kernel function mapping that trains a dataset to improve its resemblance towards a linearly separable dataset. The kernel function mapping increases the dimensionality of the data thereby improving the detection accuracy. The commonly used kernel functions are linear, quadratic, multilayer perceptron and polynomial kernel. SVM kernel generally takes low dimensional input space and converts to a high dimensional space function. Generating and selecting the right hyper plane becomes an important task in SVM. SVM implements a kernel trick where the low dimensional data are transformed to high dimensional data by adding more dimensions to it, thus improving the performance.

We have used radial basis function (RBF) kernel in SVM as given in Eq. 1.

$$K\left(x, x'\right) = e^{-\gamma|x-x'|^2} \tag{1}$$

Here, the gamma (γ) value indicates the influence of single training example over the classification and $\|x-x'\|^2$ is the Euclidean distance between two feature vectors. Gamma value is considered in a range of 10–1 to 10–5 and 'c' is penalty parameter of the error term chosen from 0.001 to 10,000. After fine-tuning the hyperparameters for precision and recall, it is observed that $\gamma = 0.01$ and $c = 1000$ are giving optimal results.

Another model used on the dataset is convolutional neural network. The CNN consists of multiple stack of layers to perform convolution, Max pooling, activation and fully connected network. Any input image in CNN sent to the convolution layers consists of filters, receptive field, stride, padding, pooling and ReLu stack. The receptive field in a CNN keeps track of any filter that responds to each pixel. It increases linearly when more stack layers are added to convolutional layers. The next activation layer clips of to zero when negative value exists. The activation function ReLu converges faster than the sigmoid activation function but it is saturated at the negative region too which makes the gradient zero.

We have used binary CNN (BCNN) where binary weights propagate both in forward and backward direction. With enabling few multipliers in CNN, a BCNN can be built with low computation time and complexity. The method to involve the BCNN in the CNN multi-layer is done by overlapping the mth weight filter to the array of the receptive field so as to enable the activation where the value is restricted to $\{-1, +1\}$. The BCNN basic architecture given in Fig. 2 contains the concatenation of the CNN layer with binary specifications.

Inception in CNN, a 48 layered network is designed for deep computations reducing the computational loss. It consists of fully connected layers with sparsely connected architecture. The inception module contains 1×1, 3×3 and 5×5

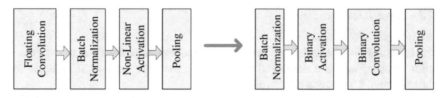

Fig. 2 Binary CNN architecture

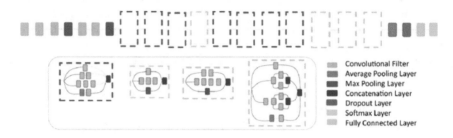

Fig. 3 Inception V3 model

convolutional layers where the output filter banks of CNN layer is concatenated to a single output vector thus incepting into the next stage for computation. The BCNN with inception V3 without pretraining is involved in the input. The proposed work architecture is described in Fig. 3. Fully connected layer block consists of two dense layers with 1024 nodes in each layer followed by a flatten and a dropout layer with 50% dropout and finally a softmax classifier of 20 classes. Other hyperparameters are listed as follows: SGD optimizer, categorical cross entropy error, 0.001 learning rate with 0.9 momentum.

4 Results

This part deals with the results through two different machine learning models and their classification report. SVM classifier had produced 79% accuracy on test dataset. The classification report indicating precision, recall and $f1$-score on test set is given in Fig. 4. After fine-tuning the hyperparameters for precision and recall, it is observed that $\gamma = 0.01$ and $c = 1000$ are giving optimal results.

On the other side, inception V3 had produced the training accuracy of 89.75% and validation accuracy of 96.79%. Training and test sets were taken in 80:20 proportion. Model was run for 15 epochs with batch size being 64. Model is neither underfit nor overfit as the same can be seen from Fig. 5. In predictive analysis, confusion matrix is given in Fig. 6 to show the misclassification of different classes.

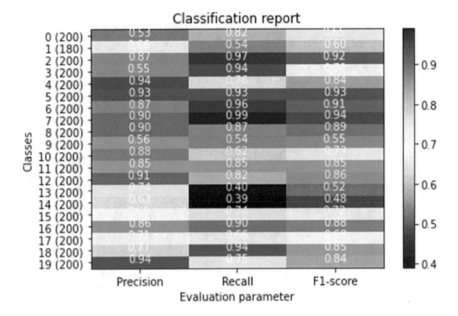

Fig. 4 Classification report using SVM

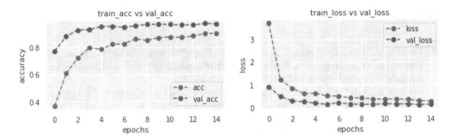

Fig. 5 Classification accuracy and error rate

5   Conclusion

To conserve plant species, their classification and recognition are necessary, an auto-mated system is required to identify plants using available information. This system can assist common people without much information regarding biology recognize them. This proposal aims at creating a database of different plant species of Karnataka region with a sample size of more than 20,000 images. Two machine learning models are proposed namely SVM and binary CNN-based inception V3 which gave a satis-factory accuracy of 79% and 89.7%, respectively, on 20 classes. It can be concluded that development of a bigger dataset and fine-tuning the hyperparameters further would definitely able to classify much more number of species in future.

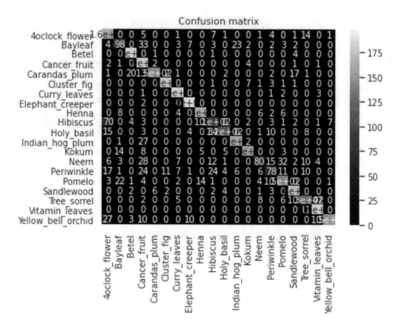

Fig. 6 Confusion matrix

References

1. Kataoka T, Kaneko T, Okamoto H, Hata S (2003) Crop growth estimation system using machine vision. In: Proceedings 2003 IEEE/ASME international conference on advanced intelligent mechatronics (AIM 2003), vol 2 1072, pp b1079–b1083
2. Ahonen T, Pietikäinen M (2007) Soft histograms for local binary patterns. In: Proceedings of the Finnish signal processing symposium, FINSIG, vol 5, pp 1–4
3. Mouine S, Yahiaoui I, Verroust-Blondet A (2013) Combining leaf salient points and leaf contour descriptions for plant species recognition. In: International conference image analysis and recognition, pp 205–214
4. Bai JY, Ren HE (2011) An algorithm of leaf image segmentation based on color features. Key Eng Mater 474:846–851
5. An N, Palmer CM, Baker RL, Markelz RC, Ta J, Covington MF, Maloof JN, Welch SM, Weinig C (2016) Plant high-throughput phenotyping using photogrammetry and imaging techniques to measure leaf length and rosette area. Comput Electron Agric 127:376–394
6. Chen Q, Zhao J, Cai J (2008) Identification of tea varieties using computer vision. Trans ASABE 51(2):623–628
7. Sun Y, Liu Y, Wang G, Zhang H (2017) Deep learning for plant identification in natural environment. Comput Intell Neurosci 2017
8. Wu SG, Bao FS, Xu EY, Wang Y-X, Chang Y-F, Xiang Q-L (2007) A leaf recognition algorithm for plant classification using probabilistic neural network. In: 2007 IEEE international symposium on signal processing and information technology, pp 11–16
9. Bakhshipour A, Jafari A (2018) Evaluation of support vector machine and artificial neural networks in weed detection using shape features
10. Rasti R, Rabbani H, Mehridehnavi A, Hajizadeh F (2017) Macular OCT classification using a multi-scale convolutional neural network ensemble. IEEE Trans Med Imaging 37(4):1024–1034

11. Petchsri S, Boonkerd T, Baum B, Karladee D, Suriyong S, Lungkaphin A (2012) Phenetic study of the *Microsorum punctatum* complex (*Polypodiaceae*). ScienceAsia 38(1):1–12

12. Tekkesinoglu S, Rahim MSM, Rehman A, Amin IM, Saba T (2014) Hevea leaves boundary identification based on morphological transformation and edge detection features. Res J Appl Sci Eng Technol 7(12):2447–2451

13. Mzoughi O, Yahiaoui I, Boujemaa N (2012) Petiole shape detection for advanced leaf identification. In: 2012 19th IEEE international conference on image processing, pp 1033–1036

14. Gill GS, Kumar A, Agarwal R (2013) Nondestructive grading of black tea based on physical parameters by texture analysis. Biosys Eng 116(2):198–204

15. Morris D (2018) A pyramid CNN for dense-leaves segmentation. In: 2018 15th conference on computer and robot vision (CRV), pp 238–245

16. Narayan V, Subbarayan G (2014) An optimal feature subset selection using GA for leaf classification. Ratio 1388:885.193

17. Mzoughi O, Yahiaoui I, Boujemaa N, Zagrouba E (2013) Advanced tree species identification using multiple leaf parts image queries. In: 2013 IEEE International conference on image processing, pp 3967–3971

18. Bhat S et al (2021) Classification of plant leaves of western Ghats using deep learning. In: 2021 IEEE international conference on distributed computing, VLSI, electrical circuits and robotics (DISCOVER). IEEE

19. Roopashree S, Anitha J (2020) Medicinal leaf dataset. Mendeley Data, V1. https://doi.org/10.17632/nnytj2v3n5.1

20. Dileep MR, Pournami PN (2019) Ayurleaf: a deep learning approach for classification of medicinal plants. In: TENCON 2019–2019 IEEE region 10 conference (TENCON). IEEE

21. Roopashree S, Anitha J (2021) DeepHerb: a vision based system for medicinal plants using Xception features. IEEE Access 9:135927–135941

22. Paulson A, Ravishankar S (2020) AI based indigenous medicinal plant identification. In: 2020 Advanced computing and communication technologies for high performance applications (ACCTHPA). IEEE

23. Habiba U et al (2019) Automatic medicinal plants classification using multi-channel modified local gradient pattern with SVM classifier. In: 2019 joint 8th international conference on informatics, electronics & vision (ICIEV) and 2019 3rd international conference on imaging, vision & pattern recognition (icIVPR). IEEE

24. Naeem S et al (2021) The classification of medicinal plant leaves based on multispectral and texture feature using machine learning approach. Agronomy 11(2):263

25. Sai Kumar TS, Prabalakshmi A (2021) Identification of Indian medicinal plants from leaves using transfer learning approach. In: 2021 5th international conference on trends in electronics and informatics (ICOEI). IEEE

26. Islam MK, Habiba SU, Masudul Ahsan SM (2019) Bangladeshi plant leaf classification and recognition using YOLO neural network. In: 2019 2nd international conference on innovation in engineering and technology (ICIET). IEEE

Implementation of All-Optical Logic Gates AND, OR, NOT, XOR Using SOA at 100 Gb/s

Sidharth Semwal, Nivedita Nair, and Sanmukh Kaur

Abstract As technology progressed, photonic wave processing and computation came which were low cost and robust. They provided a better alternative to design and combine with traditional optical network. These initial systems and architecture provided an interesting dimension in photonic wave processing and photonic computing. Here, in this work, we will be inspecting four basic logic gates, namely AND, XOR, OR, NOT based on semiconductor optical amplifiers (SOAs) devices at 100 Gb/s. Later, the assessment will be done by confirming the truth table. There are various contents which we use in designing like data signaling probes, couplers, various kinds photonic wave amplifier (e.g., SOA) and optical bandpass filter. As we advance, we can build numerous functioning schemes like a flip-flop, shift registers, and sequential circuits in the photonics aspect. All of these devices are functioning on the principle of cross-phase modulation which takes place when photonic wave field goes to the same optical medium as the change in refractive index due to one optical field causes phase modulation of the other optical fields.

Keywords Semiconductor optical amplifier · Optical bandpass filter · Cross-phase modulation

1 Introduction

There has been an extensive development in optical fiber connection since early 80 s. Systems were made and commercialized. Soon these devices which were at early stage started working at high bit rate with hundreds of varying wavelengths with

S. Semwal (✉) · N. Nair · S. Kaur
Electronics and Communication Engineering Department, Amity University Uttar Pradesh, Noida, India
e-mail: sidharth.semwal@gmail.com

N. Nair
e-mail: nnair@amity.edu

S. Kaur
e-mail: skaur2@amity.edu

© The Author(s), under exclusive license to Springer Nature Singapore Pte Ltd. 2023
A. Shukla et al. (eds.), *Computational Intelligence*, Lecture Notes in Electrical Engineering 968, https://doi.org/10.1007/978-981-19-7346-8_62

the help of time division multiplexing (TDM) systems. With the speed at which this technology is progressing and also with increasing demand of transmission systems and with ever increasing rate of information volume, a breakthrough is required in optical communication technology. In today's communication sphere with growing high-speed processing, different and varying computational functionalities, such as packet buffering, bit length conversions, header processing, switching, retiming, reshaping, pattern matching are required to overcome, speed electronics limitations [1–3].

Gates seem to be the area where solution can be achieved through various stages of solution. Due to rise in demand of data rates at a speed of terahertz electrons are replaced by the photons for photonic wave processing, gates are essential, then in case where there are various kinds photonic wave optical amplifiers (e.g., SOA) have the capability to combine with considerable number of passive and active components high operating speed, high stability, and low power consumption, the components which help the data probe with other waves to merge one of these component is Mach–Zehnder interferometers [4–6].

The functioning of the MZI is that there are coupling points in the structure the first part of it enables the coupling of light signals operating through the first two arms; these are known as the inner mode and the outer mode, and the other does exactly the opposite function as the first one [1, 2].

Another component is the optical amplifier; this component amplifies an optical signal which is coming through input the functioning of this is through the principle of stimulated emission which is the basic principle for LASERs without feedback. It basically increments the length; bits can be transmitted over an optical media. A booster amplifier is used at the opening of the data probe in the optical link to increase the amplified output. So, bits go to the architecture of the device; the optical signals get attenuated so to bring them back to its original power a regenerator is used. Among the many of such optical amplifier, one is the photonic wave optical amplifier (e.g., SOA); it has the same operating principle of optical amplifier that is through stimulated emission where an incident photon induces it. A light travel in a medium in which electron lose energy, when coming back to the original state, the electron emits photons [2, 3].

Here, we will try to stimulate the schematic design of the all-optical logic gates. All gates' schematics are stimulated on the OptiSystem 8 software. A general optical gate has two inputs one of which belongs to the data A that is coming from the primary line and the other from the second data signal probe which gives data B. Many gates have the same lines and a single output. The corresponding outputs can be compared with the inputs with gate table [7–9].

In recent research, an all-optical wavelength convertor has been implemented using various kinds of photonic wave optical amplifier (e.g., SOA) which were supposed to be cascaded and implemented. This paper talks about a counter-propagating topology which is supposed to be used in wavelength converting operation and obtained an inverted and non-inverted signal simultaneously. A feasibility study is also demonstrated over varying wavelength there are various methods of wavelength conversion such as cross-phase modulation (XPM), cross-polarization modulation (XPolM), and (XGM). Through XPoIM and interferometric wavelength converters, different wavelength are achieved [10–13].

In one of the recent research, an all-optical comparator is implemented using various kinds of photonic wave optical amplifier (e.g., SOA) which are using the nonlinear properties. This paper talks about a two-bit comparator is proposed which is able to process the return-to-zero modulated input signals at 100 Gb/s; the simple architecture makes it suitable for signal processing. due to increasing demand of data communication and ultrawide-band information processing an integrated all-optical logic comparator is indispensable optical computing systems; it is used in determining if numbers are equal or not, it is achieved by an XOR gate analyzed with a changing optical-loop mirror; also, it is implemented in by cascading a unique basic gate with XGM and XPM in a SOA [1–3].

In one of the recent research, an all-optical integrated decoder and demultiplexer architecture is made using SOA based MZI. This paper talks about a 2-to-4 decoder and a 1-to-4 demultiplexer which is integrated as one through an enable switch like signal as the output given a clear eye pattern can be seen after the simulation in OptiSystem software. We have seen that it is very easy to integrate this circuit with both varying devices. It also have a simple design structure which is desirable for processing. A decoder is a combination circuit which contains a multiple inputs and simultaneously many outputs. A demultiplexer is a combination circuit which selects output as per the select lines [4].

In one of the recent papers published, an all-optical two-bit comparator is proposed which is based on photonic wave amplifier and Mach–Zender interferometer which is showing a long-term solution with its robust structure and capability to photonic computing and signal processing. All-optical logic comparator is indispensable optical computing systems; it is used in determining if numbers are equal or not it is achieved by an XOR gate analyzed with a nonlinear optical-loop mirror; also, it is implemented in by cascading a unique basic gate with XGM and XPM in a SOA [5–9].

In one of the recent papers published, a photonic full-adder and full-subtractor which is based on various kinds of photonic wave optical amplifier (e.g., SOA) and a few all-optical logic gates. The optical amplifier, this component amplifies an optical signal which is coming through input the functioning of this is through the principle of stimulated emission which is the basic principle for LASERs without feedback. It basically increments the length over an optical media. A booster amplifier has been used at the opening of the data probe in the optical link to increase the amplified output. Working principle of optical amplifier that is through stimulated emission here an incident photon induces it. A light travels to the varying medium in which electron loses energy, when coming back to the original state the electron emits photons. A booster amplifier is also used at the opening of the data probe in the optical link to increase the amplified output. We go through the architecture of the device; the optical signals get attenuated, so to bring them back to its original power, a regenerator is used [13–16].

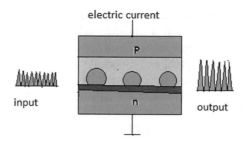

Fig. 1 Optical amplifier

2 Proposed Schematic

We are pursuing in making the schematics for photonic gates, namely AND, OR, NOR, XOR, and NOT. The functioning of phenomena, that are seen, explained here.

2.1 Optical Amplifiers

This component amplifies an optical signal which is coming through input the functioning of this is through the principle of stimulated emission which is the basic principle for LASERs without feedback. It basically increments the length of bits that are being transmitted over an optical media. A booster amplifier has been used at the opening of the data probe in the optical link to increase the amplified output [2], so bits go through the architecture of the device the optical signals get attenuated, so to bring them back to its original power, a regenerator is used. Among the many of such optical amplifier, one is photonic wave light-based amplifier (e.g., SOA); it has the same working principle of optical amplifier that is through stimulated emission here an incident photon induces it. A light goes in a medium in which electron loses energy, when coming back to the original state the electron emits photons [3] (Fig. 1).

2.2 Cross-Phase Modulation (XPM)

The principle which has been utilized cross-phase modulation which takes place, in an optical fields, which are through the same light medium as the change in refractive index due to one optical field causes phase modulation of the other optical fields; this type of modulation technique has been used for combining information to a light stream by changing phase of a coherent optical beam with the assistance of another beam when there is a nonlinear medium. This type of modulation is used in many areas like ultrafast optical switching, wavelength conversion, and pulse compression [2].

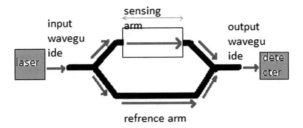

Fig. 2 Typical Mach–Zehnder interferometer

2.3 Mach–Zehnder Interferometers (MZIs)

The functioning of the MZI is there are coupling points in the structure; the first part of it enables the coupling of light signals operating through the first of two arms; these are known as the inner mode and the outer mode, and the one does exactly the opposite function as the first one. It is used in the fields of aerodynamics, to measure pressure, density, and temperature changes in gases [2] (Fig. 2).

Those logic gates which are designed and have a unique architecture are NOT, OR, AND, XOR; many gates are designed on these basic gates usually combination of two or more gates give a new gates that is NAND, NOR, and XNOR. These gates have a working principle which is based on Boolean identity.

2.4 NOT Gate

The main operation of a NOT gate is to invert the input and give it as output when a bit sequence is send as optical data signal. The coupler couples the bits with the CW laser. Then, the combine signal is send to the photonic wave optical amplifier (e.g., SOA); the output is extracted by using a bandpass filter (BPF). The input in data probe, wavelength which is chosen, is round 1550 nm with power of around 1 mW. The input bit sequence is 101010101. These data bits pass through both SOAs as injection current and are out through bandpass filter which is set with the wavelength of around 1500 nm and bandwidth of 0.4 nm (Fig. 3).

Truth Table

Signal A	Output
1	0
0	1

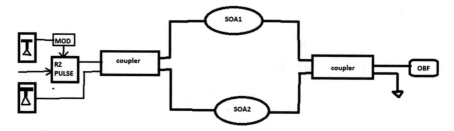

Fig. 3 Typical NOT gate

2.5 OR Gate

The operation of an OR gate is like the NOT gate, but the difference is in the input, instead of one, three input streams through which a bit sequence is send. The input signal which has been going inside semiconductor optical amplifier, it is actually very small value which is called an injection current; the output is extracted by using a bandpass filter (BPF). The input in data probe, wavelength which is chosen, is around 1550 nm with power of around 1 mW. The input bit sequence is 101010101. These data bits pass through single SOAs as injection current and give output through bandpass filter which is set with the wavelength of around 1500 nm and bandwidth of 0.4 nm (Fig. 4).

Truth table

Signal A	Signal B	Output
1	1	1

(continued)

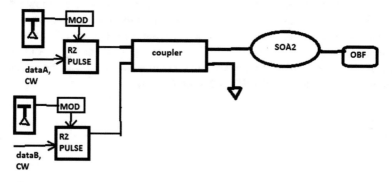

Fig. 4 Typical OR gate

(continued)

Signal A	Signal B	Output
1	0	1
0	1	1
0	0	0

2.6 AND Gate

The operation of an AND gate was two input streams through which a bit sequence is send. The input signal which has been going inside semiconductor optical amplifier, it is actually very small value which is called an injection current; the output is extracted by using a bandpass filter (BPF). The input in data probe, wavelength which is chosen, is round 1550 nm with power of around 1 mW. The input bit sequence is 101010101. These data bits pass through both SOAs as injection current and give output through bandpass filter which is set with the wavelength of around 1500 nm and bandwidth of 0 (Fig. 5).

Truth table

Signal A	Signal B	Output
1	1	1
1	0	0
0	1	0
0	0	0

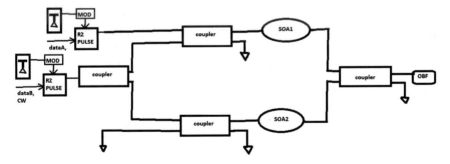

Fig. 5 Typical AND gate

2.7 XOR Gate

The operation of an XOR gate is unique from other gates; a single probe is used for both data inputs which has been going inside semiconductor optical amplifier; it is actually very small value which is called an injection current which is enough to consider all the bit sequence, and later, output is extracted by using a bandpass filter (BPF). The input in data probe, wavelength which is chosen, is round 1550 nm with power of around 1 mW. The input bit sequence is 101010101. These data bits pass through both SOAs as of injection current and give output through bandpass filter which is set with the wavelength of around 1500 nm and bandwidth of 0 (Fig. 6).

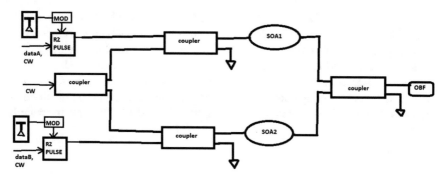

Fig. 6 Typical XOR gate

Truth table

Signal A	Signal B	Output
1	1	0
1	0	1
0	1	1
0	0	0

3 Simulation Results

When we have the logic verified that is when we give a bit sequence to the probe for which SOA, optical BPF, CW laser and optical pulse generator, parameters are varied accordingly and viewed through the visualizer that is when we confirm the output. The design shown above is stimulated in OptiSystem software. The output shown from the stimulation has been shown.

3.1 NOT Gate

Initially, the input in data probe, wavelength which is chosen, is round 1550 nm with power of around 1 mW. The input bit sequence is 101010101. These data bits pass through both SOAs as injection current and give output through bandpass filter at around 1500 nm and bandwidth of 0.4 nm. If we compare this with the gate table, we see when the value of signal that is signal A which can be eighter 1 or 0; then, in the arms which go through the couplers are going a phase shift of 180°, this is called XPM effect. SOAs are identical and recombine the signal through interference and give light indicating that output value is '1'. Hence, this is the way the output for NOT gate is obtained.

3.2 OR Gate

The input in data probe, wavelength which is chosen, is round 1550 nm with power of around 1 mW. The input bit sequence is 101010101. These data bits pass through single SOAs as injection current and gives output through bandpass at around 1500 nm and bandwidth of 0.4 nm. If we compare this with the gate table, we see when the value of both signals that is signal A and signal B is same which can be eighter 1 or 0; then, in the arms which go through the couplers which has been going a phase shift of 180°, this is called XPM effect. Lines are connected to the SOAs. which are identical and hence give no light that is the output value is zero.

3.3 AND Gate

The input in data probe, wavelength which is chosen, is round 1550 nm with power of around 1 mW. The input bit sequence is 101010101. These data bits pass through both SOAs as injection current and supply output through bandpass filter at around 1500 nm and bandwidth of 0.4. If we compare this with the gate table, we see when the value of both signals that is signal A and signal B which can be eighter 1 or

0; here, it is to know that signal *B* contains the output; then, in the arms which go through the couplers going a phase shift of 180°, this is called XPM effect. These arms are then connected to the SOAs which are identical. If both signal *A* and signal *B* are 1, i.e., output is one; if signal *A* is zero, then output will be zero since there will be no input. Hence, this is the way the output for AND gate is obtained.

3.4 XOR Gate

The input in data probe, wavelength is chosen, is round 1550 nm with power of around 1 mW; the input bit sequence is 101010101. A CW laser probe which is connected to the middle coupler, this gives the same wavelength of 1550 nm and power of 1 mV; then, these data bits pass through both SOAs as injection current and give output through bandpass filter which is set at around 1500 nm and bandwidth of 0.4 nm. If we compare this with the gate table, we see when the value of both signals that are signal *A* and signal *B* is same which can be eighter 1 or 0; then, in the arms which go through the couplers going a phase shift of 180°, this is called XPM effect. They are connected to the SOAs which are identical and hence give no light that is the output value is zero (Figs. 7, 8, 9 and 10).

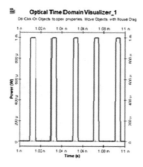 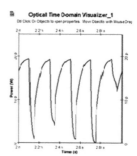

Fig. 7 Stimulated result of NOT

Fig. 8 Stimulated result of OR

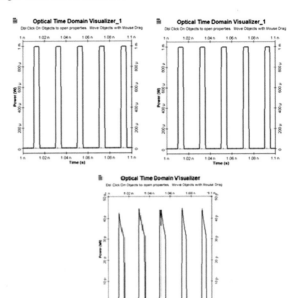

Fig. 9 Stimulated result of AND

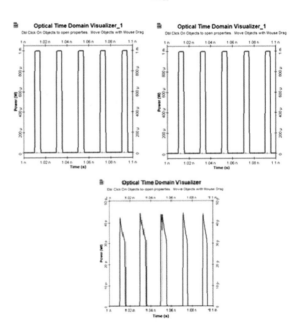

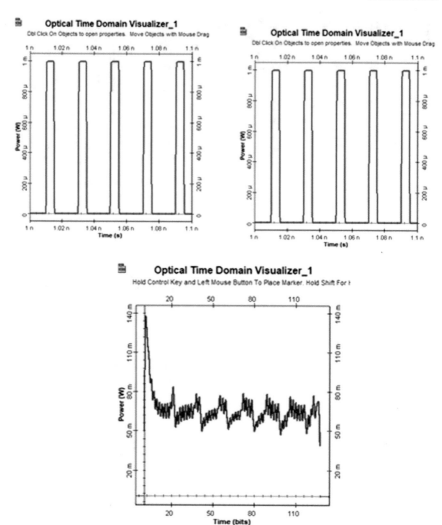

Fig. 10 Stimulated result of XOR

References

1. Wada O (2004) Femtosecond all-optical devices for ultrafast communication and signal processing. New J Phys 6:183. Received 12 July 2004. https://doi.org/10.1088/1367-2630/6/1/183.
2. Nahata PK, Ahmed A, Yadav S, Nair N, Kaur S (2020) All optical full-adder and full-subtractor using semiconductor optical amplifiers and all-optical logic gates. In: 2020 7th international conference on signal processing and integrated networks (SPIN)
3. Hamie A, Sharaiha A, Guegan M, Le Bihan J (2005) All-optical inverted and noninverted wavelength conversion using two-cascaded semiconductor optical amplifiers. IEEE Photonic Technol Lett 17(6):1229

4. Nair N, Kaur S, Singh H (2020) All-optical integrated 2-to-4 decoder and 1-to-4 demultiplexer circuit with enable using SOA based MZI
5. Nair N, Kaur S, Goyal R (2018) All-optical two-bit comparator using SOA based Mach–Zehnder interferometers. In: 2018 2nd international conference on micro-electronics and telecommunication engineering
6. Chen X, Huo L, Zhuang L, Zhao Z, Lou C (2021) 100 Gb/s all-optical multifunctional AND, XOR, NOR, OR, XNOR, and NAND logic gates in a single compact scheme based on semiconductor optical amplifiers. Guo Photonics Laboratory, Changchun Institute of Optics, Fine Mechanics, and Physics, Chinese Academy of Sciences, Changchun 130033, China
7. Muralikrishna K, Bakyalakshmi V, Palanivelan M, Sathya R (2020) Design and implementation of various optical logic gates using semiconductor optical amplifier-Mach Zehnder interferometer scheme and band pass filters. Int J Adv Sci Technol
8. Lovkesh, Marwaha A (2016) Reconfiguration of optical logics gates at 160 Gb/s based on SOA-MZI. Springer Science + Business Media New York
9. Mukherjee K, Ghosh P (2012) Alternative method of implementation of frequency encoded N bit comparator exploiting four wave mixing in semiconductor optical amplifiers. Optik 123:2276–2280
10. Bordoloi K, Theresal T, Prince S (2014) Design of all optical reversible logic gates. In: International conference on communication and signal processing. India, pp 1583–1588
11. Sribhashyam S, Ramachandran M, Prince S, Ravi BR (2015) Design of full adder and subtractor based on MZI—SOA. In: IEEE international conference on signal processing and communication engineering systems. Guntur, India, pp 19–21
12. Katti R, Prince S (2015) Implementation of a reversible all optical multiplexer using Mach–Zehnder interferometer. In: IEEE international conference on signal processing, informatics, communication and energy systems (SPICES). Kozhikode, India, pp 1–4
13. Naga Maruthi K, Manohari Ramchandran R, Prince S (2016) Design of all optical JK flip flop. In: International conference on communication and signal processing, April 6–8, 2016. India, pp 123–127
14. Ramachandran M, Prince S, Verma D (2018) Design and performance analysis of all-optical cascaded adder using SOA-based MZI. J Comput Electron 17(5):845–856
15. Swetha K, Manohari R, Prince S (2018) SOA parameters optimization for high data rate operation. In: Gnanagurunathan G et al (eds) Optical and microwave technologies. Lecture notes in electrical engineering, vol 468, pp 247–253
16. Ramachandran M, Kambham SPA, Maruthi N (2020) Design and simulation of all optical shift registers using D flip-flop. Microw Opt Technol Lett 62(7):1–12

An Efficient Hybrid Approach for Malware Detection Using Frequent Opcodes and API Call Sequences

Om Prakash Samantray ⓘ and Satya Narayan Tripathy

Abstract Malicious software attacks are increasing every day despite so many preventive measures, and many detection mechanisms are available in the literature. Most of the detection mechanisms use either static or dynamic attributes of the malicious and legitimate samples with machine learning classification methods to distinguish malware from benignware. In this article, the static and dynamic features are joined to prepare a hybrid feature set which is used with machine learning algorithms for classification. The operation code sequences of samples are extracted through static analysis, and API call sequences are extracted through dynamic analysis. Both the feature vectors are joined to form a hybrid feature set which is then passed through three machine learning algorithms for experimental evaluation. Hybrid feature set has achieved higher accuracy and low error rate in comparison with the static and dynamic datasets when used individually with all the selected algorithms.

Keywords Malware · Malware analysis · Malware detection · Opcode · API call sequence · Hybrid feature · Machine learning

1 Introduction

Malware is the abbreviation for malicious software of different types. The generic term malware can be synonymously used for virus, worm, Trojan, rootkit, botnet, adware, ransomware, and so on. These malicious programs are written intentionally to harm the integrity and security of computer systems. They get into the systems without the user's consent to execute unfamiliar deeds in the system. The number of malware attacks has surged enormously in past few years because of extensive use of Internet and computing devices in our daily life which deals with huge volumes of data. The data may comprise some sensitive information which should be kept

O. P. Samantray (✉)
Department of CSE, Raghu Institute of Technology, Visakhapatnam, Andhra Pradesh, India
e-mail: ompakash_cse@raghuinstech.com

S. N. Tripathy
Department of Computer Science, Berhampur University, Berhampur, Odisha, India

confidential in order to avoid unnecessary malware attacks on the systems in turn organizations. There are so many preventive measures adopted by the organizations, but still, the number of attacks is increasing. A recent study by AV-Test security organization [1] reveals that more than 3.5 lakhs of different malicious programs are being produced every day. Similar facts are also given by so many other security organizations which motivates the researchers toward malware research domain.

Traditional malware detection models use static features like printable string information (PSI), operation codes (Opcodes) of the samples as the signature for malware classification. Some of the models also use dynamic features like application programming interface (API) call sequences as the signature. Both these method-ologies have a set of merits and disadvantages. They may detect known malware efficiently but may not detect unknown and unseen malware variants. In this work, the static feature (opcode sequence) and the dynamic feature (API call sequence) are combined to form a hybrid dataset which can achieve the benefits of static and dynamic analysis approaches together. An opcode is the part of an instruction which denotes the operation to be carried out. Usually, instructions of a process have two parts such as operation code and one or more operands. Programs use API calls to interact with operating system and cooperating processes to perform the intended task. API calls and their sequences are used to understand how the program is going to behave in the system.

Contributions of this work are

1. Collecting recent malicious and legitimate samples for analysis.
2. Collecting opcodes of the samples through static analysis and create a static feature vector for the frequent opcodes.
3. Collecting API calls 4-g (sequences) of the samples through dynamic analysis and create a dynamic feature vector.
4. Merge the static and dynamic feature vectors to form a hybrid feature set.
5. Apply machine learning algorithms to perform feature selection and classifica-tion. Performance evaluation of the model using different ML algorithms for all the three types of feature vectors.

The next segment contains a study of some previous works based on opcode, API call, and combination of different features. Section 3 talks about the proposed system architecture. Section 4 presents details of algorithms used in the experiment and results obtained from the experiment. Section 5 concludes the article.

2 Related Work

This part of the article presents some of the previous works based on malware detection using opcode feature, API call feature, and different hybrid features.

Igor et al. [2] extracted opcode sequences using static analysis in the similar sequence they appear in the sample executable file. They tested the performance of their model using some machine learning classification algorithms. Ding et al. [3]

proposed a very identical system which takes opcode sequences as the base feature. They used a control flow-based scheme to collect opcode behaviors of the sample files. Then, they performed a comparative study of flow-based analysis with text-based opcode analysis to check the efficiency of the control flow-based approach. From the experiment, it is found that, control flow-based method is more accurate than text-based malware classification.

O'kane et al. [4] made use of histograms of opcode density obtained from dynamic analysis of the executable. SVM algorithm is used in their experiment to prove that dynamic analysis leads to more accurate and light-weight malware classification as compared to static analysis-based malware classification. Zhang et al. [5] converted the opcodes into images to perform visualized analysis for malware classification. CNN algorithm was used to assess the opcode images extracted from binary targets. Experiments stated that, accuracy of visualized analysis is 15% more than the traditional approach. An identical strategy was projected by Wang et al. [6] to convert the opcodes into images and then improve the images through histogram normalization, dilation, erosion, and principal component analysis. They found that SVM has a better accuracy rate as compared to KNN for smaller training datasets.

Alqurashi et al. [7] used API call sequences and opcode sequences as features for the ML-based malware classification. They used hidden Markov model as a classification algorithm to test and validate their proposed model and achieved satisfactory results in malware classification in case of both features considered. Ki et al. [8] used DNA sequence alignment technique to collect function call sequences from different families of malicious executable. They verified their proposed model through ML algorithms and attained good detection rate with lesser error rate. Fan et al. [9] collected API calls through hooking method and passed the features through machine learning models. They found decision tree as the best algorithm with accuracy 95% for their classification model. Liu et al. [10] used API call sequences from social networks to build the phylogenetic networks. Their method reveals the inner connection among malware families. Ma et al. [11] collected API fragments from the API execution sequence and then applied LSTM and ensemble machine learning models to achieve optimal performance in malware detection. They found CNN model as the best model with an accuracy score of 93.25%.

Amer et al. [12] used cluster of similar API functions for their experiment. They exercised Markov chain procedure to generate a transition matrix to understand relationship between API functions. Their model claimed to achieve 99.7% accuracy score with 0.010 false-positive rate value. Alazab et al. [13] proposed a model which combines permission requests and API calls used in android applications. Then, machine learning algorithms are applied on these datasets comprising 27,891 android samples. Authors claimed an F-measure of 94.3% for the proposed model.

Shijo et al. [14] suggested a cross-model using the mix of binary code and dynamic behavior of samples. The integrated approach has achieved about 98.7% accuracy with the SVM algorithm. Santos et al. [15] suggested a hybrid approach which merged the opcode frequency and execution trace information of the executable. The hybrid feature set was passed through four different classification algorithms. The SVM algorithm with normalized polynomial kernel proved to be the best algorithm with an accuracy score of 96.6% with the hybrid dataset. Su et al. [16] used the static features such as permissions, native-permissions, function and priority of the applications, and some dynamic behaviors. The hybrid dataset is passed through machine learning algorithms to get a highest accuracy value of 97.4%.

Wang et al. [17] applied static and dynamic methods to extract features available in the program codes and used correlation-based ML algorithm to select highly correlated features for the experiment. They applied ML classification algorithms on 4600 samples and accomplished an accuracy score of 96.52% with a false-positive rate of 2.38%. Maryam et al. [18] proposed a hybrid method for android malware detection. Experiment was conducted using more than 5000 android applications with a classifier named as cHybridroid. The proposed model performed well with 97% F-score value. Li et al. [19] presented a detection model called Hyda using hybrid features of malicious and authentic samples. The proposed model Hyda improved the performance as well as reduced detection time.

In this article, opcodes frequency is considered as the static feature, and API call sequences are considered as the dynamic feature. The combination of these features is then passed through the ML algorithms for classification.

3   System Outline

The malware samples can be gathered from several Web resources like VirusTotal, VirusShare, and so on. The benign samples can be collected from newly installed operating system files. Usually, EXE files or DLL files are considered as samples. The samples can be confirmed and certified by utilizing the online service Web site VirusTotal.

This proposed system uses both the analysis methods such as static analysis and dynamic analysis to collect the features for the experiment. The system outline is depicted in Fig. 1.

The code analysis process uses IDAPro tool to fetch the opcodes from the samples. The frequent opcode sequences are collected using the operation code extraction and count (OPEC) algorithm [20] and represented as a vector. On the contrary, the run-time analysis process uses Cuckoo Sandbox environment to extract the API calls of the collected samples. The n-grams ($n = 4$) or 4-g of API calls are generated to prepare the dataset. The value of n is decided based on some of the previous studies which show that 4-g are better than 1, 2, and 3-g of API calls used for malware detection. Then, both the opcode dataset and API call sequence dataset are merged to form a hybrid dataset. The hybrid dataset is then passed through different ML algorithms like decision tree (DT), support vector machine (SVM), and random forest (RF) for classification.

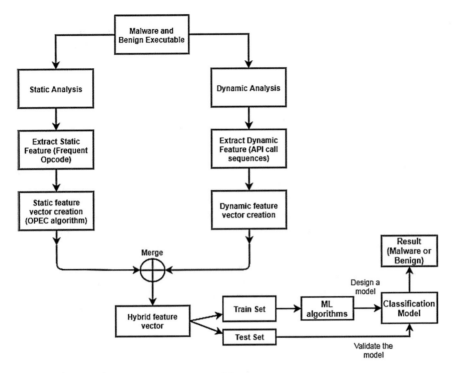

Fig. 1 Proposed hybrid malware detection model

4 Experimental Evaluation

This experiment includes 1738 malicious files and 1570 benign files in the dataset. These samples are gathered from a variety of online and offline sources. These samples are passed through the IDAPro tool which generates the assembly language format (.asm extension) of the input files. The .asm file contains all the opcodes found in the malware sample. This .asm file can be passed through the OPEC algorithm to generate a static vector containing occurrences of the most frequent opcodes in all the input samples. A portion of the static feature vector resulted from the OPEC algorithm is depicted in Fig. 2.

	mov	push	call	pop	cmp	jz	lea	test	jmp	add	...	fstcw	int	nop	pushf	rdtsc	sbb	setb	setle	shld	labels
0	3362	491	840	495	581	0	317	548	482	3477	...	0	85	340	17	0	150	1	0	0	0
1	74728	31431	14363	7866	9853	0	14338	7709	8331	45814	...	0	5453	3404	52	0	782	127	11	3	1
2	9782	12950	3422	6773	5324	0	3121	2125	1163	79531	...	0	339	1580	67	0	931	22	1	0	1
3	8224	1444	1757	1512	988	0	534	922	1203	8947	...	0	294	865	17	0	279	9	0	0	0
4	8510	2047	1241	985	1720	0	1727	505	618	2186	...	0	349	290	1	0	24	30	1	2	1

Fig. 2 Static feature vector containing frequent opcodes

File	4-gram1	4-gram2	4-gram3	...	4-gram872	Class
Malware1	1	1	0		1	Malware
Malware2	1	0	1		0	Malware
Benign1	1	1	1		1	Benign
Benign2	0	1	1		0	Benign

Fig. 3 API call 4-g binary feature vector

On the other side, the same set of samples is analyzed dynamically using Cuckoo Sandbox environment, and the API call log files are collected. The 4-g API call sequences are generated for all the samples and represented as a feature vector. The feature vector contains a collection of binary 1's and 0's to represent the presence and absence of API call grams in each sample. The process of preparing the API call 4-g feature vector [14] is given in algorithm 1. As a result of Algorithm 1, the total number of 4-g generated is equal to 872. Consequently, the dynamic feature set contains 874 columns to describe the sample name in the first column, class of the sample in the last column, and rest of the columns describe API call grams of length four. The API call 4-g binary feature vector look like as presented in Fig. 3.

Algorithm 1

API call 4-g feature vector generation algorithm

Input: Sample dataset (D)
Output: Dynamic feature vector

---

1. Start
2. repeat for each f_i in D
2.1. Generate log file and extract 4-g of API calls
2.2. repeat for each 4-g
2.3. calculate freq (4-g)
2.4. if (freq (4-g) > threshold) then
2.5. Add the n-gram to the API call feature list
2.6. end of loop
3. end of loop
4. Create a binary feature vector with an attribute 4-API-call-grams;
5. repeat for each f_i in D
5.1. repeat for each 4-g in feature list
5.2. if (API-call-gram is available in the table associated with f_i) then,
5.3. Set binary one for the feature in the vector
5.4. Else
5.5. Set binary zero for the feature in the vector
5.6. end of loop
6. end of loop
7. stop

These two datasets are merged together using Python programming language. After merging these two feature sets, the final dataset contains 3308 number of rows to represent 1738 malicious samples and 1570 benign samples. The total number of columns is 974 to represent name of the sample in the first column, opcode frequencies in the next 100 columns, API call 4-g in next 872 columns and class of the sample in the last column.

The experiment is conducted using Python language to implement machine learning-based classification algorithms. The three classification algorithms such as RF, SVM, and DT are selected for this experiment. These algorithms are selected based on their outstanding performance in the previous works for the individual datasets. The performance metrics considered are true-positive rate (TPR), false-positive rate (FPR), and accuracy score. The dataset has sent through a tenfold cross-validation method to get more accurate results. The mean values of the observed results for the evaluation metrics for all the three algorithms with the three datasets are shown in Table 1.

From Table 1, it is observed that, SVM algorithm has achieved accuracy scores 96.19%, 97.39%, and 98.21% for static, dynamic, and hybrid datasets, respectively. These values are better as compared to the other two algorithms implemented on the datasets. The accuracy score of SVM for the hybrid dataset is more than the opcode and API 4-g datasets when used individually. In case of RF and DT also, the hybrid dataset performs better as compared to individual static and dynamic datasets which is presented in Fig. 4.

Table 1 Experimental results for the three algorithms for the three datasets

Dataset	RF			SVM			DT		
	TPR	FPR	ACC (%)	TPR	FPR	ACC (%)	TPR	FPR	ACC (%)
Static	0.942	0.152	94.24	0.962	0.087	96.19	0.911	0.312	91.33
Dynamic	0.957	0.110	95.76	0.974	0.099	97.39	0.929	0.21	93.66
Hybrid	0.969	0.059	96.99	0.982	0.039	98.21	0.936	0.062	94.89

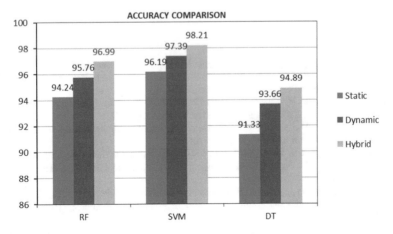

Fig. 4 Accuracy comparison of the algorithms for the three datasets

Therefore, we can state that, hybrid feature-based detection system is more accurate for malware detection as compared to either static or dynamic feature when used independently. This model performs better than many of the works studied in the related work section. Though this model gives better detection accuracy, it takes little extra time for training and testing the classification model. This little extra time can be ignored looking at the better accuracy provided by this model.

5   Conclusion

In this article, the static opcode features and dynamic API call sequences are combined together for a better representation of the malicious and legitimate samples. The purpose of preparing this hybrid dataset is to improve the accuracy of malware classification. The OPEC algorithm and API call 4-g feature vector generation algorithm are used to prepare the individual datasets for the same set of samples. Then, these static and dynamic datasets are combined to form the hybrid dataset. The hybrid dataset is then passed through the three ML algorithms such as RF, SVM, and DT using Python programming language. A tenfold cross-validation method is used to ensure accurate results of the model. From the experimental results, it is observed that, SVM performs well with the individual static and dynamic datasets and the hybrid dataset as well. It is also evident that, the hybrid feature can represent a sample more accurately as compared to either static or dynamic features. The hybrid feature set has secured better accuracy score as compared to other feature sets for all the selected algorithms. More recent samples are collected for this experiment to check the efficiency of the model with respect to known and unknown malware detection.

References

1. AV-TEST Page. https://www.av-test.org/en/statistics/malware/. Last accessed 24 Dec 2021
2. Igor S, Felix B, Xabier UP, Pablo GB (2013) Opcode sequences as representation of executables for data-mining-based unknown malware detection. Inf Sci 231:64–82
3. Ding Y, Dai W, Yan S, Zhang Y (2014) Control flow-based opcode behavior analysis for malware detection. In: Computers & security. Elsevier, pp 1–10
4. O'kane P, Sezer S, McLaughlin K (2016) Detecting obfuscated malware using reduced opcode set and optimised runtime trace. Secur Inform 5:2
5. Zhang J, Qin Z, Yin H, Ou L, Hu Y (2016) IRMD: malware variant detection using opcode image recognition. In: Proceedings of 22nd international conference on parallel and distributed systems. IEEE, pp 1175–1180
6. Wang T, Xu N (2017) Malware variants detection based on opcode image recognition in small training. In: 2nd international conference proceedings on cloud computing and big data analysis. IEEE, pp 328–332
7. Alqurashi S et al (2017) A comparison between API call sequences and opcode sequences as reflectors of malware behavior. In: Proceedings of the 12th international conference for internet technology and secured transactions. IEEE, pp 105–110

8. Ki Y, Kim E, Kim HK (2015) A novel approach to detect malware based on API call sequence analysis. Int J Distrib Sens Netw 2015(6):1–9

9. Fan CI, Hsiao HW, Chou CH, Tseng YF (2015) Malware detection systems based on API log data mining. In: Proceedings of international computer software and applications conference. IEEE, pp 255–260

10. Liu J, Wang Y, Wang Y (2017) Inferring phylogenetic networks of malware families from API sequences. In: Proceedings of international conference on cyber-enabled distributed computing and knowledge discovery. IEEE, pp 14–17

11. Ma X, Guo S, Bai W, Chen J, Xia S, Pan Z (2019) An API semantics-aware malware detection method based on deep learning. Secur Commun Netw 2019

12. Amer E, Zelinka I (2020) A dynamic windows malware detection and prediction method based on contextual understanding of API call sequence. Comput Secur 92

13. Alazab M, Alazab M, Shalaginov A, Mesleh A, Awajan A (2020) Intelligent mobile malware detection using permission requests and API calls. Futur Gener Comput Syst 107:509–521

14. Shijo PV, Salim A (2015) Integrated static and dynamic analysis for malware detection. In: Proceedings of the international conference on information and communication technologies, vol 46. Elsevier, Kochi, pp 804–811

15. Santos I, Devesa J, Brezo F, Nieves J, Bringas PG (2013) OPEM: a static-dynamic approach for machine-learning-based malware detection. In: Herrero Á et al (eds) International joint conference CISIS'12-ICEUTE'12-SOCO'12 special sessions. Advances in intelligent systems and computing, vol 189. Springer, Berlin, Heidelberg

16. Su M, Chang J, Fung K (2017) Machine learning on merging static and dynamic features to identify malicious mobile apps. In: Proceedings of ninth international conference on ubiquitous and future networks. IEEE, pp 863–867

17. Wang Y, Cai W, Lyu P, Shao W (2018) A combined static and dynamic analysis approach to detect malicious browser extensions. Secur Commun Netw

18. Maryam A, Ahmed U, Aleem M, Lin JC-W, Islam MA, Iqbal MA (2020) cHybriDroid: a machine learning-based hybrid technique for securing the edge computing. Secur Commun Netw

19. Li Z, Li W, Lin F et al (2020) Hybrid malware detection approach with feedback-directed machine learning. Sci China Inf Sci 63

20. Samantray OP, Tripathy SN (2021) An opcode-based malware detection model using supervised learning algorithms. Int J Inf Secur Privacy (IJISP) 15(4):18–30

Exploring the Emotion Recognition in Speech Using Machine Learning

Akshay Kumar, Aditya Chandrayan, and Sanjay Kumar Dubey

Abstract There is an emotion related to everything a person does, and recognizing is a much complex task. These emotions perform very important role in understanding how a person feels about something. In this world, where humans are fed what others want through speech, then analyzing these different emotions is of major importance. In this paper, we will try to combine different dataset and formulate a base methodology. Different people have used techniques ranging from CNN, SVM to Fourier parameter. Language and geography also matter. Emotions and reaction pitches and energies are different for people of different culture, nation, and perspective. 1D CNN model is studied and concluded in present study. The different dataset is used by various researchers, and such dataset plays an important contribution in the results. Datasets which have voice with loudspeakers along with some background noise can decrease the accuracy drastically with our combined dataset we hope to minimize the dataset bad effect on the model and enable them to perform better on unseen data.

Keywords Emotion recognition · Convolutional neural network · MFCC · Deep learning · Machine learning

1 Introduction

In current world of technology, human–computer interaction is increasing day by day and becoming more common. As human, we interact with computer either physically or through speech for initializing any computation. Humans are emotional being, and emotion plays an important role in our life, and to express our emotion or feeling, we use various medium as a form of communication. Speech is the method for humans to exchange, express, or interact with each other's. We all know that computer is

A. Kumar · A. Chandrayan · S. K. Dubey (✉)
Department of CSE, ASET, Amity University, Uttar Pradesh, Noida, India
e-mail: skdubey1@amity.edu

© The Author(s), under exclusive license to Springer Nature Singapore Pte Ltd. 2023
A. Shukla et al. (eds.), *Computational Intelligence*, Lecture Notes in Electrical
Engineering 968, https://doi.org/10.1007/978-981-19-7346-8_64

smarter but dumb at the same time as computer cannot understand emotion between human and computer. The demand for emotion recognition in the speech which is gaining more popularity. Emotion recognitions have wide range of application such as it can be used at hospital for sociological and mental health purposes, at call center to shield the employees from angry and toxic customers, in gaming industry or social sites to stop heat speech and decreasing toxic behavior, and many more. Speech emotion recognition (SER) aims to identify the emotion in words spoken by the speaker which enabling the computer to understand the human words more accurately and more human friendly. It can also be described as discovering the emotions of speech spoken by the person. In speech, emotion recognition is a great challenge as for understanding the emotions of a particular speaker as emotion is very subjective to person, and it is needed to extract intrinsic data from his/her speech. Then, it is converted into proper data which is later processed.

Pitches and energies of emotion are different for every person. Along with this, different nationalities and cultures have completely unique; pitches and energies are also a way of reacting to a certain thing. Same person can react to same thing with same emotions but with unique styles. Background noises can certainly be a great problem in the dataset. Tracking emotions of elderly people is even more difficult. Spontaneous reactions can be extremely hard to track. Circumstances and many other things invisible and untrickable factors also impact the emotions. Also, a lot of resources are consumed in this process. Here, both human and time cost are required. For labeling, we need humans to sit and listen to each audio and then label them. This is very costly and error-prone process. As emotions are very subjective, so human labeling other person's emotions can lead to error. In recent years, the SER has been observing the vertical advancement in various fields. Newer technologies and methodologies are coming every day to make task easier. Noises are automatically removed. Bigger and bigger datasets are being and too are being specified too much, as different nationalities, cultures, mood, ages, etc., are being categorized. All these details will help for bringing this technology on a global scale. Different sectors are contributing in this as they are realizing its importance, and increased research is being done. Huge datasets can also improve the real-time translations applications, as they do not take in account the emotions of the speakers and give the direct translation of even a sarcasm. Linguistic barriers will be removed by this. Many other previously existing phenomenal applications would also be improved a lot by SER. In this paper, we will study different research papers. All of them have used some different approach and datasets. Algorithms ranging from CNN, FP, to SVM are used. Many distinctive features including energy, pitch, MFCC, LPCC, and MEDC are also used. EMODB, CASIA, and EESDB are the major databases which are used by majority of researchers. The reason could be their huge size. Accuracy and results of all these are given. We will study all these papers and study them and then compare them. Algorithms, databases, accuracy along with time elapsed will also be studies. Then, after comparison, we will give our own research methodology with the way which we find suitable for us.

2 Proposed Method

As we know that speech is the fastest and most common way to communicate with each other and to exchange emotion with them. For us to empower the computer to better understand human communication is to start with identifying the real meaning behind what is said, and to do so, we need to identify the emotion buried under what is said human–computer interaction. We can identify the emotion in speech by identifying one or more of the classes of feature, namely the lexical features (the vocabulary used), the visual features (the expressions the speaker make), and the acoustic features (sound properties like pitch, tone, and jitter). A typical set of emotions contains 300 emotional states which are decomposed into six primary emotions like anger, happiness, sadness, surprise, fear, and neutral. Success of speech emotion recognition depends on naturalness of database [1].

Steps for detecting emotions from speech data:

(1) Emotional speech input
(2) Feature extraction and selection
(3) Classifier
(4) Emotional speech output.

3 Databases

Database is most important and necessary part in any machine learning model or research. Outcome depends greatly on the dataset. There are many datasets available for speech emotion recognition. Datasets are also important for training, testing, or validation as well as analysis of techniques or model. This paper used four dataset and created one new dataset. The purpose is to get the vast amount of data for proposed model.

The four datasets used are as follows: -

(1) Crowd-sourced emotional multimodal actors dataset (Crema-D)
(2) Ravdess dataset
(3) Surrey audio–visual expressed emotion (Savee)
(4) Toronto emotional speech set (Tess).

3.1 Surrey Audio–Visual Expressed Emotion (SAVEE)

The SAVEE database has been made by recording of four native English male speakers aged between 27 and 31 years. The dataset contains files in four folders where each folder belongs to one person, which are identified as DC, JE, JK, and KL. Emotion was created psychologically under six category that is anger, disgust, fear, happiness, sadness, and surprise. The naming of the file is such that the prefix letters

Table 1 SAVEE dataset audio file count with label

Labels (Gender_Emotion)	No. of audio file
male_neutral	120
male_angry	60
male_disgust	60
male_happy	60
male_surprise	60
male_sad	60
male_fear	60

tell the emotion which are as 'a' for anger, 'd' for 'disgust', 'f' for 'fear', 'h' for 'happiness', 'n' for 'neutral', 'sa' for 'sadness', and 'su' for 'surprise'[2] (Table 1).

3.2 Ravdess Dataset

In this dataset, there are 24 professional actors, 12 male speakers, and 12 female speakers. The lexical features (vocabulary) of the utterances are kept constant by speaking only 2 statements of equal lengths in 8 different emotions by all speakers. We got 7356 audio samples ((Total size: 24.8 GB) which were in the wav format. Speech includes calm, happy, sad, angry, fearful, surprise, and disgust expressions, and song contains calm, happy, sad, angry, and fearful emotions. The database is chosen for the verity of available speech, which is produced by trained professional under the suitable condition. The naming of the file is kept in such a way that it identifies the first digit as modality (01 = full-AV, 02 = video-only, 03 = audio-only), second one as Vocal channel (01 = speech, 02 = song), third one as Emotion (01 for neutral, 02 for calm, 03 for happy, 04 for sad, 05 for angry, 06 for fearful, 07 for disgust, 08 for surprised), fourth one as Emotional intensity (01 = normal, 02 = strong), fifth one as Statement (01 = 'Kids are talking by the door', 02 = 'Dogs are sitting by the door'), sixth one as Repetition (01 = 1st repetition, 02 = 2nd repetition) and the last one, i.e., seventh tells us about Actor (01 to 24. Odd numbered actors are male; even numbered actors are female) [3] (Table 2).

3.3 Crowd-Sourced Emotional Multimodal Actors Dataset (Crema-D)

Crowd-sourced emotional multimodal actors dataset (CREMA-D) contains 7442 original audio clips from 91 actors in which 48 were male and 43 were female actors aged between 20 to 74 belonging to various culture, race, places, and ethnicities (Table 3).

Table 2 RAVDESS dataset audio file count with label

Labels (Gender_Emotion)	No. of audio file
male_neutral	144
female_neutral	144
male_angry	96
female_happy	96
female_angry	96
female_disgust	96
male_disgust	96
female_fear	96
female_sad	96
female_surprise	96
male_sad	96
male_fear	96
male_happy	96
male_surprise	96

Table 3 CREMA-D dataset audio file count with label

Labels (Gender_Emotion)	No. of audio file
male_sad	671
male_fear	671
male_disgust	671
male_happy	671
male_angry	671
female_sad	600
female_happy	600
female_fear	600
female_angry	600
female_disgust	600
male_neutral	512
female_neutral	512

The actors for datasets spoke from a selected sentence of 12 sentences or lines. These 12 sentences were categorized under 6 different emotions, i.e., anger, disgust, fear, happy, neutral, and sad with 4 different emotion level to the emotion, i.e., low, medium, high, and unspecified.

CREMA-D is one of the best datasets as it has large number of speakers, which help with many things, viz., the model does not overfit, and there is large number of files to train model too.

The naming of the file is kept in such a way that first four digits tell the actor ID; second three alphabet tells which sentence is spoke by actor from the 12 selected

Table 4 TESS dataset audio file count with label

Labels (Gender_Emotion)	No. of audio file
female_angry	400
female_sad	400
female_surprise	400
female_neutral	400
female_happy	400
female_disgust	400
female_fear	400

sentences; third three alphabet tells the emotion the audio file, i.e., 'ANG' for Anger, 'DIS' for Disgust, 'FEA' for Fear, 'HAP' for Happy/Joy, 'NEU' for Neutral, 'SAD' for Sad, and last two alphabet tell about the emotion level of the audio file in the dataset, i.e., 'LO' for Low,'MD' for Medium, 'HI' for High, 'XX' for Unspecified [4].

3.4 Toronto Emotional Speech Set (TESS)

Toronto emotional speech set (TESS) dataset contains 2800 files which are set of 200 targeted words spoken in the phrases of 'Say the word ____' by two native speaking English actresses aged 26 and 64 years old, and they were also university educated with musical trading. Phrase in dataset was said to stimulate seven while keeping the threshold under the normal range [5] (Table 4).

3.5 Combining the Four Datasets

To obtain the final dataset, all above four datasets are combined into one dataset. In absence of this, there are chances of overfitting the model. The other reason why it is decided to use four combined dataset is because of having large variety and different circumstances for these datasets, and when proposed model is used on unseen data, it will be able to perform better than the model trained on only one dataset. The procedure was started with importing the dataset into the Python environment one by one, and then, we added the label that is emotion plus gender to each file present in dataset with the help of their nomenclature. After that we combined the four datasets into one new CSV file, which included the path of all the file with gender plus emotion label and the source of the file that is the name of the dataset to which they belong. After that features for our CNN model were extracted. We used new CSV file which contains path same as the path of all the files from all four dataset, and the label tells the emotion and gender of the audio file as similar in the original files (Table 5).

Table 5 Combined dataset audio file count under emotion and gender label

Emotion	Gender	No. of audio file
Happy	Male	827
Angry	Male	827
Sad	Male	827
Disgust	Male	827
Fear	Male	827
Angry	Female	1096
Disgust	Female	1096
Sad	Female	1096
Happy	Female	1096
Fear	Female	1096
Neutral	Male	837
Neutral	Female	1056

4 Feature Extraction and Selection

Speech signals contain many parameters which indicate emotion components of speech. Changes in these parameters mean changing the emotion component of the speech. Therefore, appropriate selection of characters or parameter is one of the most important tasks. There is need to extract useful features for our classifier from the dataset. After that it is required to load the database in Python, where we will extract the audio features from the file such as power, pitch, and vocal tract configuration with the help of Python library like librosa, pyAudio, and many more.

Most widely and simplest features used by many researchers are energy and energy-related features in speech. Energy is one of the most fundamental and necessary features in speech signal that we record. One can extract multiple energy-related characteristics in the speech sample by calculating the energy, such as mean value, max value, variance, variation range, contour of energy [6].

There are mainly three types of signal domain feature which contain most descriptive features for audio file: -

(i) Time domain: These are easily extracted from waveforms of the raw audio. Zero crossing rate, amplitude envelope, and RMS energy are examples [7].

(ii) Frequency domain: These focus on the frequency part of the audio. Signals are generally converted using the Fourier transform. Band energy ratio, spectral centroid, and spectral flux are few examples [7].

(iii) Time–frequency representation: These features combine both the time and frequency components of the audio signal. The time–frequency representation is obtained by applying the short-time Fourier transform (STFT) on the time domain waveform. Spectrogram, mel-spectrogram, and constant-Q transform are few examples [7].

A spectrogram is a chart or pictorial representation of the range of frequencies of a sound sign as it differs with time. Spectrogram can show both time and frequency part of the sound signs which is obtained by applying fast Fourier change on locally small sections of the sound file [7].

Conversion from frequency (f) to mel scale (m) is given by

$$m = 2595 \cdot \log(1 + f/500) \tag{1}$$

A mel-spectrogram is a therefore a spectrogram where the frequencies are converted to the mel scale [7].

The data of the pace of progress in ghastly groups of a sign are given through its cepstrum. A cepstrum is essentially a range of the log of the range of the time signal. The following range is neither inside the frequency space nor inside the time area, and accordingly, it changed into the name quefrency (a re-arranged word of the expression frequency) area. The mel-frequency cepstral coefficients (MFCCs) aren't anything anyway the coefficients that make up the mel-frequency cepstrum [7].

The cepstrum passes on the various qualities that develop the formants (a trademark part of the nature of a discourse sound) and tone of a sound. MFCCs accordingly are valuable for profound learning models [7] (Figs. 1 and 2).

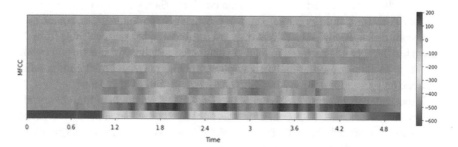

Fig. 1 Mel-frequency cestrum coefficients (MFCCs) of the speech signal

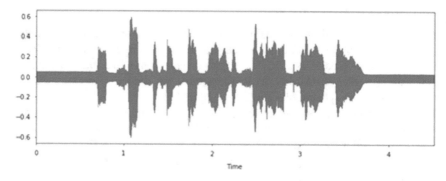

Fig. 2 Time domain plot of the speech signal

In every single speech frame, the value of pitch frequency can be calculated. Also, its statistics is derived in full speech frame. These values displayed the global properties of characteristic parameters. There is same 19 dimensions of energy in each pitch feature vector. Qualitative features of any emotional content, its utterance is very greatly related to the voice quality. One can even represent it numerically by parameters which are estimated directly from speech signal. The acoustic parameters related to speech quality are as follows: (1) Voice level: There are many measures on which we can rely upon like amplitude, energy, and duration; (2) Voice pitch; (3) phrase, feature boundary, phoneme, and word; (4) Temporal structures [6].

Linear prediction cepstrum coefficients (LPCCs): The channel of the speech can be embodied by it. Everyone who has some varying emotional speech will have different channel characteristics; then, to identify the emotions which are presented in speech, we extract these feature coefficients. The computational method of LPCC is generally a recurrence of computing the LPC. It is in accordance with the all-pole model.

Mel-frequency cepstrum coefficients (MFCCs): It is focused on human ear's hearing. It uses a nonlinear frequency unit, which simulates the hearing system of us humans. Mel-frequency scale: It is a greatly used feature of speech. It has simple calculation, ability of distinction anti-noise, and other advantages [8].

This paper used mel-frequency cepstrum coefficients (MFCCs) as our main feature to classify the speech on the bases of emotion. In negative emotion recognition using deep learning for Thai language, authors have used deep convolutional neural network (DCNN) with 6 layer, and the feature they selected is MFCC and got the accuracy of 65.83% on SAVEE dataset, accuracy of 75.83% on RAVDESS dataset, accuracy of 55.71% on TESS dataset, accuracy of 65.77% on CREMA-D dataset, accuracy of 96.60% on THAI dataset [9].

5   Result and Finding

To train and evaluate the model, we started with importing the path of all 4 datasets (Crema-D, Ravdess, Savee and Tess) into the environment and then combined them into one dataset where path of all the audio file, source dataset name and label them under male and female with emotion is saved in into new combined dataset.

For feature selection, we used mel-frequency cepstrum coefficients (MFCCs), and for calcification, 1D CNN is used to classify the MFCC array data into different emotion category that is gender and emotion or only emotion or only gender. We were able to see that our model works with the help of CNN model loss graph as the train curve is giving a smooth decreasing slope (Figs. 3, 4, 5, 6 and 7).

The gender vs emotion model accuracy score was 0.44, and the weighted average score is 0.45. The gender accuracy result of our 1D CNN model is with accuracy score of 0.80 and weighted average accuracy of 0.80. The emotion accuracy result of our 1D CNN model with accuracy score of 0.49 and weighted average accuracy of 0.50. As our combined dataset is derived from four different dataset sources and contain a

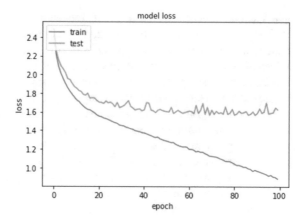

Fig. 3 CNN model loss graph

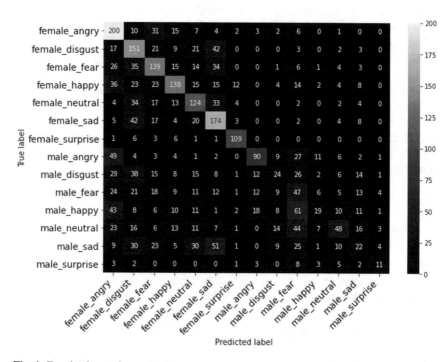

Fig. 4 Emotion by gender accuracy

large verity of file with different in parameter like length and max or min frequency which result to some files in the dataset not able to pass through the input layer of the CNN model, so they are discarded and that data file results in data loss for our model. This problem can be solved by using different data augmentation techniques

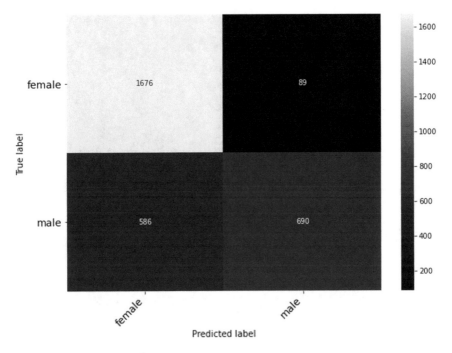

Fig. 5 Gender accuracy result

to those files which are being discarded and selecting the file after data augmentation technique are applied. So that they can pass the input layer requirement.

6 Conclusion and Future Scope

In this paper, we were able to create CNN model to classify the emotion in speech which was trained on the dataset which is combination for 4 different datasets having the weighted average accuracy of 50%. The primary objective of the paper was to create the base-level emotion recognition system with the wide dataset to train so that it can perform on new unseen data more effectively, so we trained the model on multiple combined datasets. We will say there is still chances to get better accuracy if we explore new methodology to classify the data or can manipulate the MFCC to give better and clear graph without changing the emotion. The emotion recognition is still had not been widely used in our world for consumer, and many new researches are continually happening as emotion is affected and based on multiple factors and understanding them with high accuracy is still hard. Though many companies are deriving the use emotion recognition into the new world, and we hope it to benefit many people. For further, enhancement can be done by introducing new and effective emotion model, and combining them with facial expression recognition system to the error can be minimized.

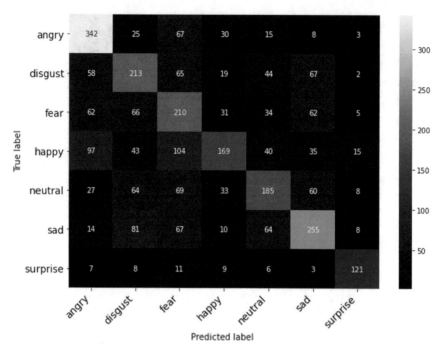

Fig. 6 Emotion accuracy

	precision	recall	f1-score	support
angry	0.59	0.70	0.64	490
disgust	0.42	0.39	0.41	468
fear	0.42	0.44	0.43	470
happy	0.51	0.41	0.46	503
neutral	0.44	0.52	0.48	446
sad	0.49	0.48	0.49	499
surprise	0.94	0.58	0.71	165
accuracy			0.50	3041
macro avg	0.54	0.50	0.51	3041
weighted avg	0.50	0.50	0.50	3041

Fig. 7 Emotion accuracy model classification report

References

1. Ayadi ME, Kamel MS, Karray F (2011) Survey on speech emotion recognition: features, classification schemes, and databases. Pattern Recogn
2. Jackson P, Haq S (2011) Surrey audio-visual expressed emotion (SAVEE) database. http://kah lan.eps.surrey.ac.uk/savee/Database.html

3. Livingstone SR, Russo FA (2018) The Ryerson audio-visual database of emotional speech and song (RAVDESS): a dynamic, multimodal set of facial and vocal expressions in North American English. PLoS ONE 13(5), Article e0196391. https://doi.org/10.1371/journal.pone. 019639
4. Cao H, Cooper DG, Keutmann MK, Gur RC, Nenkova A, Verma R (2014) CREMA-D: crowd-sourced emotional multimodal actors dataset. IEEE Trans AffectComput 5(4):377–390. https://doi.org/10.1109/TAFFC.2014.2336244
5. Dupuis K, Pichora-Fuller MK (2010) Toronto emotional speech set (TESS). Toronto, University of Toronto, Psychology Department
6. Rong J, Li G, Chen YPP (2009) Acoustic feature selection for automatic emotion recognition from speech. Information Processing and Management 45 (Elsevier)
7. https://devopedia.org/audio-feature-extraction
8. Shen P, Changjun Z, Chen X (2011) Automatic speech emotion recognition using support vector machine. In: International conference on electronic and mechanical engineering and information technology
9. Mekruksavanich S, Jitpattanakul A, Hnoohom N (2020) Negative emotion recognition using deep learning for Thai language. 71–74. https://doi.org/10.1109/ECTIDAMTNCON48261. 2020.9090768
10. Zehra W, Javed AR, Jalil Z et al (2021) Cross corpus multi-lingual speech emotion recognition using ensemble learning. Complex Intell. Syst. 7:1845–1854. https://doi.org/10.1007/s40747-020-00250-4
11. Mustaqeem, Kwon S (2020) MLT-DNet: speech emotion recognition using 1D dilated CNN based on multi-learning trick approach. Expert Syst Appl 167. https://doi.org/10.1016/j.eswa. 2020.114177
12. Xu M, Zhang F, Cui X, Zhang W (2021) Speech emotion recognition with multiscale area attention and data augmentation, In: ICASSP 2021—2021 IEEE international conference on acoustics, speech and signal processing (ICASSP), pp 6319–6323. https://doi.org/10.1109/ICASSP39728.2021.9414635
13. Meyer P, Xu Z, Fingscheidt T (2021) Improving convolutional recurrent neural networks for speech emotion recognition. IEEE Spoken Language Technology Workshop (SLT) 2021:365–372. https://doi.org/10.1109/SLT48900.2021.9383513
14. Jiang L, Tan P, Yang J, Liu X, Wang C (2019) Speech emotion recognition using emotion perception spectral feature. Concurrency Comput Pract Expert e5427. https://doi.org/10.1002/cpe.5427
15. Zisad S, Hossain M, Andersson K (2020) Speech emotion recognition in neurological disorders using convolutional neural network. https://doi.org/10.1007/978-3-030-59277-6_26
16. Li M et al (2021) Contrastive unsupervised learning for speech emotion recognition. In: ICASSP 2021—2021 IEEE international conference on acoustics, speech and signal processing (ICASSP), 2021, pp 6329–6333. https://doi.org/10.1109/ICASSP39728.2021.9413910
17. Atmaja BT, Akagi M (2021) Evaluation of error- and correlation-based loss functions for multitask learning dimensional speech emotion recognition. J Phys Conf Ser 1896:012004. https://doi.org/10.1088/1742-6596/1896/1/012004
18. Koduru A, Valiveti HB, Budati AK (2020) Feature extraction algorithms to improve the speech emotion recognition rate. Int J Speech Technol 23:45–55. https://doi.org/10.1007/s10772-020-09672-4
19. Singh P, Saha G, Sahidullah M (2021) Deep scattering network for speech emotion recognition. ArXiv abs/2105.04806 (2021): n. pag
20. https://www.analyticsinsight.net/speech-emotion-recognition-ser-through-machine-learning/
21. https://towardsdatascience.com/speech-emotion-recognition-using-ravdess-audio-dataset-ce1 9d162690
22. Mahanta SK, Khilji AFUR, Pakray P (2021) Deep neural network for musical instrument recognition using MFCCs. Computación y Sistemas 25(2). Accessed 23 May 2021

Deep Learning Framework for Compound Facial Emotion Recognition

Rohan Appasaheb Borgalli and **Sunil Surve**

Abstract Human facial expressions are an indication of true emotions. Accurately recognizing facial expressions is useful in artificial intelligence, computing, medicine, e-learning, and many more. Although facial emotion recognition (FER) can be accomplished primarily through the use of multiple sensors. However, research shows that using facial images/videos to recognize facial expressions are better because emotions can be conveyed through visual expressions that carry important information. In the past, much research was conducted in the field of FER using different approaches such as analysis through different sensor data, using machine learning and deep learning framework with static images and dynamic sequence. Previous FER research focused on studying seven basic emotions: anger, resentment, fear, excitement, sadness, surprise, and neutrality. However, humans that exhibit many more facial expressions are considered compound emotions. Recently, use of deep learning algorithms in FER has been considerable. State-of-the-art results show deep learning-based approaches are powerful over conventional FER approaches. This paper focuses on implementing deep learning frameworks for compound facial emotion recognition systems for detecting compound emotion using the facial expression image dataset compound facial expressions of emotion (CFEE).

Keywords Facial emotion recognition · Compound emotion · Deep learning · Convolution neural network

R. A. Borgalli (✉)
Department of Electronics Engineering, Fr. Conceicao Rodrigues College of Engineering, Bandra, Mumbai, India
e-mail: rohanborgalli111@gmail.com

S. Surve
Department of Computer Engineering, Fr. Conceicao Rodrigues College of Engineering, Bandra, Mumbai, India

1 Introduction

FER is a vital research area helpful in many applications. A FER system model mentioned in this paper detects the human facial expression and identifies the corresponding induced emotion for a static image or sequence of images. Recognizing basic and compound facial expressions with a high accuracy remains challenging due to the complexity and varieties of facial expressions.

Generally, human beings can convey intentions and emotions through gestures, facial expressions, and involuntary language in nonverbal ways. According to Darwin and Prodger [1], human facial expressions reflect their emotional condition and intentions. Due to the importance of the FER system, various FER systems have been developed to decode expression information from facial representations. Because of its accessibility and efficiency, FER has become one of the essential methods to detect emotions. It is also widely used in lie detection, medical assessment, human–machine interaction, driver safety, and solving other complex real-world problems with FER.

In 1978, automatic facial expression recognition was published first by Suwa [2]. Followed by, many researchers have been contributing to developing facial expression recognition methods that are robust and accurate.

At the start of the twentieth century, Ekman and Friesen [3] defined six basic emotions based on cross-culture study: anger, disgust, fear, happiness, sadness, and surprise, indicating that humans of any culture perceive these basic emotions in the same way. Most of the past works mentioned in a survey by Li and Deng [4] focused on these six basic emotions and neutral. For implementing FER on basic emotion, some databases have been proposed, such as FER2013 [5], CK + [6], MMI [7], JAFFE [8], and RaFD [9] database.

However, some facial expressions induced by humans are more complex and cannot be considered a basic emotion, leading to compound emotion [10].

Advanced research on FER focused on compound emotion by mentioning a variety set of compound emotions along with basic emotion given by Du et al. [11], Guo et al. [12], and Li and Deng [13, 14].

Initially, different image processing techniques were employed [15] to solve the FER problem, followed by machine learning techniques such as the k-nearest neighbors, multi-layer perceptron model, and support vector machines mentioned in a survey of Li and Deng [16]. These traditional methods work on the concept of extracting various features like local binary patterns (LBPs) discussed by Zhao and Pietikainen [17], histogram of gradient (HOG) features, Eigen faces, gradient and Laplacian [18] face-landmark features, edge, skin color, face muscle motion, and texture features [19]. Such features are effective for facial images. However, due to a large number of facial images in the wild and their complexity, handcrafted features alone cannot achieve optimal performance. With the advancement in research into deep learning strategies, this problem is being solved. Deep learning techniques automatically extract features specific to the task, especially the convolutional neural network (CNN) is very popular among them.

Literature survey by Li and Deng (2020) [16] showed the process of FER begins with the preprocessing stage to detect face from the image and extract features before training models either through the handcraft method or with CNN along with transfer learning can also be tried by Lim et al. [20] using standard models.

To implement the FER system effectively, it should work accurately to detect not on basic and compound emotions. This paper implemented a deep learning framework for FER systems on the CFEE dataset [11] to detect basic and compound emotions and improve performance.

2   Related Work

To move the FER system from a basic emotion to compound emotion, many researchers developed datasets such as CFEE [11] and ICV-MEFED [12]; Li and Deng [13] presented the RAF-DB database in the wild, consisting of 7 basic emotions and 12 compound emotions. Shan Li et al. [14] introduced a real-world facial expression database, RAF-ML, to make researchers explore the challenges of expression recognition in the wild and address it with a multi-label dataset. Fabian et al. [21] developed Emotion recognition in the Wild (EmotiW), to promote the transition of FER from the lab-controlled to the in-the-wild environment by collecting lots of training data from real-world scenarios to make it challenging.

Also, advanced research in FER made the transition from basic to compound emotion recognition. Du et al. [11] generally examined the people showing compound facial emotions that is a combination of more than one basic emotion (for example, the compound facial expression of happily surprised people experiencing a feel of both happiness as well as surprise simultaneously, as happens in a surprise birthday party). In their previous study [11], they reported 15 compound emotions. However, a recent survey by them [22] added two more compound emotions and came up with 17 compound emotions based on the various categories of people belonging to diverse cultures who had undergone the test.

As given by [11] indicates that the face displays compound emotion, a combination of the six basic emotions tallying as 22 emotions, including seven basic emotions and twelve compound emotions along with the three additional emotions (appall, hate, and awe) and gave CFEE database. They extracted Gabor features using Gabor filter of each image of a dataset, and then, the multi-class support vector machine (mSVM) classifier was applied to detect compound emotion. Tested with the tenfold cross-validation and leave-one-sample-out to get an accuracy of 73.61%, 70.03%, and 76.91% when using only geometric features, only appearance features, and when both components of features are combined in a single feature space, respectively.

In [13], researchers proposed a novel deep locality-preserving learning method for emotion recognition using the RBM-DB dataset. To address the problem of a non-lab-controlled database, they proposed a novel deep locality-preserving CNN (DLP-CNN) method to extract deep features, aiming to maximize the power of deep learned features to preserve the local similarities while maximizing the inter-class differences. Their experiments discriminate 11-class compound expressions Happily Surprised, Happily Disgusted, Sadly Fearful, Sadly Angry, Sadly Surprised, Sadly Disgusted, Fearfully Angry, Fearfully Surprised, Angrily Surprised, Angrily Disgusted, and Disgustedly Surprised with accuracy 44.55%. Li and Deng [14] proposed DeepBi-Manifold (DBM)-CNN architecture to learn through features from RAF-ML achieve better performance on cross-databases which indicates RAF-ML can serve as a source database for analysis of facial expression as it contains a great variety of training data. DBM-CNN adaption of source domain to target domain can help enhance the performance for cross-dataset tasks.

In 2016, EmotionNet challenge [23] included a million images of facial expressions in the wild that are automatically annotated very accurately using a real-time algorithm. This challenge has two tracks. The first track was related to the automated detection of 11 facial action units (AUs), whereas the second detected compound emotion. Only, compound emotions track was briefly reviewed here as per our work considered. The EmotionNet challenge applied to the dataset mentioned in [21]. The training set with 950 K was annotated automatically using the algorithm proposed in [23] whereas validation and test sets with 2 K and 40 K facial images were annotated manually. Finally, basic and compound emotions with 16 categories are available for detection.

Slimani et al. [24] proposed highway CNN architecture for FER. The complete system comprises different stages: preprocessing image, feature extraction, and expression classification. An experiment was performed on the CFEE database and got a test accuracy of 52.14%. They mentioned that it is difficult for a highway CNN model to discover distinguishable features to compound emotion from the basic categories that compose them.

Most FER solutions are focused on the detection of basic emotion. Hence, there is a need for an efficient and robust FER system that detects particular basic and compound facial expressions.

To correctly recognize basic and compound human emotion on the CFEE dataset [10], the proposed learning framework is based on deep learning architecture for FER systems performing better than state-of-the-art results.

3 Dataset Overview

Compound facial expressions of emotion (CFEE) database [10] contains 5,060 facial images of 230 subjects (130 women, mean age 23 years) labeled with seven basic emotions and 15 compound emotions. The CFEE database has 5,060 images, 1610 fundamental emotions, and 3450 compound ones expressed by 230 subjects, with each subject having frontal direction.

Figure 1 taken from [24] indicates sample images of 22 categories of the CFEE dataset. The emotions depicted in this database can be classified into two categories.

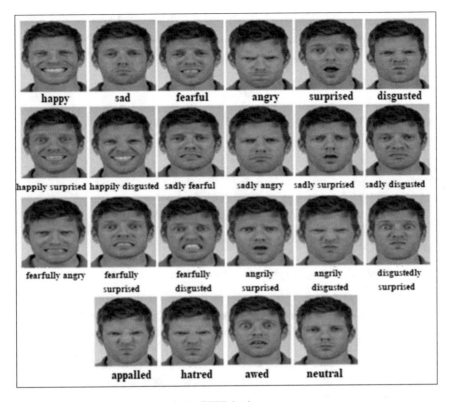

Fig. 1 Sample images of 22 categories in CFEE database

The first contains six basic emotions; "happiness," "sadness," "fear," "anger," "surprise," and "disgust" (see Fig. 1). The second relates to compound emotions. Here, compound emotions mean that the category of emotion indicates a combination of two basic emotion categories. Figure 1 shows the 22 emotions most commonly communicated by humans.

4 Proposed Methodology

For compound facial expression recognition, we use deep learning algorithms for feature extraction and classification, as shown in Fig. 2. Deep learning algorithms try to learn high-level features from data. This is a very distinctive part of deep learning and a significant step ahead of traditional machine learning. Therefore, deep learning removes the need to develop a new feature extractor for every new problem. CNN will try to learn low-level features such as edges and lines in early layers, parts of faces of people, and then a high-level representation of a face.

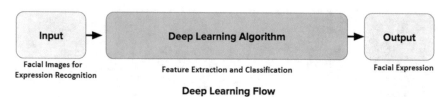

Deep Learning Flow

Fig. 2 Deep learning flow for compound facial expression recognition

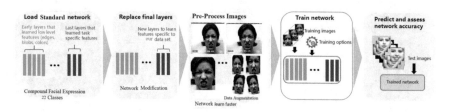

Fig. 3 Proposed methodology for compound facial emotion recognition

The main difference between deep learning and traditional machine learning concerning its performance can be found by increasing the data scale. Deep learning algorithms don't perform that well when the data are small. This is because deep learning algorithms work perfectly on a large amount of data to understand it better. On the other hand, traditional machine learning algorithms with handcrafted features work well for less data.

The proposed methodology is based on the convolution neural network-based deep learning framework given in Fig. 3. We are using standard CNN Inception v3 [24] architecture known for image classification applications. These standard networks have starting layers that are used to learn low-level features such as edges, blobs, and colors, and the last layers learn task-specific features. In our case, it is compound facial expressions.

We modify the standard network by replacing the last layers with specific features to our dataset. In the CFEE database, we have a variety of images, but to learn features faster, we are preprocessing images, so only the required part of the face is captured using the Haar Cascade algorithm to detect the frontal face.

The training set consists of different facial compound expressions; however, the training data distribution is usually uneven. The imbalance of training data distribution will harm network performance. Data augmentation and data generation operations are performed to balance the data distribution and generate more training data to eliminate this adverse effect.

To train, validate, and test the network, images are generated using image data generator, which generates batches of tensor image data with real-time data augmentation. While training the network with 4053 training samples and 1014 validation samples, accuracy and loss are calculated to get an idea of how the network is

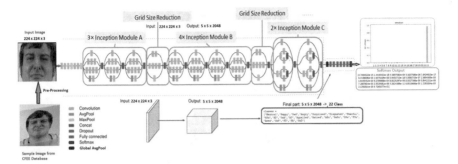

Fig. 4 Compound facial emotion recognition on CFEE dataset using standard Inception v3 architecture

training. Finally, around 895 test images are supplied, and accuracy and categorical cross-entropy loss are calculated.

4.1 Inception v3 (2015)

Inception v3 mainly focuses on reducing computational power by modifying the previous inception architectures. This idea was published in 2015 [25]. Figure 4 shows a final architecture of an Inception v3 network used for our application.

5 Results

5.1 Available Metrics

Here, we define set D of N images of a database as given in Eq. (5.1)

$$D = \{d_0, d_1, \ldots, d_N\} \tag{5.1}$$

We define $L_0, L_1, \ldots, L_{N-1}$ to be a family of compound emotion label (L) sets and $P_0, P_1, \ldots, P_{N-1}$ to be a family of prediction (P) sets where Li and Pi are the label set and prediction set, respectively, that correspond to Image di.

The set of all unique labels is given by Eq. (5.2).

$$L = \bigcup_{k=0}^{N-1} L_k \tag{5.2}$$

Mathematical representation of accuracy and f1-score is given in Eqs. (5.3) and (5.4), respectively.

$$\frac{1}{N} \sum_{i=0}^{N-1} \frac{|L_i \cap P_i|}{|L_i| + |P_i| - |L_i \cap P_i|} \tag{5.3}$$

$$\frac{1}{N} \sum_{i=0}^{N-1} 2 \frac{|P_i \cap L_i|}{|P_i| \cdot |L_i|} \tag{5.4}$$

The next step is to calculate a macro-average and weighted average

1. Macro-average accuracy

Macro-average accuracy is an arithmetic mean of the accuracy.

2. Weighted average accuracy

Weighted average accuracy is calculated by weighing the accuracy of each class by the number of samples from that class.

5.2 Confusion Matrix of CFEE Database

The confusion matrix is a cross-table that records the number of occurrences between two raters, the true/actual classification, and the predicted classification. The columns stand for model prediction for consistency throughout the paper, whereas the rows display the true classification. The classes are listed in the same order in the rows as in the columns. Therefore, the correctly classified elements are located on the main diagonal from the top left to the bottom right, and they correspond to the number of times the two raters agree. We use a confusion matrix to evaluate the accuracy between actual and predicted labels. Refer to Table 1 for label and corresponding compound facial emotion on CFEE dataset [10] to understand the normalized confusion matrix on the CFEE dataset shown in Fig. 5.

5.3 Emotion Analysis on CFEE Sample Test Image

Emotion analysis of CFEE sample test images is done at the end to check the model's working. It shows for sample test image softmax output indicating a probability of each emotion with a plotted graph the maximum will be considered as predicted emotion by the proposed model. Refer to Table 1 for label and corresponding compound facial emotion on CFEE dataset to understand predicted emotion through softmax output layer by the proposed model on sample test images of CFEE dataset shown in Fig. 6.

Table 1 Label and corresponding compound facial emotion on CFEE dataset

Label	Emotion	Label	Emotion
0	Neutral	11	Appalled
1	Happy	12	Hatred
2	Sad	13	Angrily Surprised (Asu)
3	Angry	14	Sadly Surprised (SaSu)
4	Surprised	15	Disgustedly Surprised (DSu)
5	Disgusted	16	Fearfully Surprised (FSu)
6	Fearful	17	Awed
7	Happily Surprised (Hsu)	18	Sadly Fearful (SaF)
8	Happily Disgusted (HD)	19	Fearfully Disgusted (FD)
9	Sadly Angry (SaA)	20	Fearfully Angry (FA)
10	Angrily Disgusted (AD)	21	Sadly Disgusted (SaD)

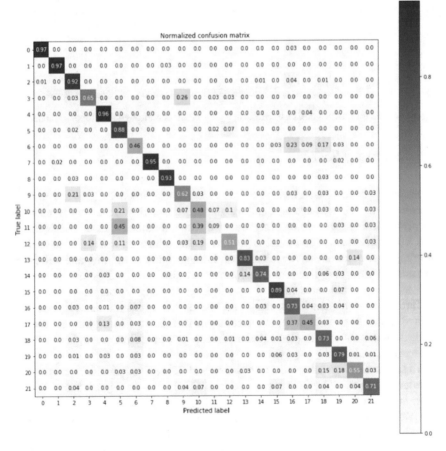

Fig. 5 Normalized confusion matrix on CFEE dataset

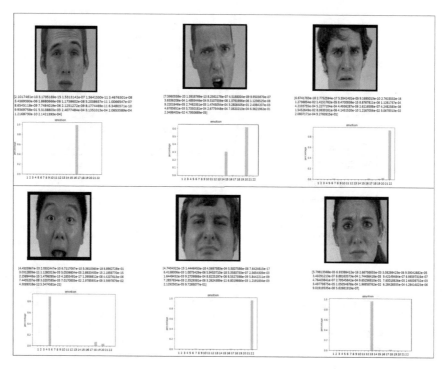

Fig. 6 Sample test images of CFEE dataset and their prediction through the proposed model

5.4 Accuracy Score

The accuracy score is calculated to evaluate the proposed model; accuracy score has overall accuracy, macro-average, and weighted average. To summarize the results of the deep learning methodology implemented with a modified standard CNN Inception V3 architecture. The model is trained from scratch on a CFEE dataset [10] of compound emotion having 22 classes. During training, training and validation datasets are used, and to measure performance, training and validation accuracy as well as loss are calculated for each epoch. Once training is completed, indicating no further improvement in accuracy and loss on the validation dataset. Finally, a trained model is used to compute accuracy performance (%) on the test dataset. Table 2 summarizes the implemented methodology's precision, recall, f1-score, and accuracy results.

To summarize, we achieved a test accuracy of **73.42%** on modified Inception v3 architecture on the 22 emotion categories of CFEE dataset. Figure 5 and Table 2 illustrate the obtained experimental results. It can be deduced that some emotions like *Neutral, Happy, Sad, Surprised, Happily Surprised, Happily Disgusted, Angrily Surprised, and Disgustedly Surprised* perform well and illustrated to be superior and simple to recognize, as they reached a successful classification rate of *97%, 97%,*

Table 2 Summary results of precision, recall, f1-score, and accuracy on compound facial emotion recognition on CFEE datasets

Emotion Class	Emotion	Precision	Recall	f1-score	Support
0	Neutral	0.97	0.97	0.97	35
1	Happy	0.97	0.97	0.97	36
2	Sad	0.83	0.93	0.88	80
3	Angry	0.79	0.65	0.71	34
4	Surprised	0.74	0.96	0.84	27
5	Disgusted	0.58	0.88	0.7	41
6	Fearful	0.52	0.46	0.48	35
7	Happily Surprised (Hsu)	1	0.95	0.98	43
8	Happily Disgusted (HD)	0.97	0.93	0.95	30
9	Sadly Angry (SaA)	0.56	0.62	0.59	29
10	Angrily Disgusted (AD)	0.38	0.48	0.42	29
11	Appalled	0.43	0.09	0.15	33
12	Hatred	0.7	0.51	0.59	37
13	Angrily Surprised (Asu)	0.8	0.83	0.81	29
14	Sadly Surprised (SaSu)	0.79	0.74	0.76	35
15	Disgustedly Surprised (DSu)	0.75	0.89	0.81	27
16	Fearfully Surprised (FSu)	0.6	0.73	0.66	67
17	Awed	0.71	0.45	0.55	38
18	Sadly Fearful (SaF)	0.71	0.73	0.72	77
19	Fearfully Disgusted (FD)	0.79	0.79	0.79	72
20	Fearfully Angry (FA)	0.75	0.55	0.63	33
21	Sadly Disgusted (SaD)	0.65	0.71	0.68	28
Accuracy				0.73	895
Macro avg		0.73	0.72	0.71	895
Weighted avg		0.73	0.73	0.72	895

88%, 84%, 98%, 95%, and 81%, respectively. At that point, some emotion categories that were recognized with classification f1-score between 60 and 79% like *"angry" (71%), "disgusted" (70%), "sadly surprised" (76%), "fearfully surprise" (66%) "sadly fearful" (72%), neutral" (71%), "fearfully disgusted (79%)"surprised" (71%), "fearfully angry" (63%), and "sadly disgusted" (68%).* On the other hand, the following emotions achieved medium classification accuracy when compared to other emotions categories: *"fearful" (48%), "sadly angry" (59%), angrily disgusted" (42%), "Hatred" (59%), and "awed" (55%).*

From Fig. 5, it can also be observed that some classes are tough to train and recognize: *"Appalled" (15%).* In the sense that the size of the CFEE database is small for CNN-based methods, it tends to have much fewer training samples compared to other databases utilized by CNN structures such as iCV-MEFED [12] database that

Table 3 Comparison with state-of-the-art results on compound facial emotion recognition on CFEE dataset

Paper	Method	Model/Classifier	Features	Accuracy (%)
[24]	CNN	Highway CNN	–	52.14
[11]	Tenfold cross-validation and leave-one-sample-out	Multi-class support vector machine (mSVM) classifier	Only geometric features	73.61,
			Only appearance features	70.03
			Both above features are combined as a single feature space	76.91
Ours	CNN	Inception v3	–	73.42

comprises 31,250 frontal face images, making it harder to train the Inception v3 architecture to recognize them. But, still, it achieved state-of-the-art performance.

Table 3 compares the performance of the proposed Inception v3 model on CFEE with other contemporary models.

6 Conclusion and Future Work

In this paper, a deep learning framework for compound facial emotion recognition was proposed to get better performance on the facial images CFEE database. Experiment shows, for the CFEE dataset, deep learning-based Inception v3 model is comparatively giving good results. Overall test accuracy of the proposed model is 73.42%, except "Appalled" that has test accuracy of 15%; all other emotions have test accuracy of greater than 40%, demonstrating our method is very robust to a wide variety of compound emotions.

In the future, to improve the accuracy of a CFEE dataset for compound emotion detection, we can consider using an unsupervised pre-training strategy from transfer learning [26], which may further decrease the recognition error rate. Deep belief network-based techniques can be used to improve facial expression analysis [27]. Transfer learning can also be tried on CFEE dataset [28]. The use of RNN and LSTM [29] can also be a good solution. FER based on facial action units (AUs) showed promising results in state-of-the-art that can also be tried [30]. Also, a separate loss function for compound emotion can be used, as mentioned in [31]. To extend the research of FER in the wild scenario, the RAF-AU dataset given by Yan et al. [32] helps to detect appropriate basic and compound facial emotion through its facial action units.

Acknowledgements We thank all the authors whose work contributed to the meta-analysis, especially Du S., Tao Y., and Martinez A. M., who took the time to respond to our requests and were given access to the CFEE dataset.

Data Availability Statement
CFEE dataset is subject to third-party restrictions. The CFEE dataset that supports the findings of this study is available from Du S., Tao Y., and Martinez, A. M. Restrictions apply to the availability of these data. The database is available for distribution for research purposes from Du S., Tao Y., and Martinez A. M. at URL link http://cbcsl.ece.ohio-state.edu/dbform_compound.html with the permission of Du S., Tao Y., and Martinez A. M.

References

1. Darwin C, Prodger P (1998) The expression of the emotions in man and animals. Oxford University Press, USA
2. Motoi S (1978) A preliminary note on pattern recognition of human emotional expression. In: Proceedings of international joint conference on pattern recognition, pp 408–410
3. Ekman P, Friesen WV (1971) Constants across cultures in the face and emotion. J Pers Soc Psychol 17(2):124
4. Li S, Deng W (2018) Deep facial expression recognition: a survey. arXiv preprint arXiv:1804.08348
5. Wolfram Research, "FER-2013" from the Wolfram Data Repository (2018)
6. Lucey P, Cohn JF, Kanade T, Saragih J, Ambadar Z, Matthews I (2010, June) The extended cohn-kanade dataset (ck+): a complete dataset for action unit and emotion-specified expression. In: 2010 IEEE computer society conference on computer vision and pattern recognition-workshops. IEEE, pp 94–101
7. Pantic M, Valstar M, Rademaker R, Maat L (2005) Web-based database for facial expression analysis. In: IEEE International Conference on Multimedia and Expo, 2005. ICME 2005. IEEE, 5 pp
8. Lyons M, Kamachi M, Gyoba J The Japanese Female Facial Expression JAFFE Dataset. Zenodo at https://doi.org/10.5281
9. Langner O, Dotsch R, Bijlstra G, Wigboldus DH, Hawk ST, Van Knippenberg AD (2010) Presentation and validation of the Radboud Faces Database. Cogn Emot 24(8):1377–1388
10. Lindquist KA, Wager TD, Kober H, Bliss-Moreau E, Barrett LF (2012) The brain basis of emotion: a meta-analytic review. Behav Brain Sci 35(3):121
11. Du S, Tao Y, Martinez AM (2014) Compound facial expressions of emotion. Proc Natl Acad Sci 111(15):E1454–E1462
12. Guo J, Lei Z, Wan J, Avots E, Hajarolasvadi N, Knyazev B, Kuharenko A, Jacques JCS Jr, Baro X, Demirel H, Escalera S, Allik J, Anbarjafari G (2018) Dominant and complementary emotion recognition from still images of faces. IEEE Access 6:26391–26403
13. Li S, Deng W (2018) Reliable crowdsourcing and deep locality-preserving learning for unconstrained facial expression recognition. IEEE Trans Image Process 28(1):356–370
14. Li S, Deng W (2019) Blended emotion in-the-wild: Multi-label facial expression recognition using crowdsourced annotations and deep locality feature learning. Int J Comput Vision 127(6):884–906
15. Revina IM, Emmanuel WS (2021) A survey on human face expression recognition techniques. J King Saud Univ Comput Inf Sci 33(6):619–628
16. Li S, Deng W (1 July-Sept 2022) Deep facial expression recognition: a survey. In: IEEE Transactions on Affective Computing, vol 13, no 3, pp 1195–1215

17. Zhao G, Pietikainen M (2007) Dynamic texture recognition using local binary patterns with an application to facial expressions. IEEE Trans Pattern Anal Mach Intell 29(6):915–928

18. Pandey RK, Karmakar S, Ramakrishnan AG, Saha N (2019) Improving facial emotion recognition systems using gradient and Laplacian images. arXiv preprint arXiv:1902.05411

19. Lekdioui K, Messoussi R, Ruichek Y, Chaabi Y, Touahni R (2017) Facial decomposition for expression recognition using texture/shape descriptors and SVM classifier. Signal Process Image Commun 58:300–312

20. Lim YK, Liao Z, Petridis S, Pantic M (2018) Transfer learning for action unit recognition. arXiv preprint arXiv:1807.07556

21. Fabian Benitez-Quiroz C, Srinivasan R, Martinez AM (2016) Emotionet: an accurate, real-time algorithm for the automatic annotation of a million facial expressions in the wild. In: Proceedings of the IEEE conference on computer vision and pattern recognition, pp 5562–5570

22. Du S, Martinez AM (2015) Compound facial expressions of emotion: from basic research to clinical applications. Dialogues Clin Neurosci 17(4):443

23. Benitez-Quiroz CF, Srinivasan R, Feng Q, Wang Y, Martinez AM (2017)Emotionet challenge: recognition of facial expressions of emotion in the wild. arXiv preprint arXiv:1703.01210

24. Slimani K, Lekdioui K, Messoussi R, Touahni R (2019) Compound facial expression recognition based on highway CNN. In: Proceedings of the new challenges in data sciences: acts of the second conference of the Moroccan Classification Society, pp 1–7

25. Szegedy C, Vanhoucke V, Ioffe S, Shlens J, Wojna Z (2016) Rethinking the inception architecture for computer vision. In: Proceedings of the IEEE conference on computer vision and pattern recognition, pp 2818–2826

26. Bengio Y (2012, June) Deep learning of representations for unsupervised and transfer learning. In: Proceedings of ICML workshop on unsupervised and transfer learning. In: JMLR workshop and conference proceedings, pp 17–36

27. Liu P, Han S, Meng Z, Tong Y (2014) Facial expression recognition via a boosted deep belief network. In: Proceedings of the IEEE conference on computer vision and pattern recognition, pp 1805–1812

28. Riaz MN, Shen Y, Sohail M, Guo M (2020) Exnet: an efficient approach for emotion recognition in the wild. Sensors 20(4):1087

29. Jarraya SK, Masmoudi M, Hammami M (2020) Compound emotion recognition of autistic children during meltdown crisis based on deep spatio-temporal analysis of facial geometric features. IEEE Access 8:69311–69326

30. Shao Z, Liu Z, Cai J, Wu Y, Ma L (2019) Facial action unit detection using attention and relation learning. IEEE transactions on affective computing

31. Li Y, Lu Y, Li J, Lu G (2019, October) Separate loss for basic and compound facial expression recognition in the wild. In: Asian conference on machine learning. PMLR, pp 897–911

32. Yan WJ, Li S, Que C, Pei J, Deng W (2020) RAF-AU database: in-the-wild facial expressions with subjective emotion judgement and objective AU annotations. In: Proceedings of the Asian conference on computer vision

Sustainably Nurturing a Plant (SNAP) Using Internet of Things

Akshayee Bharat Dhule and Divya Y.Chirayil

Abstract Plants are of great importance to us and are a vital factor in human existence on this earth. We cannot imagine a life without plants. Rapid development in economy, industries, and cities has led to an increase in the rate of deforestation. In order to have balance, the "Plant a tree" initiative is taken at the institutional level, community level, and government level. This initiative has not succeeded in a true sense as most of them just plant the trees but forget to nurture them as there is a lack of communication, feelings, and emotional quotient in people for the plants. The following paper presents a prototype that personifies the plant, helps them to create emotional quotient using IoT, monitors their health, and nurtures them. The sensor values are monitored on LCD, Cayenne dashboard, and also in the mobile app. The prototype posts customized messages on various social media like Facebook, Twitter on certain threshold values of sensors thereby personifying the plant. Moreover, the huge dataset created by the sensors lay the foundation to build algorithms in machine and deep learning, which will help to make tailor made conditions to nurture a plant and also enhance the expressions of plants through messages. The emotional connect and the awareness imparted by the prototype makes the nurturing of a plant sustainable, thus making the "Plant a Tree" initiative at the individual level successful in a true sense.

Keywords Plants · IoT · Health monitoring · Dashboard · SNAP · Plant emotions · Dataset for machine learning algorithms

1 Introduction

Everyone is aware of the benefits that we enjoy from trees. They conserve energy, save water, prevent air pollution, and prevent soil erosion. Due to industrialization and

A. B. Dhule (✉) · D. Y.Chirayil
Pillai HOC, Rasayani, New Mumbai, India
e-mail: adityatollplaza2018@gmail.com

D. Y.Chirayil
e-mail: dchirayil@mes.ac.in

urbanization, rate of deforestation is more as compared to the rate of forestation. With the benefits in mind and to maintain the balance, everyone among us support "Plant a Tree" and indulge in the tree planting activity at the individual level, community level, and at the government level. Even campaigns are organized by school such as "Go Green" and "Save Trees" so that children connect with the plants by actively taking part in tree plantation. Everyone plants on the campaign day, but reality check shows that plants are only planted by the people, but most of them forget to nurture them. Reason behind is—no strong emotional quotient develops in people for plants as in case of pet animals. Pet animals express their feelings, emotions by interacting with us, for example, they express themselves with a particular gesture when they feel hungry, feel sad, when they want to play with us, and thus, we are strongly bonded with them and nurture them. But, we do not have the same connect with the plants as they lack in communication and expressions with the humans. This gap can be bridged when plant could interact with humans as pet animals do. Our prototype personifies the plant by interacting it with the humans. IoT [5] is the technology that can assist the plants to personify themselves which ultimately will impart them the emotional quotient thus making the nurturing of a plant sustainable.

2 Literature Review

Mr. Saleem et al. [1] propose the system designed using LPC2148 microcontroller to overcome limitations of agriculture farming related to irrigation of plants by drip system with the available water tables. An electrical fence is made to keep strangers away from field by its slight shock. Authorized person's presence is known by using RFID module interfaced to controller. Sreelakshmi et al. [2] discuss about the design and execution of a sensor node in a plant monitoring system which functions even in low power. By using various sensors, each plant is monitored. The sensed information is sent to the MCU and then transmitted to various platforms through XBee S2C which consumes low power. Subashini et al. [3] discuss monitoring plant growth in real time that helps us to provide proper treatment to plants and increase the yield of our crop. The device gathers data from sensors and then compares it with ideal data. From this comparison, a check is kept on the need of plants for proper growth. Panicker et al. [4] propose the method called hydroponics for plant growth where mineral nutrient solutions are used instead of soil. By using this method, we can save water and space, control climate conditions, etc. The system is a valuable tool to support decision-making on possible interventions to increase productivity. Iwan Fitrianto Rahmad et al. [5] propose pre-programmed control and monitoring system for aeroponic agriculture. The aeroponic environmental system includes equipment that helps to maintain appropriate levels of lighting, heating, air circulation, cooling, and ventilation. Aman Jaiswal et al. [6] discuss the system that accumulates the information of environmental parameters that are favorable for various micro-organism to develop and cause diseases in crops. IoT helps the farmers to access the information and control their actual farm as a virtual firm farm anywhere and at any time around the globe. All these attribute values are then sent to the server of ThingSpeak via the Wi-Fi-enabled ESP-32. The server displays these values and sends them to the Webpage, and on the GUI. Roy et al. [7] present the design of IoT-based dynamic irrigation scheduling system (AgriSens) for efficient water management of irrigated crop fields. The AgriSens provides real time, automatic, dynamic as well as remote manual irrigation treatment for different growth phases of a crop's life cycle using IoT. The AgriSens has a farmer-friendly user interface, which provides field information to the farmers in a multimodal manner—visual display, cell phone, and Web portal. Lakshmi et al. [8] present an IoT-based agriculture monitoring system that helps to gather the data from the WSN network and sending to the mobile phone for notifying the user via IoT. Blynk dashboard is used to display the data parameters such as temperature, soil moisture level, humidity, and water quality. Farmers can access this information on their smart devices. Manikandan et al. [9] propose and test a cloud-based wireless communication framework for tracking and managing a series of sensors and actuators for determining plant water needs and environmental parameters that impact plants. Plants can be watered through a pre-planned irrigation system which is controlled by NODE-MCU. Soil moisture sensor records the moisture present in soil and actuates motor according to the specific condition in a program. Air quality sensor MQ135 is used to keep a check on air quality. The DHT11 sensor efficiently monitors temperature and humidity. Blynk mobile app is used to analyze all wireless sensor data through the cloud. Ultimately, production of the crops is increased.

3 Methodology

Every step taken like the—planning a green space in the concrete jungles, campaigns about saving and planting a tree, education and awareness among the young generations—is a good initiative, but not a single one is enough to overcome the problem. Our approach focuses more on nurturing, caring, and loving the plants. Our approach is to create an emotional quotient in humans for the plants. Our prototype is named SNAP. It has a nurturing system installed for the plant. It has a Webcam to take the images [9]. It also has a water storage tank. Plant has enough space, natural, and favorable conditions to grow. So, SNAP is a prototype that can feed and nurture the plant placed inside it.

3.1 Understanding Plant Health for Prototype Design

Soil Moisture. Soil moisture [6] measures the water content available in the soil. Our prototype serves the purpose of balancing the required amount of water. Soil moisture sensor measures the value of the moisture content in the soil. If the saturation condition is achieved, then excess water will drain off from the bottom of the pot. If the moisture sensor senses management-allowable depression condition, then it will actuate the water feeder module (irrigation system) [12] to water the plant, connected to the water tank. Scheduled watering can also be done to avoid the imbalance supply of water, and IoT helps to control the soil moisture from anywhere–anytime.

Intensity. Plant prepare their food through light [6] and produce energy. This process is called as photosynthesis. Thus, the quantity of light received by the plant is directly connected to its growth and development. Improper intensity light can affect the growth and development of the plant. An increase in the light intensity [13] may cause a decrease in flowering time and leaf number [3]. In our prototype, a threshold value of light intensity is defined for the optimum rate of photosynthesis. LDR sensor records the value of the intensity of light [13]. If the intensity of light is more than the threshold value of intensity, it triggers the actuator to cover the plant with a shade net to reduce the intensity of light. If the intensity of light is less than the threshold value, it switches on the grow light LEDs to ensure proper intensity.

Humidity. In plant, water is absorbed by the roots and evaporated through the leaves into the air. This process of transpiration cools the plant or maintains the temperature of the plant. Humidity [6] in air can affect the flow of water through plants. If humidity increases, the transpiration of water occurs slowly; the plant temperature is not maintained at the optimum level, and if the humidity level is less, it causes water stress in the plant which may affect the plant growth [10].

DHT 11 [5] (Temperature and humidity sensor) measures the moisture content. If the humidity content is less than the threshold value, then the hybrid water irrigation system actuates the sprinkler to increase the humidity.

3.2   Design and Protocol

Hardware. It consists of an Arduino Yun microcontroller [6]. Along with its normal processor, it also has a Linux processor that makes it extra powerful as it runs a Python script to click, process, and then upload the images in to the DropBox using the Webcam. It has various sensors such as DHT11 [5]—to sense the temperature and humidity values, LDR sensor—to measure the Intensity of light [15], moisture sensor—to measure the moistness of the soil [1]. It has four servo motors to rotate the plant base, to position the Webcam at various angles. It has motor relays that control drip irrigation [5] and sprinkling water on the plant. It has Webcam that clicks the images as per instructions. It has a motion sensor [1] to detect a person standing nearby or any other prominent motion. It has a TWI LCD [16] to display the readings and messages. The advantage of this LCD is it uses only two pins of the microcontroller thereby releasing other pins for more inputs.

Software. Arduino IDE is used for coding. Arduino Yun [6], the microcontroller uses the Temboo services and Python SDK to upload the pictures to DropBox. The IFTTT

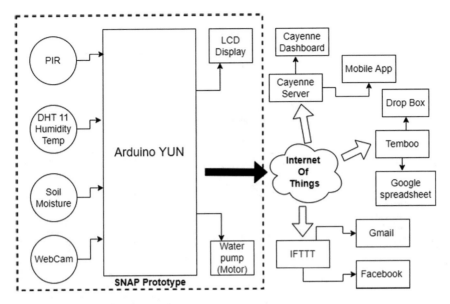

Fig. 1 Design of the SNAP prototype

service posts the images from the DropBox on to Twitter and Facebook as well as other social media. IFTTT services also help in automating various notifications of threshold values that are sent to the end user via SMS or Gmail. Google spreadsheet is used to store all the sensor data with dates and timings using IFTTT services. Cayenne [10] is the real-time IoT dashboard used in this project. Cayenne displays the sensor values and also controls the actuators by widgets on the go. The dashboard can be accessed on the PC, laptop, and mobile app.

Communication Protocol. MQTT (Message Queuing Telemetry Transport) [8] is a publish-subscribe-based messaging protocol. It is an extremely simple and lightweight messaging protocol, designed for constrained devices and low-bandwidth, high-latency, or unreliable networks. The sensors publish their values to the server. The actuators in the prototype subscribe for some specific sensor events. So, if that event occurs, the actuator gets notified and acts accordingly. So, no polling is required to continuous check whether the sensor event is occurred or not. This method avoids the fast draining of the battery which is the foremost prerequisite of an IoT project (Figs. 1 and 2).

3.3 Branding

A journey of being a prototype to emerge as a final finished product is not completed by mere hardware and software requirements. The product should have a selling

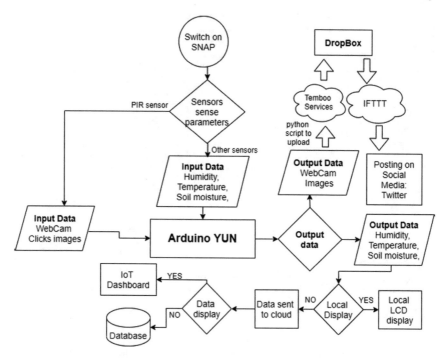

Fig. 2 Flowchart of working of the prototype

proposition. It should be advertised and branded in a unique manner. A well-branded product triggers the emotional chord in consumers. Branding can be divided in to three parts—initiation, propagation, and termination [11].

Initiation. The name of the product should be short, sweet, and catchy and somewhere around should suit your product. Magic rule is—no matter the consumer love it or hate it initially, but they should not ignore it. Our prototype is known as SNAP with a tag line "Are you ready for SNAP?" It has the interaction value most probably a "Yes" or a "No," but seldom ignorance.

Propagation. Successful initiation gives you the chance to propagate your idea and your product. The propagation should be short, exciting, innovative, attractive, and of entertainment value. Our prototype has not only the content value as it can monitor the overall health system on a dashboard. But, it also has the interaction values as it clicks the images with the people and share those images on various social platforms with customized messages using IoT. So, it can score on the initial "interest value" shown.

Termination. The most important part of branding is whether it can close a deal for itself. The branding should make the product appeal to the consumers. It should convey the need of the products existence. Our prototype has that finishing touch. The product name "SNAP—ready to SNAP?" has a very deep meaning within. Just were in the consumer thinks, that the product is just one more luxurious item, the tag line moves him completely for the second time when one is been revealed the meaning of the name. "SNAP" stands for sustainably nurturing a plant. So, the tag line is very touchy in the sense that it directly questions our bonding toward the plants. The tag line asks us whether you are ready to sustainably nurture a plant?

3.4 Features of Prototype

The prototype has multiple dimensions to it. The system senses various parameters to keep the health at check. It can also help to pick up the symptoms of diseases through images and make a timely alert. Thus, it acts as a health monitoring system.

The images are taken at various occasions, changing weathers, different time of the day. All those images are posted on various social media with customized messages. This adds to the emotional quotient. The interaction of plants with young generation across social platforms will increase the bonding. So, the people and especially the youth will not only plant but sustainably nurture a plant using the medium of IoT. So, it plays a role as a good teacher. The prototype accumulates a very diversified and detailed dataset. So, it thereby lays a foundation to build machine learning algorithms to make the plant smart and further enhance the mode of their expression.

4 Results

See Tables 1, 2 and Fig. 3.

SNAP can really click snaps and upload to the social media flawlessly. We tested the postings on Twitter and Facebook. Temboo services saved the clicked images on DropBox. IFTTT [10] and Zapier services worked successfully for all the threshold notifications. Cayenne dashboard [7] helped to implement both the IoT patterns— to control the actuators and display the sensor values with no time lag. Cayenne dashboard is accessible both as a Web app and mobile app [7] making it easily

Table 1 If soil moisture is less than threshold value, an alert message is sent, and water supply is switched on for 3–5 min

Time	Soil moisture readings	Is less than threshold	Alert message sent	Motor status
9.00 AM	60	Yes	Yes	On for 3 min
11.00 AM	79	No	No	OFF
1.00 PM	53	Yes	Yes	On for 3 min
3.00 PM	75	No	No	OFF
5.00 PM	50	Yes	Yes	On for 4 min
7.00 PM	81	No	No	OFF

Table 2 Reading from various sensors—i.e., LDR, DHT11

Day	Parameters	9 AM	11 AM	1 PM	3 PM	5PM	7 PM
Day 1	LDR	840	978	982	980	650	320
Day 1	Humidity	73	70	77	72	78	73
Day 1	Temperature	27	30	33	32	30	27
Day 2	LDR	852	970	985	976	663	298
Day 2	Humidity	74	69	78	70	80	75
Day 2	Temperature	27	30	34	33	30	28
Day 3	LDR	847	965	983	981	655	317
Day 3	Humidity	73	67	78	70	77	72
Day 3	Temperature	28	31	33	32	29	27

Fig. 3 Conversation of SNAP on Twitter account

accessible. So, the plant health gets monitored, and the environment parameters are maintained to their optimum. Over all, SNAP proved the ability to nurture the plant and at the same time strengthen the bonding with the humans (Figs. 4 and 5).

5 Conclusion

SNAP is not just "Sustainably Nurturing A Plant," its more than that. The idea of implementation of SNAP should not come out of sympathy; otherwise, it will meet the same fate as the current approaches. SNAP should be seen as a token of respect for the plants. We are not helping the plants but ourselves as we are "Sustainably

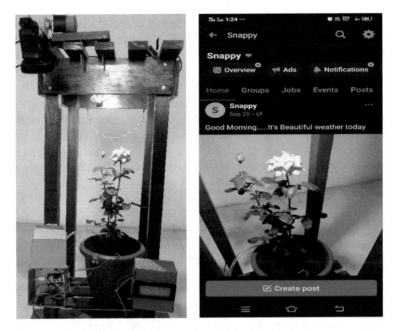

Fig. 4 Actual prototype image and the Facebook post

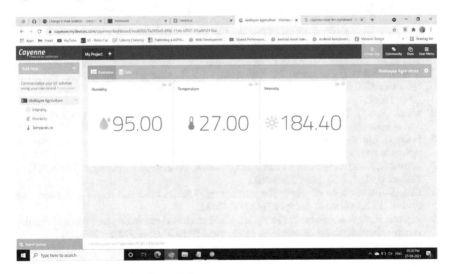

Fig. 5 Cayenne dashboard on laptop/PC

Nurturing A Producer." Finally, the implementation of SNAP ideology can save the mankind as ultimately this will lead to "Sustainably Nurturing A Planet." Also, the stored data through various sensors can help us study the various behavioral patterns of the plants. We can get an insight in to the optimum values of some parameters as when the plant health was good and as when the plant was susceptible to the disease. The plant images can be processed to check whether there are symptoms of any disease and treated accordingly. So, a huge diversified dataset is created for machine learning algorithms that will ultimately help to grow the plants in a perfect environment. Also, the datasets will personify the plants even further and impart emotional intelligence to them. We will identify and care for their needs even more.

References

1. Saleem TA, Rao KS (2017) Automatic crop monitoring using embedded system. Int Res J Eng Technol (IRJET) 4(7). ISSN 2395-0072
2. Ahmed N, De D, Hussain I (2018) Internet of Things (IoT) for smart precision agriculture and farming in rural areas. IEEE Internet of Things J 5(6):4890–4899
3. Suciu G Jr, Ijaz H, Zatreanu I, Drăgulinescu A-M (2019) Real time analysis of weather parameters and smart agriculture using IoT. In: Springer LNICST, vol 283, pp 181–194
4. Mukherjee A, Misra S, Raghuwanshi NS, Mitra S (2019) Blind entity identification for agricultural IoT deployments. IEEE Internet of Things J 6(2):3156–3163
5. Subashini P, Pandey A, Jaiswal S, Sharma A (2019) Real time plant health monitoring system using sensors and clouds. Int Res J Comput Sci (IRJCS) 6(4):2393–9842
6. Kaviyaraj R, Balakumaran A, Avinankudi M, Dhayalan A (2019) IoT enabled precision crop field monitoring system. Int Res J Eng Technol (IRJET) 6(10). ISSN 2395-0072
7. Muangprathuba J, Boonnama N, Kajornkasirata S, Lekbangponga N, Wanichsombata A, Nillaorb P (2019) IoT and agriculture data analysis for smart farm. Elsevier Comput Electron Agric 156:467–474
8. Mandal S, Ali I, Saha S (2020) IoT in agriculture: smart farming using MQTT protocol through cost-effective heterogeneous sensors. In: Springer AISC, vol 1255, pp 903–913
9. Horng G, Liu M, Chen C (2020) The smart image recognition mechanism for crop harvesting system in intelligent agriculture. IEEE Sens J 20(5):2766–2781
10. Jaiswal A, Jindal R, Verma AK (2020) Crop health monitoring system using IoT. Int Res J Eng Technol (IRJET) 7(6). ISSN 2395-0072
11. Jiang JA (2020) A novel sensor placement strategy for an IoT-based power grid monitoring system. IEEE Internet of Things J 7(8):7773–7782
12. Rajaram K, Sundareswaran R (2020) IoT based crop-field monitoring and precise irrigation system using crop water requirement. In: Springer IFIPAICT, vol 578, pp 291–304
13. Panicker VV, Koshy A, Salim S, Philip S (2020) Enhanced plant monitoring system for hydroponics farming ecosystem using IOT. GRD J Glob Res Develop J Eng 5(2). ISSN 2455-5703
14. Tyagi K, Karmarkar A, Kaur S, Kulkarni S, Das R (2020) Crop health monitoring system. In: International conference for emerging technology (INCET), pp 1–5
15. Gujar A, Joshi R, Patil A, Aranjo S (2020) Indoor plant monitoring system using NodeMCU and deep learning. Int Res J Eng Technol (IRJET) 7(11). ISSN: 2395-0072

16. Roy SK, Misra S, Raghuwanshi NS, Das SK (2021) AgriSens: IoT-based dynamic irrigation scheduling system for water management of irrigated crops. IEEE Internet of Things J 8(6):5023–5030
17. Jiang X (2021) Hybrid low-power wide-area mesh network for IoT applications. IEEE Internet of Things J 8(2):901–915

Monitoring Senior Citizens Using IoT and ML

Sunil Kumar Chowdhary, Basheer ul Hassan, and Tushar Sharma

Abstract Everyone wants their elder parents or guardians to be taken care of, it is very hard task to take care of them as they may require emergency assistance at any given point of time. So, in this Paper, a system is developed with the help of IOT sensors and python programming to monitor them closely and to take care of them. In this paper it is tried to achieve some level of monitoring of senior citizens. OpenCV library is used for the object tracking and fall detection. If it is seen that person in the view of camera fell for any reason it gives message as fall detected. It can really be useful in emergency situations, if for any miss happening like a fall emergency treatment can be provided so the damage to the person can be very less compared to not attending them for a long period of time. It can also be life saver for some situations. For this initial work of research webcam of the laptop is used to detect any fall and it can be used in various other situations to track children fall detection. This is done on python and OpenCV.

Keywords Monitoring · IoT · Senior Citizens · Sensors · Image processing

1 Introduction

Everybody want to care for their senior parents or guardians, it is really hard task to deal with them as they might call for emergency assistance at any given point of time. So, in this paper a system with taking help of IOT sensors as well as python programming and OpenCV is made, to look after them. In this research paper are attempting to achieve some degree of monitoring of senior citizens. OpenCV library of programming function is used here for object tracking as well as fall discovery. If the programme is seen that any individual fell in the view of camera, it alerts us as fall detected. Senior citizens can be monitored by using the new technology of IOT sensors, Cameras and many other technologies. If they somehow for any reason behave abnormally and they were unable to call anyone for help, this technology can

---

S. K. Chowdhary (✉) · B. Hassan · T. Sharma
Amity University Uttar Pradesh, Noida, India
e-mail: skchowdhary@amity.edu

© The Author(s), under exclusive license to Springer Nature Singapore Pte Ltd. 2023
A. Shukla et al. (eds.), *Computational Intelligence*, Lecture Notes in Electrical Engineering 968, https://doi.org/10.1007/978-981-19-7346-8_67

give them a second chance as it alerts the person that has responsibility of taking care of them. If any abnormal voice or abnormal movement whilst sleeping, or not getting up on the usual time, it can really help for any emergency assistance required. This technology can be revolutionary as it can monitor much better and efficient than a person, as a human have limitations. A real person cannot take care of old age people 24 hours, but sensors, AI and cameras can monitor 24/7. They do not need any breaks like humans.

Monitoring senior citizens is quite a difficult task as an individual. Unfortunately, any miss happening can be caused and it can lead to loss of life or a serious permanent damage, so to come up with a solution on how to prevent these losses from every individual life.

Monitoring system must be created to prevent these miss happenings. By using fall detection, these accidents from happening or if they in case happened, the caretaker of that person will get alerts on them so immediate help can provided and it can be prevented.

This Monitoring System aims to solve the problem of taking care of senior citizen via the help of IOT and Ml. The user can see and monitor their respected loved ones based on their daily routine; this includes a fall detection feature also which will help them get alert in case of any miss happening.

The various methodologies used in the project are:

To provide end-point users with an interactive and responsive UI that not only looks elegant and interesting but is also smooth and functioning.

The integration of back-end and front-end of this project will be as smooth and as compelling to the user so that he can, or she can always monitor their loved ones without any hassle and get alerts whilst anything unfortunate happens to them.

2   Methodology

For back-end development, real-time computer processing is made, so OpenCV library of programming is used for that. Computer vision is a process by which the machine can understand the images as well as videos exactly how they are saved as well as get data from them. Computer system Vision is the base, mainly utilized for Expert system. Computer-Vision is playing a significant role in self-driving vehicles, robotics as well as in image adjustment applications [8].

OpenCV is the big open-source library for the computer vision, machine learning, as well as photo handling and now it plays a significant duty in real-time operation which is extremely vital in today's systems. By using it, one can process pictures and videos to determine items it can even understand handwriting of a human. When it incorporated with numerous collections, such as NumPy, python can process the OpenCV array framework for evaluation it can determine picture pattern and its different functions and can make use of vector space as well as carry out mathematical procedures on these features [8].

The first OpenCV version was OpenCV 1.0. OpenCV is delivered under a BSD licence it is free for both academic and business use. It has C++, C, Python, Java UIs and furthermore it supports Windows, Linux, Mac OS, iOS and Android. At the point when OpenCV was made the essential accentuation was continuous applications for computational execution. Everything was formed in expanded C/C++ to exploit multi-Centre handling [9].

Applications of OpenCV:

There are bunches of applications which are addressed utilizing OpenCV, a couple of them are recorded here [1].

- Face Recognition
- Automated inspection as well as monitoring
- Number of people– count (foot traffic in a shopping mall, etc.).
- Car trusting highways together with their speeds.
- Interactive art setups.
- Anomaly (flaw) discovery in the manufacturing procedure (the odd defective items).
- Street sight picture sewing.
- Video/image search as well as retrieval.
- Robotic as well as driver-less cars and truck navigation as well as control.
- Object recognition.
- Medical image analysis.
- Flicks– 3D framework from motion.
- Television Channels promotion recognition.
- Some OpenCV Functionality.
- Picture/video I/O, handling, screen (centre, imgproc, highgui).
- Object/highlight recognition (objdetect, features2d, nonfree).
- Geometry-based monocular or sound system PC vision (calib3d, sewing, videostab).
- Computational photography (photograph, video cut, superres). Device finding out & clustering (ml, flann).
- CUDA acceleration (gpu).

Image-Processing is a process on an image, to get an upgraded photograph or to eliminate some waste data from it. "Image processing is the analysis and also manipulation of a digitized picture, especially in order to enhance its quality".

Digital-Image is an image might be specified as a two-dimensional feature f (x, y), where x and y are spatial, as well as the amplitude of any set of coordinates (x, y) is called grey level of the image at that point. A picture is absolutely nothing more than a two-dimensional matrix (3D in case of coloured pictures) which is specified by the mathematical function f (x, y) [2].

In Image-Processing generally input is a picture and outcome is picture with features according to demand related to that image. Image processing essentially consists of three steps:

1. Importing the picture.
2. Evaluating and manipulating the image.
3. Result in which result can be altered picture or report that is based on image analysis.

How does a computer read an image? Consider the picture people can quickly make it out that favours an individual. However, the computer cannot say anything since the computer is not finding out it all on its own. The computer system checks out any photo as a range of values in between 0 as well as 255. For any type of colour photo, there are 3 main colours from which almost all colours can be made which are red, green and blue.

The most recent improvements in computer vision technology are making everyone's life more and more automated-- its algorithms can run houses, drive vehicles, make medical diagnoses, and assist individuals in lots of various other ways. OpenCV library of programming that provides a rich selection of algorithms utilized for face acknowledgement, image reconstruction, and several other applications [3].

Computer system vision (CV) is a subset of artificial intelligence that manages identifying, processing and differentiating items in digital pictures and video clips. Building computer vision applications features a range of options and innovations, including libraries, structures and systems. OpenCV is just one of these services. It is a bunch of assortments with more than 2500 functions that shift from traditional ABA(ML) calculations, like direct relapse, just as decision trees, to profound finding just as semantic organizations.

OpenCV technology is an open-source library that can be straightforwardly utilized, adjusted, and disseminated under the Apache permit. The library was at first introduced by Intel in 1998, and the organization has really been supporting OpenCV and furthermore adding to it from that point.

The assortment is viable with a scope of running frameworks, including Windows Os, Linux Os, macOS, FreeBSD, Android, iOS, BlackBerry 10, and supports programming written in C/C++, Python, and Java. It has solid cross-stage limit and similarity with different systems. OpenCV consists of a wide series of modules meant to process pictures, find and track things, describe attributes, as well as carry out several various other jobs. Below are several of the modules:

- Core. Core Functionality.
- Picture Processing.
- Video Clip I/O.
- Video. Video Clip Evaluation.
- Object detection Things Discovery.
- Ml. Artificial Intelligence.

OpenCV library is outfitted with GPU module that gives high computational power to record videos, images, as well as manage other procedures in real-time. This module makes it possible for programmers to create innovative formulas for high performance computer system vision applications. OpenCV is used globally. Lots of countless AI researchers, scientists, and engineers have been providing important understandings into the collection for more than 20 years [12].

OpenCV in IoT House Automation.

Savvy homes are Internet of Things frameworks that help individuals in running family works. The organizations of IoT gadgets can handle lights, control indoor temperature, just as switch on the TV. Computer vision is one of the cutting-edge innovations that make homes shrewd–for instance, an astute cooler can utilize camcorders to analyse food materials. Architects can use an enormous assortment of computer vision calculations just as gadgets promptly accessible in the OpenCV library to make smart house tasks (Figs. 1 and 2).

Fall discovery has ended up being an important steppingstone in the study of action recognition—which is to educate an AI to categorize basic activities such as walking as well as taking a seat. What humans take an evident action of an individual falling face level is but a sequence of jumbled up pixels for an AI. To make it possible for the AI to understand the input it gets, it is needed to teach it to discover certain patterns as well as forms, as well as develop its own regulations.

To develop an AI to find drops, it is determined not to experience the torment of generating a big dataset and training a model specifically for this function. Instead, pose evaluation as the foundation is used.

Position Estimate: Posture estimate is the localization of human joints, generally called key points in pictures and video structures. Commonly, each person will be made up of a few key points. Lines will certainly be attracted in between key point

Fig. 1 Smart computer vision-based system fills missing information

Fig. 2 Posture estimation discovery

sets, successfully mapping a rough form of the individual. There is a range of posture estimation approaches based on input and detection method.

To make this design conveniently obtainable to everybody, the input as RGB images and processed by OpenCV is picked. This means it works with common cams, video documents, and HTTP/RTSP streams [11].

Multi-Stream Input: A lot of open-source versions can just process a solitary input at any once. To make this even more flexible and scalable in the future, the multi-processing library in Python to refine numerous streams concurrently using sub-processes is utilized. This permits us to totally utilize numerous processors on makers with this capability (Fig. 3).

In video frames with multiple people, it can be tough to identify an individual who falls. This is since the formula requires to correlate the exact same person between successive frameworks. But how does it recognize whether it is taking a look at the very same person if he/she is relocating frequently?

The service is to apply a several person trackers. It does not need to be elegant; simply a basic things tracker will certainly be enough. Exactly how tracking is done is rather simple and can be detailed in the following actions:

1. Calculate centroids (taken as the neck factors)
2. Assign distinct ID to each centroid.
3. Calculate new centroids in the following structure.
4. Determine the Euclidean range in between centroids of the existing and also previous frame and associate them based upon the minimum distance.
5. If the connection is discovered, update the new centroid with the ID of the old centroid.
6. If the connection is not found, provide the new centroid a one of a kind ID (beginner gets in the structure).

Fig. 3 Shows that position evaluation design can run concurrently on the two streams

7. If the individual heads out of the frame for a set quantity of structures, get rid of the centroid and also the ID [13].

The preliminary fall detection formula that was conceptualized was reasonably simplistic. Next off, the perceived elevation of the individual based upon bounding boxes that defined the entire person is calculated. Then the upright distance between neck factors at intervals of structures is calculated. If the vertical distance went beyond half the perceived elevation of the individual, the algorithm would signify a fall.

Nonetheless, after encountering several YouTube videos of people falling, it now knew there were various ways and positioning of dropping. Some falls were not detected when the field of view was at an angle, as the victims did not show up to have a radical modification moving. The design was likewise not robust enough and maintained tossing false positives when people bent to tie their shoelaces or ran right down the video frame.

So, it is decided to apply even more features to refine our formula:

- As opposed to analysing one-dimensional activity (y-axis), two-dimensional activity (both x as well as y-axis) to include different camera angles is evaluated.
- Added a bounding box check to see if the width of the person was larger than his elevation. This thinks that the person is on the ground as well as not upright. It had the ability to get rid of false positives by fast-moving people or bicyclists using this technique.
- Added a two-point check to just keep an eye out for falls if both the individual's neck and ankle joint factors can be identified. This protects against imprecise calculation of the person's elevation if the person cannot be completely determined because of occlusions [12] (Figs. 4, 5 and 6).

Fig. 4 Drunk fall

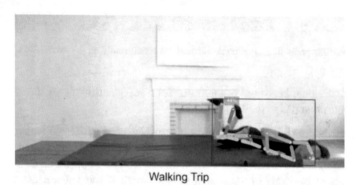

Fig. 5 Walking trip fall

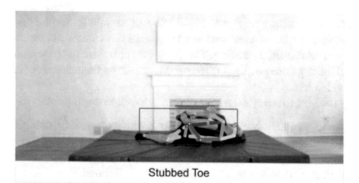

Fig. 6 Stubbed Toe Fall

Applications:
Fall discovery can be applied in several scenarios to provide support. A non-exhaustive list consists of:

- Drunk people
- The elderly
- Kids in the play area
- People that struggle with medical problems like heart attacks or strokes
- Negligent people that trip and fall

The precision of fall detection is greatly based on the pose evaluation precision. Normal pose estimation designs are educated on tidy images with a full-frontal view of the subject. However, falls trigger the subject to be bent in unusual positions, and many position evaluation models are not able to accurately define the skeletal system in such circumstances. Additionally, the models are not durable adequate to get over occlusions or picture noise [16].

To attain a human-level detection precision, existing position evaluation models will need to be re-trained on a bigger variety of postures and include lower-resolution pictures with occlusions.

Present hardware constraints also impede the capability of present estimate versions to run smoothly on video clips with high frame rates. It will be some time before these versions will be able to run conveniently on any type of laptop with a fundamental GPU, and even only with a CPU.

Besides posture estimate, a deep knowing version trained specifically on drops would likely execute also or even much better. The version must be educated thoroughly to differentiate drops from various other fall-like activities. This, certainly, have to be coupled with considerable, openly offered fall datasets to train the model on. Of course, such a model is restricted in extent as it just can determine one specific action, as well as not a selection of actions.

An additional possible strategy would certainly likewise be knowledge-based systems, which is developing a version such that it is able to learn the means humans do. This can be accomplished through a rule-based system where it makes decisions based upon specific rules, or a case-based system where it applies resemblances in previous cases it has actually seen to make a notified judgement about a new case [19].

3 Results

In the proposed model, we have used Tiny-YOLO oneclass to recognize each person in the frame, AlphaPose to obtain skeleton-pose, and ST-GCN model to forecast action from every 30 frames of each person's track. Standing, Walking, Sitting, Lying Down, Standing, Sitting, Fall Down are the seven actions presently supported in this model.

The goal was to train a new Tiny-YOLO oneclass model to recognize only human objects whilst also lowering model size. For more robust person detection in a version of angle posture, train with the rotation enhanced COCO person keypoints dataset. For training the ST-GCN model, we have used data from the Le2i Fall detection dataset (Coffee room, Home), extracted skeleton-pose using AlphaPose, and manually labelled each action frame.

As a multi-feature cross-platform collection with high handling speed, OpenCV was a suitable option to complete the above tasks. The task was to utilize OpenCV human discovery formulas as well as semantic networks to track as well as examine people's activities throughout the day.

After fragmenting a body, the human disclosure framework distinguished its biomechanical data, for example, body calculation and exercises. These standards were determined and recognized by OpenCV development following calculations. To estimate the pose of an individual, we relied on artificial information generation. By utilizing the simulation collections, proposed system produced a physical model of a body which is based upon actual percentages, biometric as well as biomechanical data. The model adjusted for various digital settings to create likely scenarios of human activities. As an example, a person might stroll, stagger and may drop overall. So, there had to train the model with such hundreds of different situations. Based upon that the approximate parameters for the algorithm are identified to approximate the pose.

The largest challenge of such type of task associated with real-time human detection in computer vision is to decreasing incorrect positives and accomplishing high discovery precision.

At first, in the model straightforward OpenCV machine discovering formulas was used to set apart positions, whether an Standing, Sitting, Walking, Lying or Fell. Applying Hidden Markov Versions aided us anticipate an autumn based upon the previous habits of an individual. The more specific the discrepancies, the greater the chance of drops.

Prima facie, this technique appeared to work, but it prove to be ineffective as the system detected a fall when a person was sitting down or simply leaning ahead. Additionally, contour detection depended upon the lights in the space. This could not detect individuals making use of computer system vision in poor illumination.

To accomplish much better precision in human body discovery, it is determined to develop neural networks based on biometric criteria and biomechanical data. To educate the models, artificial datasets are generated with the assistance of simulation solutions. Unity, Gazebo, and other simulator systems provide test settings that simplify as well as accelerate the training process.

Executing individuals' location using PC vision can get significantly more made complex in indoor spaces, like houses. An individual can be taken cover behind a furniture piece, so the camcorder will not get the whole body and furthermore hence the framework will unquestionably not get the full biometric data. After trying various strategies, it is incorporated neural networks with choice trees—classical device learning algorithms consisted of in OpenCV motion detection. These formulas have considerable benefits: they are rapid and understandable; they can gain from percentages of information or when some information is missing (Figs. 7 and 8).

Fig. 7 Representation of programme running (standing person)

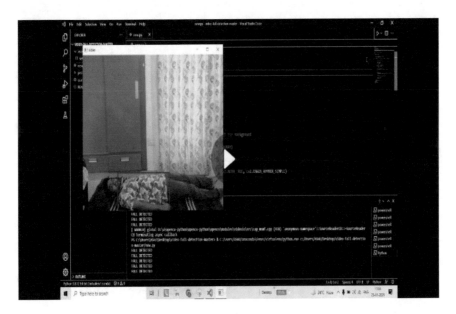

Fig. 8 Representation of programme running (fall)

4 Conclusion

Those who want to look after their elderly parents or guardians, proposed system using IoT sensors, machine learning, python programming and OpenCV can be referred. Object tracking and fall detection are both done using the OpenCV library of programming functions. If the computer detects that someone has fallen in front of the camera, it sends us an email with the subject "fall detected," as well as a link to a live stream where the concerned party may engage with the person in real-time because the camera includes a microphone and speakers. IoT-enabled lights are put in the room for the comfort of elderly residents since they detect motion and come on automatically, which may help prevent tragic situations such as injury in the dark or a fall, amongst other things. Individually, keeping track of older persons is a challenging undertaking. Unfortunately, any mishap can occur, resulting in the loss of life or major permanent harm. Therefore finding a way to avoid these losses from occurring in everyone's life is essential. By using Fall Detection, Motion detection and alert systems these accidents can be prevented from happening. In case, it happened, the caretaker of that person will get alerts on their smart gadgets. So, immediate/timely help can be provided and it can be treated as early as possible to avoid/defer the damage to the person. This monitoring system aims to solve the problem of taking care of senior citizen via the help of IoT and ML. The user can see and monitor their respected loved ones.

References

1. Abbate S, Avvenuti M, Bonatesta F, Cola G, Corsini P, Vecchio A (2012) A smartphone-based fall detection system. Pervas Mobile Comput 8:883–899. https://doi.org/10.1016/j.pmcj.2012.08.003
2. Adhikari K, Bouchachia H, Nait-Charif H (2017) Activity recognition for indoor fall detection using convolutional neural network. In: 2017 Fifteenth IAPR international conference on machine vision applications (MVA) (Nagoya: IEEE), pp 81–84. https://doi.org/10.23919/MVA.2017.7986795
3. Akagündüz E, Aslan M, Şengür A, Wang H, İnce MC (2017) Silhouette orientation volumes for efficient fall detection in depth videos. IEEE J Biomed Health Inform 21:756–763. https://doi.org/10.1109/JBHI.2016.2570300
4. Alamri A, Ansari WS, Hassan MM, Hossain MS, Alelaiwi A, Hossain MA (2013) A survey on sensor-cloud: architecture, applications, and approaches. Int J Distrib Sens Netw 9:917923. https://doi.org/10.1155/2013/917923
5. Amini A, Banitsas K, Cosmas J (2016) A comparison between heuristic and machine learning techniques in fall detection using kinect v2. In: 2016 IEEE international symposium on medical measurements and applications (MeMeA) (Benevento: IEEE), pp 1–6. https://doi.org/10.1109/MeMeA.2016.7533763
6. Aslan M, Sengur A, Xiao Y, Wang H, Ince MC, Ma X (2015) Shape feature encoding via fisher vector for efficient fall detection in depth-videos. Appl Soft Comput 37:1023–1028. https://doi.org/10.1016/j.asoc.2014.12.035
7. Auvinet E, Multon F, Saint-Arnaud A, Rousseau J, Meunier J (2011) Fall detection with multiple cameras: an occlusion-resistant method based on 3-D silhouette vertical distribution. IEEE Trans Inform Technol Biomed 15:290–300. https://doi.org/10.1109/TITB.2010.2087385

8. Available online at https://www.integrasources.com/blog/opencv-computer-vi-sion-algori thms-iot-home-automation/
9. Available online at https://www.ncbi.nlm.nih.gov/pmc/articles/PMC5751723/
10. Aziz O, Musngi M, Park EJ, Mori G, Robinovitch SN (2017) A comparison of accuracy of fall detection algorithms (threshold-based vs. machine learning) using waist-mounted tri-axial accelerometer signals from a comprehensive set of falls and non-fall trials. Med Biol Eng Comput 55:45–55. https://doi.org/10.1007/s11517-016-1504
11. C. for Disease Control, Prevention, Falls in nursing homes. http://www.cdc.gov/HomeandRe creationalSafety/Falls/nursing.html. Accessed on 23rd April 2015 (2014)
12. Delahoz Y, Labrador M (2014) Survey on fall detection and fall prevention using wearable and external sensors. Sensors 14(10):19806–19842
13. Hijaz F, Afzal N, Ahmad T, Hasan O (2010) Survey of fall detection and daily activity monitoring techniques. In: 2010 international conference on information and emerging technologies (ICIET), pp 1–6
14. Kangas M, Vikman I, Nyberg I, Korpelainen R, Lindblom J, Jämsä T (2012) Com parison of real-life accidental falls in older people with experimental falls in middle-aged test subjects. Gait Posture 35(3):500–505
15. Khan SS (2016) Classification and decision-theoretic framework for detecting and re-porting unseen falls, Ph.D. thesis, University of Waterloo
16. Perry J, Kellog S, Vaidya S, Youn J-H, Ali H, Sharif H (2009) Survey and evaluation of real-time fall detection approaches. In: 2009 6th international symposium on high-capacity optical networks and enabling technologies (HONET), pp 158–164
17. Schwickert L, Becker C, Lindemann U, Maréchal C, Bourke A, Chiari L, Helbostad J, Zijlstra W, Aminian K, Todd C, Bandinelli S, Klenk J (2013) Fall detection with body-worn sensors: a systematic review. Zeitschrift f ur Gerontologie und Geriatrie 46(8):706–719
18. Stone E, Skubic M (2015) Fall detection in homes of older adults using the microsoft kinect. IEEE J Biomed Health Inf 19(1):290–301. Noury N, Fleury A, Rumeau P, Bourke A, Laighin G, Rialle V, Lundy J (2007) Fall detection-principles and methods. In: Engineering in medicine and biology society. EMBS 2007. 29th Annual International Conference of the IEEE, IEEE, pp 1663–1666
19. Yu X (2008) Approaches and principles of fall detection for elderly and patient. In: 10th international conference on ehealth networking applications and services, HealthCom'08, IEEE, pp 42–47

Analysis of Indian Rice Quality Using Multi-class Support Vector Machine

S. Harini, Saritha Chakrasali, and G. N. Krishnamurthy

Abstract In this research, quality of Indian rice kernels is accessed using image processing and the techniques of machine learning. Analyzing rice quality through human inspection method is laborious and is prone to error based on subjectivity. Automatic machine vision solution for rice quality analysis is a breakthrough in this research area. The quality parameters considered to classify the rice grains are Degree of Milling (DOM) and Percentage of Broken kernels (PBK). The rice kernels with different DOM are collected from the commercial rice mill. The images of the collected rice samples are captured using an experimental setup. Each image captured has the dimension of 2448 × 3264 pixels. The captured images are pre-processed and features are extracted. The statistical features extracted using GLCM are fed into Machine learning model. The identification of DOM based on these extracted features is carried out using Decision Tree, K-Nearest Neighbor and Support Vector Machine algorithms. This model achieves $\geq 98\%$ of accuracy in identifying PBK. The results obtained shows that SVM with gamma $= 0.8$ and C $= 5$ or greater gives a promising results of $\geq 92\%$ for DOM.

Keywords DOM · PBK · Rice kernels · GLCM · SVM · Decision tree · KNN

1 Introduction

Analyzing grain quality continues to be an important topic of research from nearly century. Cereal grains play an imperative role in meeting the nutrient needs of the human population. Like any food, they are excellent sources of nutrients. Measuring

S. Harini (✉)
Department of ISE, B.M.S. College of Engineering, Bengaluru, India
e-mail: hharinis8@gmail.com

S. Chakrasali
Department of ISE, BNMIT, Bengaluru, India

G. N. Krishnamurthy
Department of CSE, BNMIT, Bengaluru, India

individual kernel is tedious but is still a very significant factor for quality measurement. Currently, the quality of grain is analyzed using manual method. This process is complicated and requires more time. There are more chances of human errors based on subjectivity of their perception. Hence machine-based techniques are evolved to achieve standard quality uniformly and precisely [1], it has a prime advantage of reproducing the same qualitative result efficiency upon repetition. Latest developments in the area of image processing have given a wide scope as a sample analysis tool [2].

Rice being one of the leading food crops of the world is produced in almost every continent. India being the largest rice exporter, rice is grown in around 43 million hectares (m ha) in the country- largest acreage in the world [3]. There are possibly up to 82,700 varieties of rice extant in India [4]. Among the different rice varieties, non-Basmati rice like Sona Masoori and RNR 15,048 are popular in southern India. They are moderate sized, lightweight and scented rice grown in large parts of Andhra Pradesh, Karnataka and Telangana. Since this variety of rice is widely consumed in our locality (Bangalore), this research work concentrates in analyzing the quality of RNR rice.

One of the quality analyses of the given rice kernels samples are based on amount broken and whole grain. This process is laborious and time consuming. The standard methods used for grading of rice kernels using image as inputs have used statistical methods [5]. The paddy grown undergoes a set of processes for the conversion of paddy to rice through mechanical systems. The major processing includes threshing, handling, de-husking, milling and whitening of grains. The final whitened rice kernels have high commercial value [6]. Generally, in a rice mills, the inspection of the rice quality is carried out at every 2–3 h by an experienced quality control personnel. This is a visual inspection method. In this method the person who operates, assesses the quality of the grains based on his/her experience and proficiency. This method gives subjective results and is time consuming.

A non-destructive and rapid solution is much needed work of the hour. A solution with image as an input would provide a great result. The approaches based on image processing and machine learning are expected to give promising results and help in conducting automatic analysis. This approach has yield to proven results based on the work conducted by Brosnan et al. [7], Zheng et al. [8]. In the last decade, researchers have investigated various techniques based on machine vision and digital image processing for assessing the quality of rice kernels which is non-destructive, accurate, fast and cost-effective as compared to traditional techniques [9, 10].

Hence in this work we aim in developing an image processing and machine based model for automatic analysis of rice quality.

Further section of this paper includes literature survey, sample collection and dataset preparation, methodology, results and discussion, conclusion, future work and references.

2 Related Work

Machine vision techniques provide a quick and objective ways for measuring or evaluating the visual features of grains. Researchers reported using these techniques for peach defect detection corn kernel breakage classification, corn kernel stress crack detection, wheat classification and grain classification [11]. However, there is very less work using machine vision to measure DOM of rice kernels. Fant et al. [12] proposed a technique to measure DOM using gray scale intensity. Although they classified rice into the DOM grades recognized by the United States Standards for Milled Rice, they did not attempt to quantify DOM on a linear scale [11]. In commercial practice, milling meters or color meters is used to measure DOM rapidly and objectively [13]. Measurement of milling degree of rice on basis of image analysis is dependent on the color of milled rice kernels. More white kernels correspond to higher degree of milling and are measured by pixel determination [14]. Experimental results shows that the algorithms perform well in evaluating the percentage of broken rice with overall accuracy of 92%. Shiddiq et al. [15] suggested an Adaptive Neuro-fuzzy Inference System (ANFIS) in which the relationships between rice color characteristics and its quality is expressed using Fuzzy logic and artificial neural network. Zareiforoush et al. [16] developed an intelligent system using fuzzy logic and computer vision for measuring rice quality based on DOM and PBK. The work is carried out on a popular rice variety, Hashem of Iran. In this work an overall results of 89.8% is obtained from the developed system and the results determined by the experts. Chen et al. [17] showed that head rice and broken rice could be effectively identified by Least Squares Support Vector Machine. Similarly, Fayyazi et al. [18], Gujjar and Sidappa [19], Shantaiya and Ansari [20], Kaur and Singh [21] and Prajapati and Patel [1] for classification of different rice varieties using SVM and ANN. Based on the above literature survey, we inferred the need of the work for automatic quality analysis of Indian rice variety. Thus this experiment is conducted using SVM, Decision tree and KNN to classify three different degree of the milled rice.

3 Materials and Methods

The abstract system design of the proposed research work is as shown in Fig. 1.

The abstract system design includes the four major steps. The initial and most important phase is Image acquisition and/or Data collection.

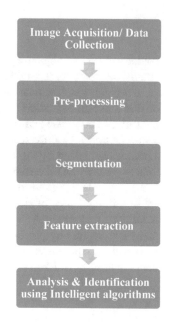

Fig. 1 Abstract System Design of the methodology

3.1 Sample Collection

The rice grain samples are collected from a commercial rice mill, S. G. Rice Mill, Antharasanahalli, Tumkur. The paddy will be cleaned and the husk is removed. This brown unpolished rice is further polished at three different stages. The third level polished rice will have no bran and will have poor nutritional value and is readily available in market. For this work, the rice samples after stage 1, 2 and 3 milling are collected from the above mentioned commercial Rice mill (S G Rice Mill, Tumkur, Karnataka, India) (Fig. 2).

The below figure shows samples of Rice kernels obtained after different milling stages. Currently the rice kernels which undergo third level of milling is available in market. The laboratory analyses are carried out at Sri Bhagyalakshmi Agro Foods Pvt.

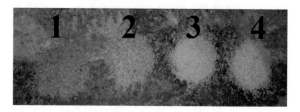

Fig. 2 Different stages of milled rice-(1) De-husked rice kernels, (2) stage one milled rice (3) stage two milled rice (4) stage three milled rice kernels (left to right)

Ltd. The evaluated rice variety, RNR (also called as "TelanganaSona") is commonly grown and consumed rice variety in southern part of India. This variety is cold tolerant rice with good cooking quality [22]. This rice variety is low in Glycemic Index (GI) and is good for Diabetic patients when compared to Sonamasoori rice [23]. The Whiteness, Transparency and Milling degree of the sample rice grain are measured by using Milling Meter (SATAKE Corporation, Milling Meter MM1D) and the dimensions of the rice kernel is measured by Vernier Calipher.

3.2 Image Acquisition

To acquire and create the image dataset for our work, we have developed an experimental setup. This standard experiment setup is fabricated for the uniform and well suited environment for capturing images of the rice kernels. The digital image of the rice kernel is acquired using the set up.

The setup is devised of non-transparent material. The whole setup has two chambers. The lower chamber is used for placing the samples. The upper has a grove which is used for placing the smart phone/ camera for capturing the images. As this work concentrates in finding a wieldy solution, we have used a mobile camera with the resolution of 8 Mega pixel and LED flash with 3.2X zooming. Each image has a dimension of 2448 × 3264 pixels. There was no source of light in both the chambers. The images are captured using the smart phone in flash mode. A black surface was used as background to simplify the segmentation process. Average of 30 to 40 rice kernels are of each class (Different levels of milling) are spread over the tray with black background. The rice kernels in each samples were manually separated and placed under the camera. This process is repeated over 25 times for each class of rice kernels. A uniform distance of 30cms in preserved between the sample and the camera. The sample images taken by the camera, are transferred to the through a USB chord and are then stored for further preprocessing. The acquired images are then pre-processed and textural features are extracted.

The sample images captured are as shown in Fig. 3.

The Fig. 3a, shows the images of rice kernel samples which have undergone stage one level of milling. The sample also includes broken kernels which will be detected and Percentage of Broken Kernels is calculated. Similarly Fig. 3b and c, shows the images of rice kernel samples which have undergone stage two and stage three level of milling, respectively.

3.3 Image Preprocessing Module

The acquired images are converted to gray scale. The converted gray images are filtered using Gaussian Blur to reduce the noise. The sample images after preprocessing are as shown in Fig. 4.

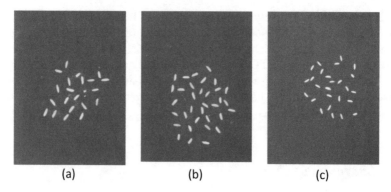

Fig. 3 Sample images of rice kernels **a** degree 1 milled **b** degree 2 milled **c** degree 3 milled

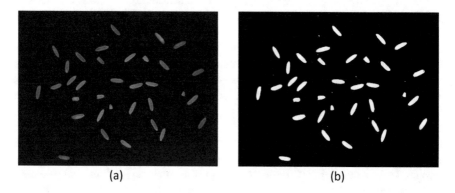

Fig. 4 Stages of preprocessing

The image in Fig. 4a, is the image after converting to gray scale and Fig. 4b, is the image obtained after filtering using Gaussian Blurring.

3.4 Image Segmentation

To find the PBK, each rice kernel in the image should be segmented and measured. Thresholding is the simplest method of segmenting images [24]. Thresholding creates a binary image based on setting a threshold value on the pixel intensity of the original image. In this work, Otsu's Thresholding is applied on the filtered image to separate rice kernels from the background [25]. Each rice kernel in the obtained binary image is separated by finding contours. After finding (as shown in Fig. 5) the contours of each rice kernel in the given sample, the contour area is considered. The major and minor axes of each rice kernel are extracted from this ellipse. The percentage of broken kernels was obtained by measuring the major axis length of the rice kernel.

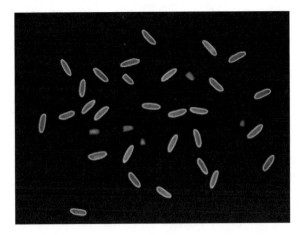

Fig. 5 Image after Contouring

The given rice kernel is considered as broken if the major axis length is less than three-quarters of the average length of whole kernels (defined by a certain pixel value) were considered as broken kernels (ISO, 2011). Based on this, if the major axis length of any rice kernel is below 75% of the average major axis length of all rice kernels present in the given image, then that rice kernel is considered as broken and it is contoured in red color.

3.5 Feature Extraction Module

The contour area obtained is then used to draw a rectangle box bounding each rice kernel. This bounding box gives minimal up-right bounding rectangle for the specified point set. The bounding box for few rice kernels are shown in Fig. 6.

Both the contours and bounding box are applied on the original image but not on the binary image. Rice kernel DOM can be measured by image texture analysis [26]. The textural properties of rice kernels in images are obtained using the Gray-gradient co-occurrence matrix (GLCM). Fang et al. [27] have used the gray-level-intensity parameter in image processing for assessing rice kernels' DOM. Zareiforoush et al. [16], have used GLCM to extract textural features and measure the rice kernel quality based on DOM and PBK. Based on the above mentioned work, this work has also considered GLCM features for quality analysis of rice kernels. The Width, Height, Aspect Ratio, Contrast, Dissimilarity, Homogeneity, Energy and Correlation features extracted using GLCM.

The features are extracted for each rice kernel, i.e., a total of 1733 rice kernels are examined for this process. The algorithms used for the analysis are: Support Vector Machine (SVM), K-Nearest Neighbors (KNN) and Decision trees are considered for classification of rice kernels based on DOM. The SVM algorithm is implemented

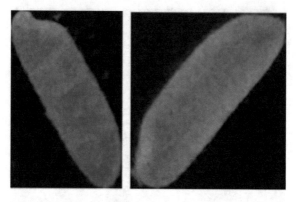

Fig. 6 The bounding box over rice kernels

in practice using a kernel. The learning of the hyperplane in linear SVM is done by transforming the problem using some linear algebra, which is out of the scope of this introduction to SVM. The different kernel functions used in SVM are:

- **Linear Kernel SVM**: The equation for Linear kernel is shown below:

$$K(x, xi) = \text{sum}(x * xi) \tag{1}$$

- **Polynomial Kernel SVM**: The equation for Polynomial kernel is as shown below.

$$K(x, xi) = 1 + \text{sum}(x * xi)^{\wedge} d \tag{2}$$

- **Radial Basis Function (RBF) Kernel SVM**: The equation for this function is as given below:

$$K(x, xi) = \exp(-\text{gamma} * \text{sum}((x - xi^2))) \tag{3}$$

4 Results and Discussion

The Machine learning algorithms like Multi-class SVM, KNN and Decision tree are used in this work and the results obtained are discussed below. The dataset of total 1733 rice kernel images are considered. Among this total 1733 images, 80% of them that is 1386 images are considered for training and 20%, i.e., 347 are used for testing. By considering all eight features, i.e., Width, Height, Aspect Ratio, Contrast, Dissimilarity, Homogeneity, Energy and Correlation, initially the work was conducted on an un-scaled data with linear kernel function of SVM. The work is carried out on all three kernel functions along with KNN and Decision trees. The results obtained for this condition is given Table 1.

Table 1 Accuracy table for multi-class SVM with Linear, RBF and polynomial kernel functions, KNN and decision tree

Algorithms	Accuracy (%)
SVM (linear Kernel function)	72.88
SVM (polynomial Kernel function)	82.33
SVM (RBF Kernel function)	60.44
Decision tree	80.09
KNN	55.22

The work has considered all features and later selected features to compare the result.

The values in the Table 2 are obtained from SVM with RBF kernel function. The work is carried out for different 'c' and 'gamma' values of RBF kernel function. It is observed that a highest accuracy of 92% is achieved for minmax normalized data with c = 4 and gamma = 0.8. The experiment is further continued on lesser number of features obtained from GLCM. The features selected are Contrast, Dissimilarity, Energy, Homogeneity and Correlation. The experiment has given noticeable results for Multi-class SVM. The results shows that an approximate of 92% accuracy is obtained for both selected and complete set of features. The accuracy of 92.03% is achievement for non-overlapping normalized test datasets with all eight features of rice kernels.

5 Conclusions and Future Work

An intelligent system with image processing and machine learning is used in this work to identify the quality of rice. The DOM and PBK are the quality parameters considered. PBK is calculated based the major and minor axis of each rice kernels. This model achieves nearly $\geq 98\%$ of accuracy in identifying PBK. The identification of DOM based on these extracted features is performed using Decision Tree, SVM and KNN algorithms. The results obtained shows that SVM with gamma = 0.8 and C = 5 or greater gives a promising results of $\geq 92\%$. This work is intended to use advanced algorithms like CNN for the same dataset in future.

Table 2 Showing results of multi-class SVM using RBF kernel function with different Gamma and C values for all GLCM features

	All features							Selected features				
	Un-normalized data			Normalized data				Un-normalized data		Normalized data		
C	Gamma = 0.05	Gamma = 0.2	Gamma = 0.5	Gamma = 0.8	C	Gamma = 0.8	Gamma = 1	Gamma = 0.8	Gamma = 1
1	61.94	58.45	86.31	87.56	1	58.95	59.2	88.3	89.55
2	61.19	58.95	89.05	89.8	2	58.95	58.95	90.04	90.04
3	61.69	59.45	89.8	91.29	3	59.7	60.94	90.29	91.54
4	60.69	59.45	91.04	92.03	4	61.44	63.68	91.79	91.29
5	61.19	59.2	91.29	92.03	5	64.92	67.41	91.29	92.03
6	61.19	59.7	91.29	91.79	6	68.15	70.64	**92.03**	91.79
7	60.44	59.2	92.03	91.54	7	70.89	72.63	**92.03**	91.79
8	60.44	58.45	92.03	91.29	8	73.13	73.38	91.79	91.54
9	59.7	57.71	91.79	91.79	9	73.38	75.62	91.54	91.54
10	59.95	57.71	91.79	91.54	10	74.62	75.37	91.29	91.54
11	59.2	57.96	91.54	91.29	11	74.87	75.87	91.29	91.79

Acknowledgements We thank Sri Bhagyalakshmi Agro Foods Pvt. Ltd., Bangalore, Karnataka, India, for providing the laboratory facilities for this work. We extend our regards to S.G. Rice Mill, Tumkur, Karnataka, India, for providing the rice samples which to use in this research.

References

1. Prajapati BB, Patel S (2013) Algorithmic approach to quality analysis of Indian basmati rice using digital image processing. Int J Emerg Technol Adv Eng 3(3): 503–504. ISSN 2250-2459, ISO 90001: 2008certified journal
2. Hobson DM, Carter R, Yan Y (2007) Characterisation and identification of rice grains through digital image analysis. IEEE Instrumentation and Measurement Technology Conference – IMTC 2007, 1-4244-0589
3. ICAR Webpage. https://www.icar-iirr.org/index.php. Last accessed 30 Jan 2019
4. Wikipedia page on Indian varieties. https://en.wikipedia.org/wiki/List_of_rice_cultivars#Indian_varieties. Last accessed 30 Jan 2019
5. Mandal D (2018) Adaptive neuro-fuzzy inference system based grading of basmati rice grains using image processing technique. Appl Syst Innov 10:19. https://doi.org/10.3390/asi10010020019. https://www.mdpi.com/journal/asi
6. Mandal D (2016) Concepts of farm machinery and power. Narendra Publishing House, Delhi, India
7. Brosnan T, Sun DW (2004) Improving quality inspection of food products by computer vision—a review. J Food Eng 61:3–16
8. Zheng C, Sun DW, Zheng L (2006) Recent developments and applications of image features for food quality evaluation and inspection – a review. Trends Food Sci Technol 17(12):642–655
9. Cheng F, Ying YB (2004) Machine vision inspection of rice seed based on Hough transform. J Zhejiang Univ Sci A 5:663–667
10. Maheshwari CV, Jain KR (2013) Parametric quality analysis of Indian Ponia Oryza Sativa SSP Indica (Rice). Int J Sci Res Dev 1:114–118
11. Liu W, Tao Y, Siebenmorgen TJ, Chen H (1998) Digital image analysis method for rapid measurement of rice degree of milling, CEREAL CHEMISTRY, Publication no. C-1998-0408-05R.1998 American Association of Cereal Chemists, Inc.
12. Perdon AA, Siebenmorgen TJ, Mauromoustakos A, Griffin VK, Johnson ER (2001) Degree of milling effects on rice pasting properties. Cer Chem 78:205–209
13. Puri S, Dhillon B, Sodhi (2014) Effect of Degree of Milling (Dom) on overall quality of rice—a review. Int J Adv Biotechnol Res (IJBR) 5(3):474–489. ISSN 0976-2612, Online ISSN 2278–599X. http://www.bipublication.com
14. Wan P, Changjiang L (2010) An inspection method of rice milling degree based on machine vision and gray-gradient co-occurrence matrix. In: International conference on computer and computing technologies in agriculture, pp 195–202
15. Shiddiq DMF, Nazaruddin YY, Muchtadi FI (2011) Estimation of rice milling degree using image processing and adaptive network based fuzzy inference system (ANFIS). In: 2nd international conference on instrumentation, control and automation 15–17 November 2011, Bandung, Indonesia
16. Zareiforoush H, Minaei S, Alizadeh MR, Banakar A (2015) A hybrid intelligent approach based on computer vision and fuzzy logic for quality measurement of milled rice. Elsevier, Measurement 66:26–34
17. Chen X, Ke S, Wang L, Xu H, Chen W (2012) Classification of rice appearance quality based on LS-SVM using machine vision. In: Information computing and applications. Springer, Berlin, pp 104–109

18. Fayyazi S, Abbaspour-Fard MH, Rohani A, Sadrnia H, Monadjemi SA (2013) Identification of three Iranian rice seed varieties in mixed bulks using textural features and Learning Vector Quantization neural network. In: 1st international e-conference on novel food processing, Mashhad, Iran
19. Gujjar HS, Siddappa DM (2013) A method for identification of basmati rice grain of India and its quality using pattern classification. Int J Eng Res Appl 3:268–273
20. Shantaiya S, Ansari U (2010) Identification of food grains and its quality using pattern classification. In: 12th IEEE international conference on communication technology (ICCT), Nanjing, China, pp 3–5
21. Kaur H, Singh B (2013) Classification and grading rice using multi-class SVM. Int J Sci Res Publ 3:1–5
22. Government of Indian upload on rice. https://kvk.icar.gov.in/API/Content/PPupload/k0331_1. pdf. Last accessed 31 Jan 2019
23. Webpage on RNR rice variety. https://agribusinessatssu.wordpress.com/2017/01/30/new-rev olutionary-rice-variety-rnr-15048/
24. Kaur D, Kaur Y (2010) Various image segmentation techniques: a review. Int J Comput Sci Mob Comput. ISSN 2320–088X
25. Gonzalez RC, Woods RE (2009) Digital image processing, 2nd edn. Prentice-Hall, 779 pp
26. Wan P, Long C (2011) An inspection method of rice milling degree based on machine vision and gray-gradient co-occurrence matrix. In: Computer and computing technologies in agriculture IV, Springer, Berlin, pp 195–202
27. Fang C, Hu X, Sun C, Duan B, Xie L, Zhou P (2014) Simultaneous determination of multi rice quality parameters using image analysis method. Food Anal Methods 1–9

COVID Detection Using Cough Sound

Jeffrey Rujen, Parth Sharma, Rakshit Keshri, and Purushottam Sharma

Abstract Due to the impact of COVID-19 on its emerging strains, there is now a greater need for quick identification and containment to prevent further superfluous cases. We aim to make a machine learning model which can distinguish an audio file/signal and categorize it as COVID likely or unlikely and identify the virus' infection by analyzing the user's cough sounds. By usage of real-time detection and a precisely trained ML model with verified data, the user can further assess their infection in conjunction with other available tools, which would instruct him/her to either seek medical attention or provide reassurance for a negative or false positive diagnosis provided by the other tools.

Keywords COVID-19 · COVID test · COUGHVID · COVID detection · Cough sound · COVID cough · COVID cough sound · COVID test cough sound

1 Introduction

These days everyone has access to smartphones and Internet connections. Hence, a smartphone device can be considered as a good cheaper option for detection of COVID. According to various research papers and research works, it is possible to detect COVID-19 infection through the person's cough sound. Hence, this method can be easily used and implemented.

A sound has various distinctive characteristics and can be represented as an audio signal. Characteristics can be frequency, bandwidth, decibels, etc. A person infected with COVID-19 has a distinctive cough that cannot be distinguished by the human ear, but it can be distinguished using computers/devices. Then we can use our simple audio input devices to receive the input cough sound, process it, and classify it into COVID likely or not.

COVID-19 broke out in China in November of 2019 and since then it has spread to various countries and eventually was spread across the globe forcing all the nations

J. Rujen · P. Sharma · R. Keshri · P. Sharma (✉)
Amity University Uttar Pradesh, Noida, India
e-mail: Psharma5@amity.edu

© The Author(s), under exclusive license to Springer Nature Singapore Pte Ltd. 2023   803
A. Shukla et al. (eds.), *Computational Intelligence*, Lecture Notes in Electrical
Engineering 968, https://doi.org/10.1007/978-981-19-7346-8_69

to close everything through the means of permanent lockdown. Everyone on this planet was forced to quarantine themselves to protect themselves, their families, and known ones from it. Detection of COVID was initially very difficult and led to significant loss of lives due to the lack of both knowledge and vaccine. After some time, there emerged newer detection methods along with various vaccines which gave significant control over the spread of infection of coronavirus. However, there was still a lack of scalable, inexpensive, and widely accessible modes of detection.

Through the idea presented in this paper, COVID symptoms can be verified to a certain extent and help the user decide whether he should go for proper COVID testing or not. This can result in a lesser number of people crowding up at the checkup centers and detection at these centers can be relatively faster and response can be taken faster.

2 Literature Review

The methods presented in this section were implemented, and a synthetic dataset based on a random sample of COVID-19 and non-COVID-19 coughs is freely available online [1, 2]. The source of our dataset used to train our machine learning algorithm is taken from the COUGHVID crowdsourcing dataset [3]. Classification of cough audio signals has been used successfully to identify a wide variety of respiratory conditions, and there is a high interest in the use of ML to enable widespread detection of COVID-19. This dataset consists of over 25,000 crowdsourced samples covering an extensive range of individuals' ages, gender, geographic location, and statuses. Four adept physicians labeled over 2800 recordings diagnosing medical abnormalities in coughs and contributed to the largest expertly-labeled cough dataset ever available on cough audio processing [4]. The uploaded coughing sound on the COUGHVID dataset is likely using a different device, which may lead to a variation in the caliber of the recording due to different recording hardware and software on each device, background noise levels, ambient noise, etc. To help the COUGHVID dataset users assess the quality of each signal, open-source code is provided to estimate the signal-to-noise ratio (SNR) of each cough entry.

Various sound features will be extracted from the cough sound and will be used along with various other user-inserted data to determine whether the person has COVID-19 [5] signs or not.

2.1 Developed Technology and Past Works

Various attempts have been made for the detection of COVID-19 infection. There are a lot of works and papers written for the following topic. Usually, all the taken approach utilizes raw audio in combination with different spectrogram variations [6], and then the audio is processed so the machine learning algorithm can be applied to the dataset for the training. However, the most commonly used technique is feature selection and feature extraction [1]. The functioning of all the works is similar where the person records his cough sound recording, the artificial intelligence, machine learning, and deep learning model predicts the presence of the COVID-19 from cough sound and recommends RT-PCR test (Fig. 1).

Existing research has used convolutional neural network (CNN) [6] for the determination. As preprocessing, each cough input recording is divided into 6-second audio blocks, padded as needed, processed with the MFCC feature extraction package [8], and then passed to biomarker

1. The output of these steps is converted to a CNN at the input [9] as shown and explained better in Fig. 2.

Another approach taken was by putting the extracted features into three ensemble models: random forest boosted and bagged decision trees to get the patient's COVID result. Each split in the tree is randomly permuted through 250 trees to provide the output [12].

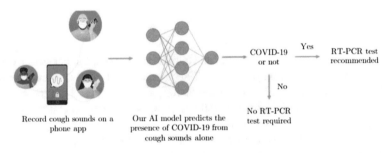

Fig. 1 Functioning of the COVID-19 detection application [7]

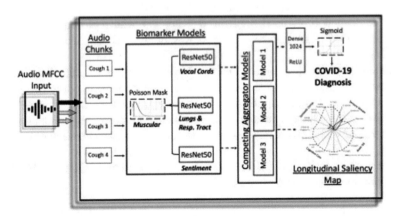

Fig. 2 Overview architecture of the COVID-19 discriminator with cough recordings as input, and COVID-19 diagnosis and longitudinal saliency map as output. A similar architecture was used for Alzheimer's [10, 11]

3 Proposed Work

The following are the proposed inputs for COVID detection using cough sound in various fields:

1. Rapid screening/diagnosis/testing in the recurrent/upcoming waves, thereby reducing the time taken for treatment/quarantine/self-isolation.
2. Contactless at-home testing without the need for medical personnel, thereby reducing the exposure of medical personnel and the user to the COVID-19 virus.
3. Reduction in manufacturing of test kits and disposal of used test kits which in return helps the environment.
4. Modification of the model for use in the diagnosis of other respiratory tract diseases.
5. A scalable model which can adapt to and incorporate newer emerging strains.
6. Lead to safer traveling environments due to everyone being screened against COVID beforehand, especially for domestic travel.

3.1 Implementation

Datasets were collected from hospitals and clinics under medical supervision. Cough sounds were recorded using the "Virufy" application, and the data was put up on "GitHub". PCR tests were also conducted on the patients, and results were appended with the cough sound data. Each sound file was subjected to segmentation, and only the cough portions were taken [13].

These cough segments were taken, and two types of processing were done.

In the first type, each segment was analyzed based on short-time Fourier transform (STFT), useful frequency bands were selected, STFT was applied on them, feature extraction was done, and then fed into an SVM classifier.

In the second type, each segment was applied with mel-frequency cepstral coefficients (MFCCs), feature extraction was done, and then fed into an SVM classifier.

The STFT-based spectrogram analysis broke the audio segments into smaller portions and applied short-time Fourier transform on these small portions.

The MFCC technique helps distinguish between frequencies that are nearly indistinguishable to the human ear and is highly efficient in recognizing sound systems. In a particular study [10], MFCC has been useful to distinguish between wet and dry coughs.

Linear and nonlinear SVM models like Leave-One-Out (LOO) and Hold-Out (HO) techniques were incorporated.

In another study [14], they used convolutional neural networks on breathing patterns and cough audio and spectrograms to classify whether a user has COVID-19 or not.

The COUGHVID Crowdsourcing Dataset

Classification of cough audio signals has been used successfully to identify a wide variety of respiratory conditions, and there is a high interest in the use of ML to enable widespread detection of COVID-19. The COUGHVID [11] dataset consists of over 25,000 crowdsourced samples covering an extensive range of individuals' ages, gender, geographic location, and statuses. Four adept physicians labeled over 2800 recordings diagnosing medical abnormalities in coughs and contributed to the largest expertly-labeled cough dataset ever available on cough audio processing.

Preprocessing of Data

We applied the following preprocessing on the dataset:

- Data observed by physicians
- Removed status-less data
- Data with cough_detected > 0.8
- Good quality data as reviewed by physicians.

Feature Extraction

We did a few data transformations before applying the model:

- Normalized
- Put a low pass filter
- Down sampled
- Selected the cough portions
- Discarded short segments
- Made all segments equal
- Rescaled the data to $[-1,1]$.

PyCaret [15]

Python library is used to perform machine learning tasks more efficiently.

We used PyCaret to compare the various models results on the dataset and found that extreme gradient boosting gave good results for our approach:

AUC - 0.722, Recall - 0.808, F1 - 0.717,

Kappa - 0.714, MCC - 0.443, TT(sec) - 0.4432
Ideas to improve model performance

- Algorithm tuning
- Data cleaning
- Ensemble methods
- Feature creation

- Feature extraction
- Feature selection
- Feature transformation
- Multiple algorithms
- Noise reduction
- Noise removal
- Treat missing/outlier values.

3.2 Algorithm Implementation

XGBOOST or extreme gradient boosting algorithm is a very strong algorithm used in machine learning field. It is used to enhance and minimize the bias error of the model. The algorithm can be used to determine both the continuous target variable (e.g., regressor) and the categorical target variable (e.g., classifier). When used as a regressor, the cost function is mean squared error (MSE) and when used as a classifier, the cost function is log loss.

Extreme gradient boosting or XGBoost is an efficient open-source implementation of the gradient boosting algorithm. The reasons for using XGBoost are execution speed and performance. The XGBoost model can also be used as a final model and makes predictions for classification.

The XGBoost ensemble is matched on to be had records, then the predict() feature may be referred to as to make predictions on new records.

Importantly, this feature expects records to be furnished as a NumPy array as a matrix with one row for every enter sample (Figs. 3, 4 and 5).

	Accuracy	AUC	Recall	Prec.	F1	Kappa	MCC
0	0.7045	0.7886	0.7324	0.6842	0.7075	0.4096	0.4105
1	0.6838	0.7903	0.6690	0.6786	0.6738	0.3671	0.3672
2	0.7285	0.8440	0.7254	0.7203	0.7228	0.4568	0.4568
3	0.7285	0.8128	0.7183	0.7234	0.7208	0.4566	0.4567
4	0.7320	0.8026	0.7113	0.7319	0.7214	0.4633	0.4634
5	0.7310	0.8117	0.7234	0.7234	0.7234	0.4617	0.4617
6	0.7448	0.8093	0.7092	0.7519	0.7299	0.4885	0.4892
7	0.6966	0.8068	0.6879	0.6879	0.6879	0.3926	0.3926
8	0.7414	0.8037	0.7324	0.7376	0.7350	0.4825	0.4825
9	0.7276	0.8094	0.7113	0.7266	0.7189	0.4547	0.4548
Mean	0.7219	0.8079	0.7121	0.7166	0.7141	0.4433	0.4435
SD	0.0190	0.0144	0.0190	0.0233	0.0182	0.0378	0.0378

Fig. 3 Build model creation

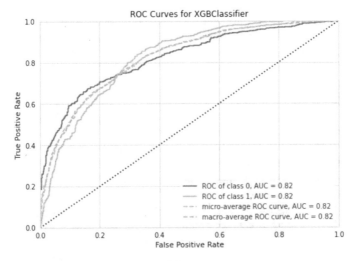

Fig. 4 Evaluate model performance (plot model)

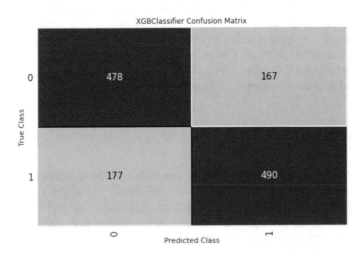

Fig. 5 Confusion matrix

4 Results

(Results co-respond to the points mentioned in the proposed work).

1. With the use of the application, an individual can quickly detect any indications leading toward COVID in real time and receive a diagnosis on the spot. This allows the person to take necessary precautions ahead of time and further proceed with self-quarantine, treatment, and other procedures. This reduces the risk of spread significantly, leading to safer surroundings for others.

2. Due to timely detection through the application, there is reduced exposure to others including the medical personnel that would be present for the at-home testing. With time and an increase in data and turn accuracy, it would eventually replace the need for another individual for testing and hence eliminate the chance of infection.

3. With gradual improvements in accuracy and detection, there will be a significant reduction in the manufacturing of test kids and their disposal leading to sustainable practice and a decrease in any pollutants.

4. As the model can identify and classify different coughs, even when afflicted with other respiratory diseases, it can also be used in the identification of said diseases with certain changes to the model. This can be used as another mode of detection for such diseases as well.

5. With the rise of newer mutations of the virus and its different strains affecting people in diverse geographical regions, there has been a greater rate of infection and challenge imposed on making suitable vaccines and more importantly their detection. With these mutations, the virus is unable to be detected by past norms as the symptoms aren't present or have changed. Once a definitive mode of detection is found, it can be added to the model making it scalable and adaptable to these newer mutations.

6. Individuals can be compelled to take a test before using any means of public transportation, which will lead to better security, limited spread, better safety, etc., for anyone else using the services. As of today, there is no such detection method in place leading to a fast and widespread of the virus.

5 Conclusion

The paper proposes a novel COVID detection technique through cough sound technology through cough analysis. This study also shows the analysis and different application for COVID detection. The research also covers the collations between the early and present technologies available in the market in comparison with the various detection methods and their processes.

The paper also shows the implementation of the proposed COVID detection through analysis of an individual's cough sounds and the different benefits in the recent pandemic scenario. The paper additionally shows the resemblances between the early and present advancements in the market by contrasting the different discovery modes and presents a clever mode of detection through the examination of individual's cough sounds.

References

1. Tena A, Clarià F, Solsona F (2022) Automated detection of COVID-19 cough. Biomed Signal Process Control 71(Part A):103175. https://doi.org/10.1016/j.bspc.2021.103175. ISSN 1746-8094

2. Tena A (2021) COVID-19 Models and Data repository. https://github.com/atenad/COVID. Date accessed 25 Aug 2021

3. Coughvid crowdsourcing dataset, 2020. https://coughvid.epfl.ch/about/

4. "Science Brief: SARS-CoV-2 Infection-induced and Vaccine-induced Immunity", October 2021. https://www.cdc.gov/coronavirus/2019-ncov/science/science-briefs/vaccine-induced-immunity.html

5. Kumar S, Viral R, Deep V, Sharma P, Kumar M, Mahmud M, Stephan T (2021) Forecasting major impacts of COVID-19 pandemic on country-driven sectors: Challenges, lessons, and future roadmap. Pers Ubiquitous Comput. https://doi.org/10.1007/s00779-021-01530-7

6. Schuller BW, Coppock H, Gaskell A (2020) Detecting COVID-19 from breathing and coughing sounds using deep neural networks. https://arxiv.org/abs/2012.14553, 29 Dec 2020

7. Bagad P, Dalmia A, Doshi J, Nagrani A, Bhamare P, Mahale A, Rane S, Agarwal N, Panicker R (2020) Cough against COVID: evidence of COVID-19 signature in cough sounds, 17 Sept 2020 (v1). Last revised 23 Sept 2020. https://arxiv.org/abs/2009.08790

8. Lyons J et al (2020, January) James lyons/python speech features: release v0.6.1. [Online]. Available https://doi.org/10.5281/zenodo.3607820

9. https://pycaret.org/guide/

10. Chatrzarrin H, Arcelus A, Goubran R, Knoefel F (2011) Feature extraction for the differentiation of dry and wet cough sounds. In: MeMeA 2011–2011 IEEE international symposium on medical measurement and application. Proceedings, IEEE Computer Society, pp 162–166. https://doi.org/10.1109/MeMeA.2011.5966670

11. Laguarta J, Hueto F, Subirana B (2020) COVID-19 artificial intelligence diagnosis using only cough recordings. IEEE J Eng Med Biol 1:275-281. https://doi.org/10.1109/OJEMB.2020.3026928

12. Despotovic V, Ismael M, Cornil M, Mc Call R, Fagherazzi G (2021) Detection of COVID-19 from voice, cough and breathing patterns: dataset and preliminary results. Comput Biol Med. https://doi.org/10.1016/j.compbiomed.2021.104944

13. Melek Manshouri N (2021) Identifying COVID-19 by using spectral analysis of cough recordings: a distinctive classification study. Cogn Neurodyn. https://doi.org/10.1007/s11571-021-09695-w

14. https://coughvid.epfl.ch/about/

15. Laguarta J, Hueto F, Rajasekaran P, Sarma S, Subirana B (2020) Longitudinal speech biomarkers for automated Alzheimer's detection. Preprint. https://doi.org/10.21203/rs.3.rs-56078/v1

Design of AMBA AHB Master and Implementing It on FPGA

Anu Mehra, Yash Chitransh, Kushaggr Sharma, and Aditya Mudgal

Abstract A microprocessor has all types of buses for transmission of data within the processor. For data transmission inside the processor, there are data buses, address buses and control buses. For data transmission outside of the processor to other external devices, we require system buses. Advanced microcontroller bus architecture is a high-performance, high-speed system bus used in high-speed microprocessors. AMBA AHB is a system bus which comes under the AMBA family. Other examples include AMBA APB and AMBA ASB. This paper is the design of AMBA AHB Master and its implementation on different FPGAs using Verilog as an HDL, ModelSim as Simulator and Vivado and Precision RTL as synthesizers.

Keywords Advanced microcontroller bus architecture · FPGA · AHB · HDL

1 Introduction

With the advancements in VLSI technologies, most designs today are based on System-on-Chips(SOC). AMBA is used in chips for the control and communication between various functional blocks of the main design. It is an optimal performance system bus used in ASIC and SOC designs. AMBA was developed by ARM in 1996 with its first buses being advanced peripheral bus(APB) and advanced system bus(ASB). In its second version in 1999, AMBA introduced advanced high-performance bus(AHB). AMBA is an open standard protocol which can be easily used on SoCs and ASICs using synthesizable HDLs [1].

The paper organization is as follows: Section 2 discusses the architecture of AMBA AHB bus. Section 3 is about the working of the protocol and Sect. 4 is all about results and discussions.

A. Mehra (✉) · Y. Chitransh · K. Sharma · A. Mudgal
Department of Electronics and Communication Engineering, Amity School of Engineering and Technology, Amity University, Noida, Uttar Pradesh, India
e-mail: amehra@amity.edu

© The Author(s), under exclusive license to Springer Nature Singapore Pte Ltd. 2023   813
A. Shukla et al. (eds.), *Computational Intelligence*, Lecture Notes in Electrical
Engineering 968, https://doi.org/10.1007/978-981-19-7346-8_70

2   Architecture of AMBA AHB Protocol

AMBA AHB Protocol basically consists of four main components: AHB Master, AHB Slave, AHB Arbiter and AHB Decoder [2-3] (Fig. 1).

A. AHB Master: Master is the component which ensures for the read/write operations. It generates the address and control signals for the slave. The bus can be accessed by one master at a time. A system can have multiple masters.
B. AHB Slave: Slave performs the operation guided by the master. Slave sends certain response to the master to indicate error or a re-transfer. There can be multiple slaves inside the AHB system.
C. AHB Arbiter: Only one arbiter is present in the AHB system. Arbiter is used to grant permission to master to access the bus. The master is allowed access to the bus on priority. The arbitration algorithm is fixed; however, any priority algorithm can be applied for arbitration.
D. AHB Decoder: It decodes the address of each transfer. It also selects the signals from the slaves.

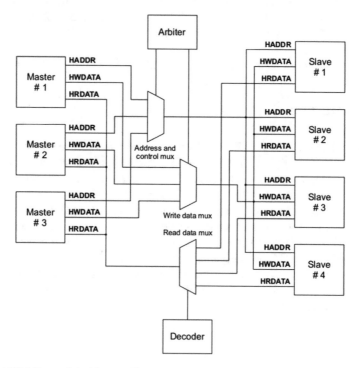

Fig. 1 AMBA Protocol Architecture [3]

3   Working of AMBA AHB

Before the transfer commences, the master must be granted access to the bus. The master asserts a request signal to the arbiter. The role of the arbiter is to indicate when this access can be given.

After the bus is accessed by the master, the master then begins the transfer and drives the address and control signals [3]. These signals indicate the address, the size of transfer and if transfer is burst type. There are two types of bursts supported:

A.   Incrementing bursts
B.   Wrapping bursts

There are two types of data buses, namely read data bus and write data bus. Write data bus transfers data from master to slave, whereas read data bus transfers data from slave to master. A transfer consists of multiple stages where the first stage is for address and control and then one or more stages for transfer of data. Slaves must sample the address during the address cycle. Address cycle cannot be extended, whereas the data stage can be extended using the HREADY signal by extending this signal LOW.

In this type of simple transfer (in Fig. 2), the master generates the control and address in the address phase and during the data phase the slave acts accordingly.

The slave can insert wait states into the transfer using the HREADY signal as shown in Fig. 3. Whenever the HREADY signal is low the data is extended till

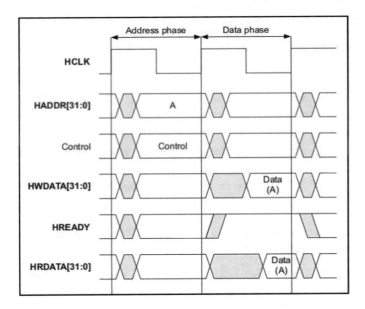

Fig. 2   A simple transfer

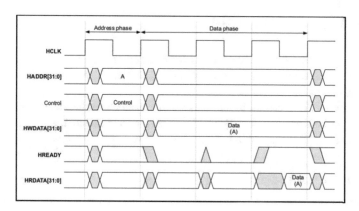

Fig. 3 A transfer with wait states

HREADY becomes high again. However, when the transfer is delayed in this manner, it also delays the address phase of the upcoming transfer [4-5].

The slave has to respond using the HRESP[1:0] signal. It can give four types of responses.

A. OKAY (HRESP = 2'b00): This response indicates that the transfer is working normally. High HREADY state indicates that transfer is completed without any error.
B. ERROR (HRESP = 2'b01): This response indicates that there has been an error in the transfer. The master has two choices when this signal is invoked - either carry on with the remaining transfer or send the data again.
C. RETRY (HRESP = 2'b10): This response indicates that there has been an error and the slave wishes to retry the transfer. The master then has to resend all the bits from the starting address.
D. SPLIT (HRESP = 2'b11): This response indicates that the slave needs to abort the current transfer and the bus must be granted to another master until slave becomes ready to receive the data from master again.

This is to be noted that AMBA bus is pipelined in nature. Thus, the occurrence of the address phase of a transfer might be same as the occurrence of data phase of the previous transfer [5]. High bus performance is achieved by the overlap of the address phase and data phase whil still providing enough time for the slave to sample the data.

There are four types of transfer in AMBA AHB protocol depending on the HTRANS [1:0] signal (Fig. 4).

A. IDLE (HTRANS = 2'b00) : This indicates that master has hold of the bus but there is no data transfer taking place. Slave must respond with OKAY to idle states and mus ignore this transfer [6].
B. BUSY (HTRANS = 2'b01): This indicates that the master wants to insert IDLE states into the transfer. When the master is carrying on bursts of transfers, it

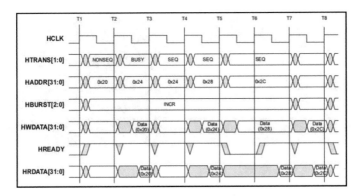

Fig. 4 Different types of transfers in a burst [4]

might want some time to process the incoming data so, master can assert the busy signal. The address and control signals must be sent in the next stage of transfer.

C. NON-SEQUENTIAL (HTRANS = 2'b10): The first transfer of a burst is always non-sequential. In non-sequential bursts, the address and data are independent of each other.

D. SEQUENTIAL (HTRANS = 2'b11): In a burst transfer, all the remaining transfers are in sequential and address and data are dependent of each other. In incrementing burst, the address of current transfer is previous address plus the incrementing size of burst. In wrapping burst, the address wraps at the address boundary.

AMBA AHB supports 4, 8 and 16 beat bursts as well as undefined—length bursts and single transfers. Bursts also have a limitation in which they must not cross a 1 kb boundary. Thus, masters should not be allowed to start a transfer from an address from where this boundary can be crossed. The total amount of data used for transmission is calculated by multiplying the data in each beat and the number of beats. For different types of wrapping bursts, the address wraps at different address boundaries. For example, in a four beat burst of HSIZE of 32 will wrap at a 16-byte address boundary. So if the transfer starts at 0x38, the address is wrapped at 0x3C followed by 0x30. Another example will be an eight beat wrapping burst of HSIZE of 32 will wrap at 32 byte address boundary. So if the transfer starts at 0x24, the address will wrap at 0x3C followed by 0x20 [7-11].

4 Results and Discussions

This section explains the results obtained after designing and implementing the bus on FPGA. Figure 5 shows the waveform of the AMBA protocol. The starting transfer address is 0 × 34. For the first transfer, the master asserts BUSY signal; thus, the

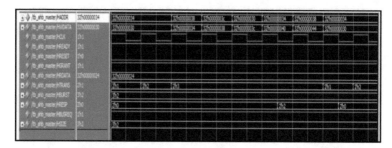

Fig. 5 Simulation Result of AMBA AHB

address and data are extended. After that the first transfer is always non-sequential so master asserts the NON_SEQ signal. In the next cycle, the master asserts NON_SEQ [8] signal high and the transfer continues in the burst mode. Since the example shown here is 4 beat wrapping, the address wraps at $0 \times 3C$ with the next address being 0×30.

Figure 6 is the further extension of Fig. 5. It shows the retry response from the slave. When the slave sends the retry signal using HRESP, master resend all the data from the starting address after two cycles. The transfer begins again from the starting address at 0×34. Figure 6 shows the concept of wait states in the protocol. When HREADY is low the data is extended for another cycle, whereas the address continues to increment. It should be noted that more than 17 consecutive wait states are not allowed in AMBA AHB. Figure 7 shows the RTL schematic of the design. The RTL schematic was derived using synthesis from precision RTL synthesis.

Figure 8 shows the tech schematic of the design. The tech schematic was again derived using synthesis from precision RTL synthesis.

For on-chip power utilization and number of components used, we synthesized the design on two different FPGAs. The first one we used is Kintex-7 KC705 Evaluation Platform (xc7k325tffg900-2) and the other one is ZedBoard Zynq.

Evaluation and Development Kit (xc7z020clg484-1). The power utilization for ZedBoard Zynq is shown in Fig. 9. The dynamic power used is 0.082 W, and the static power is 0.121 W. It should be noted that all the power utilization and utilization of

Fig. 6 Simulation result of AMBA AHB

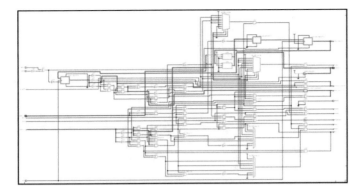

Fig. 7   RTL schematic derived from Precision RTL synthesis

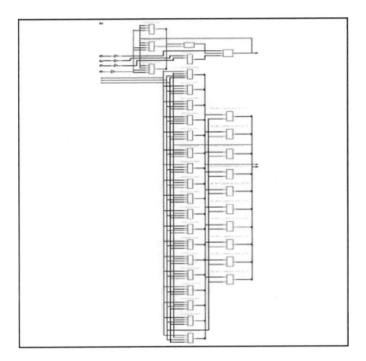

Fig. 8   Tech Synthesis from Precision RTL synthesis

components is carried out using Vivado Synthesis. Figure 10 is component utilization report. Most of the space is utilized by input/output pins which are 40%.

Now we analyse the results of Kintex-7 KC705. Figure 11 shows the power utilization report. The dynamic power for Kintex-7 is 0.075 W which is less than that of ZedBoard Zynq, and the static power is 0.157 W. For the component utilization report, we can refer to Fig. 12. Most of the space is again utilized by input/output

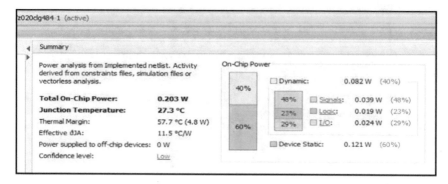

Fig. 9 Power report of AMBA AHB on ZedBoard Zynq

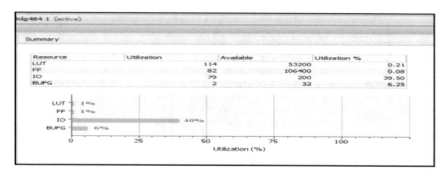

Fig. 10 Component utilization report of AMBA AHB on ZedBoard Zynq

pins which are 16%. However, it is less than that of ZedBoard Zynq. Therefore, by comparing the two results on different FPGAs, it can be concluded that the design is power efficient and utilizes less space on chip.

Table 1 shows the overall comparison of AMBA AHB between two FPGAs. It is clear that the component utilization and power utilization in Kintex-7 is less than that of ZedBoard Zynq.

5 Conclusion

The paper fulfils the objective of design of AMBA AHB bus which was verified by the simulation results carried out on ModelSim. The design is also synthesized on two different FPGAs using Vivado and precision RTL. The paper also displays the comparison of the bus running on two different FPGAs with reduced power and reduced on-chip utilization. The future scope of the paper would be design of a more comple AMBA bus with UARTs, memories and multiple masters and slaves.

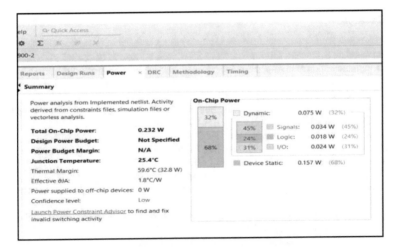

Fig. 11 Power report of AMBA AHB on Kintex-7 KC705

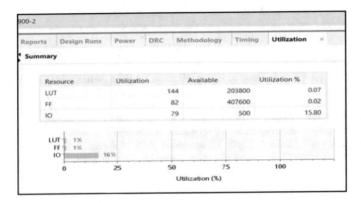

Fig. 12 Component utilization report of AMBA AHB on Kintex-7 KC705

Table 1 Comparison of AMBA AHB on two FPGAs

FPGA	On-chip process	Component utilization
ZedBoard Zynq	Dynamic: 0.082 W Static: 0.121 W	LUT: 114 FF: 82 IO: 79 BUFG: 2
Kintex-7 KC705	Dynamic: 0.075 W Static: 0.157 W	LUT: 144 FF: 82 IO: 79 BUFG: 0

References

1. Ramagundam S, Das SR, Morton S, Biswas SN, Groza V, Assaf MH, Petriu EM (2014) Design and implementation of high-performance master/slave memory controller with microcontroller bus architecture. In: 2014 IEEE international instrumentation and measurement technology conference (I2MTC), no May, pp 10–15
2. Simon D, Guruprasad U (2016) Design and implentation of AMBA-memory controller for image tansfer applications. [9] IJRET J Res Eng Technol 5(4):290–293
3. Rao S, Phadke AS (2013) Testing of AMBA compliant memory controller using pattern generator/logic analyser. Int J Adv [10] Res Electr Electron Instrum Eng 2(6):2227–2233
4. Int J Adv Eng Technol 4(2):2053–2063 (2013); Kurmi RS, Somkuwar A (2011) Design of AHB protocol block for advanced microcontrollers. Int J Comput Appl (0975) 32(8):23–29
5. Jayapraveen D, Priya TG (2012) Design of memory controller based on AMBA AHB protocol. Elixir Comp Sci Eng 51A 2:11115–11119
6. Shete PS, Oza S (2014) Design of an efficient FSM for an implementation of AMBA AHB master. Int J Adv Res Comput Sci Softw Eng 4(3):267–271
7. Khan Z, Arslan T, Erdogan ET (2005) A novel bus encoding scheme from energy and crosstalk efficiency perspective for AMBA based generic SoC systems. In: Proceedings of IEEE International Conference on VLSI Design, no January, pp 751–756
8. Deeksha L, Shivakumar B (2019) Effective design and implementation of AMBA AHB bus protocol using verilog, R. Proceedings of the international conference on intelligent sustainable systems, ICISS 8907975:1–5
9. Usha A, Jadhav MM, Jadhav (2010) A high throughput AMBA AHB Protocol. Int J Eng Sci Technol 2(5):1233–1241
10. Shete PS, Oza S (2014) Study of high performance AMBA AHB reconfigurable arbiter for on—chip bus architecture. Int J Electr Electron Data Commun 2(3)
11. Bhulania P, Chitransh Y, Kharma K, Mehra A (2020) Optimization of physical layer modules og USB 3.0 using FPGA. In: Proceedings of the 4th international conference on electronics, communication and aerospace technology, ICECA 2020, pp 345–349
12. Gaur N, Kapur S, Mehra A (2020) Application of Vedic Multiplier: Design of a FIR Filter. In: Proceedings of the 4th international conference on electronics, communication and aerospace technology, ICECA 2020, pp 234–237

An Empirical Study on the Future of Publication Repositories and Its Adaptability in Public universities—A Case Study of Shaqra University, Saudi Arabia

Nayyar Ahmed Khan⬡, Omaia Mohammed Al-Omari⬡, and Saeed Masoud Alshahrani⬡

Abstract The paper represents a dataset towards the acceptability and adaptability of publication repository system in a University environment. Data collection was over a period of two months, ranging from April 2021 till May 2021. The process was concluded using an online survey. It comprises of certain questions based on which derivations can be made. Authors made the survey in the form of a questionnaire comprising of following parts: personal details (04 Questions), acceptability about publication repositories (04 Questions), and general awareness about the current publication repository mechanism of University (05 Questions) and publications criteria / track (04 Questions). Two basic protocols were applied in this research survey, according to which any respondent must be above 30 years age, and he should be a part of University Academic curriculum. A snowball technique in accordance with purposive strategy in order to collect the information from University. The information collection commenced using emails and personalized WhatsApp messages to all the employees of University. A total of 97 observations were collected across at different centres. This data is useful for the authorities of the government universities in Saudi Arabia to find out the academic intentions to use a publication repository in a University. The data which is collected as a part of this survey is useful for various other universities seeking the deployment for their personalized publication repository systems in order to maintain the information about research being held at the University level.

Keywords Intelligent systems · Publications · Repository · Information systems · User acceptance · Adaptability · Data analysis

N. A. Khan (✉) · O. M. Al-Omari · S. M. Alshahrani
College of Computing and IT, Shaqra University, Riyadh, Saudi Arabia
e-mail: nayyar@su.edu.sa

© The Author(s), under exclusive license to Springer Nature Singapore Pte Ltd. 2023
A. Shukla et al. (eds.), *Computational Intelligence*, Lecture Notes in Electrical Engineering 968, https://doi.org/10.1007/978-981-19-7346-8_71

1 Introduction

Various universities have their own journals/publication areas. However, very few of them have a complete access to the publications by their authors and researchers. Many universities depend upon publication bodies to submit their research and evaluate. A recent study was done by [1], which clearly specifies the utility of the publication repositories at University level. Description of repository-based publication archival was presented in [2], the motivation to find out the information from the previous researches, started employing the user acceptability towards the desired information system deployment. The use of technology acceptance model proposed by [3] was done to find the adaptability criteria towards the publication repository system. The questionnaire guided by [2, 4]. Online survey comprises of the questions into languages to facilitate the readers in the Middle East region. Evaluation was done by technical experts.

2 Specifications Table

2.1 Data Description

Survey conducted for a period of two months from April 2021 till May 2021 collected 97 responses. These responses were given by the academic members. These respondents belong to different levels of teaching curriculum, including engineering, science, business, medical and humanities. The entire survey contains four parts [5]. Each part comprises of certain questions based on which ethical derivations can be done. Personal information including skills with 05 Questions, including various groups for age, different levels of education accepted by the respondent, including bachelors, masters or Ph.D. level, the status of the employment of the respondent, including full-time, part-time or contractual work conditions, gender of the person, email address for the contact, and the area of research that the respondent is interested in. Table 1 comprises of the questions for this part. Acceptability of the publication repository aims at respondent acceptance. There is a further factor which is created that gives the solution towards the improvement is expected from the respondent. This section comprises of 04 basic questions which will be helpful to identify the expectation is along with the acceptability of the publication repository. Table 2 comprises of the questions for this part. Feature's rating of the system: the level of acceptability is measured with the pointer scale out of 10. Total 05 questions were implied in which the respondents were asked, the ratings towards various points. These questions include recommendation level, belonging level, integrity level, operability level and acceptance by the respondent. These questions were on a 10-point Linkert scale from 1 to 10. Table 3 comprises of the questions for this part. Repository evaluation; this section helped to get the publication frequencies of various members in pronounced citation and indexing values like Scopus, web of science, science Citation index, emerging Citation index, ProQuest details gives a significant idea about the level of research being conducted at the University. It also included a value, which is capable enough of representing the level of publication repository expected at the University. Table 4 comprises of the questions for this part.

Table 1 Specifications for completed project on publication repositories

Subject	Computer science, publications, repository, information systems
Specific area	Information systems, research and publications
Data form	
Data collection	Using online forms (Forms.any website) as a survey. All the questions are available in a different file on the archive of the dataset
Data character	Raw, Analysed and Filtered
Properties for data collection	Total 97 University employees above the age of 30 years submitted between two months from April 2021 till May 2021
Description of data collection	The questionnaire was barely dedicated to the employees of the University. The study makes use of snowball and purposive technique for reaching towards the respondents using online forms. The distribution was done by personalized emails as well as WhatsApp messages
Data source location	Riyadh Region of Saudi Arabia
Data accessibility	The dataset is uploaded to the repository as: KHAN, NAYYAR (2021), "PubRepo", Mendeley Data, V1, https://doi.org/10.17632/5kz7vgfk34.1
Related research article	There is no related article

Table 2 Observations for the sociodemographic factors for the respondents

Variable	Description	Total	%
Age group	From 30 to 45 years old	54	55
	From 46 to 55 years old	35	36
	From 56 to 60 years old	08	8
Gender	Male	68	70
	Female	29	30
Educational level	Bachelor's degree	12	12
	Master	37	38
	PhD	48	49
Employment status	Full time	66	68
	Part time/contractual	31	32

Table 3 Acceptability factor for the staff members

Variable	Description	Total	%
What do you like most about Publication Repositories?	Interfaces	72	74
	Functionality	25	25
In your opinion, what kind of improvements would make Publication Repositories better?	Workflows	36	37
	Online access	32	33
	Membership	17	17
	Archiving	12	12
What other products may be competing with Publication Repositories?	E-Prints	8	8
	Invenio	2	2
	Zenodo	2	2
	CKAN	6	6
What would make Publication Repositories stand out among its competitors?	Public listing	8	8
	Administration	6	6
	Interface	13	13
	Financial integration	8	8
	Citations	12	12
	Indexing	12	12
	Global connect	13	13
	Workflow management	12	12
	Reviewing	6	6
	Data archiving	9	9

2.2 Data Value

The section elaborates on the use of information by various organizations. Respondent data is of great importance for various working bodies in academics. The data which is collected in this survey is useful to identify the acceptance level of the staff towards the publication repository system in a University network. It provides the intention of the academic members towards the adaptability of such system in the reality of the University [6]. The data is also helpful enough for the government universities, processing their own publication repository systems at University level [7]. The adaptability level can be predicted from the research for such universities before deploying any publication repository system at their site. The data also represents socio-demographic acceptance and adaptability of staff members towards the publication repository and their attitude for using such a system to contribute the research at University level.

Table 4 Publication repositories ratings

Variable	Description	Total	%
Would you recommend Publication Repositories to a friend or colleague?	Strongly agree	43	44
	Agree	30	31
	Neutral	24	24
	Disagree	0	0
	Strongly disagree	0	0
How likely is it for a University to hold its own Publication Repositories?	Strongly agree	67	69
	Agree	17	17
	Neutral	13	13
	Disagree	0	0
	Strongly disagree	0	0
Publication Repositories are an important for University Research	Strongly agree	72	74
	Agree	18	18
	Neutral	5	5
	Disagree	2	2
	Strongly disagree	0	0
Rate the ease of using the Publication Repositories	Strongly agree	66	68
	Agree	20	20
	Neutral	8	8
	Disagree	3	3
	Strongly disagree	0	0
How many publications usually you do annually?	Number ranging from 5 till 100		

3 Experimental Details

To find acceptability of a publication repository system at various sections of the University a survey was done. A sum total of 97 observations were collected using this survey. The questionnaire was distributed for a period of two months from April 2021 till May 2021. Online forms from "forms.any" are used. To protocols were applied to find and submit the survey to the respondents. These protocols were, respectively, the snowball and purposive [8]. The main protocol for being a respondent of this survey was, the respondent must be a University staff above 30 years age. Data collected includes the email address of the respondent with the restriction of one response per candidate. We deployed the use of personal emails and personalized sharing of form links using WhatsApp messenger. During the first phase of sharing the information to fill the form a list of members of the staff was collected and personalized email was sent. The similar link was forwarded using messenger on a basis of personal relations with different members and further their connections at University level. Friendly reminders once in 10 days for the complete period of time was sent. Collected data from all the respondents was on voluntary basis. There was no compulsion for participating in the survey. If any participant was ready to respond, he was sent a welcome message, and after the survey was complete, he was greeted with a thank you message. Just in case if a participant denies to submit, he was still sent a thank you message for not proceeding further ahead [9]. The forms.any link was released and stopped by 25 April 2021. No responses were accepted after this period of time.

3.1 Results of the Analytics on the Data Obtained

4 Conclusion

A total of 97 responses were processed. The results were then downloaded by requesting forms.any, and the responses were received an email which is included in the archived files. The user acceptability ratios were calculated based on the acceptability of different types of the academic streams. The results that were opting from the survey were used in order to calculate the content validity index [10]. Result value was 0.781. Finally, the calculation of the reliability for the survey calculated using spatial statistical analysis. Cronbach's Alpha value calculated for overall acceptance level, and it was found out to be 0.81. This value of the result represents that the overall acceptability is very high, and a publication repository system is well-accepted by the results. Various open source and other solutions are available in these regards and many universities are coming up with their repositories to monitor the publications and handling the manuscript under their copyrights. This reduces the effort to manage the research and also saves from financial investments in publications.

Ethics Statement Survey approval was taken from the Department of Computer Science and Information Technology, Shaqra University. Data collected was purely confidential, and the participant identity was kept anonymous in nature.

Acknowledgements Authors acknowledge the College of Computing and Information Technology for giving permission and ethical help to collect the information for the study. Finally, authors extend our thanks to all the people who have directly or indirectly contributed to this research.

Declaration of Competing Interest The author further declares that the study, which is conducted does not hold any conflicts of interest with any other author from another organization. There are no competing financial sources or interests towards this research.

References

1. Zervas M, Kounoudes A, Artemi P, Giannoulakis S (2019) Next generation institutional repositories: the case of the cut institutional repository KTISIS. Procedia computer science. 1(146):84–93
2. Okon R, Eleberi EL, Uka KK (2020) A Web Based Digital Repository for Scholarly Publication. J Softw Eng Appl 13(4):67–75
3. Davis FD (1989) Perceived usefulness, perceived ease of use, and user acceptance of information technology 319–340
4. Jaiswal B, Arya R (2020) Web presence of repositories of Indian institutes of technology: a webometric study. J Lib Inf Sci 8(1):83–99
5. Chisita CT, Chiparausha B (2021) An institutional repository in a developing country: security and ethical encounters at the Bindura University of Science Education. Zimbabwe 27(1): 130–143
6. Delgado JE (2021) University policies and arrangements to support the publication of academic journals in Chile, Colombia, and Venezuela, in international perspectives on emerging trends and integrating research-based learning across the curriculum. Emerald Publishing Limited
7. Romero D et al (2021) A first prototype of a new repository for feature model exchange and knowledge sharing. In: Proceedings of the 25th ACM international systems and software product line conference, vol B
8. Etikan I et al (2016) Comparison of snowball sampling and sequential sampling technique 3(1):55
9. Sieber JE (2012) The ethics of social research: surveys and experiments. Springer Science & Business Media
10. Polit DF, Beck CTJR, and health, the content validity index: are you sure you know what's being reported? Critique and recommendations. Res Nursing Health 29(5):489–497

The POPIA 7th Condition Framework for SMEs in Gauteng

Lehlohonolo Itumeleng Moraka and Upasana Gitanjali Singh

Abstract South Africa has passed its data privacy legislation dubbed the Protection of Private Information Act (POPIA). The POPIA is a new regulation that was built using the existing General Data Privacy Regulation (GDPR) a European data legislation. The objective of the POPIA is to protect the data privacy of those who reside in South Africa. The POPIA is made of eight conditions, which vary, this paper investigates the 7th condition known as the Security and Safeguards. The 7th condition is designed and drafted in a legal manner but requires technical measures to be put into place. In addition to this, organizations affected are unable to implement the POPIA without a technical guide and framework. Therefore, the key to the study was to develop a frame of reference that seeks to assist SMEs in Gauteng to implement the technical measures required in the 7th condition of the POPIA and remove the risk non-compliance. The findings indicate that SMMEs needed this frame of reference, and it will aid toward the successful integration between SMEs operations and the legislation. The purpose of this research is to also outline the Information Security in relation with the POPIA, to investigate how SMEs employees align with the POPIA regulation, to assess the benefits and weaknesses of POPIA compliance. This study serves as an awareness for South Africans as they will develop an understanding on their data privacy legislation.

Keywords COBIT 5 · ISO: International standard organization · POPIA: IR: information regulator · GDPR: General data privacy regulation · ISMS: Information security management systems

L. I. Moraka · U. G. Singh (✉)
University of Kwazulu-Natal, Durban, South Africa
e-mail: singhup@ukzn.ac.za

© The Author(s), under exclusive license to Springer Nature Singapore Pte Ltd. 2023
A. Shukla et al. (eds.), *Computational Intelligence*, Lecture Notes in Electrical Engineering 968, https://doi.org/10.1007/978-981-19-7346-8_72

1 Introduction

With the development of ICT and its benefits, businesses have found ways to operate through online platforms such as websites and mobile applications. Online-based businesses offer a convenience to the market. However, sensitive information belonging to people is shared with the businesses. It is for this reason; a data privacy legislation is designed to ensure that data belonging to individuals is protected during online transfer and storage. The POPIA legislation in its 7th condition speaks directly to the security controls required for both public and private organizations need to adopt to be compliant. Yet, the POPIA legislation in the 7th condition is drafted in a legal manner and does not provide a frame of reference for businesses to follow. Therefore, it becomes a cumbersome task for organizations to understand and implement what is required in the 7th condition of the POPIA.

The processing and dissemination of information must be done with the consent of the owner of the information, and it is the responsibility of the organization collecting the information to oversee the consent process. This is in addition to other measures where the owner of the data has rights that suppress the organization on the data. Therefore, protecting information has become a vital task for organizations, and if they are found to be in breach of their limitations, they will be sanctioned. This paper proposes a framework for SMEs to follow when implementing the 7th condition of the POPIA. The proposed frame of reference is assisted by existing Information Security conceptual and theoretical frameworks such as the ISO 27001/27002 and COBIT 5 on Information Security. Below is the structure of the paper.

Literature: This section of the paper gives a picture of literature in data privacy legislations on a local and global scale of e-commerce. This section also elaborates on SMEs, existing ISMS, and data breaches. Methods: This section covers the methodology that was adopted in the study. Results and discussion: This section covers the findings, summarizes the conclusions, reflects on the limitations of the study, and indicates the recommendations. Proposed framework: This section covers the framework, which is the main objective of this study.

1.1 Importance and Contribution of the Study

It is vital to indicate that the POPIA is a legal document, and the 7th condition requires technical implementation measures, yet it is still written in a legal format. Organizations are mandatory to implement the legislation but do not have the required guide on how this can be done, as the legal guide does not specify any technical implementation plan. As a result, organizations find it difficult. This study will allow businesses to have a technical guide on implementing the 7th condition of the POPIA and in addition, assist SMEs in the e-commerce sector to implement the legislation. The framework developed uses existing internationally recognized ISMS; as a result, it gives the SMEs an opportunity to learn and implement the policy without putting a lot of strain on their limited resources.

2 Literature

2.1 Information Security and Components

Information security refers to the process of safeguarding an information asset from illegal access, distribution, and tempering [1]. Data is a combination of symbols, characters, numbers, or unprocessed information which are stored by a computing device, processed data results in information [2]. Cybersecurity is the process adopted to protect information asset from intrusion, and cybersecurity is achieved by using software, hardware, or implementing physical measures to protect a data or information [3]. The objective behind information security and cybersecurity is to achieve integrity, confidentiality of data, and guaranteeing that the data is only available to those with the legal rights to access it.

2.2 Incidents that Affect E-Commerce Sites

Hackers have realized an increase in the online shopping space; as a result, they have invested into intruding, spamming, spoofing, e-commerce transactions. The hackers and other malicious agents make it risky for shoppers as they may find their credit information sniffed and used without their knowledge. In addition to this, the Internet has become the biggest target for hacker [6]. Users credit information, personal information and other profile data get stolen when the organizations database is compromised by hackers. Therefore, information security is important.

2.3 Small Medium Enterprises

SMEs, which is derived from 'SMME', stand for small medium and micro-enterprises. Within the South African context, it is a business with a turnover of less than R64 Million and has less than 200 employees, owns capital assets which do not exceed a value of R10 million [10]. These businesses in general do not have the capacity to invest into data privacy and information security legislation. This is because the core of their investment lies in extending their services and widening their profit streams, while ensuring that they service their existing immediate needs.

2.4   Iso 27001/27002

ISO 27001 and ISO 27002 are standards of the international standard organizations, and these duo standards cover the needs of information security. These standards enable organizations to run their ISMS and manage their information security needs. ISO 27001 is the implementation requirements while ISO 27002 is the control [11].

2.5   COBIT 5 for Information Security

COBIT 5 is a frame of reference that was born from ICASA with the objective of enabling an organization to manage its information security. COBIT 5 allows an organization to reduce its risk profile, thus, adding more value to itself [12]. An organization would easily integrate information security into its organizational strategic goals, and governance framework, and lastly, align itself to ever-growing changes on information security. Organizations that adopt COBIT 5 for information security have a process in place to manage its information security requirements at an organizational level [12].

2.6   Data Privacy Legislation Globally

In the United States of America, there is no complete general law that governs data privacy. They have several regulations that govern specific sectors, such as telecommunication, information technology, health, banking, and but not limited to, marketing and the financial sector [14]. One of the legislations is the FTC Act, which refers to the Federal Trade Commission Act. The FTC Act carries jurisdiction over commercial entities, ensuring that they prevent deceptive business operations [14]. The FTC Act does not give limitations to what organizations in the commercial space put on its website privacy policy. However, it protects the consumer and sanctions organizations that have unfair practices which affect the consumers. The US general attorney's office oversees these policies and governs how the organizations collect, disseminate, and store information. The issue of consent is vital in the space of data privacy, every consumer or data owner must consent to any of their data which is collected by an organization.

2.7   Reported Data Breaches

A data breach refers to an incident which involves the delicate and personal data of people, accessed illegally [15]. Data breaches usually result from cyberattacks where cybercriminals access a network, computer system, or database without authorization. The information targeted is usually linked to financial data, health records, market information, and but not restricted to academic records. According to the US Department of Justice, a data breach refers to the loss, compromise, disclosure, and unauthorized access of physical and electronic data, there are several types of data breaches, namely malware, phishing, denial-of-service attacks, ransomwares. Data breaches seem to be a common trend lately, this is due to cloud computing, and how data is digitally stored. Therefore, data breaches have existed since organizations kept confidential records, this was in the early 1980s and the issue grew until the early 2000.

South Africa has seen a large amount of data breaches for various reasons, but most are linked to denial-of-service attacks and ransomware. While the POPIA is now in regulation, organizations are making provisions to allocate a reasonable amount of funds to safeguard their information resource and deter data breaches. Thus, keeping a stronger association with the information regulator, who oversees the implementation of the POPIA. The data breaches affect both public and private organizations. A financial institution called Liberty Holdings has a life insurance unit named Liberty Life, which had their data repository hacked, leading to an exposure of customers' insurance policy data.

2.8 Benefits of Compliance

The benefits are customer confidence because customers will not be hesitant to store their data on the SEMs e-commerce site since they know that their data is safe. Strong business reputation as the SMEs will not have bad publicity around non-compliance, this saves their reputation and as a result attracts and retains customers. In addition to this, the organization can conserve money by not paying non-compliance fines, and the organization's members will not be arrested as they may face imprisonment for breaching the legislation.

3   Methods

The researcher adopted a quantitative approach as it was the most suitable to give the best interpretation and outcome for the study. This approach was selected because allowed an exhaustive investigation that led to a comprehensive frame of reference for the SME to implement the 7th POPIA condition. In addition, the methodology enabled the researcher to achieve the aims of the study. A questionnaire was adopted as the main data collection tool for the study. The researcher used a self-administered questionnaire. The researcher distributed the questionnaire electronically. A simple random sampling methodology was adopted. The participants of the study are based in Gauteng province, as it is one of the biggest receivers of provincial migrants in the country and offers the better employment prospects and business opportunities in South Africa. The targeted population is Information Technology experts who are responsible for implementing IT security services, IT management, business analysis, database development, and software development for SME e-commerce businesses.

4 Results and Discussion

The outcome of this research indicates that organizations are given the heavy task of implementing the POPIA legislation. However, they are not given the much-needed technical requirements to fulfill. This is the reason why they have challenges. Furthermore, organizations are under financial pressure. The bulk of their income is linked to their immediate organizational needs such as salaries, rent, and keeping afloat. As a result, the POPIA legislation is seen as an additional expense they are not able to afford. Some organizations have indicated that they have senior management providing support, and they are given finances to implement the POPIA. However, they end up following processes they understand. Therefore, they fear being told that they are not compliant. The respondents indicated that a hack may affect an organization anytime. In addition, they have proved that they do not have a strong risk assessment plan. Risk assessments are vital as they inform an organization about its weaknesses. It is for these reasons that they be required to find approaches to include a frequent risk assessment plan into their plan and increase their ways to alleviate risks that may impact them.

5 Proposed Framework

See Fig. 1

5.1 Method Used to Develop the Framework

The framework is a hybrid of two ISMS known as the ISO 27001/2700 and the COBIT 5 for Information Security. The respondents were given a questionnaire where they were asked to select ISMS's that have the potential to address the issues indicated in the POPIA's 7th condition. As a result, the researcher selected the two most selected ISMS and aligned them to the processes in the 7th condition of the POPIA.

Input 1: This input comprises processes where an organization strategically places its human resource in roles that focus on implementing information security systems, and the POPIA tasks. Input 2: This component speaks directly to the operators of the data belonging to consumers. The operator needs to stand in for the organization

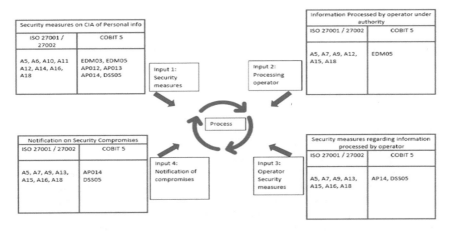

Fig. 1 POPIA 7th condition compliance framework for SMEs in Gauteng

and must ensure that data is processed in accordance with the consent which is agreed to by the data owners. Input 3: This component is linked to data access levels, confidentiality clauses, and non-disclosures. This is to ensure that data is accessed by those who are authorized and operate this data with boundaries imposed by the organization and legislation. Input 4: In this step, COBIT 5 and ISO 27001/27002 processes reflect a requirement in the POPIA, which guides how affected individuals are informed and addressed in an event of a hack or a data-related incident.

6   Conclusion

SMEs are found to employ ICT personnel or consultants on various e-commerce projects, and this is inclusive of software developers, database experts, project managers, and business analysts to name a few. The outcome in the study indicated that the professionals have demonstrated a fair understanding of the POPIA and other data privacy legislations. In addition, the SMEs respondents indicated that they get support from senior management to manage the data privacy legislation, the support includes financial resources and time. The main challenge that was displayed is the lack of understanding of the POPIA 7th condition, this is where the framework for compliance developed in this study would assist the SMEs. The results of this research study are limited to the SMEs ICT e-commerce sector, and the location where the study was done. In addition, it is prudent to indicate that this is a high-level framework for the 7th condition of the POPIA, and it does not help with the entire POPIA legislation.

It is the responsibility of the organizations at all levels to ensure they have a structure that oversees the POPIA and its administration. The information regulator must be consulted by organizations in their journey of POPIA implementation, so that they have a relationship and gather support where required. The South African government must be an enabler for compliance. They must be able to show organizations the path for maturing compliance with the data privacy legislation. This is specifically for the 7th condition of the POPIA which is drafted in a legal manner.

References

1. BER (2016) The small. University of Stellenbosch, Medium and Micro
2. Bridgwater A (2018). *The 13 Types of Data.* (Forbes) Retrieved 12 03, 2021, from https://www.forbes.com/sites/adrianbridgwater/2018/07/05/the-13-types-of-data/?sh=510dfe403362
3. Carson A (2020) Data privacy laws: what you need to know in 2021. (ONSAO) Retrieved from 2 Sept 2021. https://www.osano.com/articles/data-privacy-laws
4. Commerce Times (2007) Web 2.0 e-commerce: a new era of competition. Retrieved from 2 Dec 2021. https://www.ecommercetimes.com/story/58640.html
5. Departement of Justice USA (2018) Data breaches. Retrieved from 22 Aug 2021. https://www.doj.state.or.us/consumer-protection/id-theft-data-breaches/data-breaches/
6. Fruhlinger J (2020) What is information security? Definition, principles, and jobs. (CSO USA) Retrieved from 02 Dec 2021. https://www.csoonline.com/article/3513899/what-is-inf ormation-security-definition-principles-and-jobs.html
7. Hackers infect e-commerce sites by compromising their advertising partner (2019) (The Hackers News) Retrieved from 3 Dec 2021. https://thehackernews.com/2019/01/magecart-hacking-credit-cards.html
8. John Molson School of Business, C. U (2015) The effectiveness of COBIT 5 information security framework for reducing cyber attacks on supply chain management system. Elsevier
9. Kaspersky (2017) How data breaches happen (kaspersky). Retrieved from 27 Aug 2021 https://www.kaspersky.com/resource-center/definitions/data-breach
10. Kumar M (2019) Hackers infect e-commerce sites by compromising their advertising partner. (The hackers news) Retrieved from 12 Feb 2021. https://thehackernews.com/2019/01/mag ecart-hacking-credit-cards.html
11. Matthews T (2019) A brief history of cybersecurity. (Cyber security insiders) Retrieved from 12 Feb 2021. https://www.cybersecurity-insiders.com/a-brief-history-of-cybersecurity/
12. Mbatha A (2018) Liberty hack: a wake up call for SA firms to seriously fix up their cyber security. (BizNews) Retrieved from 5 Sep 2021 https://www.biznews.com/tech/2018/06/18/lib erty-hack-sa-fix-cyber-security
13. PixelPin (2018). Cybersecurity is becoming more and more expensive. (PixelPin) Retrieved from 12 Feb 2021. https://medium.com/@PixelPin/cybersecurity-is-becoming-more-and-more-important-381dcbab7859
14. POPIACT (2017) POPI Act Compliance. Retrieved from 19 Aug 2021. https://www.popiact-compliance.co.za/gdpr-information; Sobres R (2021) 98 Must-know data breach statistics for 2021. (VARONIS) Retrieved form 23 Aug 2021. https://www.varonis.com/blog/data-breach-statistics/
15. Tumber R (2019). 3 Compelling reasons to invest in cyber security—part 1. (Forbes) Retrieved 12 2021, from 3 compelling reasons to invest in cyber security—Part 1

Printed in the United States
by Baker & Taylor Publisher Services